The Reticuloendothelial System

A COMPREHENSIVE TREATISE

Volume 10

Infection

The Reticuloendothelial System
A COMPREHENSIVE TREATISE

General Editors:
Herman Friedman, *University of South Florida, Tampa, Florida*
Mario Escobar, *Medical College of Virginia, Richmond, Virginia*
and
Sherwood M. Reichard, *Medical College of Georgia, Augusta, Georgia*

MORPHOLOGY
Edited by Ian Carr and W. T. Daems

BIOCHEMISTRY AND METABOLISM
Edited by Anthony J. Sbarra and Robert R. Strauss

PHYLOGENY AND ONTOGENY
Edited by Nicholas Cohen and M. Michael Sigel

IMMUNOPATHOLOGY
Edited by Noel R. Rose and Benjamin V. Siegel

CANCER
Edited by Ronald B. Herberman and Herman Friedman

IMMUNOLOGY
Edited by Joseph A. Bellanti and Herbert B. Herscowitz

PHYSIOLOGY (In two parts)
Edited by Sherwood M. Reichard and James P. Filkins

PHARMACOLOGY
Edited by John W. Hadden and Andor Szentivanyi

HYPERSENSITIVITY
Edited by S. Michael Phillips and Mario R. Escobar

INFECTION
Edited by Mario R. Escobar and John P. Utz

The Reticuloendothelial System

A COMPREHENSIVE TREATISE

Volume 10
Infection

Edited by

MARIO R. ESCOBAR

Medical College of Virginia
Richmond, Virginia

and

JOHN P. UTZ

Georgetown University
School of Medicine
Washington, D.C.

Springer Science+Business Media, LLC

Library of Congress Cataloging in Publication Data

(Revised for vol. 10)

The Reticuloendothelial system.

Includes bibliographies and indexes.
Contents: v. 1. Morphology.—v. 2. Biochemistry and metabolism.—[etc.]—v. 10. Infections.
1. Reticulo-endothelial system. 2. Macrophages. I. Friedman, Herman, 1931–
II. Escobar, Mario R. III. Reichard, Sherwood M., 1928– . [DNLM: 1. Reticulo-endothelial system. WH650 R437]
QP115.R47 616.07′9 79-25933

ISBN 978-1-4899-3750-6 ISBN 978-1-4899-3748-3 (eBook)
DOI 10.1007/978-1-4899-3748-3

Originally published by Plenum Press, New York in 1988.
Softcover reprint of the hardcover 1st edition 1988

Contributors

JEFFREY D. ALDER • Department of Medical Microbiology, University of Wisconsin, Madison, Wisconsin

MAURO BENDINELLI • Institute of Epidemiology, Hygiene and Virology, University of Pisa, Pisa, Italy

FRANK M. COLLINS • Trudeau Institute, Inc., Saranac Lake, New York

RITA DHAR • Departments of Pediatrics and Microbiology, State University of New York at Buffalo, and Division of Infectious Diseases and Microbiology Laboratories, Children's Hospital, Buffalo, New York

MARIO R. ESCOBAR • Department of Pathology, Clinical Immunopathology and Virology Section, Medical College of Virginia, Virginia Commonwealth University, Richmond, Virginia

ROBERT A. FROMTLING • Department of Basic Microbiology, Merck Institute for Therapeutic Research, Rahway, New Jersey

J. TERRELL HOFFELD • Oral Biology and Medicine Study Section, Division of Research Grants, National Institutes of Health, Bethesda, Maryland

A. JULIO MARTINEZ • Division of Neuropathology, Department of Pathology, Presbyterian–University Hospital, University of Pittsburgh School of Medicine, Pittsburgh, Pennsylvania

STEPHAN E. MERGENHAGEN • Laboratory of Microbiology and Immunology, National Institute of Dental Research, National Institutes of Health, Bethesda, Maryland

SØREN C. MOGENSEN • Institute of Medical Microbiology, University of Aarhus, Aarhus C, Denmark

QUENTIN N. MYRVIK • Department of Microbiology and Immunology, Bowman Gray School of Medicine, Wake Forest University, Winston-Salem, North Carolina

PEARAY L. OGRA • Departments of Pediatrics and Microbiology, State University of New York at Buffalo, and Division of Infectious Diseases and Microbiology Laboratories, Children's Hospital, Buffalo, New York

MONIQUE PARANT • Immunopharmacologie, Centre National de la Recherche Scientifique, Unité Associée 579, Université Paris 6, 75270 Paris 06, France.

FIORENZO PARONETTO • Immunopathology Laboratory, Veterans Administration Medical Center, Bronx, New York, and Department of Pathology, The Mount Sinai School of Medicine of the City University of New York, New York, New York

RONALD F. SCHELL • State Laboratory of Hygiene and Department of Medical Microbiology, University of Wisconsin, Madison, Wisconsin

H. JEAN SHADOMY • Department of Microbiology and Immunology, Medical College of Virginia, Virginia Commonwealth University, Richmond, Virginia

STEVEN SPECTER • Department of Medical Microbiology and Immunology, University of South Florida College of Medicine, Tampa, Florida

BRET M. STEINER • State Laboratory of Hygiene and Department of Medical Technology, University of Wisconsin, Madison, Wisconsin

Foreword

This comprehensive treatise on the reticuloendothelial system is a project jointly shared by individual members of the Reticuloendothelial (RE) Society and biomedical scientists in general who are interested in the intricate system of cells and molecular moieties derived from those cells that constitute the RES. It may now be more fashionable in some quarters to consider these cells as part of what is called the mononuclear phagocytic system or the lymphoreticular system. Nevertheless, because of historical developments and current interest in the subject by investigators from many diverse areas, it seems advantageous to present in one comprehensive treatise current information and knowledge concerning basic aspects of the RES, such as morphology, biochemistry, phylogeny and ontogeny, physiology, and pharmacology, as well as clinical areas including immunopathology, cancer, infectious diseases, allergy, and hypersensitivity. It is anticipated that, by presenting information concerning these apparently heterogeneous topics under the unifying umbrella of the RES, attention will be focused on the similarities as well as interactions among the cell types constituting the RES from the viewpoint of various disciplines. The treatise editors and their editorial board, consisting predominantly of the editors of individual volumes, are extremely grateful for the enthusiastic cooperation and enormous task undertaken by members of the biomedical community in general and especially by members of the American as well as European and Japanese Reticuloendothelial Societies. The assistance, cooperation, and great support from the editorial staff of Plenum Press are also greatly valued. It is hoped that this unique treatise, the first to offer a fully comprehensive treatment of our knowledge concerning the RES, will provide a unified framework for evaluating what is known and what still has to be investigated in this actively growing field. The various volumes of this treatise provide extensive in-depth and integrated information on classical as well as experimental aspects of the RES. It is expected that these volumes will serve as a major reference for day-to-day examination of various subjects dealing with the RES from many different viewpoints.

Herman Friedman
Mario R. Escobar
Sherwood M. Reichard

Preface

The study of the reticuloendothelial system (RES) embraces much of modern allergy and immunology research, including many aspects of the perennial problems of infectious diseases and inflammation, as well as autoimmune disorders, immunodeficiency, and neoplasia. It has only been in the last quarter of this century that microbiology and immunology, which had diverged as each focused on its individual interests, have started to come together again. In phylogenetic and ontogenetic terms, it is generally thought that the immune system developed and matured in response to infectious agents and that the evolution of these agents was influenced in turn by the nature of the hosts' responses. What impact has all of this had on medicine? Most biomedical scientists and infectious disease specialists involved with the basic and clinical aspects of microbiology and immunology in one way or another and those who have watched the growth and development of microbiology and immunology during the last 30 years would agree that the impact has been extraordinary and can attest to it with some degree of astonishment.

This is the last of 10 volumes in this treatise. Although the treatise as a whole has encompassed what may appear as highly divergent areas, there is hardly any organ system or medical specialty that is not affected in some way by the immune system and the RES. Recognition of this permeation has resulted in the gradual emergence of new fields, such as psychoneuroimmunology, immunopharmacology, immunotoxicology, immunovirology, and tumor immunology.

We feel rather fortunate in having attracted contributions to this volume from experts in the field. We originally thought that the volume should provide the reader with a broad yet comprehensive view of the variety and complexity of parasite–host interactions. Nevertheless, we had to come to grips with the need for achieving a balance between the general aspects of those interactions and the rather detailed information that is available for representative microorganisms responsible for relatively common chronic human infections. Therefore, we had to choose from a large number of microbial agents within a family rather than individually reviewing everyone's favorite "bug." We intentionally decided to avoid redundancy by excluding the topic of immunoparasitology, which was covered by Drs. Mitchell and Wood in Chapter 14 in the previous volume of this series.

We have organized this volume around three areas of interest. The first four chapters deal with the major organ systems about which our knowledge of the

RES is most advanced. The second four chapters focus on bacterial and mycotic agents associated with chronic infections in humans (e.g., syphilis, tuberculosis, the mycoses). The remaining chapters are devoted to interactions between viruses and the RES. First, Dr. Mogensen focuses on natural resistance to viral infections and Dr. Specter discusses viral-induced modulation of RES function. Next, Drs. Paronetto and Escobar deal in a timely fashion with two viral diseases of significant worldwide importance—viral hepatitis (particularly hepatitis B) and AIDS. Although the tissue and cell injuries seen in these diseases have been shown to have an immunological basis, the immunopathogenetic mechanisms are entirely different. Finally, Dr. Bendinelli covers the RES and infection with RNA tumor viruses.

We would like to conclude these introductory remarks with a final note of appreciation to our contributors for their enthusiastic response to our request to review current knowledge on the general as well as specific aspects of interactions between the RES and infectious agents. We are also grateful to Mary Phillips Born of Plenum Publishing and Charlotte Phillips of the Medical College of Virginia, Virginia Commonwealth University, for their generous assistance with the administrative and editorial aspects of this project.

Mario R. Escobar
John P. Utz

Contents

1

Resistance to Respiratory Infection in the Immunocompetent Host

QUENTIN N. MYRVIK

1. INTRODUCTION

The lung is a remarkably versatile organ in terms of its normal function and its potential in the clearance of particulates and microorganisms. It has been estimated that the surface area of adult human lungs is about 90 m^2. In spite of its constant exposure to the environment, the lower respiratory tract is normally sterile, which attests to a highly efficient defense system against airborne microorganisms as well as normal flora.

To gain an understanding of the defense of the lung, it is necessary to briefly review the main anatomic systems that contribute to its resistance. The systems reviewed include (1) the mucociliary system; (2) the bronchus-associated lymphoid system; (3) the lymphatic system; and (4) the phagocytic systems.

2. ANATOMIC SYSTEMS OF DEFENSE OF THE LUNG

The lung has a highly responsive airway system that can react to all types of irritants through special irritant receptors, which, in turn, can induce the cough reflex, increased mucus secretion, and in some instances, airway constriction. These are basic physiologic responses of defense against inhaled toxicants or irritants, including gases, inanimate particulates, and infectious agents. It is the purpose of this chapter to examine the key mechanisms in the context of how the lungs of the normal host express resistance to infectious agents.

QUENTIN N. MYRVIK • Department of Microbiology and Immunology, Bowman Gray School of Medicine, Wake Forest University, Winston-Salem, North Carolina 27103.

2.1. NASOPHARYNGEAL COMPARTMENT

The nose filters particulates larger than 5 μm by inertial impaction on the nasopharynx (Hounam and Morgan, 1977). Landahl and Black (1947) and Landahl and Tracewell (1949) reported that particles greater than 20 μm in diameter were completely removed in the nose. Furthermore, there was little deposition of 1-μm particles. Dennis and Sawyer (1949) demonstrated that deposition was complete for particles of lipoidal, stearic acid, and gum arabic exceeding 6 μm in diameter. Hounam (1971) reported that deposition of condensation nuclei ranging in size from 0.001 to 0.1 μm was only about 5% at a flow rate of 35 liters min^{-1}, whereas deposition was about 20% at flow rates of 5 liters min^{-1}.

The Task Group on Lung Dynamics (1966) illustrated (Fig. 1) the effectiveness of the nose as a filter for particles with an aerodynamic diameter greater than 8μm. In contrast, deposition was negligible for particles smaller than 1 μm in diameter.

Clearance from the nose is more rapid in ciliated areas but slow from the anterior nonciliated areas. Data from Black *et al.* (1974) and Fry and Black (1973) indicate that the clearance of particles greater than 1 μm from the anterior nose had a biologic half-time of about 0.2 day. Clearance from the anterior nose is usually complete in 24 hr despite minimal mucociliary removal and is attributable to nonciliated, propelled mucus flow. The posterior ciliated areas of the nose revealed a half-time clearance of between 0.01 and 0.02 day.

Despite such efficiency, 25–60% of particulates less than 1.0 μm in diameter are not filtered. This size range encompasses much of the microbial world contained in droplet nuclei.

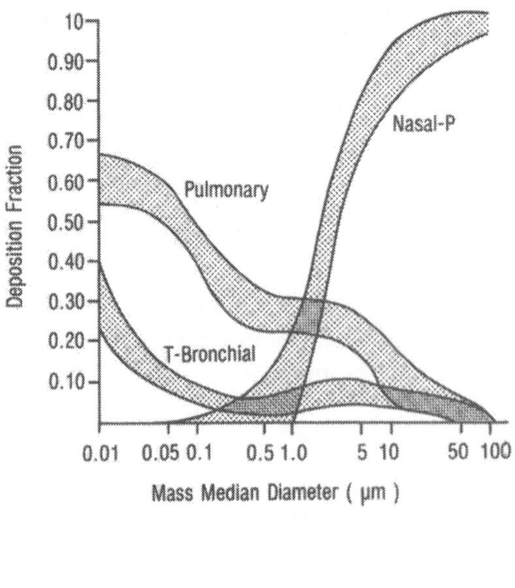

FIGURE 1. The variabilities of deposition (shaded areas) for a given mass median (aerodynamic) diameter in each lung compartment. The tidal volume was 1450 ml. (Reproduced from *Health Physics* **12**:181, by permission of the Health Physics Society.)

2.2. MOUTH

Dennis (1971) reported the deposition of particles exceeding 1 μm in the mouth of mouth breathers but at a lesser rate than in nose breathers (Fig. 2).

2.3. MUCOCILIARY CLEARANCE

One of the major clearance mechanisms in the lung involves a mucous layer lining the airways which is propelled by ciliated epithelium from the terminal bronchi to the oropharynx (Camner *et al.*, 1971). In general, inhaled particulates greater than 5 μm in diameter are impinged on the mucociliary apparatus in the upper respiratory tract and, depending on the level of penetration of the airways, are moved to the oropharynx in a matter of hours. Particles less than 10 μm in diameter penetrate the lower respiratory tract in direct relation to decreasing size. Accordingly, particles in the 0.01- to 1-μm-diameter range reach the alveoli. It has been estimated that the maximum mucociliary velocity in the tracheobronchial tree of humans is about 21.5 mm/min (Santa Cruz *et al.*, 1974).

The Task Group on Lung Dynamics (1966) concluded that about 40% of insoluble particulates deposited in the pulmonary compartment had a half-time clearance of 24 hr. Consonant with this conclusion are the studies of Thompson and Short (1969), who found that 30–50% of 5-μm-diameter particles were deposited beyond the ciliated airways. There is evidence which suggests that particles deposited in large airways are cleared in less than 1 day, whereas those

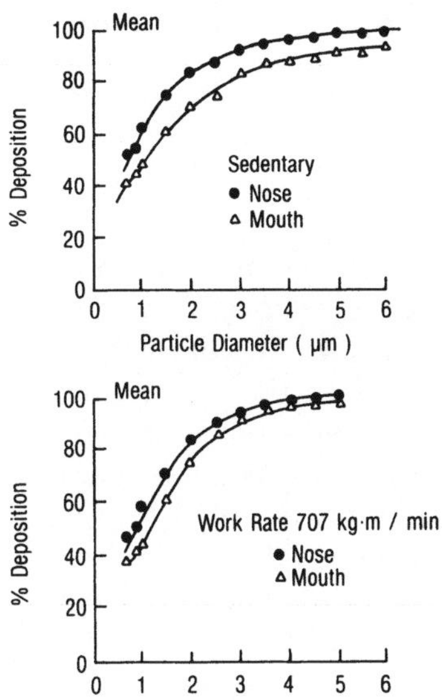

FIGURE 2. Variations in total deposition of calcium carbonate particles. Particle diameters are plotted against percent deposition and compared in nose and mouth breathers. (Adapted from Dennis, 1971, by permission of the Gresham Press.)

deposited in small airways probably exceed 1 day. Although it is assumed that the patterns of lung deposition will be different in nose-breathing subjects, definitive data on this point are limited because most of the experimental studies of deposition have used mouth-breathing subjects.

2.4. ALVEOLAR CLEARANCE

In contrast to tracheobronchial clearance, which is relatively rapid, alveolar clearance is slower. This is most likely due to the dependency on phagocytosis by resident alveolar macrophages (AM) and AM migration to the terminal bronchioles, where mucociliary clearance begins. From that site the AM are carried to the oropharynx with their phagocytic load. Alveolar clearance is also complicated by alternate routes, which can include the lymphatics, bronchial-associated lymphoid tissue, and, under special circumstances, the vascular system. These three routes are particularly important in pulmonary infections, because they provide a natural entry to the body.

Analysis of alveolar clearance kinetics is complex because of such effects as partial solubilization or breakdown of particulates, as well as the influx of other AM with their kinetic capabilities (Morrow, 1977). Accordingly, alveolar half-time clearance rates vary with the class of particle and range from 0.5 to 500 days. Kanapilly *et al.* (1973), using ^{90}Sr-labeled particles, reported an alveolar clearance half-time of 385 days.

Alveolar clearance can be perturbed when particles are endocytosed by type 1 pneumocytes or penetrate the interstitium. In the interstitium newly arrived AM can endocytose particles and transport them back to the alveoli, where alveolar clearance can begin again.

2.5. PULMONARY LYMPHATICS

The pulmonary lymphatics, like other lymphatic systems in the body, constantly remove fluids and plasma proteins that have escaped from the blood capillaries. Accordingly, the constant removal of fluids from the connective tissues of the lung serves to maintain normal fluid homeostasis by returning fluid and protein to the vascular compartment (Yoffey and Courtice, 1970). Associated with the lymphatic vessels is a complex system of lymph nodes that filter the circulating lymph. In addition to the maintenance of fluid homeostasis, the lymphatics of the lungs play an important role in the overall host defense system (Drinker and Field, 1933; Good, 1971–1972; Hall, 1967).

Pulmonary lymphatics, like lymphatics in other tissues, are valved structures. Their capillaries differ from blood capillaries in that the former have poorly developed junctions. Adjacent endothelial cells commonly overlap, but the endothelial basal lamina is discontinuous. Generally lymphatics are not found at the alveolar level but begin in the bronchioles. The pulmonary lymphatics drain peripherally to the visceral pleura and centrally to the hilar region. Accordingly, lymph from the pleural and pulmonary lymphatics traverses through the hilar, tracheal, and mediastinal lymph nodes before ultimately entering the systemic circulation.

Experiments in which colloidal ferritin or carbon was instilled into the trachea of adult mice and rabbits revealed ferritin or carbon in the interstitial areas surrounding lymphatic vessels in intracytoplasmic vacuoles of lymphatic endothelial cells 2 hr or more after instillation. At 24 hr colloidal carbon was found in draining lymph nodes. Carbon particles were commonly found in interstital and alveolar macrophages, in type 1 pneumocytes, and in the lymphatic endothelium of large lymphatic vessels (Lauweryns and Baert, 1974).

It is clear that large particles can cross squamous alveolar epithelial cells within vesicles. Since the connective tissue compartment has a negative interstitial fluid pressure, the particles move toward the lymphatics. Similar observations have been made with iron oxide particles (Sorokin and Brain, 1975).

As might be expected, the size and type of particles can affect the clearance rate by the lymphatics from the alveolar spaces. Toxic substances like quartz, which breach the alveolar-lining-cell barrier, rapidly enter the interstitium and the lymphatics (Strecker, 1967).

2.6. BRONCHIAL-ASSOCIATED LYMPHOID TISSUE

In the large airways there are specialized lymphoid structures termed bronchial-associated lymphoid tissue (BALT), similar to gut-associated lymphoid tissue. These structures are covered by a specialized epithelium that can endocytose and transport antigens and probably microorganisms from the surface of the bronchi (Racz *et al.*, 1977; Myrvik *et al.*, 1980; Tenner-Racz *et al.*, 1979).

According to Bienenstock (1984), BALT is an enriched source of potential IgA-producing cells that can be found in the submucosa. In addition, numerous IgM-marked B cells (noncommitted) are also present that are probably IgA cell precursors. Upon encounter with antigen these B cells move to the hilar regional node and expand into clones of IgA-producing cells, a switch probably mediated by helper T cells.

BALT consists of a follicular area (B cells) and parafollicular (T cells) areas (Figs. 3 and 4). It is of special interest that the specialized lymphoepithelium covering BALT can undergo hyperplasia during a cell-mediated immune reaction and can also phagocytose and transport bacteria into the dome area where numerous lymphocytes are present. BALT is distinct from classical lymph nodes in that there is no afferent circulation of lymph. It can be assumed that BALT plays a very important role in humoral immunity in the lung. There appears to be a continuous traffic of lymphoid cells from the high endothelial postcapillary venules to the BALT. The B-and-T-lymphocyte organization, however, is similar to that seen in lymph nodes.

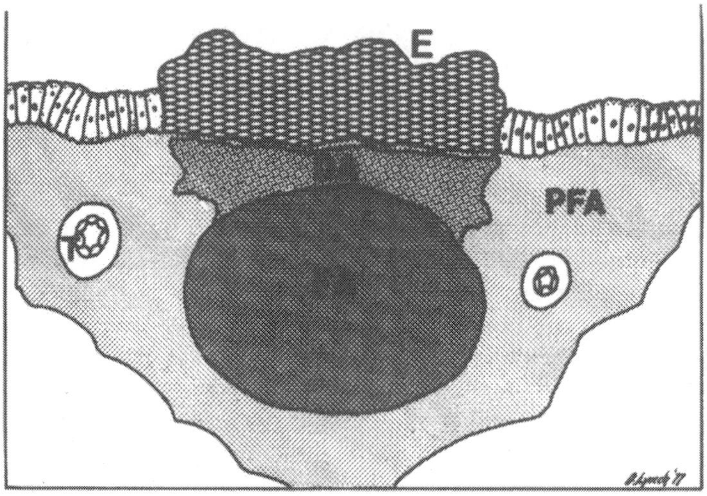

FIGURE 3. Bronchial-associated lymphoid tissue. DA, dome area; E, lymphoid epithelium; FA, follicular area; PFA, perifollicular areas; T, thymus-dependent zone. (From Racz *et al.* 1977, by permission of the Journal of RES.)

2.7. LYMPHORETICULAR AGGREGATES

Small aggregates of lymphoid cells intermixed with a few plasma cells and eosinophils are scattered throughout the lungs. These aggregates are found around bronchi, and in alveolar tissue, in interlobular septa, and in the pleura. In alveolar tissue, they are found between the terminal or respiratory bron-

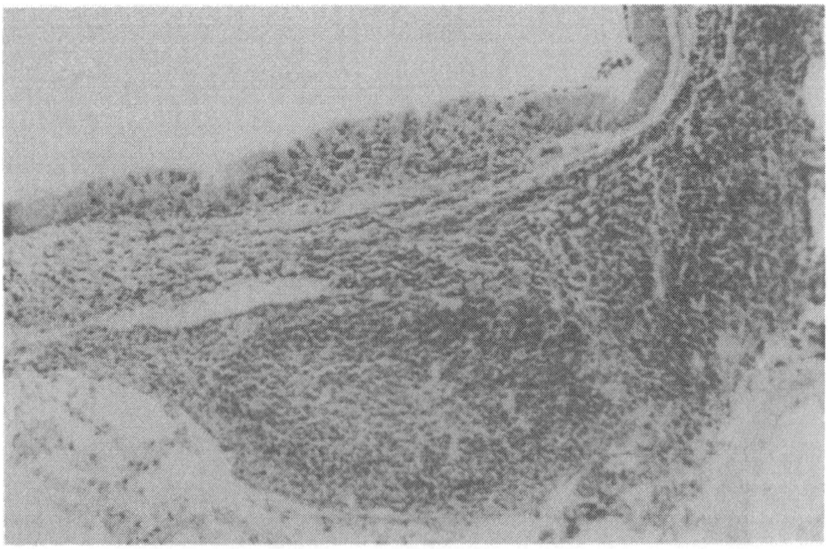

FIGURE 4. Bronchial-associated lymphoid tissue, illustrating hyperplasia following a bacille Calmette–Guérin-induced pulmonary cell-mediated response.

chioles and their associated arteries. When particles are transported to the alveolar interstitium by vesicular transport (Gonzalez-Crussi and Boston, 1972), they are carried as free particles or inside interstitial macrophages, by the flow of interstitial fluid to the lymphatics, immediately adjacent to these aggregates, and transported to the lymph nodes. It has been noted that heavy particulate deposition in alveoli leads to a lining up of particle-laden macrophages waiting their turn to enter the lymph drainage system.

3. PHAGOCYTES OF THE LUNG

Mononuclear and polymorphonuclear leukocytes combat toxic and infectious insults in the lung to varying degrees. In the normal "steady state" the free phagocytic cells in the bronchi and alveoli consist of about 95–98% macrophages, referred to as either alveolar or pulmonary. Very small numbers (<2%) of neutrophils are recoverable by lavage. The remaining small number of lavageable cells are lymphocytes. The pattern is similar in the interstitium. On the other hand, neutrophils can be seen adhering to the small blood vessels in the lung as if to migrate into the interstitium. In general terms, neutrophils respond rapidly to acute inflammatory stimuli and migrate progressively to the interstitium and into the alveoli, whereas mononuclear phagocytes respond to chronic and persisting stimuli. It becomes obvious that in the normal lung in the steady state, alveolar macrophages play the dominant role in microbial and particulate clearance, when small microbial inocula or loads are received. On the other hand, when microbial pathogens establish a clinically apparent pulmonary infection, the cellular dynamics are put in motion with first a neutrophil followed by a mononuclear phagocyte response; the latter represents the second line of phagocytic defense, which is designated for long-term, chronic host–parasite interactions.

4. LUNG DEFENSE MECHANISMS AGAINST INFECTION

The lung continuously interfaces with the environment. Because of the extensive surface area present in the lung, a staggering array of microbial agents, particulates, and gaseous pollutants must be contained and "neutralized" in the course of a lifetime. In addition to airborne exogenous microorganisms, there is the continuous interaction with the normal flora, restricting its location, controlling its numbers, and preventing its invasion of the lower respiratory tract. A consideration of the mechanisms involved follows.

4.1. FILTRATION AND MUCOCILIARY TRANSPORT

The nose represents the first line of defense by means of nasal hairs, aerodynamic design, and impaction on a mucous layer. As previously discussed, particulates greater than 10 μm are usually trapped in the nasal passages or upper airways. Smaller particles penetrate the airway system, so that particles between 2 μm and 8 μm are commonly trapped on the bifurcations of the smaller bronchial tree. Particles in the range of 0.1–2 μm generally reach the terminal airways and alveoli. In this case, the particles are ingested by alveolar macrophages and, for the most part, transported to the alveolar duct region, where they engage the mucociliary transport system.

The main mechanism for clearing particles that impinge on the mucous layer is mucociliary transport. Each ciliated cell has about 200 cilia that beat in a metachromatic fashion, propelling the mucus to the oropharynx. The ciliated cell:goblet cell ratio is about 5:1. Contrary to what was originally assumed, the mucous layer is patchy and not contiguous.

4.2. HUMORAL FACTORS EXHIBITING ANTIMICROBIAL ACTIVITY

4.2.1. Lysozyme

Mucus and fluid lining the airways are rich in lysozyme, which is secreted by epithelial cells, alveolar macrophages, and glandular cells. Lysozyme is highly active against numerous gram-positive airborne bacteria, such as staphylococci, *Sarcina lutea,* and *Bacillus* species (Konstan *et al.*, 1981). It can also enhance bacterial killing, by antibody and complement, of certain gram-negative bacteria (Hill and Porter, 1974; Crombie and Muschel, 1967; Wilson and Spitznagel, 1968; Glynn, 1969).

4.2.2. Lactoferrin

Lactoferrin, generally found in exocrine secretions including mucus, exerts a bacteriostatic effect on gram-negative and gram-positive microorganisms, as well as on *Candida albicans,* by binding iron and depriving microorganisms of this critical metabolite (Arnold *et al.*, 1977; Kirkpatrick *et al.*, 1971). It has also been reported that the bacteriostatic effect of lactoferrin may be enhanced by secretory IgA. It is alleged that secretory IgA inhibits production of iron-chelating agents (Arnold *et al.*, 1981; Bullen *et al.*, 1974; Rogers and Synge, 1978).

4.2.3. Lactoperoxidase

This enzyme, in exocrine secretions, interacts with peroxide and thiocyanate ions to generate products bactericidal for gram-negative and bacteriostatic for gram-positive microorganisms (Reiter, 1978; Pruitt *et al.*, 1979).

4.2.4. Interferon

Interferon is comprised of a family of nonantibody, host-coded proteins that induce normal cells of the same or closely related species to become resistant to most viruses. Three types of interferon are recognized, as illustrated in Table 1 (Kucera, 1984). Interferon is secreted by virus-infected cells during virus replication. For example, interferon appears in nasal secretions within 24 hr of initia-

TABLE 1. PRODUCTION AND ACTION OF THE THREE TYPES OF INTERFERON[a]

Interferon	Producer cell	Stimulus	Major action (?)
Fibroepithelial, β-type	Fibroblast Epithelial Macrophage (Other?)	Viruses Nucleic acids	Antiviral
Leukocyte, α type	Null cell	Foreign or transformed cells	Activate killer cells
	(Other?)	Viruses	Antiviral
Immune, γ type	T (and B?) lymphocyte	Mitogens antigens	Immunoregulatory

[a]Adapted from Kucera, 1984, p. 433.

tion of a virus infection (Danielscu *et al.*, 1975). It is clearly established that interferon is the first line of defense following a primary viral infection.

4.2.5. Fibronectin

This glycoprotein, with a molecular weight of 440,000 kilodaltons, is present on the surface of epithelial cells, in connective tissue, and in plasma. It is thought to play an important role in the lung by blocking adherence and colonization of certain types of bacteria. Patients who are seriously ill have decreased fibronectin levels (Woods *et al.*, 1981) and are highly susceptible to colonization and infection by gram-negative bacteria. Alveolar macrophages secrete fibronectin, although its role as an opsonin for bacteria is somewhat obscure. Undoubtedly, fibronectin functions as an opsonic protein in macrophage disposal of old effete cellular elements. Fibronectin, which is a major secretory product of fibroblasts, has domains for fibrinogen and collagen (Villiger *et al.*, 1981). It probably also promotes cell adhesion, connective tissue organization, and tissue repair.

4.2.6. Alveolar Lining Material

LaForce *et al.* (1973) and Juers *et al.* (1976) demonstrated a surfactant-containing material from the alveolar lining of rats and humans that enhanced the intracellular killing of *Staphylococcus aureus*. Subsequently, Coonrod and Yoneda (1983) examined the effect of this alveolar-lining material from rats on pneumococci and other nonstaphylococcal microorganisms. The antipneumococcal activity was associated with this surfactant-containing fraction, which was not destroyed by trypsin and was bactericidal for *Streptococcus viridans*, *Streptococcus pyogenes*, *Streptococcus bovis*, and *Bacillus* sp. Whereas diacylphospholipids had no antibacterial activity, a lysophospholipid, palmitoyl/lysophosphatidylcholine, displayed many properties resembling those of the surfactant-containing fraction, including autolysin-mediated pneumococcal lysis, altered cell membrane permeability, and antibacterial activity against other gram-positive bacteria. It is likely that LaForce *et al.* (1973) observed a similar effect in the presence of phagocytes. Notwithstanding, Coonrod and Yoneda (1983) noted that rabbit and guinea pig lavage fluids had essentially no antipneumococcal activity. Skornik *et al.* (1973, 1974) also observed that alveolar-lining material promoted killing of *Psuedomonas aeruginosa* by alveolar macrophages. More recently LaForce and Boose (1981) suggested that a peptide in rabbit lavage fluid can facilitate deoxycholate killing of *Escherichia coli in vitro.*

Collectively, these observations indicate that local nonspecific factors are important in alveolar macrophage function. The depressed antibacterial activity of the rat lung following oxygen toxicity and restrain-induced stress has been reported to be due to a failure of surfactant enhancement of alveolar macrophage bactericidal capacity (Huber *et al.*, 1974; Mullane *et al.*, 1974).

5. IMMUNOGLOBULINS IN THE AIRWAYS

Secretory IgA, IgG, and small amounts of IgM and IgE are found in bronchial secretions. Usually secretory IgA occurs in greater concentration in the upper airways, whereas IgG is in greater concentration than IgA in alveolar fluid (Kaltreider and Chan, 1976; Bell *et al.*, 1981). Plasma cells in the lamina propria are largely responsible for synthesis of IgA. The dimer of IgA, linked by a J-chain, binds to the secretory component, which is a surface membrane receptor on mucosal epithelial cells. The dimeric IgA enters the epithelial cells, and the IgA plus its secretory piece is secreted onto mucosal surfaces.

Although there is evidence that some of the IgG antibody is generated locally and secreted into the bronchial lumen, the majority of IgG appears by transudation, and if inflammation occurs, by exudation.

A summary of the immunologic components in the normal human lower-airway and alveolar-lining fluids sampled by lavage is presented in Table 2.

5.1. BIOLOGIC FUNCTIONS OF SECRETORY IgA

A major function of secretory IgA is neutralization of viruses. It is also an effective agglutinin of other microorganisms and an important neutralizer of toxins. It probably acts also as a sump for soluble and insoluble antigens. The latter property appears to be very important, because it serves to regulate the concentrations and exposures to antigens. Although it is established that IgA can agglutinate bacteria, it is a poor opsonin, in contrast to IgG and IgM. It does not activate complement via the classical pathway. Finally, IgA probably blocks adherence and colonization of some microorganisms to the epithelium.

5.2. BIOLOGIC FUNCTIONS OF IgG

Specific IgG is highly efficient in agglutinating and opsonizing bacteria, as well as neutralizing viruses and toxins (Reynolds *et al.*, 1975). Since it is an

TABLE 2. IMMUNOLOGIC SUBSTANCES IN NORMAL LOWER-AIRWAY AND ALVEOLAR-LINING FLUIDS AS SAMPLED BY BRONCHOALVEOLAR LAVAGE (BAL)[a]

Component	Estimated percentage of total protein in BAL sample
Serum-derived	
Albumin	30
Transferrin	0.1
Immunoglobulins	
IgA	5
IIS Dimeric (with bound secretory component and J-chain constitutes about 90% of IgA)	
Monomeric IgA (<10%)	
IgG	14
IgE	0.00001
IgM	<0.1
Complement components: factor B, C4, C3, C6, C1q	Trace
Epithelial cell products: free secretory component	1.0
Enzyme inhibitors	
α_1-Antitrypsin	0.7
α_2-Macroglobulin	0

[a] Adapted from Young and Reynolds, 1984, p. 160.

efficient activator of complement, generation of the chemotaxin C5a is an important means to promote chemotaxis. Because IgG is concentrated in respiratory secretions, it is highly probable that this class of antibodies is important in acquired immunity to extracellular parasites. Complement levels are very low in normal subjects. The complement factors C4, C1q, C3, C6, and factor B have been detected (Robertson *et al.*, 1976).

5.3. BIOLOGIC FUNCTIONS OF IgM

IgM usually remains in the vascular component owing most likely to its high molecular weight (Tomasi, 1976; McDermott *et al.*, 1982). Accordingly, IgM concentrations are very low in the normal lung but are increased as a consequence of inflammation. Specific IgM activates complement very efficiently and also agglutinates bacteria. In addition, it is a bactericidal and lytic for some gram-negative bacteria, when complement is present. It probably plays some role in neutralization of viruses.

5.4. BIOLOGIC FUNCTIONS OF IgE

IgE mediates anaphylactic-type hypersensitivity in the respiratory tract (Casale and Kalinger, 1984). However, its exact role in acquired immunity is still somewhat obscure (Kaltreider, 1976; McDermott *et al.*, 1982). It is unlikely that IgE could play an important role in antibacterial immunity because it is present in such low concentration. However, specific IgE could be effective in the neutralization of some viruses (Welliver *et al.*, 1980). An accumulating body of data indicates that IgE may be important in acquired immunity to certain parasitic infestations. It is also likely that IgE is involved in controlling the expression of other immune effector responses through release of mast-cell-derived vasoactive mediators, which could modulate vascular permeability.

6. EFFECTOR IMMUNE RESPONSES IN THE LUNG

The effector immune responses in the lung are designed to mediate resistance to (1) extracellular bacterial pathogens, (2) intracellular bacterial and fungal pathogens, (3) viruses, and (4) animal parasites. A diverse array of immunologic components and interactions require appropriate regulation of T helper cells, T suppressor cells, killer cells, natural killer cells, B cells, accessory cells (dendritic cells), neutrophils, eosinophils, basophils, mast cells, and macrophages. It becomes obvious that in the expression of total immunity there is an interplay between specific and nonspecific immune mechanisms.

Acquired humoral immune (antibody) mechanisms are primarily involved in immunity to extracellular parasites, whereas cell-mediated immune mechanisms are designed to handle intracellular bacterial or fungal and a few species of intracellular animal parasites. Although all categories of immune responses can be expressed in response to airway pathogens, the lung appears to produce preferentially IgA antibodies, a common characteristic of afferent immune responses in all mucocutaneous tissues. An explanation for this unique response lies in the nature of either the precursor B cells or the special helper functions in BALT T cells. Nevertheless, essentially all afferent immune clonal responses, regardless of anatomic location, are generated in regional or draining lymph nodes, except when the antigen circulates or enters the vascular compartment, in which case the spleen becomes involved. On the other hand, efferent responses are expressed locally in almost any anatomic location, when antigen is introduced the second time in an immune individual.

The remaining sections are concerned with selected examples of effector immune responses against representative microorganisms.

6.1. IMMUNE MECHANISMS AGAINST EXTRACELLULAR BACTERIAL PARASITES

By definition, extracellular parasites are highly susceptible to the normal killing activities of the two phagocyte systems (neutrophils and macrophages). Therefore, their survival depends on their being able to block phagocytosis. This is usually accomplished by a capsule having antiphagocytic properties or by a highly potent exotoxin, which destroys or inactivates local phagocytic function. The following examples illustrate these principles.

6.1.1. *Streptococcus pneumoniae*

This extracellular lung pathogen is an excellent example of a microorganism that has acquired a massive polysaccharide capsule with antiphagocytic properties (Tomasz, 1981). Immunity is type specific, and if specific opsonins are available, the microorganism is rapidly phagocytosed and killed by both neutrophils and macrophages. Immediately following primary infection, edema fluid is produced and serves as medium for the growth of the pneumococci. Chemotactic factors are generated, and neutrophils arrive at the site, followed in a few hours by macrophages. Resident alveolar macrophages fail to ingest and kill the microorganisms in the initial infecting dose in a nonimmune subject. Later, when specific opsonins appear, there is a rapid ingestion and killing of the pneumococci by the markedly increased numbers of phagocytes mobilized to the lung by microbial as well as immunologically derived mediators (Austrian, 1981). As might be expected, children with hypogammaglobulinemia have recurring bouts of pulmonary *S. pneumoniae* infections.

Viral infections, impaired ciliary function, excess mucus formation, and airway obstruction predispose to pneumococcal infections. This always raises the question to the potential role of nonspecific defense mechanisms in the normal, healthy adult. Although pneumococci are ubiquitous and are commonly carried by normal subjects, the incidence of lobar pneumonia is very low in normal, healthy adults. Coonrod *et al.* (1983) demonstrated in rats that a factor confined to the surfactant-containing fraction of lavage fluid mediated killing and lysis of types 1, 14, and 25 of *S. pneumoniae.* Decomplementation of the rats had no effect on pulmonary clearance, suggesting that activation of complement by the alternative pathway was not involved. This was further substantiated when Coonrod *et al.* could not demonstrate C3b, C3bi, or C3d receptors on the rat alveolar macrophages (Coonrod and Rehm, 1982). The alternative pathway generation of C3b thus would not have facilitated phagocytosis, but rather an extracellular nonphagocytic killing mechanism.

The possibility that surfactant or a similar component in lavage can activate the autolytic enzymes in *S. pneumoniae* is a provocative and unique antimicrobial mechanism (see Section 4.2.6).

6.1.2. *Hemophilus influenzae*

The type b strain of *H. influenzae* is an encapsulated microorganism responsible for endemic bacterial meningitis in infants and young children (Centers for Disease Control, 1979). However, two-thirds of serious adult respiratory infections have been due to nontypable strains, suggesting that noncapsulated variants have pathogenic potential.

Toews *et al.* (1984) have reported that in mice there was no difference in the clearance of a fully encapsulated *H. influenzae* type b strain and an unencapsulated variant derived from the same strain. Both multiplied for 6 hr and then cleared progressively over 18 hr, indicating that factors other than the capsule are involved in virulence. However, acquired immunity to *H. influenzae* in humans clearly correlates with serum bactericidal antibodies (Fothergill and Wright, 1933). Accordingly, resistance to encapsulated strains is probably determined by an interplay between anticapsular antibodies that promote phagocytosis and complement-dependent bactericidal antibodies directed against somatic antigens (Mpairwe, 1971). It is possible that the mice used by Toews *et al.* (1984) had low levels of bactericidal antibodies, which were natural or cross-reacting.

6.1.3. *Corynebacterium diphtheriae*

This organism colonizes the mucous membranes of the fauces, pharynx, and, occasionally, the larynx and trachea. If the organisms colonize the pharynx of a nonimmune subject, the diphtheria toxin promotes the production of fluid, pseudomembrane, and acute inflammation, which probably permits further growth of the organism in the lesion and more toxin production. In contrast, in the immune individual no disease develops (Barksdale, 1980). Thus antibody, specific for diphtheria toxin, is solely responsible for immunity against this pathogen. The lower respiratory tract and the vascular compartment of nonimmune individuals are highly resistant to infection. Antitoxin does not prevent colonization, and both IgA and IgG are antitoxic antibodies. Classically, this organism is an extracellular parasite and is highly susceptible to the killing mechanisms of the phagocyte systems. However, its superficial colonization characteristic demonstrates remarkable ecologic adaptation to avoid encounter with the phagocyte systems.

6.2. IMMUNE MECHANISMS AGAINST INTRACELLULAR PARASITES

By definition, intracellular pathogens are resistant to the killing mechanisms of the two major phagocyte systems. However, since most host–parasite relationships involving intracellular pathogens are chronic or long-term, the mononuclear phagocyte system is usually involved. Accordingly, acquired immunity is basically of the cell-mediated type, with macrophage mobilization and activation by T-cell-derived lymphokines.

Monocytes spontaneously differentiate into macrophages on leaving the vascular compartment. Furthermore, they are activated by additional exposure to a wide array of nonspecific agents of nonimmune origin, such as muramyl dipeptide and endotoxin. However, this nonspecific activation is not sufficient to activate macrophages so as to kill virulent intracellular pathogens like *Mycobacterium tuberculosis*. It is only when antigen-specific immune T cells release a series of lymphokines, which mobilize and activate macrophages at the site of antigen localization, that true immunologically activated macrophages develop (Fig. 5). Once macrophages are activated by the specific antigen-induced lymphokines, their capacity to kill intracellular pathogens is nonspecific, but even so lymphokine-activated macrophages are superior in activity to macrophages activated by nonimmune-derived stimuli (Fig. 6).

There are two theories on the resistance of intracellular pathogens to normal

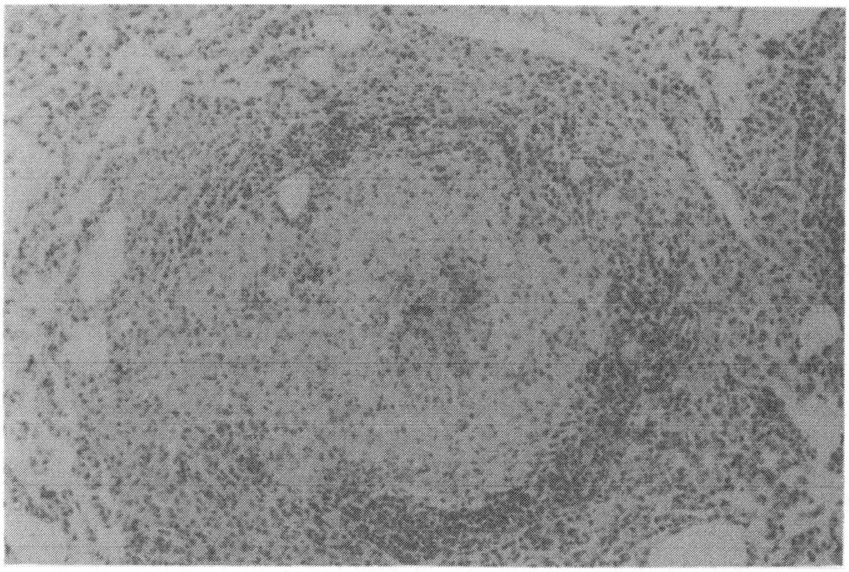

FIGURE 5. Cell-mediated immune response in the lung of bacille Calmette–Guérin-immune rabbit. Note the aggregation of macrophages rimmed by lymphocytes forming an immune structure, the immune granuloma.

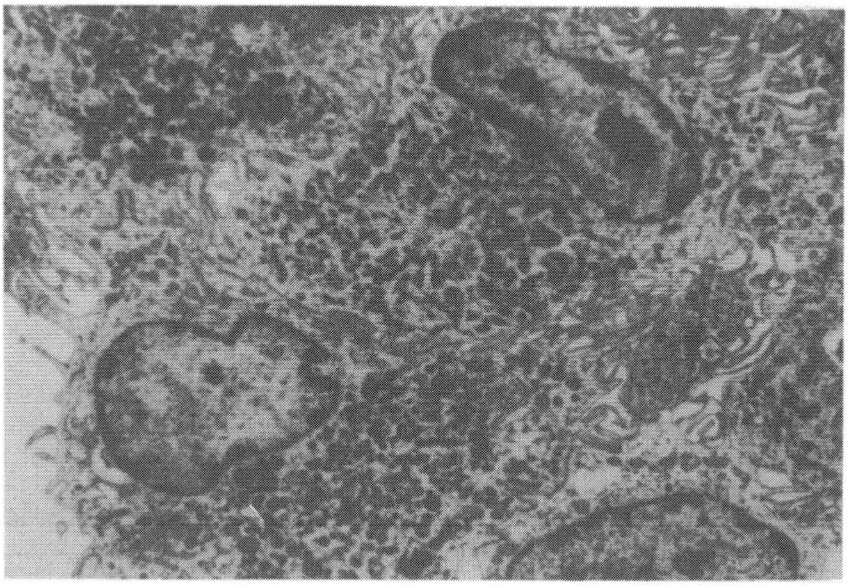

FIGURE 6. Immunologically activated alveolar macrophages. Note the marked ruffling, membrane interactions, and large array of dense lysosome structures.

macrophages: the first, by Armstrong and Hart (1975), postulates that virulent mycobacteria block lysosome–phagosome fusion, and the second states that the pathogens disrupt the phagosomal membrane so as to escape into the cytoplasm (Myrvik *et al.*, 1984).

6.2.1. *Mycobacterium tuberculosis*

This intracellular bacterium is so infectious that a single microorganism can probably initiate an infection or generate a tubercle in the lung (Lurie, 1964). However, the immunity acquired after infection can arrest infection in about 60% of infected individuals, so that disease does not develop.

Since avirulent mutants of virulent mycobacteria are readily killed by normal AM, it is necessary to postulate some type of perturbation mechanism. As mentioned previously, the most likely explanation presently available indicates a perturbation of lysosome–phagosome fusion or a total breakdown of the phagosomal membrane. Both these mechanisms could abort the normal killing of the resident macrophage.

The most convincing evidence of phagosome membrane disruption is based on the resistance of BCG-immune alveolar cells to disruption (Leake *et al.*, 1984). Although, specific humoral antibodies cannot mediate immunity, specifically immune T cells can passively transfer immunity.

6.2.2. *Francisella tularenis*

This organism is one of the most virulent pathogens known. Early studies showed that an aerosol challenge of one microorganism can produce fatal disease in rabbits. While the normal route of infection in man is via skin contact with infected rabbits, the local lesion can seed microorganisms to the lymphatics and blood with typhoidal and pulmonary tularemia.

Acquired immunity can be induced by vaccination with an attenuated LVS strain that makes rabbits resistant to 10,000-LD_{50} doses. Such immunity is largely, if not entirely, dependent on cell-mediated immune mechanisms similar to those against *M. tuberculosis.* The exact virulence mechanisms are not known, although growth is uncontrolled in nonactivated macrophages but controlled in immunologically activated ones (Claflin and Larson, 1972).

6.2.3. *Histoplasma capsulatum*

This fungus produces a chronic pulmonary disease similar to tuberculosis. Spores present in soil are inhaled and ingested by resident AM. Since most exposed individuals develop delayed cutaneous hypersensitivity, innate immunity is low or nonexistent in the normal, unexposed subject. Individuals infected with no apparent clinical disease are thereafter resistant to exogenous reinfection.

Acquired immunity is of the cell-mediated type (Domer and Carrow, 1983; Thor and Dray, 1968). The mechanisms responsible for virulence are not known, but this fungus is resistant to or perturbs the normal killing mechanisms of normal alveolar macrophages.

6.2.4. *Pneumocystis carinii*

This curious protozoan parasite infects numerous animals including humans. Serologic studies indicate that unrecognized infection with *P. carinii* is common. Meuwissen *et al.* (1977) reported that 100% of children 2 years of age and older had antibodies detectable by immunofluorescence.

The infection commonly becomes apparent when there is immunosuppression, as in transplant recipients or patients with AIDS. In animals latent infection can be activated by starvation or corticosteroid administration.

The major acquired immune mechanism is cell-mediated. Immune globulin is ineffective. Although cell-mediated immunity is effective in controlling the infection, it is ineffective in eradicating the parasite.

6.3. ACQUIRED IMMUNITY TO VIRAL AGENTS

The mechanisms of immunity in viral diseases are humoral as well as cell-mediated. In addition, the interferons are important during the early phases of infection (Kucera, 1984).

6.3.1. Respiratory Syncytial Virus

This virus is the most frequent cause of lower-respiratory-tract diseases in infants less than 6 months of age and is a frequent cause of upper-respiratory-tract illnesses in older children and adults.

Although the exact mechanisms of immunity are not in quantitative perspective, IgG and secretory IgA are important in neutralizing extracellular virus. IgG plus complement can probably destroy virus-infected cells, thus aborting viral maturation. Cell-mediated immunity probably destroys virus-infected cells and inactivates virus endocytosed by immunologically activated macrophages. Activated macrophages may not be permissive for viral replication.

Infants and young children are most susceptible probably because they have partially down-regulated immune affector responses (Murphy and Florman, 1983).

6.3.2. Influenza Virus

Three major types (A, B, and C) of influenza virus cause human disease. However, the constantly changing antigenic characteristics of influenza A virus have created undue complexity in understanding and characterizing the immunogenic antigens. Antibody responses to the hemagglutinin, neuraminidase, and viral nucleoprotein antigens after immunization and natural infection have been demonstrated (Virelizier, 1975; Biddison *et al.*, 1977; Ogra *et al.*, 1977). In addition, cell-mediated immune responses are similarly generated (Chow, 1978; Greenberg *et al.*, 1978; Habershon *et al.*, 1973). In particular, specific neutralizing IgA antibodies to influenza virus have been found in respiratory tract secretions (Couch *et al.*, 1969; Ogra and Morag, 1975). High titers of respiratory tract antibodies are best induced by administering virus directly to the respiratory mucosa (Kasel *et al.*, 1969; Waldman and Ganguly, 1974).

In the course of experimental infections, virus titer peaks at about 24 hr (McLaren and Potter, 1973; McLaren *et al.*, 1974). A decrease in virus titers at 72 hr after infection is associated with high levels of interferon (McLaren *et al.*, 1974).

The best correlate of protection against infection is the level of antibody in the respiratory tract and not the serum (Chanock and Finberg, 1957; Kasel *et al.*, 1969), although both serum and secretory antibodies reduce the frequency and severity of disease. Secretory IgA appears to block viral replication through neutralization; serum IgG appears at high titers in the lower respiratory tract when transudation occurs early as a result of inflammation. Several studies suggest that antibody to hemagglutinin prevents viral attachment, and antibody to neuraminidase limits the release of cell-to-cell spread of influenza virus (Couch *et al.*, 1975; Ogra *et al.*, 1977).

Yap and Ada (1978) reported that in mice immune T cells passively transfer immunity with clearance of virus and protection from death. Cell-mediated immunity destroys infected cells through the action of cytotoxic T cells, thus aborting replication (Effros *et al.*, 1977; Greenberg, *et al.*, 1977). However, immune cytotoxic T cells may also be responsible for the autoimmune and tissue-damaging events (Ennis *et al.*, 1977).

In summary, specific IgA and IgG block infection and specific cytotoxic T cells eradicate infected cells and, ultimately, the virus. This illustrates an important interplay between humoral and cell-mediated immune mechanisms.

6.4. SUMMARY

Host–parasite interactions involve diverse mechanisms, including colonization, toxin production, encapsulation of microorganisms, and surface virulence moieties that perturb lysosome–phagosome fusion or destroy the phagosomal membrane. It is, indeed, remarkable that the lung can tolerate and handle such a wide array of particulates, gases, and microorganisms on a continuing basis.

Although detailed analysis of all the interacting mechanisms and perturbations is beyond our reach at this time, most of the dominant immune mechanisms are evident. For example, specific opsonizing antibody is an absolute requirement for immunity to such extracellular parasites as *S. pneumoniae*, *Streptococcus pyogenes*, *H. influenzae*, and *Klebsiella pneumoniae*. These microorganisms resist phagocytosis in the absence of opsonic antibody. The effective role of antitoxins is also clearly demonstrated with toxigenic organisms like *C. diphtheriae*. Other important humoral mechanisms include the bactericidal effects of IgG or IgM and complement on certain gram-negative organisms, and IgA and IgG in blocking colonization.

Cell-mediated mechanisms are essential against intracellular bacteria and fungi. T cells orchestrate macrophage deployment and activation against these microorganisms.

Secretory IgA and IgG neutralize viruses and block uptake. Specific cytotoxic T cells assist in destroying virus-infected cells and abort virus maturation. This form of cell-mediated immunity is particularly effective in eradicating viruses and preventing cell-to-cell transmission. Interferon functions most effectively early in infection.

The most problematic issues in immune mechanisms lie in the role of nonspecific factors such as lysozyme, lactoferrin, and surfactant. These factors are undoubtedly contributory, but most likely by playing secondary roles and, therefore, interacting with specific acquired immune mechanisms.

Finally, the normal bacterial flora in the upper respiratory tract generates some resistance against colonization of pathogens, possibly by inducing cross-reacting antibodies. On the other hand, viral infections depress resistance to bacterial agents, particularly when the phagocytes are infected by the virus. As a consequence, bacterial pneumonias complicate viral pulmonary infections. Viruses can also impair mucociliary clearance by infecting epithelial cells of the airways.

During the steady rate, resident alveolar macrophages are highly efficient in maintaining sterility of the lower respiratory tract. Their function is probably supplemented by mucociliary clearance, nonspecific factors, and secretory IgA and IgG. Maintenance of sterility of the lower respiratory tract requires that microorganisms inhaled in aerosols as well as the normal flora on reaching the lower respiratory tract are destroyed on a continuous basis. Collectively, the defense mechanisms are truly a remarkable, integrated system.

ACKNOWLEDGMENTS. This work was supported in part by National Institutes of Health grants AI 17812 and HL 31624.

REFERENCES

Armstrong, J. A., and Hart, P. D., 1975, Phagosome–lysosome interactions in cultured macrophages infected with virulent tubercle bacilli, *J. Exp. Med.* **142**:1–16.

Arnold, R. R., Cole, M. F., and McGhee, J. R., 1977, A bactericidal effect for human lactoferrin, *Science* **197**:263–265.

Arnold, R. R., Russell, J. E., Champion, W. J., and Gauthier, J. J., 1981, Bactericidal activity of human lactoferrin: Influence of physical conditions and metabolic state of the target microorganism, *Infect. Immun.* **32**:655–660.

Austrian, R., 1981, Some observations on the pneumococcus and on the current status of pneumococcal disease and its prevention, *Rev. Infect. Dis.* **3**(S):S1–S17.

Barksdale, L., 1980, The immunobiology of diphtheria, in: *Immunology of Human Infection*, Part I (A. J. Nahmias and R. J. O'Reilly, eds.), *Comprehensive Immunology*, Vol. 8, Plenum Press, New York.

Bell, D. Y., Haseman, J. A., Spock, A., McLennan, G., and Hook, G. E. R., 1981, Plasma proteins of the bronchoalveolar surface of the lungs of smokers and nonsmokers, *Am. Rev. Respir. Dis.* **124**:72–79.

Biddison, W. E., Doherty, P. C., and Webster, R. G., 1977, Antibody to influenza virus matrix protein detects a common antigen on the surface of cells infected with type A influenza viruses, *J. Exp. Med.* **146:**690–697.

Bienenstock, J., 1984, Bronchus-associated lymphoid tissue, in: *Immunology of the Lung and Upper Respiratory Tract* (J. Bienenstock, ed.), pp. 96–118, McGraw-Hill, New York.

Black, A., Evans, J. C., Hadfield, E. H., Macbeth, R. G., Morgan, A., and Walsh, M., 1974, Impairment of nasal mucociliary clearance in woodworkers in the furniture industry, *Br. J. Ind. Med.* **31:**10–17.

Bullen, J. J., Rogers, H. J., and Griffiths, E., 1974, Bacterial iron metabolism in infection and immunity, in: *Microbial Iron Metabolism* (J. B. Neilands, ed.), pp. 517–551, Academic Press, London.

Camner, P., Philipson, K., Friberg, L., Holms, B., Larson, B., and Snedbert, J., 1971, Human tracheobronchial clearance studies with fluorocarbon resin particles tagged with ^{18}F, *Arch. Environ. Health* **22:**444–449.

Casale, T. B., and Kaliner, M., 1984, Allergic reactions in the respiratory tract in: *Immunology of the Lung and Upper Respiratory Tract* (J. Bienenstock, ed.), pp. 326–344, McGraw-Hill, New York.

Centers for Disease Control, 1979, Bacterial meningitis and meningococcemia—United States, 1978, *Morbid. Mortal. Wkly. Rep.* **28:**277–279.

Chanock, R. M., and Finberg, L., 1957, Recovery from infants with respiratory illness of a virus related to chimpanzee coryza agent (CCA). II. Epidemiological aspects of infection in infants and young children, *Am. J. Hyg.* **66:**291–300.

Chow, T., 1978, Immune response to hemagglutinin of neuraminidase antigens of influenza A virus in man and guinea pig, Ph.D. dissertation, State University of New York at Buffalo.

Claflin, J. L., and Larson, C. L., 1972, Infection–immunity in tularemia: Specificity of cellular immunity, *Infect. Immun.* **5:**311–318.

Coonrod, J. D., and Rehm, S. R., 1982, Complement receptors of rat alveolar macrophages, *J. Reticuloendothel. Soc.* **31:**107–115.

Coonrod, J. D., and Yoneda, K., 1983, Detection and partial characterization of antibacterial factor(s) in alveolar lining material of rats, *J. Clin. Invest.* **71:**129–141.

Coonrod, J. D., Rehm, S. R., and Yoneda, K., 1983, Pneumococcal killing in the alveolus: Evidence for a nonphagocytic defense mechanism for early clearance, *Chest* **5:**89S–90S.

Couch, R. B., Douglas, R. G., Jr., Rossen, R., and Kasel, J. A., 1969, Role of secretory antibody in influenza, in: The *Secretory Immunologic System* (D. H. Dayton, Jr., P. A. Small, Jr., R. M. Chanock, *et al.*, eds.), pp. 93–112, U.S. Government Printing Office, Washington, DC.

Couch, R. B., Kasel, J. A., Gerin, J. L., Schulman, J. L., and Kilbourne, E. D., 1974, Induction of partial immunity to influenza by a neuraminidase specific vaccine, *J. Infect. Dis.* **129:**411–420.

Crombie, L. B., and Muschel, L. H., 1967, Quantitative studies on spheroplast formation by the complement system and lysozyme on gram negative bacteria, *Proc. Soc. Exp. Biol. Med.* **124:**1029–1033.

Danielescu, G., Barbu, C., and Sorodoc, Y., 1975, The presence of interferon and type A immunoglobulins in the nasopharyngeal secretions of volunteers immunized with an inactivated influenza vaccine, *Acta. Virol. (Praha)* **19:**245–249.

Dennis, W. L., 1971, The effect of breathing rate on the deposition of particles in the human respiratory system, in: *Inhaled Particles* III (W. H. Walton, ed.), pp. 91–102, Unwin, Old Woking, Surrey, England.

Dennis, W. L., and Sawyer, K. F., 1949, *The Retention of Particles in the Human Nose*, Porton Technical Paper, no. 95, Chemical Defence Experimental Establishment, Porton, Salisbury, England.

Drinker, C. K., and Field, M. E., 1933, *Lymphatics, Lymph and Tissue Fluid*, Williams & Wilkins, Baltimore.

Domer, J. E., and Carrow, E. W., 1983, Immunity to fungal infections, in: *Host Defenses to Intracellular Pathogens* (T. K. Eisenstein, P. Actor, and H. Friedman, eds.), pp. 383–408, Plenum Press, New York.

Effros, R. B., Doherty, P. C., Gerhard, W., and Bennink, J., 1977, Generation of both cross reactive and virus specific T-cell populations after immunization with serologically distinct influenza A viruses, *J. Exp. Med.* **145:**557–568.

Ennis, F. A., Martin, W. J., Verbonitz, M. W., and Butchko, G. M., 1977, Specificity studies on cytotoxic thymus-derived lymphocytes reactive with influenza-virus infected cells: Evidence for dual recognition of H-2 and viral hemagglutinin antigens, *Proc. Natl. Acad. Sci. USA* **74**:3006–3010.

Fothergill, L. D., and Wright, J., 1933, Influenzal meningitis: The relation of age-incidence to the bactericidal power of blood against the causal organism, *J. Immunol.* **24**:273–284.

Fry, F. A., and Black, A., 1973, Regional deposition and clearance of particles in the human nose, *Aerosol. Sci.* **4**:113–124.

Glynn, A. A., 1969, The complement lysozyme sequence in immune bacteriolysis, *Immunology* **16**:463–471.

Gonzalez-Crussi, F., and Boston, R. W., 1972, The absorptive function of the neonatal lung: Ultrastuctural study of horseradish peroxidase uptake at the onset of ventilation, *Lab. Invest.* **26**:114–121.

Good, R. A., 1971–1972, Immunodeficiency in developmental perspective, *Harvey Lect.* **67**:1.

Greenberg, S. B., Criswell, B. S., Six, H. R., and Couch, R. B., 1977, Lymphocyte cytotoxicity to influenza virus infected cells. II. Requirement for antibody and non-T lymphocytes, *J. Immunol.* **119**:2100–2106.

Greenberg, S. B., Criswell, B. S., Six, H. R., and Couch, R. B., 1978, Lymphocyte cytotoxicity to influenza virus infected cells: Response to vaccination and virus infection, *Infect. Immun.* **20**:640–645.

Habershon, R. B., Molyneux, M. E., Slavin, G., Loewi, G., and Tyrrell, D. A. J., 1973, Skin tests with influenza virus, *J. Hyg. Camb.* **71**:755–761.

Hall, J. G., 1967, The response of a node to stimulation with foreign tissue, *Congies, Colloq. Univ. Liege* **45**:1–8.

Hill, I. K., and Porter, P., 1974, Studies in the bactericidal activity to *E. coli* of porcine serum and colostral immunoglobulins and the role of lysozyme with secretory IgA, *Immunology* **26**:1239–1250.

Hounam, R. F., 1971, The deposition of atmostpheric condensation nuclei in the nasopharyngeal region of the human respiratory tract, in: *Inhaled Particles* III (W. H. Walton, ed.), pp. 71–79, Unwin, Old Woking, Surrey, England.

Hounam, R. F., and Morgan, A., 1977, Particle deposition, in: *Respiratory Defense Mechanisms*, Part I (J. D. Brain, D. F. Proctor, and L. M. Reid, eds.), pp. 125–156, Marcel Dekker, New York.

Huber, G., Goodenough, S., and O'Connell, D., 1974, Oxygen toxicity and bacterial infection in the lung: Impairment of alveolar lining materials versus alveolar macrophage bactericidal activity, *Clin. Res.* **22**:506A.

Juers, J. A., Rogers, R. M., McCurdy, J. B., and Cook, W. W., 1976, Enhancement of bactericidal capacity of alveolar macrophages by human alveolar lining material, *J. Clin. Invest.* **58**:271–275.

Kaltreider, H. B., 1976, Expression of immune mechanisms in the lung, *Am. Rev. Respir. Dis.* **113**:347–379.

Kaltreider, H. B., and Chan, M. K. L., 1976, The class-specific immunoglobulin composition of fluids obtained from various levels of the canine respiratory tract, *J. Immunol.* **116**:615–618.

Kanapilly, G. M., Raabe, O. G., Goh, C. H. T., and Chimenti, R. A., 1973, Measurement of in vitro dissolution of aerosol particles for comparison to *in vivo* dissolution in the lower respiratory tract after inhalation, *Health Phys.* **24**:497–507.

Kasel, J. A., Hume, E. B., Fulk, R. V., Togo, Y., Huber, M., and Hornick, 1969, Antibody responses in nasal secretions and serum of elderly persons following local or parenteral administration of inactivated influenza virus vaccine, *J. Immunol.* **102**:555–562.

Kirkpatrick, C. H., Green, I., Rich, R. R., and Schade, A. L., 1971, Inhibition of growth of *Candida albicans* by iron-unsaturated lactoferrin: Relation to host-defense mechanisms in chronic mucocutaneous candidiasis. *J. Infect. Dis.* **124**:539–544.

Konstan, M. W., Chen, P. W., Sherman, J. M., Thomassen, M. J., Wood, R. E., and Boat, T. F., 1981, Human lung lysozyme: Sources and properties, *Am. Rev. Respir. Dis.* **123**:120–124.

Kucera, L. S., 1984, Immunity to viruses, in: *Fundamentals of Immunology* (Q. N. Myrvik and R. S. Weiser, eds.), pp. 425–443, Lea & Febiger, Philadelphia.

LaForce, F. M., and Boose, D. S., 1981, Sublethal damage of *Escherichia coli* by lung lavage, *Am. Rev. Respir. Dis.* **124**:733–737.

LaForce, F. M., Kelly, W. J., and Huber, G. L., 1973, Inactivation of staphylococci by alveolar macrophages with preliminary observations on the importance of alveolar lining material, *Am. Rev. Respir. Dis.* **108:**784–790.

Landahl, H. D., and Black, S., 1947, Penetration of air-borne particulates through the human nose, *J. Ind. Hyg. Toxicol.* **29:**269–277.

Landahl, H. D., and Tracewell, T., 1949, Penetration of air-borne particulates through the human nose, II, *J. Ind. Hyg. Toxicol.* **31:**55–59.

Lauweryns, J. M., and Baert, J. H., 1974, The role of the pulmonary lymphatics in the defenses of the diseased lung: Morphology and experimental studies of the transport mechanisms of intratracheally instilled particles, *Ann. NY Acad. Sci.* **221:**244–275.

Leake, E. S., Myrvik, Q. N., and Wright, M. J., 1984, Phagosomal membranes of *Mycobacterium bovis* BCG-immune alveolar macrophages are resistant to disruption by *Mycobacterium tuberculosis* H37Rv, *Infect. Immun.* **45:**443–446.

Lurie, M. B., 1964, *Resistance to Tuberculosis: Experimental Studies in Native and Acquired Defensive Mechanisms,* Harvard University Press, Cambridge, MA.

McLaren, C., and Potter, C. W., 1973, Immunity to influenza in ferrets, VIII. Serological response of ferrets to influenza virus vaccines after infection with heterotypic strains of the influenza. *Med. Microbiol. Immunol. (Berlin)* **159:**52–62.

McLaren, C., Potter, C. W., and Jennings, R., 1974, Immunity to influenza in ferrets. XIII. Protection against influenza infection by serum antibody to homologous haemagglutinin or neuraminidase antigens, *Med. Microbiol. Immunol (Berlin)* **160:**33–45.

McDermott, M. R., Befus, A. D., and Bienenstock, J., 1982, The structural basis for immunity in the respiratory tract, *Int. Rev. Exp. Pathol.* **23:**47–112.

Meuwissen, J. H. E., Tauber, I., Leeuwenberg, A. D. E. M., Beckers, P. T. A., and Sieben, M., 1977, Parasitologic and serologic observations of infection with *Pneumocystis* in humans, *J. Infect. Dis.* **136:**43–49.

Morrow, P. E., 1977, Clearance kinetics of inhaled particles, in: *Respiratory Defense Mechanisms,* Part II (J. D. Brain, D. F. Proctor, and L. M. Reid, eds.), pp. 491–543, Marcel Dekker, New York.

Mpairwe, Y., 1971, Immunity to *Haemophilus influenzae* type b: The nature of the bactericidal antibody in human blood, *J. Med. Microbiol.* **4:**43–49.

Mullane, J., Goodenough, S., McCarthy, C., Simmons, G., O'Connell, D., LaForce, M., Laguarda, R., and Huber, G., 1974, Alterations in alveolar lining material as the key factor in depression of interpulmonary antibacterial defenses following stress, *Chest* **66:**333.

Murphy, S., and Florman, A. L., 1983, Lung defenses against infection: A clinical correlation, *Pediatrics* **72:**1–15.

Myrvik, Q. N., Racz, P., and Racz, K. T., 1980, Ultrastructural studies on bronchial-associated lymphoid tissue (BALT) and lymphoepithelium in pulmonary cell-mediated reactions in the rabbit, in: *Macrophages and Lymphocytes,* Part B (M. R. Escobar and H. Friedman, eds.), pp. 145–153, Plenum Press, New York.

Myrvik, Q. N., Leake, E. S., and Wright, M. J., 1984, Disruption of phagosomal membranes by the H37Rv strain of *Mycobacterium tuberculosis:* A correlate of virulence, *Am. Rev. Respir. Dis.* **129:**322–328.

Ogra, P. L., and Morag, A., 1975, Immunologic and virologic aspects of secretory immune system in human respiratory tract, *Dev. Biol. Stand.* **28:**129–144.

Ogra, P. L., Chow, T., Beutner, K. R., Rubi, E., Strussenberg, J., DeMello, S., and Rizzone, C., 1977, Clinical immunologic evaluation of neuraminidase specific influenza A virus vaccine in humans, *J. Infect. Dis.* **135:**499–506.

Pruitt, K. M. Adamson, M., and Arnold, R., 1979, Lactoperoxidase binding to streptococci, *Infect. Immun.* **25:**304–309.

Racz, P., Tenner-Racz, K., Myrvik, Q. N., and Fainter, L. K., 1977, Functional architecture of bronchial associated lymphoid tissue and lymphoepithelium in pulmonary cell-mediated reactions in the rabbit, *J. Reticuloendothel. Soc.* **22:**59–83.

Reiter, B., 1978, Review on nonspecific antimicrobial factors in colostrum, *Ann. Rech. Vet.* **9:**205–224.

Reynolds, H. Y., Kazmierowski, J. A., and Newball, H. H., 1975, Specificity of opsonic antibodies to enhance phagocytosis of *Pseudomonas aeruginosa* by human alveolar macrophages, *J. Clin. Invest.* **56:**376–385.

Robertson, J., Caldwell, J. R., Castle, J. R., and Waldman, R. H., 1976, Evidence for the presence of components of the alternative (properdin) pathway of complement activation in respiratory secretions, *J. Immunol.* **117**:900–903.

Rogers, H. J., and Synge, C., 1978, Bacteriostatic effect of human-milk on *Escherichia coli:* The role of IgA, *Immunology* **34**:19–28.

Santa Cruz, R., Lauda, J., Hirsch, J., and Sackner, M. A., 1974, Tracheal mucous velocity in normal man and patients with obstructive lung disease: Effects of terbutaline, *Am. Rev. Respir. Dis.* **109**:458–463.

Skornik, W. A., Dressler, D. P., and Nathan, P., 1973, Intracellular killing of *Ps. aeruginosa* by rat alveolar macrophages stimulated by normal lung lavage fluid, *Physiologist* **16**:456.

Skornik, W. A., Dressler, D. P., and Nathan, P., 1974, Identification of a humoral alveolar-macrophage enhancement factor, *Physiologist* **17**:332.

Sorokin, S. P., and Brain, J. D., 1975, Pathways of clearance in mouse lungs exposed to iron oxide aerosols, *Anat. Rec.* **181**:581–600.

Strecker, F. J., 1967, Tissue reaction in rat lungs after dust inhalation with special regard to bronchial dust elimination and to the penetration of dust into the lung interstices and lymphatic nodes, in: *Inhaled Particles and Vapours* II (N. Davis, ed.), Pergamon Press, Oxford.

Task Group on Lung Dynamics for Committee II of the International Commission on Radiological protection, Deposition and Retention Models, 1966, *Health Phys.* **12**:173–208.

Tenner-Racz, K., Racz, P., Myrvik, Q. N., Ockers, J. R., and Geister, R., 1979, Uptake and transport of horseradish peroxidase by lymphoepithelium of the bronchus-associated lymphoid tissue in normal and bacillus Calmette-Guérin-immunized and challenged rabbits, *Lab. Invest.* **41**:106–115.

Thompson, M. L., and Short, M. D., 1969, Mucociliary function in health, chronic obstructive airway disease and asbestosis, *J. Appl. Physiol.* **26**:535–539.

Thor, D. E., and Dray, S., 1968, A correlation of human delayed hypersensitivity: Specific inhibition of capillary tube migration of sensitized human lymph node cells by tuberculin and histoplasmin, *J. Immunol.* **101**:51–61.

Toews, G. B., Viroslav, S., Hart, D. A., and Hansen, E. J., 1984, Pulmonary clearance of encapsulated and unencapsulated *Haemophilus influenza* strains, *Infect. Immun.* **45**:437–442.

Tomasi, T. B., Jr., 1976, *The Immune System of Secretions*, Prentice-Hall, Englewood Cliffs, NJ.

Tomasz, A., 1981, Surface components of streptococcus pneumoniae, *Rev. Infect. Dis.* **3**:190–211.

Villiger, B., Kelley, D. G., Engleman, W., Kuhn, C., III, and McDonald, J. A., 1981, Human alveolar macrophage fibronectin: Synthesis, secretion and ultrastructural localization during gelatin-coated latex particle binding, *J. Cell Biol.* **90**:711–720.

Virelizier, J. L., 1975, Host defense against influenza virus: The role of antihemagglutinin antibody, *J. Immunol.* **115**:434–439.

Waldman, R. H., and Ganguly, R., 1974, Immunity to infections on secretory surfaces, *J. Infect. Dis.* **130**:419–440.

Welliver, R. C., Kaul, T. N., and Ogra, P. L., 1980, The appearance of cell-bound IgE in respiratory-tract epithelium after respiratory syncytial virus infection, *N. Engl. J. Med.* **303**:1198–1202.

Wilson, L. A., and Spitznagel, J. K., 1968, Molecular and structural damage to *Escherichia coli* produced by antibody, complement and lysozyme systems, *J. Bacteriol.* **96**:1339–1348.

Woods, D. E., Straus, D. C., Johanson, W. G., Jr., and Bass, J. E., 1981, Role of salivary protease activity in adherence of gram negative bacilli to mammalian buccal epithelial cells *in vivo*, *J. Clin. Invest.* **68**:1435–1440.

Yap, K. L., and Ada, G. L., 1978, The recovery of mice from influenza virus infection: Adoptive transfer of immunity with immune T lymphocytes, *Scand. J. Immunol.* **7**:389–397.

Yoffey, J. M., and Courtice, F. C. (eds.), 1970, *Lymphatics, Lymph and the Lymphomyeloid Complex*, Academic Press, London, New York.

Young, K. R., Jr., and Reynolds, H. Y., 1984, Bronchoalveolar washings: Proteins and cells from normal lungs, in: *Immunology of the Lung and Upper Respiratory Tract* (J. Bienenstock, ed.), pp. 157–173, McGraw-Hill, New York.

2

The Role of the Reticuloendothelial System in Infections of the Central Nervous System

A. JULIO MARTINEZ

1. INTRODUCTION

1.1. HISTORICAL PERSPECTIVE

> The difficulty is not with cells which are morphologically similar to those described in classical anatomy, but rather with the many intermediate forms present.
>
> (Bornstein and Murray, 1958)

The term *reticuloendothelial system* (RES) or its alternate, *histiocytic metabolic apparatus*, was adopted by Ludwig Aschoff (1924) to define a group of diffuse collections of mesenchymal cells identified by phagocytic activity with ingestion and accumulation of particulate matter, as well as by their staining characteristics using vital dyes such as trypan blue, toluidine blue, lithium carmine, or pyrrole blue. Today some authors prefer the term *mononuclear phagocyte system* or *lymphoreticular system* as more adequate. It has been recognized that the common biologic and functional nature of these cells is of diverse ontogeny and phylogeny, but with common pathophysiologic and protective functions. They are characterized by changing their forms, proliferating and moving through fluids and tissues. The morphologic features of these cells are variable. Their origin has also been disputed by many authorities, so that several names have been coined for them. Cells are present in multiple tissues and organs and they usually appear as elements of inflammatory exudates. They have been called monocytes or

A. JULIO MARTINEZ • Division of Neuropathology, Department of Pathology, Presbyterian–University Hospital, University of Pittsburgh School of Medicine, Pittsburgh, Pennsylvania 15213.

transitional cells; endothelial leukocytes; macrophages by Metchnikoff; clasmatocytes by Ranvier; adventitial or leukocytoid cells by Marchand; histiocytes or polyblasts by Maximow; histoblasts by Kiyono; rhagiocrine cells by Renaut; endothelial sinusoidal cells; perithelial, mononuclear cells or phagocytes by Aschoff; resting wandering cells; Kupffer stellate cells of the liver; microglia of the central nervous system (CNS); or Hortega cells (Río-Hortega, 1919a).

Particular attention will be paid to the microglia, which are the RES cells in the CNS. The microglia participate in inflammatory reactions and "cleaning operations" within the CNS. They are the scavenger cells of the CNS. The blood–brain barrier separates the microglia from the remainder of the body, and for that reason these cells are particularly different, but also interesting and intriguing. They also remove cellular debris from necrotic CNS tissue, accumulate pigments from hemoglobin degradation, and engulf and phagocytize intracellular pathogens. They, like other RES cells, have an endothelial position, are intimately related to reticulum fibrils, or are lining vascular channels (Río-Hortega, 1932).

What a cell is and what it is capable of becoming cannot be determined accurately by its morphology alone. The concept of the RES is not in the definition of its cellular components, but in its role of the protective functions and the different ways in which it reacts to infections, tumors, injuries, and other pathologic conditions.

There is an abundance of publications on the RES, its cells, and their reaction to injury, some publications pertaining exclusively to the CNS (Baldwin, 1980; Baron and Gallego, 1972; Fujita and Kitamura, 1976; Johnson, 1974; Oehmichen, 1976, 1982, 1983; Oehmichen and Huber, 1976; Oehmichen *et al.*, 1979).

Elie Metchnikoff, the Russian zoologist, in 1882 became convinced that the body had an active cellular defense mechanism (Metchnikoff, 1913). Metchnikoff observed that "wandering" or mobile cells surrounded and engulfed foreign material when introduced into the body of a starfish larva. He also observed that fungal spores were engulfed and digested by leukocytes. After noting the same phenomenon with bacteria, he postulated a close relationship between immunity and phagocytosis. In this way a bridge was made linking immunology with the RES. He referred to the neutrophilic leukocytes of the blood as microphages and to the fixed-tissue histiocytes as macrophages. The terminology pertaining to the latter has been reviewed recently by Headington (1986).

Initially, it appeared that the cellular and the immunologic mechanisms were complementary but unrelated and that the concept of the RES was primarily cellular, including only the physiologic function of phagocytosis. From the morphologic point of view, it can be argued that reticulum cells are phagocytic, stain with vital dyes, form the argentaphilic reticulum, and participate in the defense mechanism of the host (Cancilla *et al.*, 1972).

It is not the purpose of this chapter to extensively discuss in detail the RES cell of origin, its morphologic features, and their different pathophysiologic interpretations or to delve into the clinical and experimental pathology of a specific CNS infection. Rather, this is an overview of the most common CNS reactions to infections and their relation to the RES and, in particular, the microglia.

1.2. PATHOPHYSIOLOGY OF THE RES

In an anatomic as well as in a functional sense, the RES can be divided into hematopoietic and lymphoreticular components. The hematopoietic component includes the bone marrow and neoplastic proliferations. The lymphoreticular component includes primitive cells and their histiocytic, lymphocytic and plasmacytic progeny, which are present in the lymph nodes, spleen, liver, tonsils, and thymus (Dunning and Furth, 1935). Many RES cells can manifest phagocytic behavior, when they are stimulated and challenged by pathogens and other noxious agents (Sheagren *et al.*, 1970).

The functions and reactions of the cells of the RES, of the undifferentiated reticular cell and its derivatives, and of the organs and specialized tissues containing these cells (the spleen, the brain, liver, lymph nodes, thymus, and bone marrow) can be classified as follows:

1. Hemopoiesis: Production, destruction, and maintenance at physiologic levels of the blood and bone marrow cells and its product
2. Phagocytosis: Engulfment of abnormal blood cells, of the cellular breakdown products, and of the particulate matter bacteria, viruses, fungi, and parasites; clearance from the blood and interstitial tissues and storage of certain chemical by-products of normal and abnormal metabolism (hemoglobin, hematoidin, hemosiderin bilirubin, lipids, and glycogen). Phagocytosis is the most important host-defense mechanism of the RES. Disorders in phagocytosis may result in often fatal viral, bacterial, protozoal, or fungal infections.
3. Production of antibodies: RES cells and their derivatives participate in the cellular and immunologic defense system, producing antibodies. Rejection of transplanted organs is characterized by significant host cell infiltration (mainly lymphocytes), indicating an important role of the RES cells in the rejection process.
4. Benign and malignant proliferation (hyperplasia and neoplasia): Cells of the RES may proliferate excessively and be transformed in tumors [hemangiopericytoma, malignant lymphoma (histiocytic type) or reticulum cell sarcoma (histiocytic lymphoma would be more accurate than reticulocell sarcoma), histiocytomas, Hodgkin's disease, and leukemias]. Immune deficiencies are sometimes associated with solid tumors, but usually these tumors may result from the use of immunosuppressive drugs with organ transplantation and the use of corticosteroids. Macrophages and lymphocytes may play a role in the surveillance and destruction of tumor cells.

1.3. IMMUNOSUPPRESSION AND OPPORTUNISTIC INFECTIONS

The immune responses, both humoral and cellular, require cells with immunologic specificity, such as T and B lymphocytes, and cells without apparent antigenic specificity, such as the macrophage. The interaction between B and T cells is necessary in the production of antibodies and also requires the participation of macrophages. The primary cell of humoral immunity is the B cell (so-called "bursa" or "bone-marrow-derived" cell). The T cell (thymus-dependent cell) is an important lymphoid cell in cell-mediated immunity (Table 1). Macrophages are also important cells that are intimately involved in the induction,

TABLE 1. T LYMPHOCYTE (THYMUS-DERIVED) DEFECTS: PREDOMINANT MICROORGANISMS, LOCATION, AND TYPE OF INFECTION

Type of CNS infection	Bacteria	Fungi	Protozoa	Viruses	Helminthic
Leptomeningitis	*Listeria monocytogenes*	*Cryptococcus neoformans*	*Toxoplasma gondii*	Varicella–zoster *herpes simplex*	*Strongyloides stercoralis*
Encephalitis	*L. monocytogenes* *Legionella pneumophila*	*C. neoformans*	*T. gondii* *Acanthamoeba* spp.	Varicella–zoster papovavirus AIDS	*S. stercoralis*
Abscess	*Nocardia asteroides*	*C. neoformans* *Aspergillus fumigatus*	*T. gondii* *Acanthamoeba* spp.		

stimulation, and expression of antibodies and cell-mediated immune responses. Macrophages are cells able to ingest and destruct antigens and intracellular pathogens. They are involved in the regulation and activation of lymphocytic functions in addition to their phagocytic capabilities (Oehmichen, 1983).

Immunoglobulin (Ig) is the effector molecule of the humoral immune response and is the product of a mature B cell, the plasma cell. A plasma cell secretes Ig of only one class and specificity. The classes of antibody include IgG, IgA, IgM, IgD, and IgE. Immunoglobulin A, IgG, and IgM are most important in viral infections. Immunoglobulin A is present in secretions on the mucosal surfaces of the respiratory and the gastrointestinal tracts where local production of virus-specific antibody can be quite effective in controlling infection. Virus-specific IgG antibodies persist for life, but those of the IgM class appear in primary infection and are short-lived.

Infection of the CNS with some bacteria (i.e., *Mycobacterium tuberculosis*) fungi (i.e., *Cryptococcus neoformans*), or protozoa (i.e., *Acanthamoeba* spp., *Toxoplasma gondii*) characteristically produces a well-developed granulomatous reaction with multinucleated giant cells (Martinez, 1982; Weiner and Fleming, 1984). These features, in some cases, may be present to a lesser degree or completely absent as a consequence of long-term treatment with immunosuppressive drugs. The host response and the immune status of the individuals in granulomatous amebic encephalitis (GAE) due to *Acanthamoeba* spp. may be influenced by prolonged immunosuppressive therapy. The mechanism of granuloma formation may be impaired when there is a deficiency in the production of histiocytes, macrophages, lymphocytes, plasma cells, giant cells, immunoblasts, neutrophils, eosinophils, or granulocytes. The histiocytes are essential cells in the formation of granulomas. The infecting organism itself may induce macrophage toxicity and therefore interfere in the production of the granulomatous reaction. When the function of the macrophages is inadequate and the irritant persists, granulomas cannot be formed. In addition, when there is a deficiency in cell-mediated immunity, a poor cellular response results and antibody production is impaired. The granulomatous reaction is an indicator of cell-mediated immunity. The essential factor in granuloma formation is a failure to eliminate the causative organism or irritant with persistence of the host defense mechanism and intact RES function (Cammermeyer, 1967).

In some of the reported cases of CNS infections due to fungi, protozoa, or bacteria the typical granulomatous reaction was not seen. In some cases with indisputable encephalitis, the evidence of immunodeficiency, the status of the cellular elements of immunity, the predisposing factors that may be responsible for development of opportunistic infections, the drug-induced immunosuppression, or the underlying disease cannot be established with certainty owing to incomplete clinical and pathologic documentation. In some CNS infections, such as primary amebic meningoencephalitis (PAM) due to *Naegleria fowleri*, it has been demonstrated that human neutrophils can kill amebas if they are exposed and stimulated by lymphokines and/or monokines (Ferrante and Mocatta, 1984). It is also well established that immunization with free-living ameba antigens may be done producing varying degrees of protection against infection in the host.

The conditions that increase the risk for infections and the clinical setting and underlying diseases associated with compromised host defense mechanisms and RES system deficiency may be multiple. They include hospitalization, malnutrition, trauma, burns, surgery, parenteral hyperalimentation, prolonged use of broad-spectrum antibiotics and corticosteroids, metabolic diseases (diabetes mellitus and liver disease), systemic lypus erythematosus, malignant tumors, blood transfusions, skin ulcers, pregancy, age-related factors such as prematurity and old age, lympho- and hematoproliferative diseases, drug addiction, organ transplantation with immunosuppressive therapy, and acquired immunodeficiency syndrome (AIDS). These conditions suppress or diminish the normal inflammatory and immune response of the host, both humoral and cell mediated, impair the RES, and increase the risk for opportunistic infections. In such cases infection with more than one pathogenic organism is possible. The clinical manifestations of the infection may mimic other diseases, because of the host's altered inflammatory responses (Johnson, 1974; Martinez, 1982; Martinez *et al.*, 1976; Weiner and Fleming, 1984).

2. RETICULOENDOTHELIAL CELLS OF THE CNS

2.1. MICROGLIA AND MACROPHAGES: ORIGIN AND TOPOGRAPHY

Río-Hortega, the Spanish pathologist, in 1919 discovered that the phagocytic cells in the CNS could be stained with his silver impregnation technique using silver carbonate. He then described these cells, also called the "third element," and named them microglia. Following injury to the CNS the microglia increase in number and engulf cellular debris and fat droplets. Río-Hortega stated that microglia cells are derived from the mesodermal or mesenchymal tissue or are adventitial in origin. Therefore, he stated that these cells belong to the RES and have no direct connection or genetic relationship to astrocytes, neurons, or oligodendroglial cells (Río-Hortega, 1919a,b, 1921, 1932). In spite of new ideas and concepts regarding the microglia, its derivatives, and their functions, the original contributions of Río-Hortega are of paramount importance and still remain the cornerstone in the study of the RES of the CNS.

The most primitive, still undifferentiated mesenchymal and multipotential cell may be called an "undifferentiated reticulum cell." This cell is neither phagocytic nor stainable with vital dyes. It is capable of differentiating into various types of connective tissue cells (osteocytes, lipocytes, chondrocytes, histiocytes, and fibrocytes) as well as into the "typical" cell of the RES, the fixed phagocytic cell, the phagocytic or nonphagocytic histiocyte, and into fibroblasts and other connective tissue cells. It is also able to differentiate in the hematopoietic direction, at which time it may be considered a hematopoietic reticulum cell.

Microglia first appear in the brain, mainly at the base, during the last stages of embryogenesis (Ferrer and Sarmiento, 1980a,b; Kershman, 1939; Stensaas and Reichert, 1971). They accumulate in large numbers in areas where the pia mater covers the white matter. From there, they spread over the entire CNS, in both the gray and white matter. They enter the brain with the invasion of the blood vessels at the moment of embyronic vascularization from the leptomeninges and choroid plexus and remain as resting elements of the RES. Some may arise by detachment of cells from the adventitial connective tissue of the neural blood vessels (Stensaas and Reichert, 1971).

These cells are "activated" when injury or damage occurs (Figs. 1 and 2). In the resting state, the microglia remain in place and immobile. They are cells of relatively simple structure. The bipolar microglia are called "rod cells," because the nucleus assumes a rodlike, elongated form. They are capable of responding to a variety of stimuli which provoke them into a state of intense physiologic and phagocytic activity. They are the source of some of the macrophages or "gitter cells." They become spherical and accumulate large amounts of detritus and tissue debris and are transformed into globoid, foamy cells, known also as "compound granular corpuscles, gitterzellen" or macrophages (Río-Hortega, 1919a).

Other macrophages found in lesions of the brain may enter the CNS from the blood stream, contributing as much as 60% of the cellular reaction at sites of injury (Kitamura, 1975; Kitamura *et al.*, 1972a,b; Konigsmark and Sidman, 1963; McKeever and Balentine, 1978). This is accompanied by changes in function and

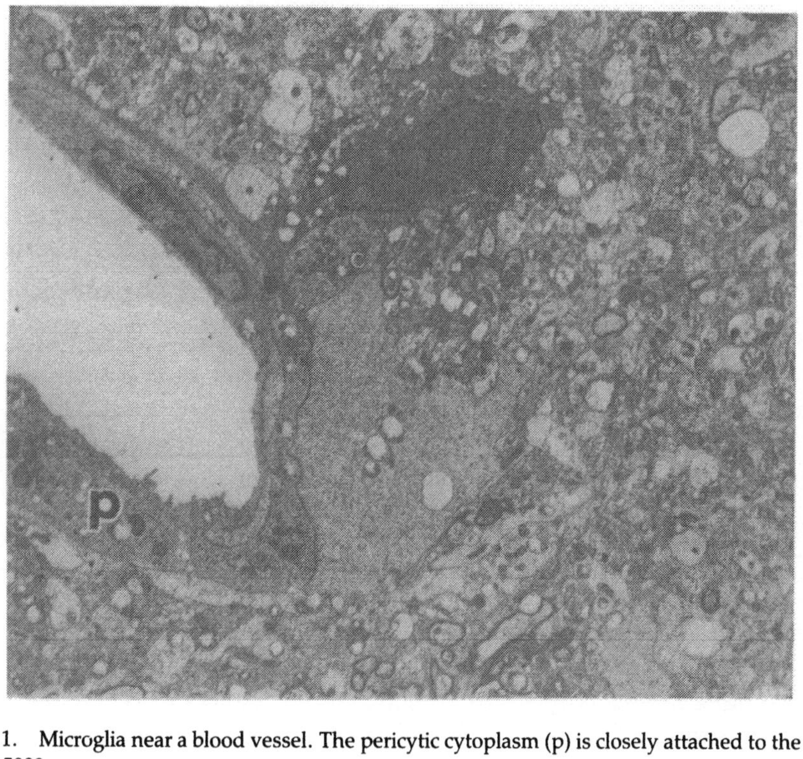

FIGURE 1. Microglia near a blood vessel. The pericytic cytoplasm (p) is closely attached to the vessel wall. ×5000.

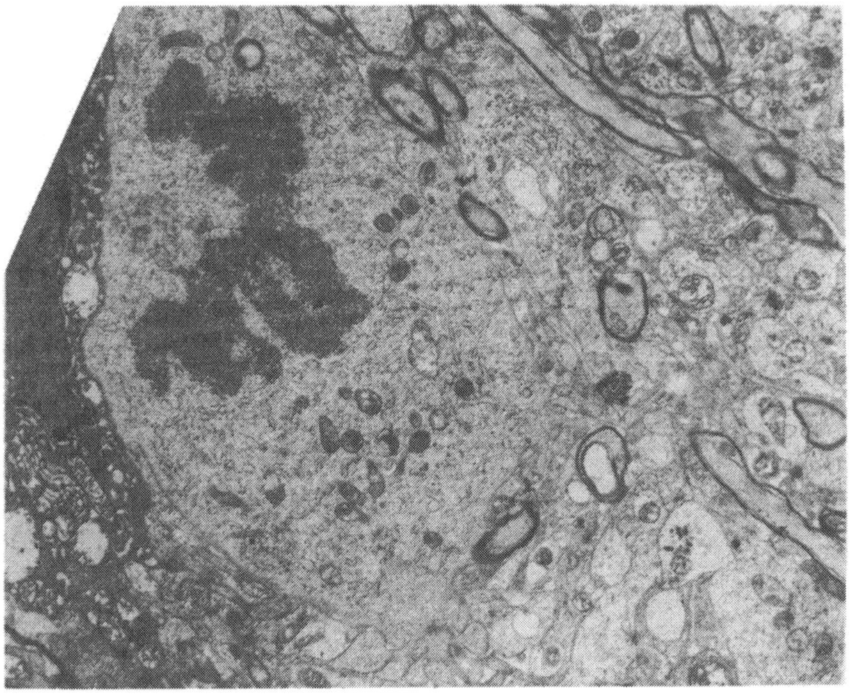

FIGURE 2. Phagocytic microglial cell in mitosis. Experimental *Acanthamoeba castellanii*, encephalitis, 20 days postinoculation. ×10,000.

morphology. The microglia are poorly branched, spine-covered cells elongated and distorted in shape, with an oval-shaped or elongated nucleus with one nucleolus. The microglial processes are short and run perpendicular to the surface of the cells. The cytoplasm of a microglia is scant and contains a moderate number of mitochondria, many vacuoles, dense granules, a well-developed rough endoplasmic reticulum, and inconspicuous Golgi membranes.

The average microglia length is 60 μm, with the nucleus measuring 12 μm. Under normal conditions, resting microglia cells do not store intravenously injected dyes because the wall of the blood vessel is encompassed by the foot processes of the astrocytes, which prevent the dye from reaching the microglia. However, if the wall of the blood vessel is injured, this allows the dye to penetrate. Then the microglia cells can store the dye. The microglia become a macrophage and lose their processes (Markov and Dimova, 1974). This is called "compound granular corpuscles," "gitterzellen," or lipophage. The cytoplasm swells and becomes granular in case of injury or disease. The chief function of the microglia appears to be phagocytosis (Ramón y Cajal, 1920; Río-Hortega, 1919a).

Autoradiographic studies have demonstrated that some of the brain macrophages in traumatic and other lesions in the CNS are derived from blood monocytes which migrate through the vessel wall (Casanova *et al.*, 1986; Schelper and Adrian, 1986). However, some of the brain macrophages are exclusively of endogenous (CNS) tissue origin (Kitamura *et al.*, 1972a; Mathew *et al.*, 1983; Matthews, 1974; Weiner and Fleming, 1984). Some authors believe that microglia are neuroectodermal and not the precursors of macrophages (Fujita and Kitamura, 1976).

Because of the silver staining techniques and the studies of Río-Hortega, it was demonstrated that the microglia in the brain and similar cells of the extracerebral sites all have argentaphilic cytoplasm and cytoplasmic processes and they are neither phagocytic nor stainable with vital dyes until altered by an inflammatory reaction. These are precisely the fundamental characteristics of the RES cells and, in particular, the microglia or Hortega cells (Río-Hortega, 1921).

2.2. MONOCYTES AND LYMPHOCYTES: RES INTERACTIONS

There are no hematogenous cells in the normal CNS parenchyma. Monocytes, lymphocytes, and polymorphonuclear leukocytes arrive in the CNS from the circulating blood (Fujita and Kitamura, 1976; Kitamura, 1975; McKeever and Balentine, 1978; Oehmichen, 1983; Stenwig, 1972). This happens when there is an attack by a noxious agent: a toxin, foreign material, or bacteria, fungus, virus, or a parasite. Immunoglobulins and other plasma proteins are produced, catabolized, or attached to RES cells, lymphocytes, macrophages, monocytes, and fibroblasts. Most brain macrophages originate from blood monocytes (Raedler and Raedler, 1984). Lymphocytes are not considered or included in the RES, but they interact with mononuclear cells and macrophages in the performance of their functions. In addition, the concept that plasma cells develop by differentiation of lymphocytes has been demonstrated, and there is no question that there is a direct relationship between the total cell population of plasma cells and the number of lymphocytes in chronic inflammatory reactions. These reasons are sufficient to include the lymphocytes as playing a role in the RES, defense mechanisms, and immunity.

2.3. PERICYTES AND PERIVASCULAR CELLS: RES INTERACTIONS

The pericytes are found throughout the CNS. They are also called Rouget cells. They lie between the two layers of the capillary basement membrane. It has been suggested that they may give rise to macrophages and microglia (Cancilla *et al.*, 1972). However, it has been demonstrated that microglia and pericytes are different cells (Stensaas, 1975; Stensaas and Reichert, 1971). They may act as phagocytes in response to tracer substances such as peroxidase when injected into the perivascular space (Cancilla *et al.*, 1972).

The pericytal microglia usually assume the immature feature in various stages and appear evenly in the normal cerebral hemispheres, cerebellum, brain stem, and spinal cord. The pericyte is a nonendothelial adventitial cell which is in close association with the capillary bed and is surrounded by the two layers of the basement membrane (Dodson *et al.*, 1976). Pericytes usually accumulate small amounts of lipofuscin in the aging brain. Activated pericytes usually contain electron-dense, lysosomallike structures (Fig. 3).

Multiple functions performed by the pericyte have been suggested, such as the synthesis of basement membrane, participation in some aspects of the blood–brain-barrier mechanism, mechanical support and stability to the vessel wall, differentiation into smooth muscle cells, and a phagocytic role. Pericytes may also participate in the capillary contractility, controlling the capillary diameter. Some investigators have regarded the pericytes as mesenchymal elements, possibly giving rise to tumors such as the hemangiopericytomas. Others believe that pericytes are involved in antibody formation and in the defense reaction and regard them as immature histiocytes or histioblasts.

The pericytes may be involved in the pathogenesis of acute and chronic inflammatory reaction and other injuries involving the CNS (Maxwell and Kruger, 1965; Stensaas, 1975; Stensaas and Reichert, 1971). It is important to study the capacity of pericytes to respond to different forms of stimuli, such as chemical, physical, infectious, allergic, and immunologic, with subsequent formation of perivascular infiltrates and the potential evolution or relationship of

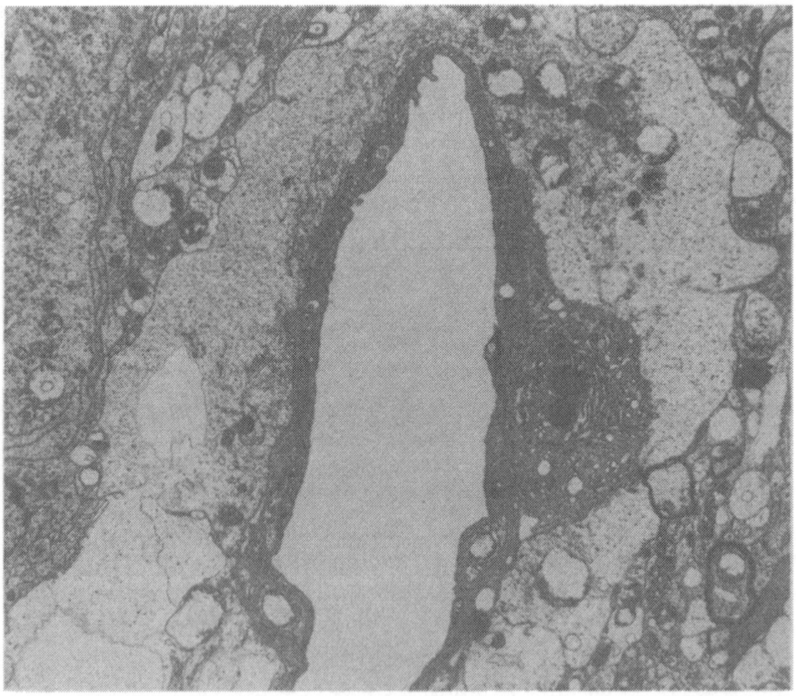

FIGURE 3. Activated pericyte. There are foci of cytoplasmic osmiophilic material, or "lysosomallike" bodies. There is swelling of the astrocytic foot processes around the vessel. Primary amebic meningoencephalitis due to *Naegleria fowleri*. Animal inoculated intranasally 5 days before death. ×5000.

these cells to lymphocytes, monocytes, plasma cells, epithelial cells, multinucleated giant cells, microglia, fibroblasts, blood vessels, and tumors.

Some of the properties of the pericytes may be attributed to their proximity and close attachment to the blood vessels, probably because of their better nutritional conditions. Pericytes as well as lymphocytes play an important function in the mechanism of defense in the CNS (Markov and Dimova, 1974).

3. INFECTION OF THE CNS: HOST RESPONSE OF THE CNS—ITS REACTION TO INFECTION

3.1. BACTERIAL

Most infectious agents are prevented from reaching the CNS by a variety of protective mechanisms, including the RES, the blood–brain barrier, the humoral and cellular immune responses, and the covering membranes (meninges) of the CNS. Otherwise, the CNS tissue is very susceptible to infections, because it is a highly specialized tissue in motor and sensory functions.

In spite of the multiplicity of infectious agents that may provoke diseases of the CNS, most infections of the CNS display similar or identical morphologic features. Therefore, infections of the CNS are characterized by nonspecific cellular reactions, but they do have some topographic sites of predilection. Interest in mechanisms by which CNS cells react to infectious agents has increased rapidly in recent years. Some studies show that susceptibility to infection is determined by the general immune system or defense mechanism of the host. It has been demonstrated that in experimental *Escherichia coli* meningitis the lining cells of the pia mater and the pericytes (adventitial cells) become phagocytic (Nelson *et al.*, 1962). These cells act as RES cells able to respond to a noxious stimulus and be transformed into macrophages. These cells are capable of responding to a variety of stimuli, changing their function and morphology. By the same token, when there is disruption of the protective mechanism and an increased vascular permeability, the CNS parenchyma may be invaded by the attacking "opportunisitic" pathogens.

Pyogenic (suppurative) infections are usually produced by bacteria. The inflammatory response is characteristically composed of polymorphonuclear leukocytes (acute or purulent exudate). These cells come from the circulating blood, and because the infection is usually fast or fulminant, there is no sufficient time for a proper RES response. Some bacteria may produce a necrotizing angiitis. The most common bacterial organisms associated with this type of reaction are *Hemophilus influenza, Streptococcus pneumonia, Neisseria meningitidis, Listeria monocytogenes, E. coli, Pseudomonas aureoginosa, Klebsiella pneumoniae, Staphylococcus aureus,* and *Streptococcus viridans.*

M. tuberculosis and *Treponema pallidum* usually produce a subacute or chronic inflammatory response in the leptomeninges, with marked vascular changes of occlusive endarteritis. The RES is particularly involved in syphilitic encephalitis, which is characterized by proliferation of rod-shaped microglia in the cerebral cortex and perivascular infiltration of lymphocytes and plasma cells.

Rickettsial infections, on the other hand, may produce perivascular histiocytic, microglial, or phagocytic nodules in the gray matter and in the brain stem. In Rocky Mountain spotted fever, rickettsiae may be found in enlarged endothelial cells. They invaded endothelial cells and smooth muscle cells of blood vessels. Acute angiitis involving small arteries, capillaries, and venules is the most significant finding in the CNS. Rickettsiae are prokaryotic, coccobacillary microorganisms.

3.2. VIRAL

Viruses usually stimulate the RES cells within the CNS parenchyma. Many viruses can replicate in CNS tissue and its covering if directly inoculated. Some viruses are neurotropic and characterized by their selective attractability to CNS structures. In addition, when the protective barriers are affected by viruses, they can reach and replicate in the CNS tissue more easily. Viruses can invade the brain by several routes, including the blood stream, the peripheral nerves, and the olfactory nerves or directly through open wounds. Passage through the placenta may occur in congenital viral infections. There is a selective vulnerability of the CNS cells, blood vessels, and meninges to different viruses. This is one of the most striking characteristics of this superspecialized tissue (Table 2).

TABLE 2. SELECTIVE VULNERABILITY OF CNS CELLS TO BACTERIA, VIRUSES, AND PROTOZOA

Cell	Disease	Infective agent
Meningeal	Viral meningitis (leptomeningitis) Meningovascular syphilis	Herpes simplex type 2; echovirus; mumps; coxasackievirus; Treponema pallidum
Neurons A. Cortical	Viral encephalitis Rabies Rubeola AIDS	Western equine encephalitis Venezuelan equine encephalitis viruses; paramyxovirus; Herpes simplex; arbovirus; rhabdovirus. HIV
B. Motor	Poliomyelitis	Poliovirus
C. Sensory	Shingles; herpes zoster	Varicella–zoster; herpes simplex
Oligodendroglia	Progressive multifocal Leukoencephalopathy	Papovavirus; cytomegalovirus; rubella; JC virus
Vascular endothelial (breakdown of blood–brain barrier)	Rocky Mountain spotted fever	*Rickettsia rickettsii*
	Exanthematic thyphus	
	Sleeping sickness	*Trypansoma gambiense*
	American trypanosomiasis (Chagas' disease)	*Trypanosoma cruzi*
	Malaria	*Plasmodium falciparum*
	Meningovascular syphilis	*Treponema pallidum*

Most viral infections of the CNS display quite similar histopathologic features even though there are some topographic differences between them. They are characterized by chronic inflammatory cell infiltrates in the leptomeninges, blood vessels, cranial or peripheral nerves, and the CNS parenchyma. The inflammatory reaction in such sites is mainly composed of lymphocytes, plasma cells, histiocytes, and macrophages (Fig. 4). The polymorphonuclear leukocytes are usually transient cells, present at the onset of the infection. The landmark of viral infection in the CNS is the presence of microglial (phagocytic) nodules or Babès nodules, which are characterized by focal collections of microglia, histiocytes, few lymphocytes, and rare plasma cells (Fig. 4). Diffuse proliferation of microglia or rod cells with intranuclear inclusion bodies (Cowdry type A and B) is also a frequent morphologic feature of viral infections within the CNS. Reactive astrocytosis, loss of myelin, and damage of oligodendrocytes may also be seen.

In addition, viral diseases of the CNS are characterized by neuronal degeneration, satellitosis, or neuronophagia (Figs. 5 and 6). Chromatolysis and pyknotic changes may be present. The dead or dying neurons are usually surrounded by a few polymorphonuclear leukocytes, macrophages, lymphocytes, and microglia.

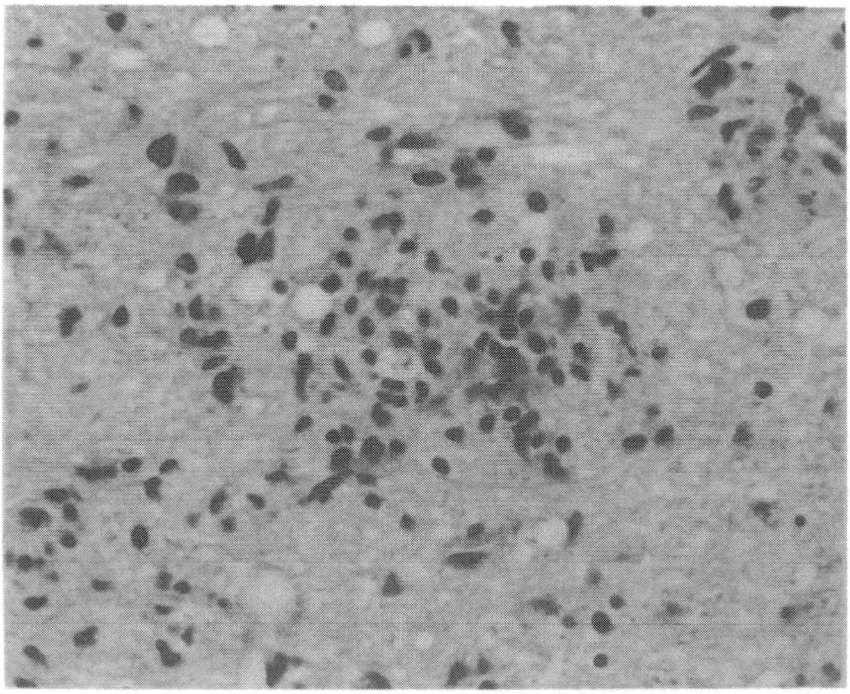

FIGURE 4. Subacute sclerosing panencephalitis. Phagocytic or microglial nodule in the medulla oblongata. There are proliferation of microglia, scattered lymphocytes, few mononuclear cells and macrophages. H&E. ×500.

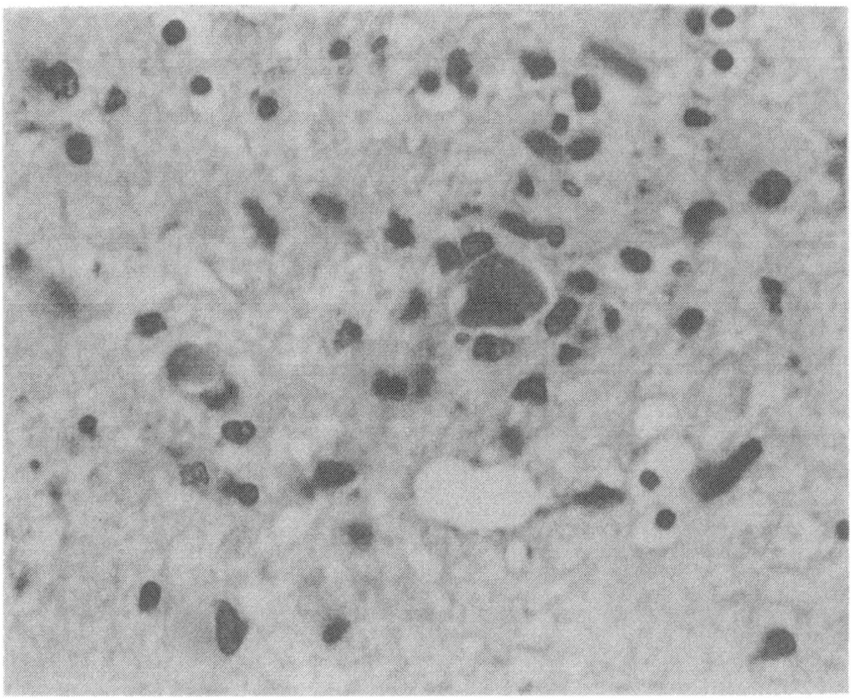

FIGURE 5. Neuronophagia in a case of subacute sclerosing panencephalitis in the medulla oblongata. Proliferating and activated microglia and mononuclear cells around a neuron. H&E. ×620.

Some of the most common viruses that may attack the CNS are the following:

1. RNA: Enteroviruses: poliomyelitis; coxsackieviruses
   Arenaviruses: lymphocytic choriomeningitis
   Arboviruses (arthropod-borne viruses): VEE; St. Louis and Japanese B encephalitis viruses
   Paramyxoviruses: mumps; measles = rubeola = subacute sclerosing panencephalitis
   Retrovirus: HIV
   Rhabdoviruses: rabies
   Togaviruses: rubella = chronic progressive panencephalitis
2. DNA: Papovaviruses: progressive multifocal leukoencephalopathy
   Herpesviruses: *herpes simplex* encephalitis
   Varicella–zoster virus: herpes zoster encephalomyelitis and varicella encephalitis
   Cytomegalovirus: cytomegalovirus necrotizing encephalitis

Virus-related CNS diseases occur in many different clinical patterns. Both the incubation period (short or long) and the period of apparent disease (acute,

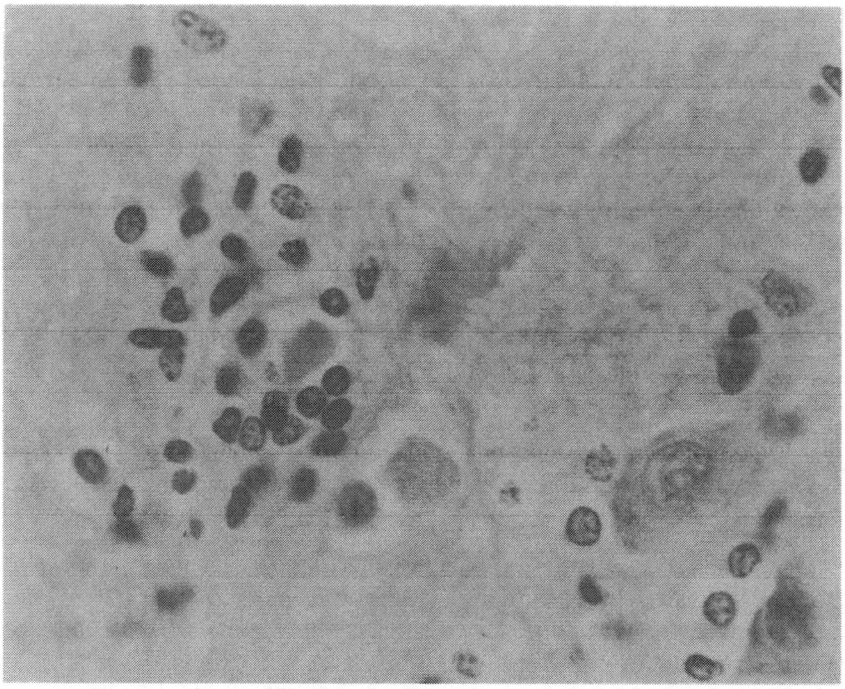

FIGURE 6. Neuronophagia in the cerebral cortex of a patient with subacute sclerosing panencephalitis. The neuron is almost destroyed. It is surrounded by microglia and mononuclear cells, probably lymphocytes. H&E. ×1000.

subacute, or chronic) are variable, and virtually all possible combinations of incubations and clinical types may occur (Weiner and Fleming, 1984).

Viruses may be divided into two main groups. In one group are viruses with a well-defined protein structure and a nucleic acid genome; these are termed "conventional" agents. In the other group are the "unconventional" agents or transmissible, slow viruses, which are viruses in which the protein structure and nucleic acid genome have not been demonstrated. The conventional viruses are the classic agents such as measles, mumps, and poliovirus, whereas the unconventional agents are those causing transmissible subacute spongiform encephalopathies or Creutzfeldt–Jakob disease and kuru.

A significant number of reports indicate that some human hematopoietic neoplasms may be caused by viruses. Viruses of the RNA and DNA groups have been associated with the production of malignant tumors. It has been demonstrated that a polyoma virus was responsible for the production of progressive multifocal leukoencephalopathy (PML), a demyelinating opportunistic viral infection of the CNS. In addition, polyoma viruses known as simian virus 40 (SV40) and JC virus are able to produce malignant tumors in hamsters. SV40–PML and JC viruses have been isolated from CNS tissue of patients with PML

(ZuRhein and Chou, 1965). Some viruses of the herpes virus family and the Epstein–Barr viruses are also oncogenic viruses.

Herpes simplex (type 1) encephalitis is associated with a high incidence of mortality and morbidity. Herpes simplex encephalitis is the most common sporadic, severe, and fatal encephalitis in the United States. The clinical picture is not specific, but the initial symptoms are usually headache and fever, progressing over a few days to confusion, personality changes, and speech difficulties. Focal neurologic symptoms and seizures may follow. The lesions are commonly located on the basal aspect, mesial surface of the temporal lobes and are characterized by a necrotizing, hemorrhagic meningoencephalitis. The open-brain biopsy is the method of choice for a definitive diagnosis. With appropriate CNS tissue, studies by light and electron microscopy and immunoperoxidase and viral cultures can be made. The replication cycle for herpes simplex virus is 18 hr. The cultures are usually positive in 24 hr. However, delays up to 1 week have been reported. Herpes simplex type 2 usually produces aseptic meningitis and spreads venereally, producing lesions in the genitalia.

Rabies may be considered the most important viral disease transmitted from animals to humans. It is a frightful disease with no known cure. Only vaccination of animals and of the victims of bites can prevent the spread of the disease. A neural pathway for entry into the CNS is used by the virus. Viral particles are transported within the axoplasm of peripheral nerves and its Schwann cells. Intracytoplasmic eosinophilic inclusion bodies or Negri bodies usually are found in neurons and Purkinje cells of patients and animals dying of rabies.

In AIDS dementia HIV-III, DNA, and RNA virus have been detected in the brain of AIDS patients with encephalopathy.

3.3. FUNGAL

Fungal infections of the CNS are more common in patients with immunosuppression, defects of cellular immunity, or AIDS. Fungi characteristically produce granulomatous reactions in the CNS. However, impaired RES may show absence or minimal granulomatous reaction and minimal chronic inflammation. Abscesses due to fungi may be seen occasionally.

Cryptococcus neoformans, Nocardia asteroides, Aspergillus fumigatus, and *Candida* species (most commonly *C. albicans* and *C. tropicalis*) are the most frequent fungi that produce infections in the CNS. Bloodstream invasion and potential dissemination into the CNS may occur as a complication of immunosupressive therapy.

The fungal infections of the CNS may produce granulomatous leptomeningitis with epithelioid and giant cell formation as well as intraparenchymal mycotic abscesses and granulomas. The blood vessels are particularly affected in invasive aspergillosis and candidiasis. In such instances, the blood vessel walls are pierced by hyphae of the fungal organism, which are usually associated with thrombosis and infarct of the adjacent CNS parenchyma. The RES reaction in such cases is absent or minimal, probably owing to the rapid development of the disease and the impaired defense mechanisms of the host.

3.4. PROTOZOAL

The role of the RES in protozoal infections of the CNS is not well understood. Arachnoidal phagocytic reaction can be elicited after intranasal inoculation with pathogenic free-living amebas *N. fowleri* and *Acanthamoeba* spp. (Martinez, 1982; Martinez *et al.*, 1976). The pia-arachnoidal cells actively phagocytize and destroy some of the invading protozoa in the subarachnoidal space. The neuropil may reveal extensive necrotizing hemorrhagic lesions during the course of the infection, mainly on the rostroventral aspect of the brain and cerebellum. The arachnoidal cells may be transformed as migratory macrophages. Other macrophages from the circulation probably participate in the host-defense mechanism as active phagocytes. These suggest that an allergic type of inflammatory response is probably produced, at least in the earliest phases of the infection, by free-living amebas. This type of allergic reaction is probably due to disintegration of neural tissue and its by-products and the enzymes and proteins of the pathogenic ameba. An antigen–antibody reaction may be initiated, thus forming the characteristic histopathologic features of free-living amebic infection in the CNS. In the experimental PAM due to *N. fowleri* and GAE due to *Acanthamoeba* spp., the arachnoidal and pial cells participate as active phagocytes, as sustentacular cells of the olfactory neuroepithelium do, engulfing and digesting foreign proteins. Arachnoid phagocytic capability might be elicited following the introduction of foreign protein from the ameba into the subarachnoid space (Martinez *et al.*, 1976).

Macrophages may have been formed by the perivascular adventitial cells, perivascular mesenchymal tissue, pericytes, pial cells, and/or arachnoidal cells (Fig. 7). The microglia may originate from the perivascular mesenchymal tissue, but it is known that some of the macrophages may enter the CNS from the bloodstream, and both of them assume comparable functions in the process of immune response. Pial cells are commonly swollen with pinocytotic vesicles, cytoplasmic vacuoles, and abundant granular cytoplasm, as in "aseptic meningitis." In bacterial meningitis the phagocytic cells appear to be arachnoid cells rather than migratory cells from the circulation.

E. histolytica is a parasitic protozoa that occasionally produces brain abscesses. *T. gondii* is often the cause of CNS infection in the immunocompromised host. It can be seen associated with underlying lymphomas, leukemias, or in individuals who are immunosuppressed by iatrogenic means or with AIDS. *Toxoplasma* encephalitis frequently is characterized by multifocal areas of necrotizing encephalitis. It can be found in newborns from healthy mothers without a history of immunoincompetence. The diagnosis rests on a demonstration of the pseudocyst or the trophozoite in the cerebral tissue. *Trypanosoma gambiense* and *Trypanosoma rhodesiense* provoke the "sleeping sickness" characterized by diffuse encephalitis or meningoencephalitis with marked RES reaction and the

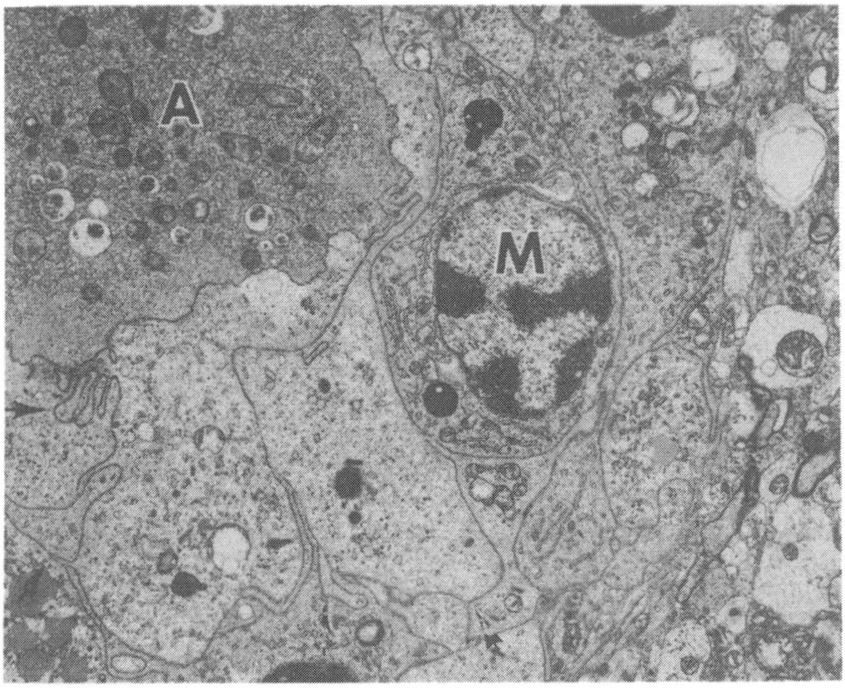

FIGURE 7. Trophozoite of *Acanthamoeba castellani* (A) encompassed by interdigitating cytoplasmic projections of macrophages (arrows). There is an "activated microglia" (M). Same experiment as Fig. 3. ×2500.

presence of Mott cells. *Trypanosoma cruzi* or *Bartonella bacilliformis* may produce meningoencephalitis with severe participation of the vascular endothelium, microglia proliferation, and neuronal loss. *Plasmodium falciparum* may produce malaria with CNS involvement when the infected erythrocytes become clogged in brain capillaries (Sheagren *et al.*, 1970).

3.5. HELMINTHIC

Examples of common helminthic CNS infections are *Strongyloides stercoaralis* and *Angiostrongylus cantonensis.* Disseminated *Strongyloides* infection has occasionally been observed in individuals with an apparently normal immune system. This disease is most often seen in immunocompromised patients, particularly those with T-lymphocyte dysfunction. Two forms of severe *Strongyloides* infection are recognized in the immunocompromised host: a "hyperinfection syndrome," in which there is massive invasion of the gastrointestinal tract and lungs (the major site of involvement in the usual life cycle of the organisms), and "disseminated strongyloidiasis," in which large numbers of filariform larvae penetrate the bloodstream and infest many organs, particularly the CNS. In disseminated strongyloidiasis there may be an associated gram-negative infection of the bloodstream and of the CNS, possibly resulting from transport of these bacteria by the parasite. *A. cantonensis* usually produces an eosinophilic leptomeningitis and multifocal areas of chronic inflammatory reaction with modest microglia proliferation around affected areas of the CNS.

A careful epidemiologic history and search for ova and adult worms are indicated in anyone with a history of travel to an endemic area, even many years previously.

Other parasitic infections in the CNS with minimal RES response are cysticercosis, due to *Taenia solium* or its larva (*Cysticercus cellulosae*); echinococcosis, due to *Echinococcus granulosus;* schistosomiasis, due to *Schistosoma* spp.; paragonimiasis, due to *Paragonimus westermani;* and trichinosis, due to *Trichinella spiralis*. These parasites may produce diffuse or nodular isolated granulomas and single or multiple cystic structures with mainly reactive astrocytosis and minimal microglia proliferation.

4. CONCLUSIONS

The RES, or the mononuclear phagocyte system, is represented in the CNS by the microglia, the pericyte, and its derivatives. Monocytes, T and B lymphocytes, and macrophages interact in the performance of the defense mechanism and immunity. These cells have a protective function in the CNS and its coverings. Microglia remain "inactive" in normal condition, but when a pathogen or other noxious agent invades the CNS, the microglia are activated and usually participate with macrophages from the blood in the defense mechanism of the host. It is possible that the perivascular pericytic cells may also be transformed in microglia or microglialike cells.

Most pathogenic or infectious organisms may be prevented from reaching the CNS, mainly by the protective role of the RES and its cells, by humoral and cell-mediated immunity, by the covering membranes of the brain, and by the blood–brain barrier. When one or all of these protective mechanisms is at fault, by either trauma, compromised host, immunosuppressive drug therapy, or any other defects of cellular immunity such as AIDS, it is then likely that infection with opportunisitic viruses, fungi, bacteria, parasites, or protozoa might rapidly and overwhelmingly develop.

The most dramatic proliferation of microglia or reactivation of the RES cells in the CNS takes place in viral diseases and in syphilis. This proliferation is less conspicuous in granulomatous amebic encephalitis due to *Acanthamoeba* spp. and basically negligible in primary amebic meningoencephalitis due to *N. fowleri*. One consequence of altered immunocompetence is a compromised capacity of the host to tolerate viral, fungal, protozoal, and parasitic or bacterial infections. Any compromised defense mechanism of the host significantly enhances the susceptibility to CNS infections.

There is an important link between the way microorganisms interact with RES cells and the cells lining the vasculature in the CNS. Perhaps, adhesions of antigens on the microviral or cellular surfaces play a role in the pathogenesis of infections, particularly CNS infections. It is possible that cells of the RES or the mononuclear phagocyte system are not the only cells or structures involved in vascular clearance and vascular change.

Infections of the CNS may be considered complications of systemic infections. The CNS may be invaded during the course of generalized or localized infections through a hematogenous route, peripheral nerves, or by contiguity.

REFERENCES

Aschoff, L., 1924, Das Reticulo-endotheliale system, *Ergebnd. Inn. Med. u. Kinderheild.* **26**:1.

Baldwin, F., 1980, Microglia and brain macrophages, in: *The Reticuloendothelial System. A comprehensive treatise*, Volume 1 (I. Car and W. T. Daems, eds.), pp. 635–669, Plenum Press, New York.

Baron, M., and Gallego, A., 1972, The relation of the microglia with the pericytes in the cat cerebral cortex, *Z. Zellforsch.* **128**:42.

Bornstein, M. B., Murray, M. R., 1958, Serial observations on patterns of growth, myelin formation, maintenance and degeneration in cultures of new-born rat and kitten cerebellum, J. Biophysical and Biochemical Cytology, **4**(5):499–504.

Cammermeyer, J., 1967, Microglia cells in diffuse and granulomatous encephalitis in the rabbit, *Acta Neuropathol. (Berlin)* **7**:261.

Cancilla, P. A., Baker, R. N., Pollock, P. S., and Frommes, S. P., 1972, The reaction of pericytes of the central nervous system to exogenous protein, *Lab. Invest.* **26**(4):376–383.

Casanova, M. F., Troncoso, J. C., and Price, D. L., 1986, Hematogenous origin of brain macrophages: A case report, *Neurology* **36**:844–847.

Dodson, R. F., Tagashira, Y., and Wai-Fong, C. L., 1976, Acute pericytic response to cerebral ischemia, *J. Neurol. Sci.* **29**:9–16.

Dunning, H., and Furth, J., 1935, Studies on the relation between microglia, histiocytes and monocytes, *Am. J. Pathol.* **11**(6):895–913.

Ferrante, A., and Mocatta, T. J., 1984, Human neutrophils require activation by mononuclear leucocyte conditioned medium to kill the pathogenic free-living amoeba, *Naegleria fowleri, Clin. Exp. Immunol.* **56**:559–566.

Ferrer, I., and Sarmiento, J., 1980a, Nascent microglia in the developing brain, *Acta Neuropathol. (Berlin)* **50**:61–67.

Ferrer, I., and Sarmiento, J., 1980b, Reactive microglia in the developing brain, *Acta Neuropathol. (Berlin)* **50**:69–76.

Fujita, S., and Kitamura, T., 1976, Origin of brain macrophages and the nature of the microglia, in: *Progress in Neuropathology*, Volume 3 (H. M. Zimmerman, ed.), pp. 1–50, Grune & Stratton, New York.

Headington, J. T., 1986, The histiocyte. In memoriam, *Arch. Dermatol.* **122**:532–533.

Johnson, R. T., 1974, Pathophysiology and epidemiology of acute viral infections of the nervous system, in: *Advances in Neurology*, Volume 6 (R. A. Thompson and J. R. Green, eds.), pp. 27–40, Raven Press, New York.

Kershman, J., 1939, Genesis of microglia in the human brain, *Arch. Neurol. Psychiatry* **41**:24–49.

Kitamura, T., 1975, Hematogenous cells in experimental Japanese encephalitis, *Acta Neuropathol. (Berlin)* **32**:341–346.

Kitamura, T., Hattori, H., and Fujita, S., 1972a, EM-Autoradiographic studies on the inflammatory cells in the experimental Japanese encephalitis, *J. Electron Microsc.* **21**(4):315–322.

Kitamura, T., Hattori, H., and Fujita, S., 1972b, Autoradiographic studies on histogenesis of brain macrophages in the mouse, *J. Neuropathol. Exp. Neurol.* **31**:502–518.

Konigsmark, B. W., and Sidman, R. L., 1963, Origin of macrophages in the mouse, *J. Neuropathol. Exp. Neurol.* **22**:643–676.

Markov, D. V., and Dimova, R. N., 1974, Ultrastructural alterations of rat brain microglial cells and pericytes after chronic lead poisoning, *Acta Neuropathol. (Berlin)* **28**:25–35.

Martinez, A. J., 1982, Acanthamoebiasis and immunosuppression, *J. Neuropathol. Exp. Neurol.* **41**:548–557.

Martinez, A. J., Astruc, J., and Escobar, M., 1976, The reticuloendothelial system (RES) in experimental amebic meningoencephalitis, in: *Advances in Experimental Medicine and Biology,* Volume 73A (S. M. Reichard, M. R. Escobar, and H. Friedman, eds.), pp. 277–286, Plenum Press, New York.

Mathew, R. C., Gupta, S. K., Katayama, I., Curtis, J., and Turk, J. L., 1983, Macrophage specific antigen is expressed by resting microglia in the CNS but not by Langerhans cells in the skin, *J. Pathol.* **141:**435–440.

Matthews, M. A., 1974, Microglia and reactive "M" cells of degenerating central nervous system: Does similar morphology and function imply a common origin? *Cell Tissue Res.* **148:**477.

Maxwell, D. S., and Kruger, L., 1965, Small blood vessels and the origin of phagocytes in the rat cerebral cortex following heavy particle irradiation, *Exp. Neurol.* **12:**33–54.

McKeever, P. E., and Balentine, J. D., 1978, Macrophage migration through the brain parenchyma to the perivascular space following particle ingestion, *Am. J. Pathol.* **93:**153–164.

Metchnikoff, E., 1913, Die Lehre von den phagocyten und deren experimentelle grundlagen, in: *Handbuch der Pathogenen Mikroorganismen* (W. Kalle and A. Wassermann, eds.), pp. 655–731, Fischer, Jena.

Nelson, E., Blinzinger, K., and Hager, H. 1962, Ultrastructural observations on phagocytosis of bacteria in experimental (*E. coli*) meningitis, *J. Neuropathol. Exp. Neurol.* **21:**155.

Oehmichen, M., 1976, Receptor activity on some mesenchymal cells in CNS of normal rabbits, *Acta Neuropathol. (Berlin)* **35:**205–218.

Oehmichen, M., 1982, Functional properties of microglia, in: *Recent Advances in Neuropathology* (W. T. Smith and J. B. Cavanagh, eds.), pp. 83–107, Churchill Livingstone, Edinburgh.

Oehmichen, M., 1983, Inflammatory cells in the central nervous system: An integrating concept based on recent research in pathology, immunology and forensic medicine, in: *Progress in Neuropathology,* Volume 5 (H. H. Zimmerman, ed.), pp. 277–335, Grune & Stratton, New York.

Oehmichen, M., and Huber, H., 1976, Reactive microglia with membrane features of mononuclear phagocytes, *J. Neuropathol. Exp. Neurol.* **35:**30–39.

Oehmichen, M., Wiethölter, H., and Greaves, M. F., 1979, Immunological analysis of human microglia: Lack of monocytic and lymphoid membrane differentiation antigens, *J. Neuropathol. Exp. Neural.* **38:**99–102.

Raedler, E., and Raedler, A., 1984, Local proliferation of brain macrophages in central nervous system tissue cultures, *J. Neuropathol. Exp. Neurol.* **43:**531–540.

Ramón y Cajal, S., 1920, Algunas consideraciones sobre la mesoglia de Robertson y Río-Hortega, *Trabajos Lab. Invest. Biol.* **18:**109–127.

Río-Hortega, P del., 1919a, Poder fagocitario y mobilidad de la microglia, *Bol. Soc. Españ. Biol.* **9:**154–171.

Río-Hortega, P del., 1919b, El tercer elemento de los centros nerviosos. I. La microglia en estado normal. II. Intervención de la microglia en los procesos patológicos. III. Naturaleza probable de la microglia, *Bol. Soc. Españ. Biol.* **9:**69.

Río-Hortega, P del., 1921, Histogenesis y evolución normal, éxodo y distribución regional de la microglia, *Mem. R. Soc. Españ. Hist. Nat.* **11:**213.

Río-Hortega, P del., 1932, Microglia, in: *Cytology and Cellular Pathology of the Nervous System,* in: Volume 2 (W. Penfield, ed.), pp. 482–534, Hafner, New York.

Schelper, R. L., and Adrian, E. K., 1986, Monocytes become macrophages: they do not become microglia: Light and electron microscopic autoradiographic study suing 125-iododeoxyuridine, *J. Neuropathol. Exp. Neurol.* **45:**1–19.

Sheagren, J. N., Tobie, J. E., Fox, L. M., and Wolff, S. M., 1970, Reticuloendothelial system phagocytic function in naturally acquired human malaria, *J. Lab. Clin. Med.* **75**(3):481–487.

Stensaas, L. J., 1975, Pericytes and perivascular microglial cells in the basal forebrain of the neonatal rabbit, *Cell Tissue Res.* **158:**517–541.

Stensaas, L. J., and Reichert, W. H., 1971, Round and amoeboid microglial cells in the neonatal rabbit brain, *Z. Zellforsch.* **119:**147–163.

Stenwig, A. E., 1972, The origin of brain macrophages in traumatic lesions, Wallerian degeneration, and retrograde degeneration, *J. Neuropathol. Exp. Neurol.* **31:**696.

Weiner, L. P., and Fleming, J. O., 1984, Viral infections of the nervous system, *J. Neurosurg.* **61**:207–224.

Wood, G. W., Gallahan, K. A., Tilzer, S. A., Vats, T., and Morantz, R. A., 1979, The failure of microglia in normal brain to exhibit mononuclear phagocyte markers, *J. Neuropathol. Exp. Neurol.* **38**:369–376.

ZuRhein, G., and Chou, S. M., 1965, Particles resembling papova viruses in human cerebral demyelinating disease, *Science* **148**:1477–1479.

3

Chronic Adult Periodontitis
An Oral Infection/Infestation

J. TERRELL HOFFELD and STEPHAN E. MERGENHAGEN

1. INTRODUCTION

The term *periodontal* refers to any of the bony and soft tissues that support the teeth. Diseases of these supporting tissues affect the majority of the adult population throughout the world and are the major cause of loss of teeth in adults. The most common form of periodontal disease and the major consideration of this chapter is chronic adult periodontitis (CAP). The pathologic process(es) involved in CAP are initiated and sustained by bacteria in the dental plaque adherent to the teeth. In the absence of these bacteria, the disease can neither begin nor progress, but the destructive process of CAP does not appear to require the presence of organisms within the periodontal tissues. Thus, semantically, CAP is more properly classified as an infestation than an infection. As with other pathological infestations, bacterial products initiate the host response in the tissues; this response is the major source of destruction. Two of the most interesting aspects of the study of this disease concern the nature of the ecologic development of the initiating mixed bacterial flora and the mechanisms involved in the breakdown of supporting tissues by the host response. Indeed, much of the knowledge gained by researchers interested in the immunopathology of CAP is applicable to the mechanisms of tissue destruction in other chronic inflammatory diseases such as rheumatoid arthritis (Snyderman and McCarty, 1982; Decker *et al.*, 1984).

In the following sections, the progressive aggregation of bacteria in the dental plaque and their association with progressive clinical disease are addressed. Following that, the nature of the host response, especially the pivotal role of the macrophage, is discussed, not only in controlling this disease process, but also in causing bone resorption in its differentiated form as an osteoclast.

J. TERRELL HOFFELD • Oral Biology and Medicine Study Section, Division of Research Grants, National Institutes of Health, Bethesda, Maryland 20892. STEPHAN E. MERGENHAGEN • Laboratory of Microbiology and Immunology, National Institute of Dental Research, National Institutes of Health, Bethesda, Maryland 20892.

2. THE NORMAL PERIODONTIUM

The slow progression of CAP can occur over many decades and range from moderate tissue damage to complete loss of teeth. The relationships of the components of the periodontium to a tooth are diagrammed in Fig. 1. Principal support of the tooth is provided by the collagenous fibers of the periodontal ligament. Some of these fibers insert into the cementum covering the root of the tooth, some into the alveolar bone of the jaw, and some into the lamina propria of the gingival tissue. Other fibers interconnect the inserted fibers in a dense meshwork. This attachment involves the entire root surface of the erupted tooth. The surface of the periodontal ligament exposed to the oral cavity is covered by a thin epithelial attachment. This epithelium prevents gross ingress of contents of the gingival crevice into the periodontal ligament space. The fragility of this epithelium is one of the weakest links in the chain of protection of the periodontium.

The gingivae cover the outer surfaces of the alveolar bone, providing adequate protection of that bone from external influences. The gingiva immediately adjacent to the crown of each tooth is free-standing, unsupported by bone. This loose cuff around each tooth forms a peripheral crevice. The gingival crevice is

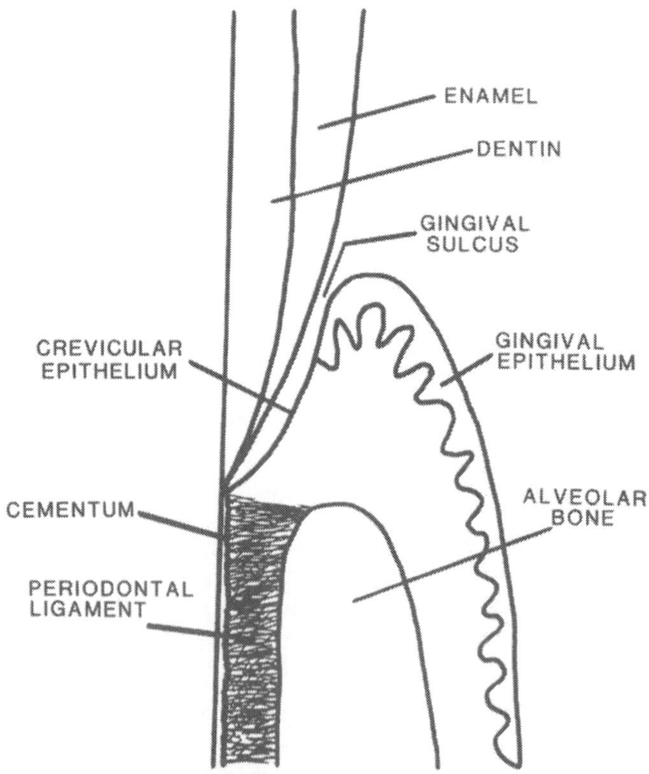

FIGURE 1. The tissues of the periodontium and their relationship to the tooth.

not only a site for the accumulation and proliferation of microorganisms, but also the repository for gingival fluid (see Fig. 1).

The gingival fluid, once considered merely a passive tissue fluid, contains not only acute-phase serum proteins such as transferrin, ferritin, ceruloplasmin, and interleukin 1 (Charon *et al.*, 1982a), but also the effector proteins of the complement system and immunoglobulins of the A. G, and M classes (Cimasoni, 1974, 1983). Additionally, this transudate of serum from peripheral capillaries through an epithelium of extremely high metabolic activity is quite depleted of oxygen (Mettraux and Graf, 1982), thus providing an excellent growth niche for anaerobic bacteria. The constant outward flow of gingival fluid prevents retrograde diffusion of oxygen from the oral cavity, thus ensuring a pO_2 no higher than that depleted by the epithelial cells at the base of the crevice. In a pathologically deepened gingival crevice (periodontal pocket), this control of pO_2 becomes increasingly more important (*vide infra*).

The cellular component of gingival fluid is composed primarily of neutrophils, with occasional lymphocytes and monocytes. These cells have historically been considered as castoff, dying cells. This viewpoint was predicated on the large number of cells and their poor viability when recovered from whole saliva. In contrast, studies over the past decade have shown excellent viability of lymphocytes and neutrophils when isolated before contamination with saliva (Skapski and Lehner, 1976). Additionally, the neutrophils are as capable as peripheral blood neutrophils of chemotactic, oxidative burst, phagocytic, and bactericidal activities (Charon *et al.*, 1982b; Miller *et al.*, 1984; Van Dyke *et al.*, 1985).

3. PLAQUE ACCUMULATION

The freshly cleaned surface of the tooth enamel, primarily an hydroxyapatite matrix, is rapidly coated by glycoproteins present in saliva (Mayhall, 1970). This acellular coating, the plaque pellicle, in turn, serves as the site of attachment of nonmotile, gram-positive cocci, primarily streptococci. These organisms have the capacity to produce viscous extracellular dextrans which facilitate colonization of the tooth. These streptococci, in turn, serve as specific attachment sites for certain fimbriated, nonmotile actinomycetes (Cisar *et al.*, 1979; Mergenhagen and Rosan, 1985). In a mutually beneficial ecological relationship, the actinomycetes provide a meshwork for the extension of the plaque colonies on the tooth surface while the streptococci provide additional adhesion to the pellicle. This basic infestation of nonmotile organisms is sufficient to provide a physical attachment for numerous sequentially colonizing microorganisms which, in turn, alter the metabolic and toxic nature of the dental plaque.

4. MICROBIAL PROGRESSION TO GINGIVITIS

The accumulation of "normal" amounts of plaque around the tooth surface and in the gingival crevice leaves the structures of the periodontium intact. This intact periodontium is characterized by a sparse, but present, lymphocytic infiltrate of the gingival tissue and a discontinuous crevicular epithelium, through which neutrophils and gingival fluid pass into the gingival crevice. In the absence of mechanical debridement of accumulating plaque, a more severe lymphocytic infiltrate appears in the gingival tissue. This infiltrate consists primarily of T cells with only a few B cells, macrophages, or plasma cells (Schroder *et al.*, 1973). Occasional discontinuities of the crevicular epithelium permit a more rapid egress of gingival fluid and neutrophils into the gingival crevice. The gingivae become red and edematous, despite the fact that no bacteria are demonstrable in the tissues. The exact nature of the change in the bacterial flora of the plaque which leads to this inflammation is not well defined. It may simply be a quantitative difference, but is more likely the early establishment of one or more periodontopathic microorganisms (*vide infra*) into the altered ecologic niche. In either case, the simple removal of the excessive plaque is sufficient to permit full resolution of the inflammation (i.e., gingivitis) with no clinically measurable damage (Loe *et al.*, 1965). This experimental model of human gingivitis (viz., cessation of oral hygiene, plaque accumulation, gingivitis, debridement, resolution) confirmed all or part of the bacterial infestation as the initiator and sustainer of the host response.

5. MICROBIAL PROGRESSION TO CHRONIC ADULT PERIODONTITIS

As already mentioned, the gross accumulation of dental plaque not only physically occupies a greater volume of the gingival crevice, but also alters local microenvironments, making new niches suitable for different additional microorganisms. When oral hygiene measures are not used regularly, the "normal" flora of predominantly aerobic, nonmotile cocci and filaments gradually can support the ingress of anaerobic gram-positive rods such as *Bacteroides* spp. and *Eikenella* spp. (Tanner *et al.*, 1979). Just as striking are the entry and colonization of motile forms, primarily spirochetes (Listgarten and Levin, 1981). As the superficial microorganisms continue to proliferate, many of the organisms deep in the plaque concentrate calcium, leading to a calcification, not only of the bacteria themselves, but also of the surrounding plaque matrix (Rizzo *et al.*, 1962). Other salivary components can also contribute to this calcification process (Slomiany *et al.*, 1980). This tenacious, calcified encrustation is known as dental calculus (often referred to as tartar). This calculus serves not only as a rigid base for further accumulation and calcification, but also as a physical barrier to access to the depths of the gingival crevice (Mandel and Gaffar, 1986). Such a physical barrier contributes to stagnation and anoxia at the base of the crevice, thereby further promoting the growth of the anaerobic pathogens. A partial listing of microorganisms associated with periodontal diseases is given in Table 1. These changes in flora are, themselves, interesting in terms of a mixed infestation, but the host response accompanying this change is even more dramatic.

The crevicular epithelium, that thin protection of the periodontium from external influence, is progressively assailed by multiple insults. The high meta-

TABLE 1. PARTIAL LISTING OF MICROORGANISMS ASSOCIATED WITH PERIODONTAL DISEASES

Actinobacillus actinomycetemcomitans
Actinomyces viscosus
Bacterionema matruchotii
Bacteroides gingivalis
Campylobacter sputorum
Capnocytophaga ochracea
Eikenella corrodens
Fusobacterium nucleatum
Hemophilus aphrophilus
Mycoplasma salivarius
Selenomonas sputigena
Treponema dentiola
Wolinella recta

bolic activity of the epithelial cells competes with the overload of aerobic organisms for the oxygen present. Under conditions of low oxygen tension, the repair, regeneration, and baseline metabolic activity of epithelial cells are decreased (Taylor and Camalier, 1982), resulting in more frequent discontinuities of the protective epithelium. Toxic bacterial products that accumulate further challenge the survival of the epithelial cells. Finally, as plaque and calculus gradually widen the space between the tooth and the gingiva, the span traversed by the epithelium appears, the regenerating epithelium reattaches at sites further toward the apex (tip) of the root.

As the closure maintained by the epithelium is progressively lost, leakage of toxic and antigenic bacterial products into the periodontium is enhanced. This expanded antigenic and toxic load causes a shift in the inflammatory cellular infiltrate from primarily resting T cells to B-cell blasts, plasma cells, and T-cell blasts. The stimulation of these inflammatory cells is probably both specific and nonspecific. Evidence for specificity is derived, not only from the measured levels in serum, gingival fluid, and saliva of antibodies specific for disease-associated bacteria (Genco and Krygier, 1972), but also from altered cellular proliferation, *in vitro,* in response to specific bacteria and bacterial products (Ivanyi and Lehner, 1971). Nonspecific and toxic reactions to bacterial products are implied by nonspecifically stimulated lymphocyte proliferation in response to endotoxins from plaque bacteria (Ivanyi and Lehner, 1974). The host response is presumed to be primarily to bacterial products or structural components, since only a few reports have demonstrated the actual entry of any microorganisms into the periodontal tissues in CAP (Frank, 1980; Saglie *et al.*, 1982). The total consequence of this continuously stimulated, never resolved, activity of immune cells is similar to that seen in other unresolved chronic inflammations—tissue damage (e.g., rheumatoid arthritis.

The array of potentially deleterious products that have been implicated locally in CAP is quite broad (see Table 2). Vasodilation, lethal or sublethal damage to fibroblasts, killing of epithelial cells, lysis of collagen fibers, and

TABLE 2. COMPONENTS OF CELL-MEDIATED IMMUNITY IMPLICATED IN CHRONIC INFLAMMATION

Antigen-specific lymphocyte proliferation
Helper T cells
Suppressor T cells
Killer T cells
Lymphokines/Cytokines
Interleukin 1
Interleukin 2
Migration-inhibition factor
Macrophage-activating factor
Lymphotoxin
Osteoclast-activating factor
Leukocyte-derived chemotactic factor
Products of activated macrophages
Complement components
Prostaglandins
Leukotrienes
Thromboxanes
Superoxide anion
Hydrogen peroxide
Hydroxyl radical
Lysozyme
Lysosomal hydrolases
Collagenase
Elastase
Plasminogen activator

resorption of bone all occur. The rates and sites of these activities vary and are dependent on the nature of the stimulus and the cellular infiltrate. This is also important in understanding the nature of the disease. Contrary to earlier clinical impression, the tissue destruction may not be continuous and gradual, but, rather, may pass through cycles of exacerbation and remission (Goodson *et al.*, 1982). Although the soft tissues have the potential to repair during a period of remission, they seldom do so unless the inciting plaque has been removed during that period. The resorption of alveolar bone is the diagnostic sign, *sine qua non*, for cumulative damage to the periodontium. The increase in numbers of osteoclasts and the enhancement of bone resorption can result, not only from bacterial products such as cell walls (Lanscraf *et al.*, 1979) or endotoxins (Hausmann *et al.*, 1970), but also from the host response products such as complement (Sandberg *et al.*, 1977), prostaglandins (Klein and Raisz, 1970), or osteoclast-activating factor (OAF) (Horton *et al.*. 1972). Both the resorption of the thin crest of alveolar bone and the destruction of periodontal ligament fibers remove the structural support of the tooth and permit further deepening of the gingival crevice. Such a pathologically deepened crevice (periodontal pocket) is a classic sign of cumulative damage in CAP. The deeper the pocket progresses, the lower is the pO_2 (Mettraux and Graf, 1982) and the more negative the redox potential (Kenney and Ash,

1969). This permits further colonization of anaerobic organisms. As the bacterial overgrowth proceeds and more bacterial products diffuse into the tissues, further loss of periodontal ligament attachment and bone occurs, until, eventually, the mechanical support of the tooth is inadequate to resist even normal masticatory forces and the tooth is lost or extracted.

6. ATYPICAL PERIODONTITIDES AS PROBES IN THE PATHOLOGIC PROCESSES OF CHRONIC ADULT PERIODONTITIS

Over the past decade, the reevaluation and better definition of the characteristics of several atypical periodontal diseases have permitted better insight into the pathologic processes involved in all the periodontitides (Page and Schroeder, 1982). Although those authors define the two most common early-onset forms of periodontal disease as juvenile periodontitis and rapidly progressive periodontitis, common clinical usage retains the earlier designations, localized juvenile periodontitis (LJP) and generalized juvenile periodontitis (GJP), respectively. In contrast to CAP, with onset commonly in the fourth decade, LJP can begin circumpubertally and GJP can occur between puberty and the end of the third decade.

Data concerning LJP were first systematically compiled by Baer (1971), who outlined the unique diagnostic criteria. This condition is characterized by rapid bone loss around the permanent incisor and first molar teeth. The bone loss is so rapid that these teeth are usually lost owing to lack of bony support before the end of the third decade. Two microorganisms have been associated with this condition: *Actinobacillus* (*Hemophilus*) *actinomycetemcomitans* (Slots *et al.*, 1980) and *Capnocytophaga* spp. (Newman and Socransky, 1977). Although these organisms have subsequently been isolated, in small numbers, from apparently healthy sites, they are isolable in large numbers only in LJP-affected sites. Additionally, many data implicate tissue invasion by these organisms in the pathogenesis of LJP (Christersson *et al.*, 1983). The capacities, not only of *Capnocytophaga* spp. to induce a serum inhibitor of neutrophil chemotaxis (Shurin *et al.*, 1979), but also of *A. actinomycetemcomitans* to directly kill neutrophils and macrophages with a leukotoxin (Taichman *et al.*, 1980), have led to a great interest in the potential protective role of phagocytic cells, particular neutrophils (Van Dyke *et al.*, 1980), in the pathophysiology of LJP, and, thence, other periodontitides.

GJP is similar to LJP in many ways and can be a later stage of LJP. It is characterized by widespread loss of alveolar bone at an early age and can result in complete loss of teeth. The prevalent organism isolated from these patients is *Bacteroides gingivalis* (Mouton *et al.*, 1981). Limited entry of bacteria into the tissue has been implicated in these cases. In both GJP and LJP, the role of this tissue infection (c.f., infestation) is further implicated by the lack of efficacy of standard conservative periodontal therapy (viz., debridement by scaling and root planing), which is effective in mild to moderate CAP. The rapid recurrence of isolable organisms after debridement suggests a reservoir in the gingival tissues (Slots and Rosling, 1983). Gingival resection and systemic treatment with tetracycline, therapeutic measures that can eliminate tissue reservoirs of infecting organisms, are effective in the treatment of these atypical conditions (Slots and Rosling, 1983; Mandell *et al.*, 1986). These unusual disease conditions have led to a reevaluation of the possible role of limited tissue infection and its prevention by neutrophils in CAP. In contrast to LJP and GJP, mild to moderate CAP can be effectively treated by debridement. Advanced CAP can often be more effectively treated surgically and the adjunctive use of antibiotics. These clinical observations suggest that the atypical periodontitides may be extremes in the spectrum of periodontal diseases, but that they may provide a useful probe in the evaluation of host response mechanisms in CAP (Engel *et al.*, 1984).

7. EXPERIMENTAL APPROACHES TO DEFINING HOST FACTORS IN CHRONIC ADULT PERIODONTITIS

Several problems have prevented firm definition of the precise role of the host response in CAP, not unlike other inflammatory diseases. First, clinical measurement of the disease, in humans, is inadequate for most research purposes (Haffajee *et al.*, 1983). Measures such as the plaque index (Loe, 1967), the bleeding index (Suomi, 1969), pocket depth (WHO Scientific Group, 1978), and loss of attachment (Suomi *et al.*, 1969) must be assessed qualitatively, or, at best, nonparametrically (WHO Scientific Group, 1978). Unfortunately, there is, at present, no good, noninvasive measure (Niekrash and Patters, 1986; Offenbacher *et al.*, 1986; Smith *et al.*, 1986) of acute exacerbation (active disease), in contrast to the cumulative damage observed during remission (Fine and Mandel, 1986; Lisgarten, 1986). Second, the best way to study a chronic disease process is through a longitudinal program. Unfortunately, the slow, intermittent process of CAP over decades of life has limited appropriately designed longitudinal studies to a very few institutions with very stable populations. Third, although many animal species display periodontal tissue destruction, either induced or spontaneous, similar to the human condition, none is precisely the same. The use of different animal models in studies of CAP has been compiled and reviewed (Page and Schroeder, 1982). Since differences among animal models have led to many contradictions and disputations through the years, that review is aiding in retrospective interpretation of results in different species and the design of appropriate prospective studies. Indeed, known host response differences among the species, in some cases, explain differences in the experimental disease processes and their extrapolability to human CAP.

Studies of host factors have centered on almost every known responding and effector cell, type of immune response, and effector molecule. Many of those basic studies of host factors are reviewed in the proceedings of a symposium (Genco and Mergenhagen, 1982). In the next section, some of the findings implicating different host responses in the tissue destruction in CAP are briefly discussed.

8. THE ROLE(S) OF HUMORAL IMMUNITY IN CHRONIC ADULT PERIODONTITIS

Humoral immunity appears to be a two-edged sword in the host response to periodontopathic microorganisms. Timing of initial immunization and class of antibody appear to make the difference between deleterious and beneficial effects. The early studies of Lehner *et al.* (1974) used the experimental gingivitis model in humans (Loe *et al.*, 1965). In this model, patients voluntarily discontinue oral hygiene procedures for a specified period of time. As already explained, the gradual accumulation of masses of dental plaque initiates a gingivitis. Among the parameters studied longitudinally during such a period, serum IgM levels were found to be increased. This was attributed to a general nonspecific adjuvant effect of endotoxins from gram-negative bacteria and dextrans produced by streptococci. The effect of that antibody, *in vivo,* was not defined. Several other investigators have assessed humoral response to periodontopathic organisms, both locally and systemically (Martin *et al.*, 1986). Lally *et al.* (1980) found that the capacity of gingival biopsy specimens to produce IgG or IgA was proportional to the severity of clinical inflammation, thus implicating local synthesis in the immunopathologic mechanism. Mouton *et al.* (1981) showed increased serum IgA, IgM, and IgG levels, specific for *Bacteroides gingivalis.* The increases were correlated to both increasing age and prevalence of CAP. It is difficult and probably inappropriate to assign a protective or destructive role to such local or systemic antibodies based on a simple correlation in a cross-sectional study. Longitudinal studies (already described as almost prohibitively lengthy and requiring large, stable populations) are needed to confirm that issue.

Some animal studies support a protective role for serum antibody. Germ-free rats, monoinfected orally with *Eikenella corrodens,* demonstrate a rapid bone loss with a pattern similar to human periodontal bone loss. If the germ-free animals are repeatedly inoculated with a sonicate of *E. corrodens* in Freund's incomplete adjuvant, prior to monoinfection, there is a significant decrease in bone loss following monoinfection (Behling *et al.*, 1981). The enhancement of salivary IgA specific for plaque organisms is similarly protective, but by a totally different mechanism. Inoculation of rat parotid glands with *Streptococcus mutans* or *Actinomyces naeslundii* or *Actinomyces viscosus* in Freund's incomplete or complete adjuvant produces a decrease of eventual bone loss, proportional to salivary IgA titers (Crawford *et al.*, 1978). In this case, the antibody presumably disrupts the interaction of these bacteria in the plaque substructure, thereby preventing formation of a periodontopathic plaque.

A more specific protective role for serum antibody can also be inferred from other investigations. These studies were based on the earlier work of Cisar (1982), in which fimbriae of *A. viscosus* were found to serve as specific attachment sites for streptococci. Those data suggested that the sequential colonization of the tooth surface by actinomycetes, after the primary colonization by streptococci, is mediated by a specific carbohydrate–lectin interaction. Furthermore, other fimbriae mediate attachment of the organisms directly to saliva-coated hydroxyapatite, *in vitro*. Clark *et al.* (1984) have shown that immunization of mice with purified *A. viscosus* fimbriae elicits a serum IgM and IgG titer against those fimbraie. These serum antibodies are found in the exuded gingival fluid. The concomitant failure of *A. viscosus* to colonize the tooth surface is presumed to be the result of these antibodies. Studies such as these (reviewed in Mergenhagen and Rosan, 1985) show promise for the disruption of the microbial ecosystem and the prevention of plaque development by immunization.

In contrast, local production of antibody (Lally *et al.*, 1980) can lead to the accumulation of immune complexes (Clagett and Page, 1978). The presence of these potentially damaging complexes and the prevalence of plasma cells in the advanced stages of the disease have led some to describe CAP as a B-cell lesion (Seymour *et al.*, 1979). Certainly the plasma cells are secreting specific antibody in the local tissue and that activity is probably potentiated by endotoxins and dextrans, but whether that activity is protective, destructive, or both requires further definition.

9. THE ROLE(S) OF CELL-MEDIATED IMMUNITY IN CHRONIC ADULT PERIODONTITIS

The implication of cell-mediated immunity in CAP has followed the same history as in many other diseases. As each new *in vitro* assay of cell-mediated immunity has been developed, it has been tested on diseased and control patients (Engel *et al.*, 1984). Through this approach, it can easily be concluded that CAP involves at least the following cell-mediated immune-associated processes: antigen-specific lymphocyte proliferation, lymphokine/cytokine production, and macrophage activation (see Table 2).

In the same longitudinal studies of experimental gingivitis cited earlier, Lehner *et al.* (1974) also considered the blastogenic response of peripheral blood cells to either whole plaque or certain specific oral organisms. They found increased proliferation with each of these stimuli and an even greater enhancement if the same patients repeated the cessation of oral hygiene a second time. Studies correlating the *in vitro* activities of peripheral blood cells, exposed to sonicated whole plaque from patients, with clinical measurements of CAP showed unexpected relationships. Proliferation of cells from patients with either gingivitis (alone) or mild periodontitis increased over normal controls, whereas cells from patients with severe periodontitis were intermediate (Ivanyi *et al.*, 1972). Again, based on such a cross-sectional study of CAP, it is not clear whether an etiologic or bystander role can be appropriately assigned to any observed, associated phenomena. Furthermore, methodologic confounders have raised further disputations about the significance of such data (Engel *et al.*, 1984).

The role of lymphokines and cytokines in the periodontal lesion was initially assumed by their implication at any chronic inflammatory site where activated macrophages are present and lymphocytes are proliferating. Several studies (O'Neill and Woodson, 1986) have shown that plaque sonicates and several specific suspected periodontopathogens stimulate peripheral blood cells *in vitro* to produce migration inhibitory factor (Ivanyi *et al.*, 1972), leukocyte-derived chemotactic factor (Mackler *et al.*, 1974), lymphotoxin (Horton *et al.*, 1973), and OAF (Horton *et al.*, 1972). The latter lymphokine is particularly relevant, since bone resorption is the invariable clinical sign of CAP. Understanding the mechanisms of production and action of OAF may be one way of specifically modifying bone resorption in CAP. Indeed, the studies of Dewhirst *et al.* (1985) have shown OAF to be one of the many acronymically designed immune factors now known as interleukin-1-beta. The demonstrated identity of these two molecules should facilitate more rapid progress in defining the role(s) of various mediators in the pathophysiology of CAP.

The presence and functions of activated macrophages at sites of periodontal tissue damage are intriguing, because of the many destructive products that the cells can release. The production of collagenase by activated macrophages can not only contribute to the direct breakdown of periodontal ligament fibers, but also provide for both ingress and diffusion of bacterial products (Wahl *et al.*, 1975). Similarly, elastase produced by activated macrophages has been proposed to play a role in tissue degradation (Werb and Gordon, 1975). Prostaglandins, thromboxanes, and leukotrienes have all been proposed as mediators of continued cell stimulation and bone resorption (Klein and Raisz, 1970). As mentioned earlier, bacterial endotoxins stimulate eicosanoid production by macrophages (Humes *et al.*, 1977), and this, in turn, stimulates osteoclastic bone resorption (*vide infra*). Lysosomal hydrolases, plasminogen activator (Yoshida and Ward, 1981), and reactive oxygen species (Hoffeld, 1982) have all been implicated in tissue destruction at chronic inflammatory sites. Macrophages ensure adequate levels of C1q, C2, C4, and alternate-pathway complement components in the inflamed tissues (Whaley, 1980). Macrophage-derived lysozyme, in addition to its antibacterial properties, may modify activity of neutrophils by decreasing superoxide anion production and chemotactic activity while enhancing phagocytosis (Gordon *et al.*, 1979).

The role of the macrophage in not only causing direct tissue damage, but also modifying the lymphocytic responses at chronic inflammatory sites has been underestimated. At least two animal experiments warrant mention in that regard. When beagle dogs (which spontaneously develop a CAP-like condition) were treated with antilymphocyte serum, no change in gingivitis was observed (Nobreus *et al.*, 1974). In contrast, when the animals were treated with antineutrophil serum (notoriously cross-reactive to macrophages, but not tested in those studies), the development of gingivitis was greatly slowed (Attstrom *et al.*, 1971). Other animal experiments strongly implicate cell-mediated immunity as a major destructive process in CAP. Freund's complete adjuvant, a classic inducer of local cell-mediated immunity, was injected into the gingivae of marmosets or capuchins, resulting in bone-destructive lesions (Levy *et al.*, 1976). All these data strongly suggest a destructive role for cell-mediated immunity in human CAP.

10. BONE RESORPTION—THE MONOCYTE/MACROPHAGE/OSTEOCLAST AXIS

The enhanced and apparently destructive cell-mediated immunity associated with CAP involves the macrophage, as a cell both controlling the immune response and being activated by it to release degradative products. Additionally, the macrophage is involved in CAP as the cell precursor for the osteoclast. Whether the inflammatory macrophage, itself, becomes a precursor to the osteoclasts, which lead to inflammatory bone resorption, is somewhat speculative. It can be definitively stated that the bone marrow monocyte has, at least, the potential to become a bone-resorbing osteoclast (Burger *et al.*, 1982).

The progenitor of the osteoclast was a source of contention for many years. The several lines of evidence conclusively demonstrating the progression from bone marrow to monocyte to osteoclast have been reviewed in detail (Chambers, 1980a; Loutit and Nisbet, 1982). The presence of the precursor in circulating blood was shown by parabiosis experiments, in which one animal was lethally irradiated while the other was not. Tritiated thymidine injected into the nonirradiated animal (with cross-circulation blocked during labeling) appeared in osteoclasts at a fracture site in the irradiated rat after reestablishment of cross-circulation (Gothlin and Ericsson, 1973). Parabiosis of osteopetrotic mice (which fail to remodel bone) with normal mice or injection of normal spleen or bone marrow cells corrects the remodeling defect (Walker, 1975). If the "normal" cells are derived from a beige mouse, whose phagocytic cells contain giant lysosomes, the "cured" osteopetrotic mouse demonstrates giant lysosomes in the osteoclasts of its long bones (Ash *et al.*, 1980). Finally, the grafting of quail limb buds, not containing marrow cavities, to a chick embryo results in a chimera with osteoclast nuclei morphologically characteristic of the chick. In contrast, if the quail limb bud contained a hematopoietically active marrow cavity, the chimera has osteoclasts of both morphologically distinguishable nuclear types (Jotereau and le Douarin, 1978). Thus, blood circulation is sufficient to transport progenitor monocytic cells to bone for the initiation of remodeling.

The actual process of bone resorption has been a focus for much attention, not only in CAP, but also in rheumatoid arthritis (Decker *et al.*, 1984), multiple myeloma, osteoporosis, and other calcium-labilizing conditions. There is a growing list of both inflammatory and physiological agents that stimulate osteoclastic bone resorption (Table 3). It has been suggested that some of these act through cAMP-dependent mechanisms and that the cAMP-independent agents may act by directly entering the cell, thereby obviating the need for second messengers (Raisz *et al.*, 1969). Definition of the pathways of stimulation should allow pharmacologic manipulation of the process. Inhibitors of prostaglandin synthesis should inhibit bone resorption initiated by cyclooxygenase products. Indomethacin does inhibit such resorption, *in vivo*, in an animal model of CAP (Weaks-Dybvig *et al.*, 1982). cAMP phosphodiesterase agonists reduce cAMP concentrations and should thereby inhibit bone resorption mediated through cAMP action. Imidazole does inhibit cAMP-mediated bone resorption, *in vitro* (Goldhaber and Rabadjija, 1982). Since the mechanism of action of the cAMP-

TABLE 3. AGENTS THAT STIMULATE OSTEOCLASTIC BONE RESORPTION

Parathyroid hormone
Prostaglandin E_2
Bacterial lipopolysaccharide
Osteoclast activating factor
Muramyl dipeptide
Actinomyces amphiphile
Bacterial lipoteichoic acid
cAMP agonists
cAMP phosphodiesterase antagonists
β-Adrenergic agonists
Dopaminergic agonists
Mechanical deformation of bone

independent agents is unknown, either broader-spectrum inhibitors or agents specific to the undefined pathways are needed. One broader-spectrum inhibitor is a group of compounds known as diphosphonates. These agents intercalate into the hydroxyapatite crystals, inhibiting both resorption and further new crystallization (Fleisch *et al.*, 1969). The inhibition of bone resorption by these agents appears to be due to toxicity to osteoclasts during the abortive resorption process (Chambers, 1980b). In any event, pharmacologic inhibition of bone resorption, systemically, may not be justifiable in a regionalized disease such as CAP, but local administration and development of more specific inhibitors of both the activation process and the resorptive process hold some promise for therapeutic benefits.

11. SUMMARY AND CONCLUSIONS

The erupted tooth provides a unique ecologic niche for bacterial colonization. The only other, somewhat analogous structures of the human body, which are continuous from inside the body, through the epithelium to the outside, are the nails and the hair shafts. Each of the latter structures is continuously erupting, thereby limiting ingress of microorganisms unless the niche is altered (e.g., ingrown hair or nail). The long-term immobility of the tooth and the presence of the gingival crevice not only encourage colonization by microorganisms, but also select for unique microflorae. The host response is at a much greater disadvantage in CAP than with paronychial or parafollicular infestations. The latter can become actual infections and the host is more capable of eliminating the inciting agents. In CAP, the organisms only rarely enter the soft tissues. Instead, the bacterial products diffuse insidiously into the tissues, eliciting a multitude of inflammatory reactions. The macrophage participates in these events, not only as an osteoclast precursor, but also as an immunoregulator. Ultimately, the immune response accomplishes over years and decades the only way in which the host can truly defend itself against the ingress of microorganisms: the removal of the discontinuity of the gingival epithelium—the tooth itself. Following the loss of the tooth, the infestation and the infection both resolve with the healing of the epithelium and the exclusion of the microorganisms.

REFERENCES

Ash, P., Loutit, J. F., and Townsend, K. M. S., 1980, Osteoclasts derived from haematopoietic stem cells, *Nature* **283:**669.

Attstrom, R., Tynelius-Bratthall, G., and Egelberg, J., 1971, Effects of experimental leukopenia on gingival inflammation in dogs. II. Induction of leukopenia by heterologous anti-neutrophil serum, *J. Periodont. Res.* **6:**200.

Baer, P. N., 1971, The case for periodontosis as a clinical entity, *J. Periodontol.* **42:**516.

Behling, U. H., Sallay, C., Sanavi, F., Pham, P. H., and Nowotny, A., 1981, Humoral immunity and reduced periodontal bone loss in *Eikenella corrodens*-monoassociated rats. *Infect. Immun.* **33:**801.

Burger, E. H., van der Meer, J. W. M., van de Gevel, J. S., Gribnau, J. C., Thesingh, C. W., and van Furth, R., 1982, *In vitro* formation of osteoclasts from long-term cultures of bone marrow mononuclear phagocytes, *J. Exp. Med.* **156:**1604.

Chambers, T. J., 1980a, The cellular basis of bone resorption, *Clin. Orthoped. Rel. Res.* **151:**283.

Chambers, T. J., 1980b, Diphosphonates inhibit bone resorption by macrophages in vitro, *J. Pathol.* **132:**255.

Charon, J. A., Luger, T. A., Mergenhagen, S. E., and Oppenheim, J. J., 1982a, Increased thymocyte-activating factor in human gingival fluid during gingival inflammation, *Infect. Immun.* **38:**1190.

Charon, J. A., Metzger, Z., Hoffeld, J. T., Oliver, C., Gallin, J. I., and Mergenhagen, S. E., 1982b, An *in vitro* study of neutrophils obtained from the normal gingival sulcus, *J. Periodont. Res.* **17:**614.

Christersson, L. A., Albini, B., Zambon, J., Slots, J., and Genco, R. J., 1983, Demonstration of *Actinobacillus actinomycetemcomitans* in gingiva of localized juvenile periodontitis, *J. Dent. Res.* **62:**198.

Cimasoni, G., 1974, *The Crevicular Fluid*, Karger, Basel.

Cimasoni, G., 1983, *Crevicular Fluid Updated*, Karger, Basel.

Cisar, J. O., 1982, Coaggregation reactions between oral bacteria: Studies of specific cell-to-cell adherence mediated by microbial lectins, in: *Host–Parasite Interactions in Periodontal Diseases* (R. J. Genco and S. E. Mergenhagen, eds.), pp. 121–131, American Society for Microbiology, Washington, DC.

Cisar, J. O., Kolenbrander, P. E., and McIntire, F. C., 1979, Specificity of coaggregation reactions between human oral streptococci and strains of *Actinomyces viscosus* or *Actinomyces naeslundii*, *Infect. Immun.* **24:**742.

Clagett, J. A., and Page, R. C., 1978, Insoluble immune complexes and chronic periodontal disease in man and dog, *Arch. Oral Biol.* **23:**153.

Clark, W. B., Wheeler, T. T., and Cisar, J. O., 1984, Specific inhibition of adsorption of *Actinomyces viscosus* T14V to saliva-treated hydroxyapatite by antibody against type 1 fimbriae, *Infect. Immun.* **43:**497.

Crawford, J. M., Taubman, M. A., and Smith, D. J., 1978, The effects of local immunization with periodontopathic microorganisms on periodontal bone loss in gnotobiotic rats, *J. Periodont. Res.* **13:**445.

Decker, J. L., Malone, D. G., Haraouli, B., Wahl, S. M., Schreiber, L., Klippel, J. H., Steinberg, A. D., and Wilder, R. L., 1984, Rheumatoid arthritis: Evolving concepts of pathogenesis and treatment, *Ann. Intern. Med.* **101:**810.

Dewhirst, F. E., Stashenko, P., and Mole, J. E., 1985, Purification and partial sequence of human osteoclast-activating factor: Identity with interleukin 1-beta, *J. Immunol.* **135:**2562.

Engel, D., Monzingo, S., Rabinovitch, P., Clagett, J., and Stone, R., 1984, Mitogen-induced lymphoproliferation response of peripheral blood mononuclear cells from patients with severe generalized periodontitis: Lack of correlation with proportion of T cells and T cell subsets, *Clin. Immunol. Immunopathol.* **30:**374.

Fine, D. H., and Mandel, I. D., 1986, Indicators of periodontal disease activity: An evaluation, *J. Clin. Periodontol.* **13**:533.

Fleisch, H., Russell, R. G. G., and Francis, M. D., 1969, Diphosphonates inhibit hydroxyapatite dissolution *in vitro* and bone resorption in tissue culture and *in vivo*, *Science* **165**:1264.

Frank, R. M., 1980, Bacterial penetration in the apical pocket wall of advanced human periodontitis, *J. Periodont. Res.* **15**:563.

Genco, R.J., and Krygier, G., 1972, Localization of immunoglobulins, immune cells and complement in human gingiva, *J. Periodont. Res.* **10**:30.

Genco, R. J., and Mergenhagen, S. E., 1982, *Host–Parasite Interactions in Periodontal Diseases*, American Society for Microbiology, Washington, DC.

Goldhaber, P., and Rabadjija, L., 1982, Influence of pharmacological agents on bone resorption, in: *Host–Parasite Interactions in Periodontal Diseases* (R. J. Genco and S. E. Mergenhagen, eds.), pp. 363–375, American Society for Microbiology, Washington, DC.

Goodson, J. M., Tanner, A. C. R., Haffajee, A. D., Sornberger, G. C., and Socransky, S. S., 1982, Patterns of progression and regression of advanced destructive periodontal disease, *J. Clin. Periodontol.* **9**:472.

Gordon, L. I., Douglas, S. D., Kay, N. E., Yamada, O., Osserman, E. F., and Jacob, H. S., 1979, Modulation of neutrophil function by lysozyme, *J. Clin. Invest.* **64**:226.

Gothlin, G., and Ericsson, J. L. E., 1973, On the histogenesis of the cells in fracture callus: Electron microscopic autoradiographic observations in parabiotic rats and studies on labeled monocytes, *Virchows Arch. Abt. B Zellpath.* **12**:318.

Haffajee, A. D., Socransky, S. S., and Goodson, J. M., 1983, Clinical parameters as predictors of destructive periodontal disease activity, *J. Clin. Periodontol.* **10**:257.

Hausmann, E. L., Raisz, L. G., and Miller, W. A., 1970, Endotoxin stimulation of bone resorption in tissue culture, *Science* **168**:862.

Hoffeld, J. T., 1982, Oxygen radicals in inflammation and immunity, in: *Host–Parasite Interactions in Periodontal Diseases* (R. J. Genco and S. E. Mergenhagen, eds.), pp. 343–353, American Society for Microbiology, Washington, DC.

Horton, J. E., Raisz, L. G., Simmons, H. A., Oppenheim, J. J., and Mergenhagen, S. E., 1972, Bone resorbing activity in supernatant fluid from cultured peripheral blood leukocytes, *Science* **177**:793.

Horton, J. E., Oppenheim, J. J., and Mergenhagen, S. E., 1973, Elaboration of lymphotoxin by cultured human peripheral blood leukocytes stimulated with dental plaque deposits, *Clin. Exp. Immunol.* **13**:383.

Humes, J. L., Bonney, R. J., Pelus, L., Dahlgren, M. E., Sadowski, S., Kuehl, F. A., Jr., and Davies, P., 1977, Macrophages synthesize and release prostaglandins in response to inflammatory stimuli, *Nature* **269**:149.

Ivanyi, L., and Lehner, T., 1971, Lymphocyte transformation by sonicates of dental plaque in human periodontal disease, *Arch. Oral Biol.* **16**:1117.

Ivanyi, L., and Lehner, T., 1974, Stimulation of human lymphocytes by B cell mitogens, *Clin. Exp. Immunol.* **18**:347.

Ivanyi, L., Wilton, J. M. A., and Lehner, T., 1972, Cell-mediated immunity in periodontal disease: Cytotoxicity, migration inhibition and lymphocyte transformation studies, *Immunology* **22**:141.

Jotereau, F. V., and le Douarin, N. M., 1978, The developmental relationship between osteocytes and osteoclasts: A study using the quail-chick nuclear marker in endochondral ossification, *Dev. Biol.* **63**:253.

Kenney, E. B., and Ash, M. M., Jr., 1969, Oxidation reduction potential of developing plaque, periodontal pockets and gingival sulci, *J. Periodontol. Periodontics* **40**:630.

Klein, D. C., and Raisz, L. G., 1970, Prostaglandins: Stimulation of bone resorption in tissue culture, *Endocrinology* **86**:1436.

Lally, E. T., Baehni, P. C., and McArthur, W. P., 1980, Local immunoglobulin synthesis in periodontal disease, *J. Periodont. Res.* **15**:159.

Lehner, T., Wilton, J. M. A., Challacombe, S. J., and Ivanyi, L., 1974, Sequential cell-mediated immune responses in experimental gingivitis in man, *Clin. Exp. Immunol.* **16**:481.

Lenscraf, E. J., Greenblatt, J. J., and Bawden, J. W., 1979, Effect of group A streptococcal peptidoglycan and group A streptococcal cell wall on bone in tissue culture, *Arch. Oral Biol.* **24**:495.

Levy, B. M., Robertson, P. B., Dreizen, S., Mackler, B. F., and Bernick, S., 1976, Adjuvant induced destructive periodontitis in nonhuman primates: A comparative study, *J. Periodont. Res.* **11**:54.

Listgarten, M. A., 1986, A perspective on periodontal diagnosis, *J. Clin. Periodontol.* **13**:175.

Listgarten, M. A., and Levin, S., 1981, Positive correlation between the proportions of subgingival spirochetes and motile bacteria and susceptibility of human subjects to periodontal destruction, *J. Clin. Periodontol.* **8**:122.

Loe, H., 1967, The gingival index, the plaque index and the retention index systems, *J. Periodontol. Periodontics* **38**:610.

Loe, H., Theilade, E., and Jensen, S. B., 1965, Experimental gingivitis in man, *J. Periodontol. Periodontics* **36**:177.

Loutit, J. F., and Nisbet, N. W., 1982, The origin of osteoclasts, *Immunobiology* **161**:193.

Mackler, B. F., Altman, L., Rosenstreich, D. L., and Oppenheim, J. J., 1974, Induction of lymphokine production by EAC and of blastogenesis by soluble mitogens during human B cell activation, *Nature* **249**:834.

Mandel, I. D., and Gaffar, A., 1986, Calculus revisited: A review, *J. Clin. Periodontol.* **13**:249.

Mandell, R. L., Tripodi, L. S., Savitt, E., Goodson, J. M., and Socransky, S. S., 1986, The effect of treatment on *Actinobacillus actinomycetemcomitans* in localized juvenile periodontitis, *J. Periodontol.* **57**:94.

Martin, S. A., Falkler, Jr., W. A., Suzuki, J. B., Hawley, C. E., and Mackler, B. F., 1986, Local and systemic immunoglobulins reactive to *Bacteroides gingivalis* in rapidly progressive adult periodontitis, *J. Periodont. Res.* **21**:351.

Mayhall, C. W., 1970, Concerning the composition and source of the acquired enamel pellicle of human teeth, *Arch. Oral Biol.* **15**:1327.

Mergenhagen, S. E., and Rosan, B. (eds.), 1985, *Molecular Basis of Oral Microbial Adhesion*, American Society for Microbiology, Washington, DC.

Mettraux, G., and Graf, H., 1982, pO_2 in human periodontal pockets, *J. Dent. Res.* **61**:289.

Miller, D. R., Lamster, I. B., and Chasens, A. I., 1984, Role of the polymorphonuclear leukocyte in periodontal health and disease, *J. Clin. Periodontol.* **11**:1.

Mouton, C., Hammond, P. G., Slots, J., and Genco, R. J., 1981, Serum antibodies to oral *Bacteroides asaccharolyticus* (*Bacteroides gingivalis*): Relationship to age and periodontal disease, *Infect. Immun.* **31**:182.

Newman, M. G., and Socransky, S. S., 1977, Predominant cultivable microbiota in periodontosis, *J. Periodont. Res.* **12**:120.

Niekrash, C. E., and Patters, M. R., 1986, Assessment of complement cleavage in gingival fluid in humans with and without periodontal disease, *J. Periodont. Res.* **21**:233.

Nobreus, N., Attstrom, R., and Egelberg, J., 1974, Effect of anti-thymocyte serum on chronic gingival inflammation in dogs, *J. Periodont. Res.* **9**:236.

Offenbacher, S., Odle, B. M., and Van Dyke, T. E., 1986, The use of crevicular fluid prostaglandin E_2 levels as a predictor of periodontal attachment loss, *J. Periodont. Res.* **21**:101.

O'Neill, P. A., and Woodson, D. L., 1986, Lymphokine production by human gingival lymphocytes, *J. Periodont. Res.* **21**:338.

Page, R. C., and Schroeder, H. E., 1982, *Periodontitis in Man and Other Animals: A Comparative Review*, Karger, Basel.

Raisz, L. G., Brand, J. S., Klein, D. C., and Au, W. Y. W., 1969, Hormonal regulation of bone resorption, in: *Progress in Endocrinology* (C. Gual, ed.), pp. 696–703, Excerpta Medica, Amsterdam.

Rizzo, A. A., Martin, G. R., Scott, D. B., and Mergenhagen, S. E., 1962, Mineralization of bacteria, *Science* **135**:439.

Saglie, R., Carranza, F. A., Jr., Newman, M. G., and Pattison, G. A., 1982, Scanning electron microscopy of the gingival wall of deep periodontal pockets in humans, *J. Periodont. Res.* **17**:284.

Sandberg, A. L., Raisz, L. G., Goodson, J. M., Simmons, H. A., and Mergenhagen, S. E., 1977, Initiation of bone resorption by the classical and alternative C'pathways and its mediation by prostaglandins, *J. Immunol.* **119**:1378.

Schroeder, H. E., Munzel-Pedrazzoli, S., and Page, R. C., 1973, Correlated morphometric and biochemical analysis of gingival tissue in early chronic gingivitis in man, *Arch. Oral Biol.* **18**:899.

Seymour, G. J., Powell, R. N., and Davies, W. I. R., 1979, Conversion of a stable T cell lesion to a progressive B cell lesion in the pathogenesis of chronic inflammatory periodontal disease: An hypothesis, *J. Clin. Periodontol.* **6**:267.

Shurin, S. B., Socransky, S. S., Sweeney, E., and Stossel, T. P., 1979, A neutrophil disorder induced by *Capnocytophaga*, a dental microorganism, *N. Engl. J. Med.* **301**:849.

Skapski, H., and Lehmer, T., 1976, A crevicular washing method for investigating immune components of crevicular fluid in man, *J. Periodont. Res.* **11**:19.

Slomiany, B. L., Slomiany, A., and Mandel, I. D., 1980, Lipid composition of human submandibular gland secretion from light and heavy calculus formers, *Arch. Oral Biol.* **25**:749.

Slots, J., and Rosling, B. G., 1983, Suppression of the periodontopathic microflora in localized juvenile periodontitis by systemic tetracycline, *J. Clin. Periodontol.* **10**:465.

Slots, J., Reynolds, H. S., and Genco, R. J., 1980, *Actinobacillus actinomycetemcomitans* in human periodontal disease: A cross sectional microbiological investigation, *Infect. Immun.* **29**:1013.

Smith, Q. T., Hinrichs, J. E., and Melnyk, J., 1986, Gingival crevicular fluid myeloperoxidase at periodontitis sites, *J. Periodont. Res.* **21**:45.

Snyderman, R., and McCarty, G. A., 1982, Analogous mechanisms of tissue destruction in rheumatoid arthritis and periodontal disease, in: *Host–Parasite Interactions in Periodontal Diseases* (R. J. Genco and S. E. Mergenhagen, eds.), pp. 354–362, American Society for Microbiology, Washington, DC.

Suomi, J. D., 1969, Periodontal disease and oral hygiene in an institutionalized population: Report of an epidemiological study, *J. Periodontol.* **40**:5.

Suomi, J. D., Greene, J. C., Vermillion, J. R., Chang, J. J., and Leatherwood, E. C., 1969, The effect of controlled oral hygiene procedures on the progression of periodontal disease in adults: Results after two years, *J. Periodontol.* **40**:416.

Taichman, N. S., Dean, R. T., and Sanderson, C. J., 1980, Biochemical and morphological characterization of the killing of human monocytes by a leukotoxin derived from *Actinobacillus actinomycetemcomitans*, *Infect. Immun.* **28**:258.

Tanner, A. C. R., Haffer, C., Bratthall, G. T., Visconti, R. A., and Socransky, S. S., 1979, A study of bacteria associated with advancing periodontitis in man, *J. Clin. Periodontol.* **6**:278.

Taylor, W. G., and Camalier, R. F., 1982, Modulation of epithelial cell proliferation in culture by dissolved oxygen, *J. Cell. Physiol.* **111**:21.

Van Dyke, T. E., Horoszewicz, H. O., Cianciolo, L. J., and Genco, R. J., 1980, Neutrophil chemotaxis dysfunction in human periodontitis, *Infect. Immun.* **27**:124.

Van Dyke, T. E., Levine, M. J., and Genco, R. J., 1985, Neutrophil function and oral disease, *J. Oral Pathol.* **14**:95.

Wahl, L. M., Wahl, S. M., Mergenhagen, S. E., and Martin, G. R., 1975, Collagenase production by lymphokine-activated macrophages, *Science* **187**:261.

Walker, D. G., 1975, Bone resorption restored in osteopetrotic mice by transplants of normal bone marrow and spleen cells, *Science* **190**:784.

Weaks-Dybvig, M., Sanavi, F., Zander, H., and Rifkin, B. R., 1982, The effect of indomethacin on alveolar bone loss in experimental periodontitis, *J. Periodont. Res.* **17**:90.

Werb, Z., and Gordon, S., 1975, Elastase secretion by stimulated macrophages: Characterization and regulation, *J. Exp. Med.* **142**:361.

Whaley, K., 1980, Biosynthesis of the complement components and the regulatory proteins of the alternative complement pathway by human peripheral blood monocytes, *J. Exp. Med.* **151**:501.

WHO Scientific Group, 1978, *Epidemiology, Etiology and Prevention of Periodontal Diseases*, World Health Organization, Geneva.

Yoshida, T., and Ward, P. A., 1981, Macrophage participation in the physiopathology of the immune response and inflammation, in: *Pathophysiology of the Reticuloendothelial System* (B. M. Altura and T. M. Saba, eds.), pp. 31–46, Raven Press, New York.

4

Host–Pathogen Interactions in the Gastrointestinal Tract

RITA DHAR and PEARAY L. OGRA

1. INTRODUCTION

During the last two decades, much attention has been devoted to the understanding of the local immune system especially in the intestinal and respiratory tracts. As a result, a large body of literature is now available concerning the role of humoral as well as cell-mediated immune responses at the mucosal surfaces against mucosal infections and in limiting systemic uptake of other luminal antigens. Since most infections are acquired via mucosal surfaces, the development of mucosal immunologic reactivity has been considered by many as the first line of defense at these sites.

Under normal conditions an ecologic balance is maintained between the host and the microorganisms, but when this balance is disrupted, the microorganisms can adhere to the luminal epithelial cells, subsequently replicate, and thus overcome the mucosal immunologic barrier. The characteristics of the interaction between the host and microorganisms in the gastrointestinal tract and the immune response it elicits are briefly reviewed in this chapter.

2. INTESTINE AS AN IMMUNOLOGIC TISSUE

There is enough histologic evidence to suggest that the gastrointestinal tract is replete with immunocompetent cells. It has been estimated that there are about 7.5×10^{10} antibody-producing cells in the mucosa of the intestinal tract (Heremans, 1974). This figure represents more than 50% of the total lymphoid tissue mass in humans, as the number of Ig-producing cells in bone marrow,

RITA DHAR and PEARAY L. OGRA • Departments of Pediatrics and Microbiology, State University of New York at Buffalo; and Division of Infectious Diseases and Microbiology Laboratories, Children's Hospital, Buffalo, New York 14222. *Present address for R.D.:* Department of Laboratories, Al-Adan Hospital, Fahaheel, Kuwait.

spleen, and lymph nodes collectively has been estimated to be about 2.5×10^{11} (Turesson, 1976). The antibodies elaborated by these cells, which are also present at other secretory sites, belong predominantly to the IgA class. Other immunoglobulin classes, IgM and IgG, are also evident in the intestine but to a lesser extent (Crabbe and Heremans, 1966). The antibodies at the intestinal mucosal level help provide protection by neutralizing viruses, inhibiting bacterial adherence and proliferation, and preventing penetration of dietary and microbial antigens and other environmental macromolecules (Walker and Isselbacher, 1977).

Exposure in the gastrointestinal tract to infections or noninfectious antigens results in the activation of precursor immunocompetent cells in gut-associated lymphoreticular tissue (GALT). It is generally believed that it is in the organized lymphoid structures, such as Peyer's patches (PP), that the primary interaction between the immunocompetent cells and the mucosally available antigens takes place, whereas the cells in the lamina propria and epithelium are the effector cells which carry out specific immune functions. The cellular components that participate in the immunologic reactions in GALT include two principal accessory cell types (macrophages and dendritic cells), regulatory T cells [T helper (T_h) and T suppressor (T_s) cells], and precursor IgA B cells. Based on the functional characteristics of the cells in the mucosal lymphoid tissue, it has been shown that 60% of the cells bear T-cell markers with a predominance of suppressor–cytotoxic phenotype. The immunoglobulin-containing cells constitute 60% of the cellular population, and of these 85% bear markers for the IgA isotype. Recent studies carried out in animal models suggest that there is a special type of T cell in PP or murine intestine, which brings about switching of IgM-bearing B cells to IgA B cells in response to an antigen. This switch is restricted to IgM-bearing B cells and is not seen with IgG B cells. It has also been shown that switch T cells do not cause proliferation and maturation of B cells containing IgA. Since these switch T cells have been found to exist in the PP, it probably explains the high numbers of B cells committed to the production of cells expressing IgA in the GALT (Kawanishi *et al.*, 1983).

2.1. COMMON MUCOSAL IMMUNE SYSTEM

After interacting with an antigen, the specifically sensitized lymphocytes in the GALT leave mucosal lymphoid follicles by way of efferent lymphatics. Following a migratory pathway that involves their passage through mesenteric lymph nodes, thoracic duct, and blood circulation, these cells settle in the lamina propria of gastrointestinal, upper-respiratory and genitourinary tracts, and in the interstitium of salivary, lacrimal, and mammary glands. These findings have brought to light the existence of a "common mucosal" immune system for induction of secretory immune response at sites away from the site of initial antigenic exposure in the GALT or possibly respiratory mucosal sites (Mestecky *et al.*, 1978). In experiments carried out in rabbits (Montgomery *et al.*, 1974) and in humans (Goldblum *et al.*, 1975), pregnant females were given microbial antigens orally. Following such exposure, specific IgA antibodies or IgA B cells were detected in milk in the absence of serum antibody of any class. The lack of serum antibody strongly suggested that antibody-forming cells or their precursors, and probably not antigen, had migrated from gut to the breast. Lamm *et al.* (1979), in their experiments with female mice using ferritin as orally administered antigen, demonstrated a similar phenomenon. However, other studies have suggested the possibility of antigenic transport as well. For example, it has been shown that by injecting radiolabeled or fluorescent antigens into donor rats it is possible to identify a subpopulation of cells in thoracic duct lymph bearing the injected antigen. These cells have been called the "antigen laden cells" and are characterized by being large in size, phagocytic, and positive for esterase (Bell and Botham, 1982).

Following intranasal administration of RA27/3 or subcutaneous injection with HPV77 live attenuated rubella vaccines, rubella-specific secretory IgA was consistently detected in the colostrum and milk. However, nasopharyngeal IgA to rubella was observed only after intranasal administration of RA 27/3 virus. This observation also suggests the circulatory transport of rubella virus antigen to the mammary gland after subcutaneous administration of HPV77 and may underlie the differential ability of HPV77 vaccine to induce secretory antibody response in the milk but not in the nasopharynx (Losonsky *et al.*, 1982).

2.2. IgA SYSTEM

IgA has been found to be the principal immunoglobulin in the secretions bathing most external mucosal surfaces. The secretory IgA (S-IgA) is composed of two molecules of IgA linked together with the J chain. This dimeric form of IgA acquires secretory component (SC), an integral part of the molecule, from the epithelial cells before S-IgA makes its appearance in the external secretions (Mestecky *et al.*, 1980). In recent studies, the fine details of the IgA transport pathway have been described by use of immunofluorescence and electron microscopic techniques. It has been proposed that SC on the epithelial cells serves as the receptor site for binding polymeric J-chain-containing IgA and represents an initial step in the eventual transport of IgA through the epithelial cells (Brandtzaeg, 1974; Brandtzaeg and Baklien, 1977; Brown *et al.*, 1976; Crago *et al.*, 1978; Nagura *et al.*, 1979). Distribution of IgA subclasses in various secretions and in serum was compared by Delacroix *et al.* (1982). Two subclasses, IgA1 and IgA2, have been demonstrated in colostrum, milk, and saliva. A preponderance of IgA2-containing plasma cells in secretory sites as compared to peripheral lymph nodes has also been observed. Although IgA2 and polymeric IgA (p-IgA) predominate in secretions, a recent study suggested a similar proportion of IgA2 in both monomeric IgA and p-IgA, indicating that the two molecular forms are not necessarily linked.

A number of publications have described the appearance of serum-derived IgA in bile (Fisher *et al.*, 1979; Hopf *et al.*, 1978; Jackson *et al.*, 1978; Lemaitre-Coelho *et al.*, 1978a,b; Orlans *et al.*, 1978, 1979). In studies conducted in a murine model, the ligation of rat bile duct resulted in an elevation of serum IgA levels and the appearance of free as well as IgA-bound SC in the circulation (Lemaitre-

Coelho *et al.*, 1978b). Interestingly, the level of IgA declined when the ligature was released. The mechanisms involved in the selective transport of p-IgA from the circulation into the bile and subsequently into the intestine are not well understood. The rat hepatocytes have been shown to synthesize and express SC on their membranes. This suggests that the murine liver may actively transport large amounts of p-IgA from plasma into bile and hence play a unique role in the gut secretory immune system (Vaerman *et al.*, 1978; Lemaitre-Coelho *et al.*, 1981). In humans, however, discordant data have been obtained. There is often lack of detectable SC in or on hepatocytes, which may explain the smaller plasma-to-bile transport of p-IgA in humans. Histochemical studies have revealed that SC in liver is found selectively in or on bile duct cells, suggesting their predominant role in p-IgA transport. Biliary obstruction, however, does not result in selective increase of p-IgA, although during parenchymal liver disease preferential increase of IgA has been observed. It is, therefore, theorized that liver could play a vital role in the selective removal of p-IgA from plasma by another SC-independent mechanism (Delacroix and Vaerman, 1983).

2.3. CELL-MEDIATED IMMUNITY AT MUCOSAL SURFACES

Although the concept of humoral immune response at secretory sites is well accepted, the role of cellular immunity in protection against microorganisms at the mucosal surfaces has remained unexplored until recently. There is now growing evidence that T cells do participate in effector mechanisms along with B cells. The bulk of the data available on cell-mediated immunity at mucosal surface is based on studies done in the respiratory tract of humans and experimental animals. The cell-mediated immune response is manifested by cell-to-cell killing by activated T cells, or by the release of cytotoxic lymphokines, which also bring about increase in vascular permeability, and recruitment of other cell types like macrophages and granulocytes to the site of antigenic exposure. Most cellular response studies to date, are based on the development of *in vitro* correlates of cellular immunity as evidenced by the appearance of lymphoproliferative response or inhibition of leukocyte or macrophage migration (Ottaway *et al.*, 1979).

There are reports suggesting that the populations of T cells observed in mucosal sites are functionally distinct and exhibit several unique characteristics. For example, the proliferative response to several viral antigens tested to date appears to be qualitatively different and somewhat hyporesponsive in colostral lymphocytes as compared with concurrent samples of peripheral blood T cells. In addition the relative proportions of suppressor–cytotoxic, T_h, and other T-cell subsets in mucosal lymphoid tissues are discernibly different from those in the peripheral blood and other systemic lymphoid tissues. Recently, specific studies on mucosal T-cell traffic have indicated that cellular components of mucosal cellular immunity may exhibit similar homing patterns as observed for IgA precursor B cells (Parrott and Ferguson, 1974). Natural killer cell activity has also been demonstrated in mucosal lymphocytes in response to experimental infection with several respiratory viruses (Kumagai *et al.*, 1985).

2.4. OTHER SPECIALIZED CELLULAR COMPONENTS OF GASTROINTESTINAL MUCOSA AND THEIR INTERACTION WITH VIRUSES AND OTHER MICROORGANISMS

Throughout the intestine, the mucosa presents several distinct cellular components. These include intraepithelial lymphocytes (IEL), M cells, mucosal mast cells (MMC), monocytes, and villus enterocytes.

Large numbers of IEL have been observed by electron microscopy to be present between the columnar epithelial cells. These cells have been shown to possess T-cell suppressor–cytotoxic phenotype markers, and their presence is attributed to the ongoing cell-mediated immune responses to various intraluminal antigens. In germ-free animals and also in grafts of fetal gut implanted under the kidney capsule the number of IEL is reduced (Parrott and Ferguson, 1974). About 40% of these cells appear to be granulated and contain sulfated mucopolysacchrides, metachromatic granules, and small but detectable amounts of histamine (Rudzik and Bienenstock, 1974; Guy-Grand *et al.*, 1978; Befus and Bienenstock, 1982). Many of these cells have specific receptors for IgE, although the number of such receptors is estimated to be 10–15 times lower than in tissue mast cells. IEL represent a heterogenous population of epithelial cells, since they participate in an array of immunologic activities. Recently, it has been suggested that like sensitized IgA B cells and T cells, IEL may also migrate to distant mucosal sites and hence participate in common mucosal immunologic functions (Guy-Grand *et al.*, 1978; Arnaund-Battandier, 1982).

The epithelium overlying the Peyer's patch lymphoid follicles is unique and specialized. It consists of low cuboidal cells called M cells. These cells are characterized by short, irregular microvilli on their luminal surface (Owen and Jones, 1974). The elegant ultrastructural studies by Owen have revealed that M cells play an important role in transporting antigens from the lumen and presenting it to immunocompetent precursor cells within the mucosal lymphoid follicles, thereby initiating early events in immune response (Owen, 1977). Recently, in studies with reoviruses it has been shown that the M cell may act as specific receptor for initial penetration of virus from the mucosal lumen to the systemic sites (Wolf *et al.*, 1981).

MMC are found among the IEL population as well as in the lamina propria. Besides having fewer IgE-specific receptors than their tissue counterparts, MMC also differ in their response to a compound like cromoglycate which causes suppression of autacoid release from tissue mast cells. IEL closely resemble MMC and share several characteristics. It has been suggested that the two mucosal cells may have a common precursor. MMC tend to accumulate in large numbers in response to parasitic diseases, as observed in mice infected with the nematode *Nippostrongylus brasiliensis* (Guy-Grand and Vassalli, 1983). Their contribution to the mechanism of protection or pathogenesis in viral infections, however, remains largely unknown.

The role of villus enterocytes in the initiation of infection has emerged from the studies of rotaviruses in humans and experimental animals. There is evidence that terminally differentiated villus enterocytes have specific receptors for initial attachment and subsequent replication of rotaviruses. In the murine model, the presence of virus-specific binding sites has been demonstrated on the enterocytes by incubation with purified mouse rotavirus (MRV)-coated sheep red cells (Riepenhoff-Talty *et al.*, 1982). Pronounced rotavirus binding was observed to the villus enterocyte in mice under 10–11 days of age as compared to mice over 75 days old (Table 1). These data suggest that the degree of adherence of MRV in the intestine may be determined by the availability of virus-specific receptors on the enterocytes. By using indirect immunofluorescence techniques, viral antigens was demonstrated in the cytoplasm of up to 30% of the enterocytes from suckling mice while little or no antigen could be seen in the enterocytes isolated from mice over 26 days of age (Riepenhoff-Talty *et al.*, 1982). It therefore appears that the initial binding and subsequent replication of the virus in the villus enterocytes is an age-dependent phenomenon (Table 1). But information is lacking regarding the mechanisms underlying such age restriction of the clinical disease and the factors that define virus intestinal epithelial cell interaction in human and other mammalian rotavirus infections.

3. HOST–PATHOGEN INTERACTION AND IMMUNE RESPONSE TO SELECT MICROORGANISMS

3.1. VIRUSES

3.1.1. Poliovirus

Prevalence of wild poliovirus in the environment and subsequent naturally acquired infection have been shown to induce a profound state of mucosal immunity in the gut as compared to that from active immunization with live poliovaccine (LPV) (Ghendon and Sankoyeva, 1961). Infection is initiated by primary replication of the poliovirus in the pharyngeal or intestinal mucosa. Subsequently, further replication takes place in the tonsils and PP. A study conducted in children with double-barreled colostomies, in whom the distal

TABLE 1. AGE-DEPENDENT INTERACTION OF MOUSE ROTAVIRUS WITH VILLUS ENTEROCYTES

Age of mice (days)	Mean percent of [a] MRV-infected enterocytes	Mean percent of [b] EC-SRBC$^{MRV c}$ rosettes
2–5	19	15
6–11	20	16
12–19	11	13
20–26	1	9
>75	0	5

[a] Percentage of enterocytes infected with mouse rotavirus (MRV) following infection per oral route, relative to the age of the animal at the time of infection.
[b] Percentage of MRV-specific receptors on the enterocytes as determined by rosetting of enterocytes with MRV-coated sheep red blood cells (SRBC), in relationship to the age of the animal at the time of collection of enterocytes.
[c] EC, enterocytes; SRBC^{MRV}, MRV-coated SRBC.

colon was immunized with LPV, resulted in implantation and replication of the virus in the distal immunized segment with no recovery from other distant nonimmunized segments of the gut or nasopharynx. Viral replication, therefore, appears to be restricted to the segments that are exposed to the antigen (Ogra and Karzon, 1971).

Exposure of nasopharynx or alimentary tract to wild virus or LPV results in the production of specific S-IgA on the mucosal surface. The viral-specific S-IgA has neuralizing activity and results in decreased implantations and shorter periods of virus excretion on reinfection challenge. Several studies have revealed that inactivated (IPV) or LPV vaccines administered by the parenteral or oral route, respectively, elicit similar serum antibody response (Table 2). In both instances the response is characterized by the appearance of serum IgM and IgG in a predictable sequence. IgA appears later at lower levels. In spite of high levels of specific antibodies in the circulation following parenteral IPV administration, the gut mucosa may remain susceptible to poliovirus infection; such immunization is often characterized by the absence of S-IgA response in the gut or respiratory tracts. However, these individuals appear to excrete the virus in the feces or nasopharynx for a shorter duration and at lower titers on subsequent reinfection with naturally acquired or LPV-induced challenge. Furthermore, repeated use of large doses of IPV administered orally result in low levels of local secretory response.

Despite the simplicity of this overview, the events leading to the development of viral-specific antibody response at the mucosal sites are complex and may be selectively amplified or suppressed under natural or experimentally induced environmental conditions. Information from tropical countries suggests that oral immunization with LPV may be associated with a high failure rate for initial mucosal viral replication and seroconversion. Such unsuccessfully immunized subjects remain highly susceptible to subsequent infection with wild virus. The precise mechanism for the observed failure of immunization by the mucosal route for a replicating agent has not been determined. Recently, in a study of Pakistani women who were presumably previously infected with wild poliovirus, oral administration of LPV in conjunction with systemic immunization with killed *Vibrio cholerae* resulted in significant decrease in the preexisting secretory IgA antibody to poliovirus in the breast milk and nasopharynx. These findings suggest a profound suppressive effect on mucosal antibody response after prior mucosal exposure to viral antigens under certain natural or induced environmental conditions (Hanson *et al.*, 1979).

The effects of prior mucosal or systemic exposure to poliovirus antigen on the development of antibody response in serum and nasopharynx following subsequent reexposure were carefully evaluated in groups of individuals immunized by the oral or intramuscular route with LPV or IPV, in the United States. Primary oral immunization followed by booster immunization 6–8 months later with LPV resulted in predictable serum and nasopharyngeal response (Table 2). However, parenteral administration of IPV during primary immunization, followed by similar booster immunization subsequently, resulted in heightened serum IgG and the appearance of nasopharyngeal secretory IgA antibody re-

TABLE 2. ANTIBODY RESPONSE TO TYPE I POLIOVIRUS AFTER MUCOSAL OR PARENTERAL IMMUNIZATION WITH LIVE TRIVALENT (LPV) OR INACTIVATED TRIVALENT (IPV) POLIOVIRUS VACCINE[a]

Routes of immunization: Primary/Booster (no. subjects)	Type of vaccine: Primary × 3/Booster × 1	Serum IgG antibody response to poliovirus type I: Peak titer mean ± SD after[b]			NPS SIgA antibody response to poliovirus type I: Peak titer mean ± SD after[b]		
		Primary immunization	Booster immunization	p value	Primary immunization	Booster immunization	p value
p.o./p.o. (7)	LPV/LPV	86 ± 3.2	84 ± 3.4	NS	32 ± 3.0	30 ± 2.3	NS
i.m./i.m. (7)	IPV/IPV	66 ± 5.0	87 ± 4.3	<0.01	0	5 ± 1.5	<0.01
p.o./i.m. (8)	LPV/IPV	79 ± 3.1	90 ± 4.5	NS	28 ± 2	19 ± 1.3	NS
i.m./p.o. (6)	IPV/IPV	78 ± 2.1	156 ± 1.8	<0.01	0	62 ± 2.4	<0.001

[a]NS, not significant; p.o., per oral; i.m., intramuscular.
[b]Data based on samples obtained 2 months after the last immunization.

sponse after booster immunization (Table 2). Intramuscular primary immunization with IPV resulted in no secretory antibody response in the nasopharynx. However, subsequent reimmunization of these subjects with orally administered LPV had a significant booster effect on nasopharyngeal IgA antibody activity, compared to that of subjects who had received primary immunication with orally administered LPV (Table 2) (Ogra, 1984). These observations suggest that nonmucosal priming may result in enhancement of mucosal antibody response on subsequent mucosal reexposure to the antigen.

Insufficient information is available regarding cellular immune response elicited during infection with poliovirus. However, interaction between poliovirus and lymphocytes has been examined *in vitro*. It was shown that poliovirus produces an inhibitory effect on the phytohemagglutinin response of leukocytes. This inhibitory effect has been suggested to be mediated by macrophages (Soontiens and Van der Veen, 1973).

3.1.2. Rotaviruses

Rotaviruses are a major cause of gastroenteritis in human infants and other young mammalian species (Bishop *et al.*, 1974; Kapikian *et al.*, 1974; Middleton *et al.*, 1974; Davidson *et al.*, 1977; Riepenhoff-Talty *et al.*, 1982).

The initial interaction between rotaviruses and the villus enterocytes in the gut epithelium has already been described in Section 2.4.

The temporal characteristics of the response of rotavirus-specific IgM, IgG, and IgA antibodies in serum and intestine were studied in hospitalized patients with rotavirus-induced gastroenteritis. The response in the serum was characterized by an initial rise in IgM antibody, followed by an increase in IgG and IgA levels. The titers of IgG antibody in convalescent sera were found to be lower in patients with severe gastroenteritis, and prolonged shedding of virus was observed as compared to patients with mild to moderate disease. On the other hand, virus-specific S-IgA antibody was found in higher concentration in the former group. These findings suggest a relationship between the severity of rotavirus infection and the nature of systemic and secretory antibody responses (Riepenhoff-Talty *et al.*, 1981). The characteristics of cell-mediated immune responses in suckling mice have been studied by employing *in vitro* lymphocyte transformation technique using spleen cells exposed to MRV. Viral-specific cell mediated immunity (CMI) was observed in spleen lymphocytes as early as 2 days after infection. The peak activity (around 10 days postinfection) closely followed the termination of diarrhea, suggesting that CMI may function as a limiting factor in disease (Riepenhoff-Talty *et al.*, 1983).

3.1.3. Other Viruses

Immunologic response to live enteric-coated adenovirus (ADV) type 21 vaccine was studied in human volunteers by Scott *et al.* (1972). An increase in the levels of serum IgM, IgA, and IgG antibodies was observed following vaccine-induced infection, which was similar to the response seen following natural infection. Virus specific S-IgA activity was observed in the intestinal tract but not in the nasal secretions of the vaccinees. Similar findings were reported by Smith *et al.* (1970), who were unable to demonstrate S-IgA in the nasal secretions of volunteers immunized with ADV type 4 enteric vaccine, although a neutralizing antibody response was observed in the sera. ADV type 4 vaccine has been shown to be highly protective against ADC type 4–associated disease, and it has, therefore, been suggested that serum antibody alone may afford protection against adenoviral disease (Scott *et al.*, 1972). In contrast, the mode of protection in other upper-respiratory virus infections such as rhino, respiratory syncytial, or parainfluenza viruses has been suggested to be mediated by S-IgA in the nasal secretion (Chanock, 1970).

Other studies have examined the immunologic response to the enteric infection using a mouse adenovirus strain K87 as the infecting agent in the conventional mouse model. Following oral challenge with this strain, the host developed resistance to subsequent infections, which has been attributed to virus-specific IgA antibodies in the gastrointestinal tract. A concurrent rise in serum IgG antibodies was also observed in these mice. Although the specific IgG antibodies in the serum had neutralizing capability, they were unable to confer resistance to subsequent oral challenges in animals after passive transfer of serum containing specific antibodies. These observations indicate that the serum IgG antibody plays no role in the protection and that the intestinal antibody may have a role in this process (Hashimoto *et al.*, 1973).

S-IgA antibody response in the gut has also been observed following infection with echovirus types 6, 9, and 11 and coxsackie virus type A-1, but concurrent rise in serum IgA was observed only occasionally (Ogra, 1970).

3.2. BACTERIA AND BACTERIAL ANTIGENS

A strong inflammatory response is generated when the microorganisms proliferate in the lymphoid follicle. Failure of local cellular responses to contain the replicating organisms often leads to severe local pathology. For example, perforation of PP may occur in typhoid fever, and narrowing of the intestine may result from tuberculosis.

Induction of local and systemic antibody response to cholera vaccine candidate immunogen (purified B subunit) was studied in Bangladesh in lactating women. Primary immunization with the oral vaccine resulted in a significant rise in secretory IgA antibodies in the intestine. Although intramuscular injection (i.m.) produced a similar magnitude of S-IgA response in the intestinal tract, the antibodies lasted longer after per oral (p.o.) than i.m. administration. A second dose by either route did not significantly increase the titers. An increase in salivary IgA antibodies was observed in most intestinal responders following p.o. but not i.m. vacciniation. On the other hand, specific IgA antitoxin antibody in milk was present in low titers in only 25% of the lactating women. Study of antibody response in serum revealed an increase in IgG antibody following parenteral as well as p.o. administration, although the titers were somewhat lower after the oral vaccination. A secondary administration by either route did not result in any increase in the antibody titer obtained by an initial i.m. vaccination, whereas a parenteral dose increased in IgG titers induced by an initial p.o. vaccination (Svennerholm *et al.*, 1981b).

Cholera toxin is an effective mucosal immunogen because of two of its properties. First, its antigenic B subunit binds specifically to receptors on the cell membrane to promote trapping of absorbed antigen by lymphoid cells in PP. Second, activation of adenyl cyclase, presumably in lymphoid cells, appears to enhance the immune response. Study of mucosal immune response to cholera toxin in rats has revealed that production of S-IgA is confined to the site of antigen exposure in the gut, while the memory cells disseminate to distant mucosal sites. Evidence came from the observation that the primary antibody response restricted to the primed site was modest; restimulation of the primed site led to a rapid and enhanced local secondary immune response. Secondary stimulation at a distant mucosal site evoked a secondary-type response at the challenged site but not at the site of priming. Parenteral administration of cholera toxin modulated the mucosal immune response by provoking suppressor mechanisms that interfere with the mucosal immune response subsequent to oral administration (Pierce, 1981).

Recently, Inman and Cantey (1984) have shown that *Escherichia coli* (strain RDEC-1), owing to the presence of an 85×10^6 D plasmid that codes for pili, can adhere to gut absorptive epithelium in rabbits. They compared the adherence of the RDEC-1 strain with that of *Shigella flexneri* (ShD 15) that also contained the 85×10^6 D plasmid and expressed the adherence pili. Electron microscopic and direct fluorescent microscopic examinations of PP tissue revealed that both piliated bacteria adhere to M cells of rabbit PP lymphoid follicle epithelium and that the adherence of RDEC-1 strain persisted, while ShD 15 strain was rapidly taken into M cells. The RDEC-1 strain resulted in loss of microvilli, but it was rarely taken into the follicle epithelium and it did not precipitate an acute inflammatory response in the follicle dome as did the ShD 15 strain. These observations suggest that certain bacteria may have additional virulence factors enabling them to persist on the epithelial surface in large numbers, without being taken into the follicle and without eliciting any inflammatory response.

Oral vaccination of pregnant swine with live virulent enterotoxigenic *E. coli* (ETEC) conferred protection on suckling pigs against ETEC-induced fatal diarrhea. ETEC-specific antibodies were detected in the colostrum, and since antigen and antibody do not normally cross the placenta in swine, the protection in the suckling pig was attributed to colostral antibodies (Moon and Issacson, 1981).

The kinetics of the IgA antibody response to *Shigella* has been studied in an isolated loop model in rabbits. Direct stimulation of isolated ileal loops with *Shigella* antigen resulted in considerable local S-IgA but not systemic IgG response. Furthermore, a mucosal memory response was elicited using only a live *Shigella* organism for priming and rechallanging by the oral route. The explanation may be based on a dose-related phenomenon, as live bacteria are able to multiply and invade the mucosal tissues, whereas killed bacteria are not. Parenteral priming with *Shigella* antigen shows no mucosal IgA memory response in spite of antigen-specific IgG (Keren *et al.*, 1983).

Natural sensitization occurs in normal subjects with *Streptococcus mutans*, a common commensal microorganism of human oral cavity. The mechanism by which *S. mutans* adheres to and colonizes tooth surfaces and eventually leads to caries has been studied in detail. The colonization of teeth by *S. mutans* was inhibited in monkeys by immunization with *S. mutans*, whereas nonimmunized monkeys were susceptible to implantation of the organism (Lehner *et al.*, 1976a,b). In the immunized group, the antibodies found in the saliva belonged to the S-IgA class and were believed to play a role in interfering with colonization. Dental caries is more severe in patients with IgA deficiency. Studies on cellular immunity in patients with periodontitis have revealed an increase in the number of antigen-specific lymphocytes in the peripheral blood. Local production of lymphokines by the stimulated lymphocytes may be responsible for periodontal disease (Genco, 1979). In another study it was shown that immune response to streptococcal antigens (SA) consists of IgG antibodies to SA in the serum and IgA-specific antibodies in colostrum and saliva. It is suggested that both systemic and local antibody response are regulated by the presence of helper and suppressor subsets of T lymphocytes *in vitro* and *in vivo* (Lehner, 1983).

Natural antibacterial and antibody-dependent cellular cytotoxic (ADCC) activities are important defense mechanisms at the mucosal level. In a murine model, cell-mediated immune response studies using *Salmonella typhimurium* have shown the presence of natural antibacterial activity in GALT as well as in peripheral lymphoid organs, although the activity was higher in the former. ADCC was observed in cells from IEL, spleen, and peripheral blood lymphocytes, while it was absent from PP lymphocytes and mesenteric lymphocytes. The effector cells for natural antibacterial, natural killer (NK), and ADCC activities differed from each other in their phenotype markers (Nencioni *et al.*, 1983).

3.3. PARASITES

Following parasitic infection in humans and in animals, the effector mechanisms of host resistance in the form of nonspecific inflammation and antigen-specific immune response are generated. IgA antibody secreted by the gastrointestinal epithelium appears to play an important role by inhibiting adherence to the epithelium, and by interfering with feeding, reproduction, and sensorimotor functions of the parasite (Davis and Porter, 1979). Both parasite-specific and heterologous IgE responses have been identified in many mucosal parasitic infections. IgE antibody primarily acts by activating mast cells, basophils, eosinophils, macrophages, and lymphocytes following interaction with specific antigen. These events result in enhancement of inflammatory response, which leads to release of parasiticidal oxidants, amines, mucus, and prostaglandins. Activated T lymphocytes also localize at the mucosal site and contribute by controlling lymphocyte and macrophage activities and stimulating allergic reactions which create an unfavorable environment for the parasite (Befus and Bienenstock, 1982).

Parasitic protozoa and helminths, which result in various disease syndromes in humans and animals, represent a heterogenous group. Most of the pathologic lesions arise, presumably, from the host immune response.

In *Schistosoma mansoni* infection in mice, it has been suggested that the immune response evoked around the eggs of the parasite in the tissues resulting in granuloma formation may be associated with the develoment of delayed hypersensitivity and the manifestations of the disease (Warren, 1977). Two T-cell subsets—(1) effector T lymphocytes, with Lyt 1.1. membrane markers, and (2) suppressor T lymphocytes, with Lyt 2.3 markers—have been shown to exert an active immunomodulatory effect on such granuloma formation. On the other hand, in *Schistosoma japonicum*–infected mice humoral antibodies, and not T cells, may play an important role in immunomodulation (Mahmoud, 1983).

Owen *et al.* (1981) examined the interaction between *Giardia muris* and lymphoid cells in experimentally infected normal and nude mice, to define the mechanism for clearance from the mammalian intestinal tract. Electron microscopic studies revealed that the parasites invade the epithelium over PP follicles at the site of desquamating columnar cells and not through the M cell (Owen *et al.*, 1979). The macrophages in the lamina propria extend pseudopods into the epithelium and phagocytose *G. muris.* It is speculated that both humoral and cellular immune processes are active during clearance of *Giardia* infection. Patients with immunoglobulin deficiences are particularly susceptible to symptomatic giardiasis compared to patients with cellular immune deficiencies. Nude mice were protected by thymus-independent defense mechanisms, one of them being enhanced phagocytosis by increased proliferation of activated macrophages. Similar mechanism may occur in patients with deficient T-cell functions. Other studies by Stevens *et al.* (1978) and Robert-Thomson and Mitchell (1978) have shown that thymus-deficient nude mice display an impaired immune response to *G. muris* and little resistance to reinfection.

4. MODULATION OF IMMUNE RESPONSE VIA INTESTINE

There is enough evidence to suggest that exposure to an antigen via the gastrointestinal tract results in immunologic unresponsiveness in systemic sites subsequent to systemic administration of the same antigen (Chase, 1946; Vaz *et al.*, 1977; Richman *et al.*, 1978; Stokes and Swarbrick, 1977). Most studies describing this phenomenon of systemic suppression of immune response are summarized in Table 3 and have included the use of chemicals like picryl chloride (Chase, 1946) or soluble proteins like dinitrophenylated human γ-glubulin (Tomasi, 1980); ovalbulin (Challacombe and Tomasi, 1980), and bovine serum albumin (Peri and Rothberg, 1981). Other antigens also resulting in systemic unresponsiveness following sensitization by oral route include sheep red cells (Andre *et al.*, 1975; Mattingly and Waksman, 1978); ragweed antigen (Challacombe and Tomasi, 1980); and transplantation antigens. Based on these observations, Tomasi (1980) introduced the concept of "oral tolerance." The antigen-specific tolerance has been observed in the IgM, IgG, and IgE antibody classes as well as in T cells.

The mechanisms of such tolerance are not fully understood. It has been

TABLE 3. EFFECT OF THE ROUTE OF PRIMING ON THE OUTCOME OF IMMUNE RESPONSE TO SUBSEQUENT CHALLENGE WITH INFECTIOUS MICROORGANISMS AND OTHER ANTIGENS[a]

Route of priming / Challenge	Agent	Effects on immune response		Reference
		Systemic IgG	Mucosal sIgA	
1. Mucosal / Systemic	LPV / KPV	B	NE, B(S)	Svennerholm *et al.*, 1981a
	KPV	NE	NE, B	Ogra, 1984
	Strep. mutans	S	NE	Challacombe and Tomasi, 1980
	V. cholerae-LPS	B	B, NE	Svennerholm *et al.*, 1980
	KPV	NE	B	Ogra, 1984
	Picryl chloride	S	—	Chase, 1946
	SRBC	S	—	Andre *et al.*, 1975
	OVA	S	NE	Challacombe and Tomasi, 1980
	Ragweed - E	S	NE	Challacombe and Tomasi, 1980
	DNP-HGG	S	—	Tomasi, 1980
	Transplantation Ag	S	—	Tomasi, 1980
2. Mucosal / Mucosal	KPV	NE	NE	Ogra, 1984
	KPV / LPV	NE	B	Ogra, 1984
	LPV	B	B	Ogra, 1984
	LP(W)	NE	S[b]	Svennerholm *et al.*, 1981a
	LPV	NE	S, B[c]	Svennerholm *et al.*, 1981a
	RSV[d]	B	B	Fishaut *et al.*, 1981
	Rubella	B	B	Ogra *et al.*, 1974
	V. cholerae	NE, B	B	Pierce, 1978
	OVA	S	NE	Challacombe and Tomasi, 1980
	BSA	S	NE	Peri and Rothberg, 1981

[a] B, booster effect; NE, no effect; —, no available data; S, suppression (tolerance); KPV, killed poliovaccine; LPV, live poliovaccine; LP(W), live wild poliovirus; ( ), possible.
[b] LPV used in conjuction with cholera vaccine.
[c] LPV used alone.
[d] Data based on milk and serum, after natural infection.

suggested that antigen-specific IgA T_h and IgG T_s are generated simultaneously in PP and that T_s emigrate to spleen and other peripheral lymphoid tissues and produce systemic tolerance. This notion is further strengthened by an earlier study that showed that immunization with soluble proteins leads in a predictable manner to induction of secretory immune response and concomitant systemic tolerance. The role of T_s in oral tolerance is rather dubious, since it has been observed that tolerance to parenterally administered deaggregated human γ-globulin may be induced in the absence of suppressor cells (Tomasi *et al.*, 1983).

Although development of oral tolerance to soluble proteins and other noninfectious agents has been observed on many occasions, an extreme degree of variation seems to exist for the induction of systemic hyporesponsiveness to infectious agents. The available data on the effects of mucosal versus systemic priming on the outcome of subsequent reexposure challenge with infectious organisms or specific microbial antigens are reviewed in Table 3. These observations suggest that with most infectious agents the effect of mucosal or systemic priming on subsequent challenge is in fact in favor of a booster effect on the systemic immune response rather than tolerance, exceptions being *S. mutans* and cholera toxoid. As reviewed earlier, booster effect or no effect was observed on preexisting S-IgA response after LPV vaccination, with the exception of mucosal challenge with LPV, in subjects previously infected presumably with wild poliovirus.

5. CONCLUSIONS

The specialized mucosal immune system appars to act as a mechanism of defense against colonization by pathogens. Exposure of the gut mucosal surface to a variety of infections and noninfectious agents results in a complex chain of immunologic processes, which determine the outcome of the host–pathogen interaction. In most instances, the enterically acquired viral or bacterial infections result in the appearance of antigen-specific S-IgA in the intestine, which confers varying degrees of protection against subsequent reinfection challenges. For example, cholera B subunit antigen has been found to give rise to antitoxin antibody levels in the intestine of similar magnitude and duration as seen after clinical cholera (Svennerholm *et al.*, 1980). Although LPV induces virus specific S-IgA in the gut, the antibody level is lower than that observed following infection with wild poliovirus. Under certain conditions challenge with LPV in previously naturally infected subjects may result in a decrease in the preexisting S-IgA antibody to poliovirus (Hanson *et al.*, 1979). It has been well documented that antibodies play an important role as effectors of host defense against parasites, as has been observed in *Schistosoma mansoni* and *Taenia taeniaeformis* infection in rats and mice. The parasite-specific antibodies may react with surface antigens on the parasite, thereby interfering with the parasite's functions essential for pathology in the host and survival of the parasite. The other possible mechanism may be antibody-dependent complement-mediated lysis, as well as various forms of antibody-driven cellular toxicity. Based on a number of experiments in *N. brasiliensis*–infected mice, a two-step mechanism of worm expulsion has been formulated, according to which the worms are first damaged by antibodies and then expelled by mechanisms involving a variety of cells (Mauel, 1982).

The cell-mediated immune response evoked at the mucosal level also helps in protection, except in certain situations where it may play a role in the pathogenesis of the disease. It has been shown that the T-cell subsets in the mucosal lymphoid follicle contribute to immunoregulatory functions at the mucosal as well as systemic levels. The development of serum antibody and secretory IgA appears to be regulated by the activation of an isotype-specific T-cell subset at the mucosal site. Finally, macrophages, NK cells, and granulocytes mobilized to the site of bacterial, viral, or other parasitic replication in the intestinal mucosa may also play an important role. A combination of such nonspecific and immunologically specific events may ultimately determine the outcome of host–pathogen interaction in the gastrointestinal tract.

REFERENCES

Andre, C., Heremans, J. F., Vaerman, J. P., and Cambiaso, C. L., 1975, A mechanism for the induction of immunological tolerance by antigen feeding. Antigen–antibody complexes, *J. Exp. Med.* **142:**1509.

Arnaud-Battandier, F., 1982, Immunologic characteristics of isolated gut mucosal lymphoid cells, in: *Recent Advances in Mucosal Immunity* (W. Strober, L. A. Hanson, and K. W. Sell, eds.), pp. 289–299, Raven Press, New York.

Befus, A. D., and Bienenstock, J., 1982, Factors involved in symbiosis and host resistance at the mucosa–parasite interface, *Prog. Allergy,* **31:**76.

Bell, E. B., and Botham, J., 1982, Antigen transport I. Demonstration and characterization of cell laden with antigen in thoracic duct lymph and blood, *Immunology* **47:**477.

Bishop, R. F., Davidson, C. P., Holms, I. H., and Ruck, B. J., 1974, Detection of a new virus by electron microscopy of fecal extracts from children with acute gastroenteritis, *Lancet* **1:**149.

Bradtzaeg, P., 1974, Mucosal and glandular distribution of immunoglobulin components, differential localization of free and bound SC in secretory epithelial cells, *J. Immunol.* **112:**1553.

Bradtzaeg, P., and Baklien, K., 1977, Intestinal secretion of IgA and IgM: A hypothetical model, in: *Immunology of the Gut,* pp. 77, Ciba Foundation Symposium 46, Elsevier, Excerpta Medica, Amsterdam.

Brown, W. R., Isobe, Y., and Nakane, P. K., 1976, Studies on translocation of immunoglobulins across intestinal epithelium. II. Immunoelectron microscopic localization of immunoglobulins and secretory component in human intestinal mucosa, *Gastroenterology* **71:**985.

Challacombe, S. J., and Tomasi, T. B., 1980, Systemic tolerance and secretory immunity after oral immunization, *J. Exp. Med.* **152:**1459.

Chanock, R. M., 1970, Control of acute mycoplasmal and viral respiratory tract disease, *Science* **169:**248.

Chase, M. W., 1946, Inhibition of experimental drug allergy by prior feeding of the sensitizing agent, *Proc. Soc. Exp. Biol.* **61:**257.

Crabbe, P. A., and Heremans, J. F., 1966, The distribution of immunoglobulin-containing cells along the human gastrointestinal tract, *Gastroenterology* **51:**305.

Crago, S. S., Kulhavy, R., Prince, S. J., and Mestecky, J., 1978, Secretory component on epithelial cells as a surface receptor for polymeric immunoglobulins, *J. Exp. Med.* **147:**1832.

Davidson, G. P., Gall, D. G., Petric, M., Butler, D. G., and Hamilton, J. R., 1977, Human rotavirus enteritis induced in conventional piglets. Intestinal structure and transport, *J. Clin. Invest.* **60:**1402.

Davis, P. J., and Porter, P., 1979, A mechanism for secretory IgA-mediated inhibition of the cell penetration and intracellular development of *Eimeria tenella, Immunology* **36:**471.

Delacroix, D. L., Dine, C., Ramband, J. C., and Vaerman, J. P., 1982, IgA subclasses in various secretions and in serum, *Immunology* 47:383.

Delacroix, D. L., and Vaerman, J. P., 1983, Function of the human liver in IgA homeostasis in plasma, in: *The Secretory Immune System* (J. R. McGhee and J. Mestecky, eds.), pp. 383–401, New York Academy of Sciences, New York.

Fishaut, M., Murphy, D., Neifert, M., McIntosh, K., and Ogra, P. L., 1981, Bronchomammary axis in the immune response to respiratory syncytial virus. *J. Pediatr.* **99:**186.

Fisher, M. M., Nagy, B., Bazin, H., and Underdown, B. J., 1979, Biliary transport of IgA: Role of secretory component, *Proc. Natl. Acad. Sci. U.S.A.* **76**:2008.

Genco, R. J., 1979, Immune responses to oral organisms: Implications for dental caried and periodontal disease, *J. Clin. Periodontol.* **6**:22.

Ghedon, Y. U. Z., and Sankoyeva, L. I., 1961, Comparison of the resistance of the intestinal tract to poliomyelitis virus (Sabin's strain) in persons after naturally and experimentally acquired immunity, Acta Virol. **5**:265.

Goldblum, R. M., Ahlstedt, S., Carlsson, B. Hanson, L. A., Jodal, U., Lidin Janson, G., and Sohl-Akerlund, A., 1975, Antibody forming cells in human colostrum after oral immunization, *Nature* **257**:797.

Guy-Grand, D., Griscelli, C., and Vassalli, P., 1978, The mouse gut T lymphocytes, a novel type of T cell: Nature, origin and traffic in mice in normal and graft-versus-host conditions, *J. Exp. Med.* **148**:1661.

Guy-Grand, D., and Vassalli, P., 1983, Gut mucosal lymphocyte subpopulations and mast cells, in: *Regulation of the Immune Response* (P. L. Ogra and D. M. Jacobs, eds.), pp. 122–130, 8th Int. Convoc. Immunol., Amherst, NY, Karger, Basel.

Hanson, L. A., Carlsson, B., Cruz, J. R., Garcia, B., Holmgren, J., Khan, S. R., *et al.*, 1979, Immune response in the mammary gland, in: *Immunology of Breast Milk* (P. L. Ogra and D. M. Dayton, eds.), pp. 145–157, Raven Press, New York.

Hashimoto, K., Onishi, N., Tajiri, T., Okada, Y., Maeda, R., Sugihara, Y., and Saski, S., 1973, Viral enteric infection and local IgA antibody: Studies in the germfree mice thymectomized and/or infected with a mouse adenovirus, in: *Germ Free Research—Biological Effects of Gnotobiotic Environment* (J. B. Heneghan, ed.), pp. 501–506, Academic Press, New York, London.

Herberman, R. B., 1983, Immunoregulation and natural killer cells, in: *Regulation of the Immune Response* (P. L. Ogra and D.M. Jacobs, eds.), pp. 38–47, 8th Int. Convoc. Immunol., Amherst, NY, Karger, Basel.

Heremans, J. F., 1974, Immunolgobulin A, in: *The Antigens*, Volume II (M. Sela, ed.), pp. 365–522, Academic Press, New York.

Hopf, U., Brandtzaeg, P., Hutteroth, T. H., and Meyer zum Bushenfelde, K. H., 1978, *In vivo* and in vitro binding of IgA to the plasma membrane of hepatocytes, *Scand. J. Immunol.* **8**:543.

Inman, L. R., and Cantey, J. R., 1984, Peyer's patch lymphoid follicle epithelial adherence of a rabbit follicle epithelial adherence of a rabbit enteropathogenic *Escherichia coli* (strain RDEC-1): Role of plasmid-mediated pili in initial adherence, *J. Clin. Invest.* **74**:90.

Jackson, G. D. F., Lemaitre-Coelho, I., Vaerman, J. P., Bazin, H., and Beckers, A., 1978, Rapid disappearance from serum of intravenously injected rat myeloma IgA and its secretion into bile, *Eur. J. Immunol.* **8**:123.

Kapikian, A. Z., Kim, H. W., Wyatt, R. G., Rodriquez, W. J., Ross, S., Cline, W. L., Parrott, R. H., and Chanock, R. M., 1974, Reovirus-like agent in stools: Association with infantile diarrhoea and development of serologic tests, *Science* **185**:1049.

Kawanishi, H., Saltzman, L. E., and Strober, W., 1983, Mechanism regulating IgA class-specific immunoglobulin production in murine gut associated lymphoid tissues. I. T cells derived from Peyer's patches that switch SIgM cells to SIgA B cells *in vitro*, *J. Exp. Med.* **157**:433.

Keren, D. F., Scott, P. J., McDonald, R. A., and Kern, S. E., 1983, Local IgA-memory response to bacterial antigens, in: *The Secretory Immune System* (J. R. McGhee and J. Mestecky, eds.) pp. 734–744, New York Academy of Sciences, New York.

Kumagai, T., Wong, D., and Ogra, P. L., 1985, Development of cell–mediated cytotoxic activity in the respiratory tract after experimental infection with respiratory syncytial virus, *Clinic. Exp. Immunol.* **61**:351–359.

Lamm, M. E., Weisz-Carrington, P., Estela Roux, M., McWilliams, M., and Phillips-Qualiata, J., 1979, Mode of induction of an IgA response in the breast and other secretory sites by oral antigen in: *Immunology of Breast Milk* (P. L. Ogra and D. H. Dayton, eds.), pp. 105–114, Raven Press, New York.

Lehner, T., 1983, Regulation of the immune response to a streptococcal antigen by helper and suppressor functions in man, in: *The Secretory Immune System* (J. R. McGhee and J. Mestecky, eds.), pp. 273–283, New York Academy of Sciences, New York.

Lehner, T., Challacombe, S. J., and Caldwell, J., 1976a, Immunologic basis for vaccination against dental caries in rhesus monkeys, *J. Dent. Res.* **55**(spec. no.):C166.

Lehner, T., Challacombe, S. J., Wilton, J. M. A., and Caldwell, J., 1976b, Cellular and humoral immune responses in vaccination against dental caries in monkeys, *Nature* **264**:69.

Lemaitre-Coelho, T., Jackson, G. D. F., and Vaerman, J. P. 1978a, Relevance of biliary IgA antibodies in rat intestinal immunity, *Scand. J. Immunol.* **8**:459.

Lemaitre-Coelho, I., Jackson, G. D. F., and Vaerman, J. P., 1978b, High levels of secretory IgA and free secretory component in the serum of rats with bile duct obstruction, *J. Exp. Med.* **147**:934.

Lemaitre-Coelho, I., Altamirano, G. A., Barranco-Acosta, C., Meykens, R., and Vaerman, J. P., 1981, *In vivo* experiments involving secretory component in the rat hepatic transfer of polymeric IgA from blood into bile, *Immunology* **43**:261.

Levantal, G. S., Kulkarni, S. S., Meistrich, M. L., Newland, J. R., and Zander, A. R., 1983, Isolation of murine small bowel intraepithelial lymphocyte, *J. Immunol. Meth.* **63**:35.

Losonsky, G. A., Fishaut, J. M., Strussenberg, J., and Ogra, P. L., 1982, Effect of immunization against rubella on lactation products. I. Development and characterization of specific immunologic reactivity in breast milk, *J. Infect. Dis.* **145**:654.

Mahmoud, A. A. F., 1983, Regulation of immunopathology in parasitic infections, in: *Regulation of Immune Responses* (P. L. Ogra and D. M. Jacobs, eds.), 8th Int. Convoc. Immunol., Amherst, NY, pp. 267–277, Karger, Basel.

Mattingly, J. A., and Waksman, B. H., 1978, Immunologic suppression after oral administration of antigen I. Specific suppressor cells formed in rat Peyer's patches after oral administration of sheep orythrocytes and their systemic migration, *J. Immunol.* **121**:1878.

Mauel, J., 1982, Effector and escape mechanisms in host–parasite relationships, *Prog. Allergy* **31**:1.

Mestecky, J., McGhee, J. R., Michalek, S. M., Arnold, R. R., Crago, S. S., and Babb, J. L., 1978, Concept of the local and common mucosal immune response, in: *Secretory Immunity and Infection* (J. R. McGhee, J. Mestecky, and L. Babb, eds.), pp. 185–192, Adv. Exp. Med. Biul., Volume 107, Plenum Press, New York.

Mestecky, J., McGhee, J. R., Crago, S. S., Jackson, S., Kilian, M., Kiyono, H., Babb, J. L., and Michalek, S. M., 1980, Molecular–cellular interactions in the secretory IgA response, *J. Reticuloendothel. Soc.* **28**(S):45S.

Middleton, P. J., Szymanski, M. T., Abbott, G. D., Bortolussi, R., and Hamilton, J. R., 1974, Orbivirus of acute gastroenteritis of infancy, *Lancet* **1**:1241.

Montgomery, P. C., Rosner, B. R., and Cohn, J., 1974, The secretory antibody response. Anti-DNP antibodies induced by dinitrophenylated type III pneumococcus, *Immunol. Commun.* **3**:143.

Moon, H. W., and Issacson, R. E., 1981, Pili of enterotoxigenic *E. coli* as protective antigens in live oral vaccines, in: *Symposium on Cholera* (S. Kuwahara and T. Zinnaka, eds.), pp. 335–341, Proceedings of the 16th Joint Conference US–Japan Cooperative Medical Science Program, Government Printing Office, Washington, D.C.

Nagura, H., Nakane, P. K., and Brown, W. R., 1979, Translocalization of dimeric IgA through neoplastic colon *in vitro*, *J. Immunol.* **123**:2359.

Nencioni, L., Villa, L., Boraschi, D., Berti, B., and Tagliabue, A., 1983, Natural and antibody-dependent cell mediated activity against *Salmonella typhimurium* by peripheral and intestinal lymphoidcells in mice, *J. Immunol.* **130**:903.

Ogra, P. L., 1970, Distribution of echovirus antibody in serum, nasopharynx, rectum and spinal fluid after natural infection with echovirus type 6, *Infect. Immun.* **2**:150.

Ogra, P. L., 1984, Mucosal immune response to poliovaccines in childhood, *Rev. Infect. Dis.* **6**(Suppl):S361.

Ogra, P. L., and Karzon, D. T., 1971, Formation and function of poliovirus antibody in different tissues, *Prog. Med. Virol.* **13**:156.

Ogra, P. L., Wallace, R. G., Umana, G., Ogra, S. S., Kerr Grant, D., and Morag, A., 1974. Implications of secretory immune system in viral infections, in: *The Immunoglubulin A System* (J. Mestecky and A. R. Lawton, eds.), pp. 271–282, Plenum Press, New York.

Orlans, E., Peppard, J., Reynolds, J., and Halls, J., 1978, Rapid active transport of immunoglobulin A from blood to bile, *J. Exp. Med.* **147**:588.

Orlans, E., Peppard, J., Fry, J. R., Hinton, R. H., and Mullock, B. M., 1979, Secretory components as the receptor for polymeric IgA on rat hepatocytes, *J. Exp. Med.* **150:**1577.

Ottaway, C. A., Rose, M. L., and Parrott, D. M. V., 1979, The gut as an immunological system, in: *International Review of Physiology*, Gastrointestinal Physiology III, Volume 19 (R. K. Crane, ed.), pp. 323–356, University Park Press, Baltimore.

Owen, R. L., 1977, Sequential uptake of horseradish peroxidase by lymphoid follicle epithelium of Peyer's patches in the normal unobstructed mouse intestine: An ultrastructural study, *Gastroenterology* **72:**440.

Owen, R. L., and Jones, A. L., 1974, Epithelial cell specialization within human Peyer's patches: An ultrastructural study of intestinal lymphoid follicles, *Gastroenterology* **66:**189.

Owen, R. L., Lennel Allen, C., and Stevens, D. P., 1981, Phagocytosis of *Giardia muris* by macrophages in Peyer's patch epithelium in mice, *Infect. Immun.* **33:**591.

Owen, R. L., Nemanic, P. C., and Stevens, D. P., 1979, Ultrastructural observations on giardiasis in a murine model. I. Intestinal distribution, attachment, and relationship to the immune system of *Giardia muris*, *Gastroenterology* **76:**757.

Parrott, D. M. V., and Ferguson, A., 1974, Selective migration of lymphocytes within the mouse small intestine, *Immunology* **26:**571.

Peri, B. A., and Rothberg, R. A., 1981, Specific suppression of antibody production in young rabbit kits after maternal ingestion of bovine serum albumin, *J. Immunol.* **27:**2520.

Pierce, N. F., 1978, The role of antigen form and function in the primary and secondary intestinal immune responses to cholera toxin and toxoid in rats, *J. Exp. Med.* **148:**195.

Pierce, N. F., 1981, Experimental evidence favours oral immunization against cholera, in: *Symposium on Cholera* (S. Kuwahara, and Y. Zinaka, eds.), pp. 319–326, Proceedings of the 16th Joint Conference US–Japan Cooperative Medical Science Program, Government Printing Press, Washington, D.C.

Richman, L. K., Chiller, J. M., Brown, W. R., Hanson, D. G., and Vaz, N. M., 1978, Enterically induced immunologic tolerance. I. Induction and suppressor T lymphocytes by intragastric administration of soluble proteins, *J. Immunol.* **121:**2429.

Riepenhoff-Talty, M., Bogger-Goren, S., Li, P., Carmody, P. J., Barrett, H. J., and Ogra, P. L., 1981, Development of serum and intestinal antibody response to rotavirus after naturally acquired rotavirus infection in man, *J. Med. Virol.* **8:**215.

Riepenhoff-Talty, M., Lee, P. C., Carmody, P. J., Barrett, H. J., and Ogra, P. L., 1982, Age-dependent rotavirus-enterocyte interactions, *Proc. Soc. Exp. Biol. Med.* **170:**146.

Riepenhoff-Talty, M., Suzuki, H., and Ogra, P. L., 1983, Characteristics of the cell-mediated immune response to rotavirus in suckling mice, *Dev. Biol. Standard.* **53:**263.

Robert-Thomson, I. C., and Mitchell, 1978, Giardiasis in mice. I. Prolonged infections in certain mouse strains and hypothymic (nude) mice, *Gastroenterology* **75:**42.

Rudzik, O., and Bienenstock, J., 1974, Isolation and characteristics of gut mucosal lymphocytes, *Lab. Invest.* **30:**260.

Scott, R. M., Dudding, B. A., Romano, S. V., and Russell, P. K., 1972. Enteric immunization with live adenovirus type 21 vaccine II. Systemic and local immune response following immunication, *Infect. Immun.* **5:**300.

Smith, T. J., Buescher, E. L., Top, F. E., Altemeier, W. A., and McCown, J. M., 1970, Experimental respiratory infection with type 4 adenovirus vaccine in volunteers: Clinical and immunological responses, *J. Infect. Dis.* **122:**239.

Soontiens, F. J. C. J., and Van der Veen, J., 1973, Evidence of a macrophage-mediated effect or poliovirus on the lymphocyte response to phytohemagglutinin, *J. Immunol.* **5:**1411.

Stevens, D. P., Frank, D. M., and Mahmoud, A. A. F., 1978, Thymus dependency of host resistance to *Giardia Muris* infection: Studies in nude mice, *J. Immunol.* **120:**680.

Stokes, C., and Swarbrick, E. T., 1977, Induction of tolerance after and feeding of soluble protein antigen, *Biochem. Soc. Trans.* **5:**1573.

Svennerholm, A. M., Hanson, L. A., Holmgren, J., Lindblad, B. S., Nilsson, B., and Quereshi, F., 1980, Different secretory immunoglobulin A antibody response to cholera vaccination in Swedish and Pakistani women, *Infect. Immun.* **30:**427.

Svennerholm, A. M., Hanson, L. A., Holmgren, J., Jahl, F., Lindblad, B. S., Khan, S. R., Nilsson, A., and Svennerholm, B., 1981a, Antibody response to live and killed poliovirus vaccines in the milk of Pakistani and Swedish women, *J. Infect. Dis.* **143:**707.

Svennerholm, A. M., Sack, D. A., Bardhan, P. K., Hudas, Jertborn, M., and Holmgren, J., 1981b, Local and systemic antibody responses in man after immunization with cholera B subunit by various routes, in: *Symposium on cholera* (S. Kuwahara and Y. Zinaka, eds.), pp. 310–318. Proceedings of the 16th Joint Conference US–Japan Cooperative Medical Science Program, Government Printing Press, Washington, D.C.

Tomasi, T. B., 1980, Oral tolerance (overview), *Transplantation* 29:353.

Tomasi, T. B., Ban, W. G., Challacombe, S. J., and Curran, G., 1983, Oral tolerance and accessory cell function of Peyer's patches, in: *The Secretory Immune System* (J. R. McGhee and J. Mestecky, eds.), pp. 145–163, New York Academy of Sciences, New York.

Turesson, J., 1976, Distribution of immunoglobluin containing cells in human bone marrow and lymphoid tissues, *Acta Med. Scand.* **199:**293.

Vaerman, J. P., Lemaitre-Coelho, I., and Jackson, G. D. F., 1978, Role of the liver in the rat intestinal S-IgA system, *Adv. Exp. Med. Biol.* **107:**233.

Vaz, N. M., Maia, L. C. S., Hanson, D. G., and Lynch, J. J., 1977, Inhibition of homocytotropic antibody responses in adult inbred mice by previous feeding of the specific antigen, *J. Allergy Clin. Immunol.* **60:**110.

Walker, W. A., and Isselbacher, K. J., 1977, Intestinal antibodies, *N. Engl. J. Med.* **297:**767.

Warren, K. S., 1977, Modulation of immunopathology and disease in schistosomiasis, *Am. J. Trop. Med. Hyg.* **26**(Suppl)**:**113.

Wolf, J. L., Rubin, D. H., Finberg, R., Kauffman, R. S., Sharpe, A. H., Trier, J. S., and Fields, B. N., 1981, Intestinal M cells: A pathway for entery of reovirus into host, *Science* **212:**471.

5

Host Resistance against Bacterial Infections

Immunomodulation by Antimicrobials and Synthetic Immunoenhancers

MONIQUE PARANT

1. INTRODUCTION

Understanding the interactions between bacteria, antibiotics, and host defense mechanisms has increased relevance, as there are more and more patients particularly susceptible to severe infectious complications. Bacteria can modulate regulation of the immune system, though there are few studies of the possible role of bacterial immunosuppressive factors in acute infections (Schwab, 1983). A number of drugs—including antibiotics—are immunosuppressive. The role of corticoids in conditioning susceptibility to infections is well known, and most of the agents employed in the palliation of various types of cancer may also interfere with normal defense mechanisms. The immunosuppressive therapy used in organ transplantation is an outstanding example of the process leading to increased susceptibility to infections. In addition, more and more patients in an immunocompromised or frankly immunodeficient state are being subjected to major surgery. Despite the refinement in aseptic techniques, there is little evidence that these methods have any impact on the majority of surgical wound infections, which now derive most often from endogenous bacterial contaminants. In surgical patients, there has been an increasing incidence of infections from gram-negative organisms and an increase in the number of superimposed or secondary infections induced by bacterial overgrowth in patients on antibiotic therapy. Therefore, particular host factors are of great importance in determining the outcome of the antimicrobial treatment. Chemotherapy of infections can be unsatisfactory in immunocompromised hosts, since without adequate cellular

MONIQUE PARANT • Immunopharmacologie, Centre National de la Recherche Scientifique, Unité Associée 579, Université Paris 6, 75270 Paris 06, France.

defenses, even the most potent antibiotic may be useless, no matter how sensitive the responsible organisms are. Therefore, one of the most significant determinants of the effectiveness of antimicrobial agents is the functional state of the host immune system (Weinstein and Dalton, 1968).

As awareness of the limitations of antibiotics increases, so the concept of host defense stimulation comes under renewed and increasing consideration. Actually, immunopotentiating products may turn out to be of major importance in the case of problem infections, mainly in the immunocompromised host, or for the treatment of immunodeficiencies concomitant to certain chronic infectious diseases and occurring under various conditions. Such agents can be used either to enhance immunologically specific responses and/or to activate mechanisms of nonspecific host resistance. With the development of synthetic biologic response modifiers (BRMs), immunotherapy has entered the era of modern biotechnology. Microorganisms and microbial products had been utilized intensively as potential BRMs in a great number of animal models. In the last decades, chemical studies have led to isolation, purification, and analysis of bacterial cell wall components endowed with immunostimulant properties. As a result, the chemical synthesis of structural parts could be approached, in order to produce a number of analogs. In many cases various biologic effects of the original compound can be dissociated depending on the chemical structure, e.g., synthetic lipid A (Galanos *et al.*, 1984), trehalose dimycolate—also called cord factor (Lederer, 1984)—peptidolipids (Migliore-Samour *et al.*, 1980; Floc'h *et al.*, 1981; Kitaura *et al.*, 1982; Sone *et al.*, 1984), and muramyl dipeptide or MDP (Lefrancier and Lederer, 1981). Since this chapter is intended to give a picture of the therapeutic usefulness of antibiotic–immunostimulant combination against infections, it deals mainly with the synthetic BRMs that have been used in such experimental models to enhance nonspecific resistance, namely, muramyl peptides and peptidolipids.

2. IMPORTANCE OF HOST FACTORS IN THERAPEUTIC EFFECTIVENESS

There is a great deal of evidence that patients with qualitative and quantitative immunologic defects experience infectious diseases more often than normal individuals. Moreover, in the presence of ineffective host defense mechanisms, cure of infection cannot be accomplished even with highly active antibiotics. Systemic immunostimulants may therefore be used for the correction of generalized immunodeficiencies resulting, for example, from the use of drugs or from various chronic diseases, and to enhance the efficacy of chemotherapeutics.

2.1. IMMUNE SUPPRESSION PRODUCED BY BACTERIAL FACTORS

Bacteria per se or their products and certain factors associated with the infectious process have been found to either enhance or inhibit the functioning of immune cells (Schwab, 1977, 1983). It is probable that the bacterial flora is essential for efficient functioning of the immune system and has a normal regulatory role (Dubos *et al.*, 1963). Infectious agents may influence specific and nonspecific resistance to a variety of agents, as well as affect the integrity of the immune system itself (for review see Friedman *et al.*, 1983). There are several examples of suppression during the course of chronic infections involving facultative or obligatory intracellular bacteria. Infections inducing specific and nonspecific anergy include tuberculosis, brucellosis, and streptococcal diseases. In experimental models, *Propionibacterium acnes* (Scott, 1972), *Listeria monocytogenes* (Kongshavn *et al.*, 1977), and bacille Calmette Guérin (BCG) (Bennett *et al.*, 1978) induced suppression of T-cell responses through stimulation of macrophages. Prostaglandins have often been involved in these mechanisms (see Schwab, 1983). The loss of cell-mediated immunity may be ascribed to the massive accumulation of bacteria cell walls, which are poorly degradable by mammalian cells. This accumulation of toxic immunogenic debris in the tissues may initiate pathogenic processes which could be greatly influenced by an impairment of the immune system (Schwab, 1983).

Host factors such as very young or old age, pulmonary diseases, or combined drug regimens may predispose to superinfections (Weinstein and Dalton, 1968). In some cases, the superinfecting species are derived from the external environment but, for the most part, the organisms responsible for the new disease are members of the indigenous microflora. The invading species are those that have not been affected by the antibiotic treatment. Superinfecting organisms may also produce substances that inactivate antibiotics, such as penicillase-producing staphylococci responsible for the therapeutic failure of *Streptococcus pyogenes* infections that have been treated with penicillin G (see Bernstein *et al.*, 1964).

2.2. IMMUNE STATUS OF THE HOST

Gaps in antibacterial chemotherapy, especially for chronic bacterial diseases and infections of immunodeficient patients, have become more and more evident in recent years. That the activity of defense mechanisms is an absolute requirement for therapeutic effectiveness of antibiotics is constantly apparent though frequently overlooked (Weinstein and Dalton, 1968). Nonspecific stimulation of the host defense system is therefore of major importance in the effective treatment of infection with antimicrobials.

In immunodepressed patients, immunologic reconstitution therapy, such as immunoglobulin administration, bone marrow transplanation, fetal thymus graft, and polynuclear transfusions, has been shown to be effective in helping the antibiotic cure or in preventing recurrent infections (for review see Young, 1980; Lagrange, 1983). The importance of intact host defense mechanisms for successful antimicrobial therapy was investigated in animal models. In mice made leukopenic by exposure to x radiation, destruction of the bacteria in pneumococcal myositis lesions was 1000-fold lower than in comparable lesions with normal leukocytic exudates. Thus, in fully established infections of muscle, pneumococci are more rapidly destroyed by the combined effect of penicillin and phagocytes than by penicillin alone (Smith and Wood, 1956). In rats, depletion of plasma C3 by repeated injections with cobra venom factor markedly impaired the opsonic capacity of serum. They were incapable of producing phagocytosis of pneumococci in the alveoli and did not show signs of recovery after inoculation of *Streptococcus pneumoniae* in a lung lobe despite therapy with penicillin (Bakker-Woudenberg *et al.*, 1979). Again, penicillin allowed a loss of efficiency against infection by susceptible organisms in animals with selectively impaired host defense mechanisms.

3. IMMUNOMODULATING EFFECTS OF ANTIBIOTICS

Apart from their antimicrobial activity, antibiotics may exhibit immunologic side effects, either inhibiting or stimulating immune response mechanisms (Gillissen, 1980; Finch, 1980). The influence of antibiotics on defense mechanisms and the antibacterial chemotherapy of compromised hosts have therefore attracted increasing attention as patients with marginal defective host immunity may need more intensive therapy.

3.1. STIMULATING EFFECTS

A number of antibacterial agents showed potentiation of phagocytosis and the effect was due solely to the influence of drugs on the neutrophil rather than on the bacteria, as already demonstrated with human cells incubated with antibiotics such as penicillin G and cephalothin (Mandell, 1982). Some antibiotics such as cefaclor can influence intracellular killing by polymorphonuclear cells and concomitantly increase levels of myeloperoxidase, which is an important component of the oxidative bactericidal system of neutrophils (Kowolik *et al.*, 1982). Amphotericin B and the less toxic amphotericin-methyl-ester are potent stimulants of humoral and cell-mediated immunity in mice. Many reports have provided evidence that T cells are the predominant effectors in the enhancement of immune responses (see Stein *et al.*, 1978). Several cephalosporins can enhance the formation of IgM-producing spleen cells to sheep red blood cells (Gillissen, 1981).

3.2. INHIBITION OF IMMUNE FUNCTIONS

It is becoming increasingly apparent that a number of drugs used in antimicrobial chemotherapy may adversely affect the immune response and phagocytic cell function. These adverse effects should be considered in long-term application during chronic illness or when the patient is already immunocompromised.

The suppressive activity of antibiotics was often detected at doses higher than the human therapeutic dose. Thus, the inhibition of antibody production and of delayed-type hypersensitivity was obtained with high dosage levels of rifampicin and is reversible *in vivo* (Paunescu, 1970). A tetracycline depression of *in vivo* cell-mediated responses has also been found (Thong and Ferrante, 1980), whereas only a slight effect of doxycycline has been shown on allograft survival and antibody titers in mice (Bellahsene and Forsgren, 1985). A cyclosporin, cephalothin, and a cephamycin, cefoxitin, have been reported to inhibit the formation of IgM-producing cells in mice immunized with sheep red blood cells (Gillissen, 1982).

Considering the nonspecific immune responses, several antibiotics were shown to impair cell functions *in vitro*. Whether these effects result in any impairment of *in vivo* responses remains to be determined. Discrepencies in the influence on human polymorphonuclear leukocytes (PMN) may be due to the widely different methods used to evaluate cell functions. Some results suggest that even therapeutic concentrations of tetracycline, trimethoprim, and cephalothin may inhibit optimal PMN microbicidal capacity (Welch *et al.*, 1981). Lymphocyte responsiveness to nonspecific mitogens can be inhibited, among others, by rifampicin, rifamycin SV, tetracycline, and cyclosporins (Osoba, 1974; Forsgren and Banck, 1978; Thong and Ferrante, 1980; Chaperon, 1982). Doxycycline, cefadroxil, for example, exhibited a dose-dependent suppressive effect at concentrations that are consistent with serum levels (Clot and Pelous, 1987). The susceptibility of lymphocytes to suppression can be ascribed to the inhibitory role of prostaglandins (Webb *et al.*, 1982) and was shown to be partially reversed by adding indomethacin to the cultures (Chaperon, 1982; Clot and Pelous, 1987). Tetracyclines have also been reported to produce a remarkable inhibition of migration inhibitory factor production in human cells pulsed with concanavalin A (Ganguly *et al.*, 1983). Among several tetracyclines, doxycycline was the only one to inhibit the natural killer cell-mediated cytotoxicity (Goh and Ferrante, 1984).

Antibiotics may therefore predispose to infection due to their suppressive effect on components of cell-mediated immunity. Marrow suppression has been noted during and/or following the use of many antibiotics, and nonreversible bone marrow aplasia from the use of chloramphenicol has been recognized for years (Sharma and Remington, 1980). These observations may caution against the indiscriminate use of some antibiotics, especially when long-term therapy is prescribed.

4. INFLUENCE OF LOW CONCENTRATIONS OF ANTIBIOTICS ON THE HOST–BACTERIA RELATIONSHIP

Antibiotics at concentrations equal or inferior to the minimal inhibitory concentration (MIC) produce visible changes on bacteria in the morphology and ultrastructure with serious consequences on their virulence as well as on their susceptibility to the host immune defenses. Abnormal forms of bacteria produced by antibiotics *in vitro* are observed in specimens obtained from patients (Lorian, 1985). Thus, exposure of *Escherichia coli* to ampicillin and of *Pseudomonas aeruginosa* to azlocillin and cefsuladin causes the bacilli to elongate into filaments (Lorian and Atkinson, 1984). Electron microscopy revealed that streptococci exposed to clindamycin were largely denuded of the hairlike structures bearing M protein (Gemmell *et al.*, 1981).

4.1. ADHERENCE OF BACTERIA

Bacterial adhesion to eukaryotic cells is thought to represent an important factor in the pathogenesis of certain infectious diseases. Bacteria attach to host tissues with a high degree of specificity. This appears to be one of the dominating factors that determine whether or not they may colonize (Gibbons, 1977). Several approaches tend to inhibit the attachment of pathogens, which is mediated by specific molecules of recognition (Beachey *et al.*, 1981). Certain antibiotics may change the adhesiveness in different ways (Sandberg *et al.*, 1979; Beachey *et al.*, 1981; Vosbeck, 1982). For example, exposure of *E. coli* during the growing phase to sublethal concentrations of penicillin, streptomycin, or tetracycline markedly suppressed adhering ability (Eisenstein *et al.*, 1979). The addition of penicillin at sub-MIC resulted in diminished adherence of *Streptococcus pyogenes*, and this was associated with a loss of lipoteichoic acid, which normally binds the organism to specific receptor sites on the epithelial cell (Ofek *et al.*, 1979). Clindamycin was shown to decrease the adherence of *Klebsiella pneumoniae*, but to promote the adhesion of *Pseudomonas aeruginosa* and *Proteus mirabilis*. However, the three strains became more susceptible to PMN phagocytosis and killing. Therefore clindamycin potentiated host defense mechanisms against gram-negative aerobic organisms, though it has virtually no bactericidal or bacteriostatic activity (Lianou *et al.*, 1985).

4.2. PHAGOCYTOSIS AND KILLING

Antibiotics in subinhibitory concentrations may enhance both the phagocytic activity and the intracellular killing of PMN and of macrophages. Pruul *et al.* (1981) were able to demonstrate that human PMNs killed chloramphenicol-pretreated *E. coli* more efficiently than they did untreated controls, and similar findings have been described by Root (1981) using *Staphylococcus aureus* and β-lactam agents. Similarly, clindamycin rendered *E. coli* organisms more susceptible to phagocytosis and killing by human PMN (Bassaris *et al.*, 1984). In this review, Root (1981) remarks that for delineation of the mechanism involved, experimental conditions must be such that effects on bacteria and those on the cells can be clearly distinguished. Pretreatment with antibiotics may result in a loss of antiphagocytic surface components inducing a higher ingestion rate, probably through an increment in opsonin deposition (Gemmell *et al.*, 1981). In the case of *S. aureus*, killing was enhanced only be treatment with cell wall–

active antibiotics. Killing does not appear as a function of increased cellular uptake. Moreover, PMNs totally deficient in myeloperoxidase or incapable of generating O_2- and H_2O_2 in chronic granulomatous disease killed penicillin-pretreated but not untreated staphylococci normally (Root, 1981). Results obtained with mouse or human macrophages are comparable; for instance, *E. coli* pretreated with semisynthetic penicillins, i.e., cyclacillin and ampicillin, are killed more effectively by mouse macrophages (Friedman and Warren, 1982). Antibiotics in subinhibitory concentrations may enhance the phagocytic activity of human macrophages. This has been determined with different antibiotics (ampicillin, tetracycline, chloramphenicol) and unrelated organisms, *Listeria monocytogens* and *E. coli* (Adam, 1982). It has been shown in this study that the effect was not due to action on the macrophages themselves. Alterations of the outer membrane of the bacteria appeared in many cases responsible for the increased susceptibility of bacterial pathogens (Pruul *et al.*, 1981; Adam, 1982; Lorian and Atkinson, 1984).

4.3. ADDITIVE EFFECT WITH SERUM FACTORS

S. pyogenes organisms incubated with clindamycin at levels less than the MIC were able to activate complement more readily and to fix more C3 on their surface (Gemmell *et al.*, 1981). Studies concerning effects of antibiotics and normal serum have produced diverse results that may be related to technical difficulties concerning the two-component systems: killing by complement and growth of cells promoted by the broth. However, the killing of gram-negative bacterial cells by normal human serum has been reported to be either enhanced or inhibited by chloramphenicol or tetracycline. These effects may be the result of alterations at more than one location on the cell surface (Hart *et al.*, 1982). An increased susceptibility of *E. coli* to killing by specific antibody and complement has also been observed after treatment with semisynthetic penicillins (Friedman and Warren, 1976). Several antibiotics were used to test phagocytic reaction of mouse peritoneal cells against *P. aeruginosa* with the addition of specific γ-globulin. The bactericidal activity was much more enhanced in the presence of specific antibodies. Subinhibitory concentrations of gentamicin or cephalothin did not render these opsonized bacteria more susceptible to phagocytosis, but the number of viable organisms was significantly reduced in the presence of the antibiotics (Brunner and Undeutsch, 1982).

4.4. *IN VIVO* SYSTEMS

The indication that *in vitro* antibiotic-enhanced killing of *E. coli* by PMN may have clinical significance has been shown by *ex vivo* and *in vivo* experiments (Lam *et al.*, 1982). The pyrazole cephalosporin derivative strongly enhanced bacterial clearance when administered by a single intravenous dose to infected mice. In rabbit, the viable *E. coli* contained in diffusion chambers implanted in the peritoneum were reduced after an intravenous injection of this antibiotic only in the presence of PMNs. Of direct relevance to this study is the possibility that this cephalosporin may weaken the integrity of the bacterial membrane, allowing easy access for PMN level oxidative factors. Such enhanced activity may explain in part the high *in vivo* efficacy of this antibiotic, even when its tissue fluid concentration has dropped to an undetectable amount (Lam *et al.*, 1982). Similarly, subinhibitory concentrations of mezlocillin and azlocillin rendered *P. aeruginosa* or *K. pneumoniae* organisms more susceptible to an antimicrobial action of rat leukocytes in both *in vitro* and *ex vivo* models. *In vivo*, using the granula pouch model, low levels of the same antibiotics caused a significant increase in the rate of phagocytosis (Dalhoff and Stübner, 1985). In a mouse model of surgical infection with *E. coli*, Lamont *et al.* (1983) have demonstrated that sub-MICs of clindamycin *in vivo* induced a significant reduction in bacteremia when combined with stimulation of host defenses by muramyl dipeptide (MDP).

5. ENHANCEMENT OF HOST RESISTANCE BY SYNTHETIC LOW-MOLECULAR-WEIGHT BRMs

As mentioned earlier, two kinds of low-molecular-weight microbially derived compounds are examined in this chapter: glycopeptides and peptidolipids, prepared by analogy with peptidoglycans. Therapeutic attempts to enhance overall host nonspecific resistance to infections and to restore impaired defense mechanisms have been carried out in many experimental models, including immunodepressed animals.

Over the past decade, identification and synthesis of the smallest cell wall structure having the full adjuvant activity of peptidoglycan were decisively achieved (reviewed in Lederer, 1980). MDP (AcMur-L-Ala-D-iGln) is a small-molecular-weight glycopeptide, nonimmunogenic, which is the first of a series of analogs and derivatives. In addition to its adjuvant effect on specific immune responses, MDP was shown to stimulate nonspecific resistance in experimental infections when given in saline before the challenge with *K. pneumoniae* (Chedid *et al.*, 1977). Since then, many reports on the induction of resistance to bacterial, fungal, or parasitic infections have been published, reflecting the ever-increasing efforts made by several laboratories (Parant, 1979, 1983; Kotani *et al.*, 1982; Dietrich *et al.*, 1983b). The requirement of the AcMur moiety in the nonspecific antibacterial resistance has been questioned with the preparation of tetrapeptides linked to a fatty acyl chain (Parant *et al.*, 1980; Floc'h *et al.*, 1981; Kitaura *et al.*, 1982).

5.1. GENERAL FEATURES OF MDP TREATMENT

Some characteristics of the antiinfectious activity of MDP and of related peptidolipids have been delineated from numerous studies conducted in acute marine infections. These small-molecular-weight products are less potent than more complex and heterogeneous agents, such as BCG or lipopolysaccharide (LPS), in inducing an enhanced resistance to infection. They are nonimmunogenic unless they are coupled to a carrier (Chedid *et al.*, 1978). Moreover, MDP and certain of its derivatives are capable of stimulating nonspecific resistance when administered by the oral route (Chedid *et al.*, 1977). Muramyl dipeptides are less potent than LPS in increasing host defense mechanisms in that the treatment is not capable of reducing an overwhelming bacterial challenge and depends on the experimental infection model, i.e., the microorganisms used and their virulence (Dietrich *et al.*, 1983b), the route of injection, and the mouse strain. The role of timing in MDP treatment is always critical; thus it is protective against *K. pneumoniae* if administered before, simultaneously, or shortly after the challenge, whereas therapeutic administration after onset of the disease has no effect (Chedid *et al.*, 1977; Dietrich *et al.*, 1983b). In contrast to other immunostimulants (BCG, *Propionibacterium acnes*, LPS) a simultaneous injection of MDP with the bacterial inoculum does not produce a negative phase of higher susceptibility even with a large range of dose (Polk *et al.*, 1982; Parant and Chedid, 1984).

5.2. HYDROPHILIC MDP

The synthesis of MDP was followed by preparation of a large number of analogs (Lefrancier and Lederer, 1981). In several studies, chemically different MDP were comparatively evaluated and their effect assessed in a variety of experimental test systems. Although most of the MDP analogs that induce an increased resistance to infections are also active in the adjuvants tests, both the stimulating properties can be dissociated in some experimental models. Thus, animals that are low responders to the adjuvant effect of MDP, such as Lewis rats or C57B1/6 mice, are protected against a *Klebsiella* infection as effectively as high-responder animals (Parant, 1979). Actually, hydrophilic MDPs are effective against an infectious challenge under timing conditions that may lead to a depression of the immune response (Parant, 1979). Moreover, like most immunoadjuvants, MDP has been shown to generate a T-cell-mediated immunosuppressive activity when given in repeated large doses (Leclerc *et al.*, 1982), but such a treatment does not impair its protective activity in mice subsequently infected with *K. pneumoniae* (Parant and Chedid, 1984). Many adjuvant-active MDP analogs do not stimulate nonspecific resistance (see Parant, 1979), but there are only a few examples of compounds devoid of adjuvant effect and capable of protecting against an infectious challenge.

Several of the side effects inherent to the administration of bacterial peptidoglycan can be partially reproduced by MDP (reviewed by Chedid *et al.*, 1982b; Kotani *et al.*, 1982). Minor chemical modification has already provided compounds that do not exhibit these pharmacologic reactions, and up to now, they have been found among the water-soluble MDP analogs (Chedid *et al.*, 1982a; Byars, 1984). Murabutide [MDP(Gln)-nO*n*Bu] has thus been selected after a comparative evaluation of the biologic properties of a great number of analogs, which showed that in saline it was as effective as MDP in enhancing both specific and nonspecific immune responses, although it did not retain any side effect (Parant and Chedid, 1984). Its safety has been assessed by animal toxicologic assays and in a phase I clinical trial (Chedid, 1983), and it is undergoing clinical applications as adjuvant of vaccines (Oberling *et al.*, 1985; Telzak *et al.*, 1986). Against a bacterial infection, the efficacy of murabutide has been demonstrated in normal mice, under conditions that make the animals very susceptible to the challenge, as is the case in newborn mice or in adults splenectomized at 7 days of age (Chedid *et al.*, 1982a). Since, in contrast to MDP (Kotani *et al.*, 1976), murabutide is free of any pyrogenicity and cannot induce the release of endogenous pyrogen (EP), either *in vivo* or *in vitro*, its protective effect has shown that an increase in body temperature is not a prerequisite for the immunostimulant properties. *In vitro* murabutide can induce monocytic cells to produce interleukin-1 (IL-1), which is likely to be related to the adjuvant effect (Staruch and Wood, 1983). This result has provided the first indication that IL-1 and EP may correspond to different factors that can be obtained separately (Damais *et al.*, 1982), in contrast to what had been postulated. Moreover, murabutide does not provoke the pharmacologic reactions observed with MDP, such as induction of neutropenia, lowering of serum iron levels, and increase of SAA production (see Parant and Chedid, 1984).

The degree of antiinfective activity of the various hydrophilic MDP analogs able to enhance resistance has usually been found quite similar when determined in comparative assays, unless they had been conjugated to a carrier. However, in certain studies, non-MDP produced an increase greater than the original MDP in the survival time of mice infected with *P. aeruginosa* (Fraser-Smith and Matthews, 1981; Fraser-Smith *et al.*, 1982).

5.3. LIPOPHILIC MDP AND PEPTIDOLIPIDS

Several lipophilic MDP derivatives have been prepared to reproduce the antitumor activity of bacterial cell walls which have a high lipid content (Petit and Lederer, 1984). Among them, some compounds were significantly effective in causing tumor regression, such as 6-*O*-acyl-MDP analogs (Kotani *et al.*, 1982), MDP-dipalmitoyl-sn-glycerol (MDP-GDP) (Phillips *et al.*, 1985), and muramyl tripeptide-dipalmitoyl phosphatidyl-ethanolamine (MTP-PE) (Dietrich *et al.*, 1983a). Against bacterial challenges, lipophilic derivatives have often been shown to exhibit a protective activity stronger than the original MDP (see Table 1).

A series of 6-*O*-acyl-MDP derivatives containing various lengths of linear fatty acids indicated that increasing lipophilicity has influenced the protective activity. Thus, the introduction of a stearoyl group (L18-MDP) resulted in an increased protective effect as evaluated by comparative assays in mice infected subcutaneously with *E. coli* (Matsumoto *et al.*, 1981). Both L18-MDP and MDP-Lys(L18) were then found more effective than MDP in enhancing the mouse resistance to *P. aeruginosa, S. aureus*, and *K. pneumoniae* (Kotani *et al.*, 1982). Our data indicate that the presence of a lipophilic chain at the C-terminal peptide end

TABLE 1. EXAMPLES OF SYNTHETIC COMPOUNDS PROVIDING A BETTER PROTECTION THAN MDP AGAINST SOME BACTERIAL INFECTIONS

Compound	References
MDP analogs	
Non-MDP	Fraser-Smith and Matthews, 1981
[Abu^1]-MDP	Fraser-Smith and Matthews, 1981
Lipophilic derivatives	
MDP-L-Ala-glycerol mycolate	Parant *et al.*, 1980
6-*O*-stearoyl-MDP, L18-MDP, and MDP-Lys(L18)	Matsumoto *et al.*, 1981
MTP-PE	Dietrich *et al.*, 1983b
Peptidolipids	
DP-L-Ala-glycerol mycolate	Parant *et al.*, 1980
Lauroyl tetrapeptide	Floc'h *et al.*, 1981
Lactoyl tetrapeptide (FK-156)	Mine *et al.*, 1983
Oligomers and conjugates	
Pol(PAP-MDP)	Parant *et al.*, 1978a
MDP(A--L)	Chedid *et al.*, 1979

(MDP-L-Ala-glyceryl mycolate) results in increased effectiveness of MDP (Parant *et al.*, 1980). The lipophilic compound was more effective than MDP against *K. pneumoniae* and *P. aeruginosa* infections and was shown to elicit a significant protection when mice were challenged with a virulent strain of *S. pneumoniae*, whereas MDP was inactive. This compound is of particular interest since it was shown to induce delayed-type hypersensitivity to a protein antigen in the absence of Freund's incomplete adjuvant (FIA) (Parant *et al.*, 1980). Other lipophilic compounds such as MTP-PE also appeared to be somewhat more potent (Dietrich *et al.*, 1983b) (see Table 1).

Desmuramyl peptides related to bacterial peptidoglycans have been reported to have immunostimulating properties and to be more effective than MDP in some bacterial infection experiments (as summarized in Table 1). The "desmuramyl" derivative of the lipophilic MDP-L-Ala-glyceryl mycolate mentioned earlier was as active as the AcMur derivative in stimulating antibacterial resistance, i.e., more active than the original MDP. The DP-L-Ala-glyceryl mycolate is poorly active as adjuvant of specific immune responses, but the absence of pyrogenic effect even when given intravenously to rabbits may increase the interest of such a compound. A lauroyl-tetrapeptide prepared by *N*-laurylation of a tetrapeptide derived from *Streptomyces stimulosus* was found to be quite effective against *Listeria* infection, whereas MDP failed to induce a significant protection. Its efficacy against a *Klebsiella* challenge was similar (Floc'h *et al.*, 1981). A lactyl-tetrapeptide, FK-156, isolated from culture filtrates of *Streptomyces olivaceogriseus*, then synthesized (Kitaura *et al.*, 1982), is the first of a series of derivatives endowed with interesting immunostimulant properties (Mine *et al.*, 1983; Sone *et al.*, 1984). FK-156 and FK-565, an acyl-tripeptide, significantly enhance mouse resistance against various infectious challenges such as *E. coli*, *K. pneumoniae*, *Serratia marcescens*, *Pseudomonas aeruginosa*, and *L. monocytogenes*.

They are active when given parenterally with a higher efficacy than that obtained with MDP, and the analog FK-565 is also effective by the oral route (Mine *et al.*, 1983). In the same study, a significant decrease in the viable cell counts has been reported in a local abscess produced by subcutaneous inoculation of *Staphylococcus aureus*.

5.4. OLIGOMERS AND CONJUGATES

A change in the physicochemical properties by polymerization may serve as a means for enhancing antiinfectious activities by reducing the rapid excretion of the muramyl dipeptide (Parant *et al.*, 1979) and modifying affinity for the target cells. Previous studies had shown that most activities of MDP were abolished upon the formation of glycosidic linkage with the *para*-aminophenyl group. Surprisingly, after cross-linking with glutaraldehyde the oligomer (26,000 mol. wt.) was more active than MDP in protecting mice nonspecifically against bacterial challenge by *Klebsiella* or *Listeria* (Parant *et al.*, 1978a).

The conjugation procedure with an appropriate carrier, natural or synthetic, immunogenic or nonimmunogenic, was also used for increasing the size and efficiency of MDP and of some analogs. Thus, antiinfectious activity has been potentiated to a considerable extent when MDP or MDP(Lys) was linked to a synthetic carrier, multipoly(DL-Ala)–poly(L-Lys) (A–L), to bovine serum albumin, tetanus toxoid, or viral subunits (Chedid *et al.*, 1979; Parant *et al.*, 1987, unpublished data). With the A–L chain, it has been shown that an inactive MDP stereoisomer, MDP(D,D) could acquire a good potency against various infections (*Klebsiella, Pseudomonas,* and *Listeria*) after coupling, although it remained inactive as immunoadjuvant (Chedid *et al.*, 1979). Other carriers such as influenza viral subunits have been used to prepare conjugates endowed with a similar protective activity, i.e., effective at a very low dosage level of MDP. Even more interesting is the fact that such a conjugate (with viral subunits) administered intranasally may enhance the mouse resistance against an unrelated viral infection, whereas free MDP was inactive (Parant *et al.*, 1987, unpublished data). This type of investigation has a potential for developing antibacterial or antiviral agents endowed with both specific and nonspecific activities (Chedid and Audibert, 1985). In some cases the pyrogenicity of the glycopeptide has been increased after conjugation (Chedid *et al.*, 1982b), but the effective dose is far smaller than with free MDP, and nonpyrogenic MDP analogs could also be used.

5.5. EFFECTIVENESS OF BRMs IN IMMUNODEFICIENT HOSTS

Several experimental models of immunodeficient animals have been used to investigate the protective effect of MDPs. These studies are summarized in Table 2. In most examples, treatment with MDP conferred a significant protection against gram-negative pathogens, which are quite frequent in patients with

TABLE 2. PROTECTIVE EFFECT OF MURAMYL PEPTIDES AGAINST BACTERIAL INFECTIONS IN IMMUNOCOMPROMISED ANIMALS

Animal group	Organism (route)	References
Young mice	*K. pneumoniae* (s.c.)	Parant *et al.*, 1978b
Aged mice	*E. coli* (s.c.)	Osada *et al.*, 1982a
CBA/N mice	*S. enteritidis* (i.p.)	Onozuka *et al.*, 1984a
Beige mutant mice	*S. enteritidis* (i.p.)	Onozuka *et al.*, 1984b
Starved mice	*K. pneumoniae* (i.m.)[a]	Galland *et al.*, 1982
Splenectomized mice	*K. pneumoniae* (i.v.)	Chedid *et al.*, 1982a
B mice	*K. pneumoniae* (i.v.)	Parant, 1979
X-ray-irradiated mice	*E. coli* (s.c.)	Osada *et al.*, 1982a
Corticoid-treated mice	*K. pneumoniae* (i.v.)	Chedid, 1978
Cyclophosphamide-treated mice	*K. pneumoniae* (i.v.)	Chedid, 1978
Cyclophosphamide-treated mice	*K. pneumoniae* (i.p.)	Dietrich *et al.*, 1980
Cyclophosphamide-treated mice	*K. pneumoniae* (i.m.)[a]	Galland *et al.*, 1983
Cyclophosphamide-treated mice	*E. coli* (s.c.)	Osada *et al.*, 1982a
Cyclophosphamide-treated mice	*E. coli* (i.p.)	Dietrich *et al.*, 1983b
Cyclophosphamide-treated mice	*P. aeruginosa* (i.p.)	Fraser-Smith *et al.*, 1982
Cortisone-treated guinea pig	*P. aeruginosa* (itr)[b]	Osada *et al.*, 1982b

[a]i.m. = intramuscular insertion of a suture laded with bacteria.
[b]itr = by intratracheal inoculation.

defective host defenses. Several interesting examples have also been provided in animals challenged with fungal infections that do not come within the scope of this review (Fraser-Smith *et al.*, 1982; Osada *et al.*, 1982a).

Most immunodeficient animals are highly susceptible to bacterial infections, and treatment with immunostimulants may turn out to be of great importance. In very young, 7-day-old animals, agents such as LPS are poorly effective, inducing a short delay in mortality, whereas (Table 2) they can be fully protected by MDP or derivatives (Parant *et al.*, 1978b; Chedid *et al.*, 1979; Chedid *et al.*, 1982a). Genetically defective mice may also provide useful models for studies on the mechanisms of bactericidal dysfunctions (Onozuka *et al.*, 1984a, 1984b). Splenectomy in young animals may correspond to a situation analogous to the clinical setting of postoperative infections, though the mechanisms of their high susceptibility to certain pathogens remain to be elucidated.

Various forms of immunosuppressive treatment have been used to evaluate the potency of these immunostimulants in systemic or local infections (Table 2) and even against combined bacterial challenge (Ausobsky *et al.*, 1982). The immunologic defect thus obtained caused a greatly increased susceptibility to the bacteria examined. In some cases, such as in burned guinea pigs challenged with *P. aeruginosa* (Stinnett *et al.*, 1983), no beneficial effect has been obtained with MDP treatment, although in another study a subcutaneous injection of L18-MDP significantly delayed death from pneumonia caused by *P. aeruginosa* in guinea pigs immunosuppressed by cortisone treatment (Osada *et al.*, 1982b). To assess the restorative effects of FK-156 and FK-565 on resistance to infections, various immunosuppressive drugs, such as cyclophosphamide, carrageenan, mitomycin

C, and hydrocortisone, have been assayed, and immunocompromised mice have also been prepared by implantation of sarcoma 180 tumor cells. Survival rates were significantly higher when mice had been treated with FK-156 or FK-565 during the suppressive therapy and subsequently challenged with *P. aeruginosa*, *Salmonella enteritidus*, or *L. monocytogenes* (Yokota *et al.*, 1983).

Therefore, in many cases reported here, treatment with MDP was capable of reversing the detrimental effects of immunodepressive treatments on host defenses. A similar efficacy has been demonstrated in animals infected with a bacterial strain rendered resistant to the bactericidal effect of antibiotics by plasmid transfer (Parant, 1979). However, an accumulating body of evidence suggests that certain antibiotics at subinhibitory doses act to alter the susceptibility of some pathogens to host defense mechanisms (see Section 4). These observations have aroused interest in the potential role of combined defense stimulation and antibiotic therapy, particularly in immunocompromised hosts.

6. IMMUNOSTIMULATION COMBINED WITH ANTIBIOTICS

Several studies have been undertaken to evaluate the therapeutic usefulness of antibiotic–immunostimulant combination against infection. The potential importance of these treatments has stimulated the evaluation of MDP with regard to their possible synergistic and/or additive effect with antibiotics (Table 3).

In normal mice the first observation of synergism has been reported with soluble glycopeptides administered subcutaneously, MDP or non-MDP, and a cephalosporin derivative given orally. It was thus shown that the median effec-

TABLE 3. EXAMPLES OF PROTECTIVE EFFECT IN INFECTED MICE OF A COMBINED TREATMENT WITH IMMUNOSTIMULANT AND ANTIBIOTICS

Stimulant	Antibiotic	Organism	References
MDP, non-MDP	Cefroxadine	*K. pneumoniae* *P. multocida* *S. typhimurium*	Dietrich *et al.*, 1980
Non-MDP	Gentamicin	*P. aeruginosa*	Fraser-Smith and Matthews, 1981
MDP	Cephaloridine Chloramphenicol	*K. pneumoniae*	Polk *et al.*, 1982
MDP-Lys (L18)	Cefazolin Gentamicin	*E. coli* *P. aeruginosa*	Osada *et al.*, 1982a
FK-156	Ticarcillin	*P. aeruginosa*	Nishida and Mine, 1982
MDP	Clindamycin	*K. pneumoniae*	Lamont *et al.*, 1983
MDP-Lys(L18)	Aminobenzyl penicillin	*S. aureus*	Kotani *et al.*, 1982
MDP-CPEH	Erythromicin	*S. aureus*	Dietrich *et al.*, 1983b
Murabutide MDP-GDP	Gentamicin	*K. pneumoniae*	Parant and Chedid, 1987

tive dose of antibiotic was lowered by MDP or non-MDP treatment in many types of acute infections (Dietrich *et al.*, 1980). When non-MDP was administered before an intraperitoneal infection of *P. aeruginosa*, a synergistic effect was observed with gentamicin administered subcutaneously (Fraser-Smith and Matthews, 1981). Similar effects have been described with other antibiotics and other systemic infectious challenges (Table 3). An implantation of a bacterially contaminated suture into the thigh of mice has been used in Polk's laboratory to stimulate surgical infection. This model was used with *K. pneumoniae*, and it was shown that MDP consistently enhanced the effect of a bacteriostatic antibiotic, chloramphenicol, and a bactericidal antibiotic, cephaloridine. The effect of the combined treatment was significant with small doses of antibiotics (ED_{25} or ED_{50}) but MDP failed to enhance the effect of large doses (Polk *et al.*, 1982). With a dose of clindamycin ineffective by itself in mice challenged with a suture infected with *E. coli*, the reduction in bacteremia was higher in animals pretreated with MDP than in mice given MDP alone (Lamont *et al.*, 1983).

As in normal mice, MDPs enhance the efficacy of antibiotics also in immunosuppressed animals, provided that the bacterial inoculum does not exceed an approximately 10-fold LD_{50} (Dietrich *et al.*, 1983b). Thus, in cyclophosphamide-treated mice, non-MDP had a synergistic effect with the antibiotic against a *K. pneumoniae* challenge. With the lactoyl tetrapeptide FK-156, the ED_{50} values of ticarcillin and gentamicin were significantly decreased even in immunosuppressed mice, which are poorly protected against a pseudomonal infection by the stimulating agent alone (Yokota *et al.*, 1983).

7. CONCLUSIONS

There are now a great number of active immunomodulator molecules which should contribute to a better knowledge of the underlying mechanisms in the control of bacterial infections. They represent potential drugs in the field of veterinary and human medicine, and there are actually several clinical situations in which immunomodulation will have a good change to help active antibacterial chemotherapy. Indeed, control of infection by antibiotics is difficult in hosts with impaired immunity. It is well known that even when large doses of potent antibiotics are directed against highly sensitive organisms, reduction of the humoral and cellular factors necessary to the elimination of infectious agents may lead to a therapeutic failure. Antibiotic efficacy may therefore be dependent in part on an intact host defense system. It is hoped that further studies on the possible collaboration of antibiotics with host defenses will contribute to the establishement of a better strategy of antibiotic dosage. However, though it is clear that antimicrobials could favorably affect phagocyte–bacterium interactions at each step of the phagocytic process, in most cases the mode of action of antibiotics in subinhibitory concentrations remains to be elucidated. The capacity to affect these interactions favorably, particularly under conditions of host impairment or ineffective host defense, may be a property worth exploiting.

The synergistic combination of immunostimulants such as MDPs or related peptidolipids and antibiotics represents a positive contribution leading to a decrease in ED_{50} values of antibiotics, and to an increase in the efficacy of both the synthetic BRM and subinhibitory doses of the chemotherapeutic agent (see Table 3). It is important to note that muramyl peptides enhance the efficacy of antibiotics also in immunosuppressed animals (Dietrich *et al.*, 1983b). Many studies indicate that modification of the bacterial surface by low concentrations of antimicrobials substantially alters the result of their encounter with phagocytic cells (for review see Lorian, 1985). In addition, a nonspecific enhancement of phagocytic cell function by the glycopeptides is involved in the mechanism(s) of enhanced resistance against infections. This is illustrated by results reported by Friedman and Warren (1984). They show that peritoneal macrophages from mice treated with MDP have an enhanced ability to phagocytize and kill *E. coli in vitro,* and that preincubation of organisms with a subinhibitory dose of cyclacillin or ampicillin—but not penicillin—still enhances the uptake and killing by stimulated macrophages.

The macrophage is an important target cell for the biologic effects of MDP as demonstrated by a number of *in vitro* studies (reviewed in Leclerc and Chedid, 1982). There is also evidence that *in vivo* treatment with MDPs results in macrophage stimulation and may at least partly be responsible for enhanced resistance to infection. Thus, MDPs have been shown to enhance the carbon clearance capacity (Tanaka *et al.*, 1979), and to protect against infection with *L. monocytogenes* (Fraser-Smith and Matthews, 1981), *Toxoplasma gondii* (Krahenbuhl *et al.*, 1981), and *Trypanosoma cruzi* (Kierszenbaum and Ferraresi, 1979). The enhancement of resistance against these agents is generally regarded as involving activation of macrophages. In addition, a linear correlation has been seen between carbon clearance and protection against *Candida albicans* infection induced by several MDP analogs in both normal and immunosuppressed mice (Fraser-Smith *et al.*, 1982). An important biochemical event apparently related to the enhanced microbicidal potential of activated macrophages is their increased ability to generate the superoxide anion (O_2^-) and hydrogen peroxide. *In vivo* priming by MDP enhances the ability of macrophages to generate O_2^- in response to opsonized zymosan or to phorbol ester stimulation. Simultaneously, the killing of *C. albicans in vitro* by these cells is increased (Cummings *et al.*, 1980). Several MDP analogs have then been used *in vitro* to evaluate both the release of O_2^- and the tumoricidal activity of monocytic cells, and the lipophilic derivatives such as 6-*O*-stearoyl MDP were consistently more efficacious than the hydrophilic ones (Fogler and Fidler, 1984). The influence of MDPs on polymorphonuclear leukocytes has not yet been examined in detail. *In vitro,* hydrophilic MDPs enhance the respiratory burst in human PMNs (Damais *et al.*, 1984). In normal and immunosuppressed guinea pigs, the stimulation of protective mechanisms against *Pseudomonas* pneumonia by a stearoyl MDP, L18-MDP, may be attributed at least partly to the increased chemotactic and phagocytic activity of PMNs (Osada *et al.*, 1982a).

The diversity and the flexibility of synthetic BRMs may lead to the production of agents displaying only the desired activity. The lipophilic compounds have often been shown to exhibit a higher protective activity, but they seem to produce stronger side effects than most of the hydrophilic molecules. They presently have not reached the stage of clinical trials. However, the toxicologic manifestations that may be elicited by lipophilic compounds is dependent on the route of administration (Dietrich *et al.*, 1983a). Various constructs containing MDP may have particularly interesting properties. Besides being used as synthetic vaccines, such conjugates could induce in addition a broader spectrum of antibacterial and antiviral nonspecific immunity (Chedid and Audibert, 1985).

The superiority of a combined therapeutic approach with immunostimulating agents may be considered against various types of infections to lower the dose of the chemotherapeutic drug that may possess significant inherent toxicity or immunosuppressive effect, and to enhance its efficacy. Several interesting examples have also been provided against fungal infection (Sackmann and Dietrich, 1982), and against *Leishmania donovani* challenge (Adinolfi *et al.*, 1985). In these cases, the combined therapy with MDP was more effective than either treatment by itself. Future validation of such a strategy may encourage trials of a potentially superior approach. Moreover, the synergistic action of macrophage-activating factor and MDPs in therapeutic studies against cancer metastases (see Fogler and Fidler, 1984) may open an exciting field of research by taking advantage of the opportunity offered by modern technology to produce cytokines.

REFERENCES

Adam, D., 1982, Enhanced *in vitro* phagocytosis of different pathogens by human monocytes in the presence of antibiotics, in: *The Influence of Antibiotics on the Host–Parasite Relationship* (H. U. Eickenberg, H. Hahn, and W. Opferkuch, eds.), pp. 67–73, Springer-Verlag, Berlin.

Adinolfi, L. E., Bonventre, P. F., Van Der Pas, M., and Eppstein, D. A., 1985, Synergistic effect of glucantime and a liposome-encapsulated muramyl dipeptide analog in therapy of experimental visceral leishmaniasis, *Infect. Immun.* **48**:409.

Ausobsky, J. R., Trachtenberg, L. S., and Polk, H. C., 1982, Enhancement of non specific host defenses against combined local bacterial challenge, *Surg. Forum* **33**:46.

Bakker-Woudenberg, I. A. J. M., de Jong-Hoenderop, J. Y. T., and Michel, M. F., 1979, Efficacy of antimicrobial therapy in experimental rat pneumoniae: Effects of impaired phagocytosis, *Infect. Immun.* **25**:366.

Bassaris, H. P., Lianou, P. E., and Papavassiliou, J. T., 1984, Interaction of subminimal inhibitory concentration of clindamycin and *Escherichia coli:* Effects on adhesion and polymorphonuclear leukocyte function, *J. Antimicrob. Chemother.* **13**:361.

Beachey, E. H., Eisenstein, B. I., and Ofek, I., 1981, Sublethal concentrations of antibiotics and bacterial adhesion, in: *Adhesion and Microorganism Pathogenicity* (K. Elliott, M. O'Connor, and J. Whelan, eds.), pp. 288–300, Ciba Foundation Symp., Turnbridge Wells, London.

Bellahsene, A., and Forsgren, A., 1985, Effect of doxycycline on immune response in mice, *Infect. Immun.* **48**:556.

Bennett, J. A., Rao, V. S., and Mitchell, M. S., 1978, Systemic bacillus Calmette Guerin (BCG) activated natural suppressor cells, *Proc. Natl. Acad. Sci. USA* **75**:5142.

Bernstein, S. H., Stillerman, M., and Allerhand, J., 1964, Demonstration of penicillin inhibition by pharyngeal microflora in patients treated for streptococcal pharyngitis, *J. Lab. Clin. Med.* **63**:14.

Brunner, H., and Undeutsch, C., 1982, Effect of cephalothin and gentamicin on phagocytosis of *Klebsiella pneumoniae* by guinea pig alveolar macrophages, in: *The Influence of Antibiotics on the Host-parasite Relationship* (H. U. Eickenberg, H. Hahn, and W. Opferkuch, eds.), pp. 129–138, Springer Verlag, Berlin.

Byars, N. E., 1984, Two adjuvant-active muramyl dipeptide analogs induce differential production of lymphocyte-activating factor and a factor causing distress in guinea pig, *Infect. Immun.* **44**:344.

Chaperon, E. A., 1982, Suppression of lymphocytes by cephalosporins, in: *The Influence of Antibiotics on the Host–Parasite Relationship* (H. V. Eickenberg, H. Hahn, and W. Opferkuch, eds.), pp. 22–32, Springer Verlag, Berlin.

Chedid, L., 1978, Immunopharmacological activities of muramyl dipeptide, a synthetic mycobacterial analog, in: *Advances in Pharmacology and Therapeutics, Prostaglandins—Immunopharmacology,* Volume 4 (B. B. Vargaftig, ed.), pp. 187–196, Pergamon Press, Oxford.

Chedid, L., 1983, Muramyl peptides as possible endogenous immunopharmacological mediators, *Microbiol. Immunol.* **27**:723.

Chedid, L., and Audibert, F., 1985, New approaches for control of infections using synthetic or semisynthetic constructs containing MDP, *Springer Semin. Immunopathol.* **8**:401.

Chedid, L., Parant, M., Parant, F., Lefrancier, P., Choay, J., and Lederer, E., 1977, Enhancement of non specific immunity to *Klebsiella pneumoniae* infection by a synthetic immunoadjuvant (*N*-acetylmuramyl-L-alanyl-D-isoglutamine) and several analogs, *Proc. Natl. Acad. Sci. USA* **74**:2089.

Chedid, L., Audibert, F., and Johnson, A. G., 1978, Biological activities of muramyl dipeptide, a synthetic glycopeptide analogous to bacterial immunoregulating agents, *Prog. Allergy* **25**:63.

Chedid, L., Parant, M., Parant, F., Audibert, F., Lefrancier, P., Choay, J., and Sela, M., 1979, Enhancement of certain biological activities of muramyl dipeptide derivatives after conjugation to a multi-poly (DL-alanine)–poly (L-lysine) carrier, *Proc. Natl. Acad. Sci. USA* **76**:6557.

Chedid, L., Parant, M., Audibert, F., Riveau, G., Parant, F., Lederer, E., Choay, J., and Lefrancier, P., 1982a, Biological activity of a new synthetic muramyl peptide adjuvant devoid of pyrogenicity, *Infect. Immun.* **35**:417.

Chedid, L., Parant, M., and Riveau, G., 1982b, Immunopharmacological activities of MDP, in: *Immunopharmacology and the Regulation of Leukocyte Function* (D. R. Webb, ed.), pp. 159–182, Marcel Dekker, New York.

Clot, J. P., and Pelous, C., 1987, Stimulation of suppressor cell activity by cefadroxil, in: *Antibiosis and Host Resistance* (A. Szentivanyi, H. Friedman, and G. Gillissen, eds.), pp. 237–242 Plenum Press, New York.

Cummings, N. P., Pabst, M. J., and Johnston, R. B., 1980, Activation of macrophages for enhanced release of superoxide anion and greater killing of *Candida albicans*, *J. Exp. Med.* **152**:1659.

Dalhoff, A., and Stübner, G., 1985, Comparative analysis of the antimicrobial action of polymorphonuclear leucocytes *in vitro ex vivo*, and *in vivo*, *J. Antimicrob. Chemother.* **15**:283.

Damais, C., Riveau, G., Parant, M., Gerota, J., and Chedid, L., 1982, Production of lymphocyte-activating factor in the absence of endogenous pyrogen by rabbit or human leukocytes stimulated by a muramyl dipeptide derivative, *Int. J. Immunopharmacol.* **4**:451.

Damais, C., Jupin, C., Riveau, G., and Parant, M., 1984, Murabutide-induced production of non-pyrogenic LAF/IL1 by monocytic cells. *Responses of PMN, Immunobiology* **167**:269.

Dietrich, F. M., Sackmann, W., Zak, O., and Dukor, P., 1980, Synthetic muramyl dipeptide immunostimulants: Protective effects and increased efficacy of antibiotics in experimental bacterial and fungal infections in mice, in: *Current Chemotherapy and Infectious Diseases,* Volume 2 (J. D. Nelson and G. Grassi, eds.), pp. 1730–1732, American Society of Microbiology, Washington, DC.

Dietrich, F. M., Lukas, B., and Schmidt-Ruppin, K. H., 1983a, MTP-PE (synthetic muramyl peptide): Prophylactic and therapeutic effects in experimental viral infections, in: 13th Intern. Congr. Chemotherapy, Volume 91, pp. 50–53, American Society of Microbiology, Washington, DC.

Dietrich, F. M., Sackmann, W., Mitchell, G. H., and Voller, A., 1983b, Modulation of resistance to infections by synthetic low-molecular-weight compounds endowed with immunostimulatory properties: Experimental results and potential role in human medicine, in: *Chemotherapy and Immunology in the Control of Malaria, Filariasis, and Leishmaniasis* (N. Anand, and A. B. Sen, eds.), pp. 277–303, Tata McGraw-Hill, New Delhi.

Dubos, R. J., Schaedler, R. W., and Costello, R., 1963, Composition, alteration and effects of the intestinal flora, *Fed. Proc.* **22**:1322.

Eisentein, B. I., Ofek, I., and Beachey, E. H., 1979, Interference with the mannose binding and epithelial cell adherence of *Escherichia coli* by sublethal concentration of streptomycin, *J. Clin. Invest.* **63**:1219.

Finch, R., 1980, Immunomodulating effects of antimicrobial agents, *J. Antimibrob. Chemother.* **6**:691.

Floc'h, F., Bouchaudon, J., Werner, G. H., Migliore-Samour, D., and Jolles, P., 1981, Augmentation par des lipopeptides immunostimulants de la résistance de la souris à l'infection par *Listeria monocytogenes, Ann. Immunol. (Paris)* **132D**:265.

Fogler, W. E., and Fidler, I. J., 1984, Modulation of the immune response by muramyl dipeptide, in: *Immune Modulation Agents and their Mechanisms* (R. L. Fenichel, and M. A. Chirigos, eds.), pp. 499–512, Marcel Dekker, New York.

Forsgren, A., and Banck, G., 1978, Influence of antibiotics on lymphocyte function *in vitro, Infection* **6**(Suppl 1):91.

Fraser-Smith, E. B., and Matthews, T. R., 1981, Protective effect of muramyl dipeptide analogs against infections of *Pseudomonas aeruginosa* or *Candida albicans, Infect. Immun.* **34**:676.

Fraser-Smith, E. B., Waters, R. V., and Matthews, T. R., 1982, Correlation between in vivo anti-*Pseudomonas* and anti-*Candida* activities and clearance of carbon by the reticuloendothelial system for various muramyl dipeptide analogs, using normal and immunosuppressed mice, *Infect. Immun.* **35**:105.

Friedman, H., and Warren, G. H., 1976, Antibody mediated bacteriolysis: Enhanced killing of cyclacillin-treated bacteria, *Proc. Soc. Exp. Biol. Med.* **153**:301.

Friedman, H., and Warren, G., 1982, Increased phagocytosis of *Escherichia coli* pretreated with subinhibitory concentration of cyclacillin or ampicillin, *Proc. Soc. Exp. Biol. Med.* **169**:301.

Friedman, H., and Warren, G., 1984, Muramyl dipeptide-induced enhancement of phagocytosis of antibiotic-pretreated *Escherichia coli* by macrophages, *Proc. Soc. Exp. Biol. Med.* **176**:366.

Friedman, H., Klein, T. W., and Specter, S., 1983, Infection and Immunosuppression, in: *Advances in Immunopharmacology,* Volume 2 (J. W. Hadden, L. Chedid, P. Dukor, F. Spreafico, and D. Willoughby, eds.), pp. 773–783, Pergamon Press, Oxford.

Galanos, C., Lehman, U., Lüderitz, O., Rietschel, E. T., Westphal, O., Barde, H., Barde, L., Freudenberg, M. A., Hausen-Hagge, T., Lüderitz, T., Mc Kenzie, G., Schade, U., Strittmatter, W., Tanamoto, K., Zähringer, U., Imoto, M., Yoshimura, H., Yamamoto, M., Shimamoto, T., Kusumoto, S., and Shiba, T., 1984, Endotoxic properties of chemically synthesized lipid A part structures. Comparison of synthetic lipid A precursor and synthetic analogues with biosynthetic lipid A precursor and free lipid A, *Eur. J. Biochem.* **140**:221.

Galland, R. B., Trachtenberg, L. S., Rynerson, N., and Polk, H. C., 1982, Non-specific enhancement of resistance to local bacterial infection in starved mice, *Arch. Surg.* **117**:161.

Galland, R. B., Heine, K. J., and Polk, H. C., 1983, Non-specific stimulation of host defenses against bacterial challenge in immunosuppressed mice, *Arch. Surg.* **118**:333.

Ganguly, R., Pennock, D., and Kluge, R. M., 1983, Suppression of the production of migration inhibitory factor in humans by tetracycline, *J. Infect. Dis.* **148**:611.

Gemmell, C. G., Peterson, P. K., Schmeling, D., Kim, Y., Mathews, J., Wannamaker, L., and Quie, P. G., 1981, Potentiation of opsonization and phagocytosis of *Streptococcus pyogenes* following growth in the presence of clindamycin, *J. Clin. Invest.* **67**:1249.

Gibbons, R. J., 1977, Adherence of bacteria to host tissue, in: *Microbiology, 1977* (D. Schlessinger, ed.), pp. 395–406, American Society of Microbiology, Washington, DC.

Gillissen, G., 1980, Antibiotika und Immunanwort. Begleiteffekte der Chemotherapie, *Immun.* **Infekt. 8**:79.

Gillissen, G., 1981, Nouveaux antibiotiques beta-lactam et réponse immunitaire, *Med. Hyg.* **39**:122.

Gillissen, G. J., 1982, Influence of cephalosporins on humoral immune response, in: *The Influence of Antibiotics on the Host-parasite Relationship* (H. U. Eickenberg, H. Hahn, and W. Opferkuch, eds.), pp. 5–11, Springer Verlag, Berlin.

Goh, D. H. B., and Ferrante, A., 1984, *In vitro* inhibition of natural killer cell activity by doxycycline, *Int. J. Immunopharmacol.* **6**:51.

Haranaka, K., Satomi, N., Sakurai, A., and Kunii, O., 1982, The combined use of antibiotics and specific antibodies against mouse *Pseudomonas aeruginosa* infection *in vivo* and the phagocytosis of peritoneal exudate cells *in vitro,* in: *The Influence of Antibiotics on the Host–Parasite Relationship* (H. U. Eickenberg, H. Hahn, and W. Opferkuch, eds.), pp. 161–167, Springer Verlag, Berlin.

Hart, R. F., Reynard, A. M., and Cunningham, R. K., 1982, Effects of antibiotics on the bactericidal activity of normal human serum, in: *The Influence of Antibiotics on the Host–Parasite Relationship* (H. U. Eickenberg, H. Hahn, and W. Opferkuch, eds.), pp. 255–262, Springer Verlag, Berlin.

Kierszenbaum, F., and Ferraresi, R. W., 1979, Enhancement of host resistance against *Trypanosoma cruzi* infection by the immunoregulatory agent muramyl dipeptide, *Infect. Immun.* **25**:273.

Kitaura, Y., Nakaguchi, O., Takeno, H., Okada, S., Yonishi, S., Hemmi, K., Mori, J., Senoh, H., Mine, Y., and Hashimoto, M., 1982, N^2-(γ-D-glutaminyl)-meso-2 (L), 2′(D)-diaminopimelic acid as the minimal prerequisite structure of FK-156: Its acyl derivatives with potent immunostimulating activity, *J. Med. Chem.* **25**:335.

Kongshavn, P. A., Ho, A., and Sebaldt, R. J., 1977, Suppression of *in vitro* antibody responses by *Listeria* primed spleen cells, *Cell. Immunol.* **28**:284.

Kotani, S., Watanabe, Y., Shimono, T., Harada, K., Shiba, T., Kusumoto, S., Yokogawa, K., and Taniguchi, M., 1976, Correlation between the immunoadjuvant activities and pyrogenicities of synthetic *N*-acetylmuramyl-peptides or -aminoacids, Biken J. **19**:9.

Kotani, S., Takada, H., Tsujimoto, M., Kubo, T., Ogawa, T., Azuma, I., Ogawa, H., Matsumoto, K., Siddiqui, W. A., Tanaka, A., Nagao, S., Kohashi, O., Kanoh, S., Shiba, T., and Kusumoto, S., 1982, Nonspecific and antigen specific stimulation of host defence mechanisms by lipophilic derivatives of muramyl dipeptides, in: *Bacteria and Cancer* (J. Jeljaszewicz, G. Pulverer, and W. Roszkowski, eds.), pp. 67–107, Academic Press, London.

Kowolik, M. J., Raeburn, J. A., and Grant, M., 1982, In vivo effect of ampicillin and cefaclor on blood and gingival crevicular neutrophil myeloperoxidase activity, in: *The Influence of Antibiotics on the Host–Parasite Relationship* (H. U. Eickenberg, H. Hahn, and W. Opferkuch, eds.), pp. 56–66, Springer Verlag, Berlin.

Krahenbuhl, J. L., Sharma, S. D., Ferraresi, R. W., and Remington, J. S., 1981, Effects of muramyl dipeptide treatment on resistance to infection with *Toxoplasma gondii* in mice, *Infect. Immun.* **31**:716.

Lagrange, P. H., 1983, Clinical immunomodulation of bacterial infection, in: *Advances in Immunopharmacology*, Volume 2 (J. W. Hadden, L. Cehdid, P. Dukor, F. Spreafico, and D. Willoughby, eds.), pp. 195–204, Pergamon Press, Oxford.

Lam, C., Georgopoulos, A., Laber, G., Schütze, E., Wenzel, A., 1982, *in vitro* and *in vivo* modulation of *Escherichia coli* susceptibility to PMN leukocyte microbicidal activity by a pyrazole cephalosporin derivative, in: *The Influence of Antibiotics on the Host–Parasite Relationship* (H. U. Eickenberg, H. Hahn, and W. Opferkuch, eds.), pp. 242–254, Springer Verlag, Berlin.

Lamont, P. M., Trachtenberg, L. S., West, C. S., and Polk, H. C., 1983, Non-specific host defence stimulation and antibiotic-resistant infection, *J. Antimicrob. Chemother.* **12**(Suppl C):117.

Leclerc, C., and Chedid, L., 1982, Macrophage activation by synthetic muramyl peptides, in: *Lymphokines*, Volume 7 (E. Pick, ed.), pp. 1–21, Academic Press, New York.

Leclerc, C., Bourgeois, E., and Chedid, L., 1982, Demonstration of muramyl dipeptide (MDP)-induced T suppressor cells responsible for MDP immunosuppressive activity, *Eur. J. Immunol.* **12**:249.

Lederer, E., 1980, Synthetic immunostimulants derived from the bacterial cell wall, *J. Med. Chem.* **23**:819.

Lederer, E., 1984, Chemistry of mycobacterial cord factor and related natural and synthetic trehalose esters, in: *The Mycobacteria*, Part A (G. P. Kubica, and L. G. Wayne, eds.), pp. 361–378, Marcel Dekker, New York.

Lefrancier, P., and Lederer, E., 1981, Chemistry of synthetic immunomodulant muramyl peptides, in: *Progress in Chemical and Organic Natural Products*, Volume 40 (W. Herz, H. Grisebach, and G. W. Kirby, eds.), pp. 1–47, Springer Verlag, New York.

Lianou, P. E., Bassaris, H. P., Votta, E. G., and Papavassiliou, J. T., 1985, Interaction of subminimal inhibitory concentrations of clindamycin and gram-negative aerobic organisms: Effects on adhesion and polymorphonuclear leukocyte function, *J. Antimicrob. Chemother.* **15**:481.

Lorian, V., 1985, Low concentrations of antibiotics, *J. Antimicrob. Chemother.* **15**(Suppl A):15.

Lorian, V., and Atkinson, B., 1984, Bactericidal effect of polymorphonuclear neutrophils on antibiotic-induced filaments of gram-negative bacilli, *J. Infect. Dis.* **149**:719.

Mandell, L. A., 1982, The effects of antibacterial, antiviral and antifungal drugs on the phagocytic, microbicidal, and chemotactic functions of the human polymorphonuclear leukocytes, in *The Influence of Antibiotics on the Host–Parasite Relationship* (H. U. Eickenberg, H. Hahn, and W. Opferkuch, eds.), pp. 40–55, Springer Verlag, Berlin.

Matsumoto, K., Ogawa, H., Kusama, T., Nagase, O., Sawaki, N., Inage, M., Kusumoto, S., Shiba, T., and Azuma, I., 1981, Stimulation of non-specific resistance to infection induced by 6-*O*-acyl muramyl dipeptide analogs in mice, *Infect. Immun.* **32**:748.

Migliore-Samour, D., Bouchaudon, J., Floc'h, F., Zerial, A., Ninet, L., Werner, G. H., and Jolles, P., 1980, A short lipopeptide representative of a new family of immunological adjuvants devoid of sugar, *Life Sci.* **26**:883.

Mine, Y., Yokota, Y., Wakai, Y., Fukada, S., Nishida, M., Goto, S., and Kuwahara, S., 1983, Immunoactive peptides, FK-156 and FK-565. I- Enhancement of host resistance to microbial infection in mice, *J. Antibiotics (Japan)* **36**:1045.

Nishida, M., and Mine, Y., 1982, Effect of antibiotics and an immunostimulant on the ability of immunosuppressed mice to defend against pseudomonal infection, in: *The Influence of Antibiotics on the Host–Parasite Relationship* (H. U. Eickenberg, H. Hahn, and W. Opferkuch, eds.), pp. 148–159, Springer Verlag, New York.

Oberling, F., Morin, A., Duclos, B., Lang, J. M., Beachey, E. H., and Chedid, L., 1985, Enhancement of antibody response to a natural fragment of streptococcal M protein by Murabutide administered to healthy volunteers, *Int. J. Immunopharmacol.* **7**:398.

Ofek, I., Beachey, E. H., Eisenstein, B. I., Alkan, M. L., and Sharon, N., 1979, Suppression of bacterial adherence by subminimal inhibitory concentrations of beta-lactam and amino-glycoside antibiotics, Rev. Infect. Dis. **1**:832.

Onozuka, K., Saito-Taki, T., and Nakano, M., 1984a, Augmentation of protective and antibacterial activity induced by muramyl dipeptides in CBA/N defective mice with X-linked immunodeficiency for *Salmonella enteritidis* infection, *Infect. Immun.* **45**:424.

Onozuka, K., Saito-Taki, T., and Nakano, M., 1984b, Effect of muramyl dipeptide on *Salmonella enteritidis* in beige mice with Chediak–Higashi syndrome, *Microbiol. Immunol.* **28**:1211.

Osada, Y., Mitsuyama, M., Matsumoto, K., Une, T., Otani, T., Ogawa, H., and Nomoto, K., 1982a, Stimulation of resistance of immuno-compromised mice by a muramyl dipeptide analog, *Infect. Immun.* **37**:1285.

Osada, Y., Otani, T., Une, T., Ogawa, H., and Nomoto, K., 1982b, Enhancement of non-specific resistance to *Pseudomonas pneumoniae* by a synthetic derivative of muramoyl dipeptide in immunosuppressed guinea pigs, *J. Gen. Microbiol.* **128**:2361.

Osoba, D., 1974, Effect of rifamycin SV derivatives on immunologic responses in culture, *J. Immunol.* **112**:844.

Parant, M., 1979, Biologic properties of a new synthetic adjuvant, muramyl dipeptide (MDP), *Springer Semin. Immunopathol.* **2**:101.

Parant, M., 1983, Bacterial immunoregulatory agents, in: *Regulation of the Immune Response* (P. L. Ogra and D. M. Jacobs, eds.), pp. 195–204, Karger, Basel.

Parant, M., and Chedid, L., 1984, Stimulation of nonspecific resistance to infections by synthetic immunoregulatory agents, *Infection* **12**:230.

Parant, M., and Chedid, L., 1987. Muramyl dipeptides, host immunity and enhancement, in: *Antibiosis and Host Resistance* (A. Szentivanyi, H. Friedman, and G. Gillissen, eds.), pp. 291–300, Plenum Press, New York.

Parant, M., Damais, C., Audibert, F. A., Parant, F., Chedid, L., Sache, E., Lefrancier, P., Choay, J., and Lederer, E., 1978a, *In vivo* and *in vitro* stimulation of nonspecific immunity by the β-D-*p*-aminophenyl glycoside of *N*-acetylmuramyl-L-alanyl-D-isoglutamine and an oligomer prepared by cross-linking with glutaraldehyde, *J. Infect. Dis.* **138**:378.

Parant, M., Parant, F., and Chedid, L., 1978b, Enhancement of the neonate's nonspecific immunity to *Klebsiella* infection by muramyl dipeptide, a synthetic immunoadjuvant, *Proc. Natl. Soc. Acad. Sci. USA* **75**:3395.

Parant, M., Parant, F., Chedid, L., Yapo, A., Petit, J-F., and Lederer, E., 1979, Fate of the synthetic immunoadjuvant, muramyl dipeptide (^{14}C-labeled) in the mouse, *Int. J. Immunopharmacol.* **1**:35.

Parant, M., Audibert, F. A., Chedid, L., Level, M., Lefrancier, P., Choay, J., and Lederer, E., 1980, Immunostimulant activities of a lipophilic muramyl dipeptide derivative and of desmuramyl peptidolipid analogs, *Infect. Immun.* **27**:826.

Paunescu, E., 1970, *In vivo* and *in vitro* suppression of humoral and cellular immunologic response by rifampicin, *Nature* **228**:1188.

Petit, J. F., and Lederer, E., 1984, The structure of the mycobacterial cell wall, in: *The Mycobacteria* (G. P. Kubica, and L. G. Wayne, eds.), pp. 301–313, Marcel Dekker, New York.

Phillips, N. C., Moras, M. L., Chedid, L., Lefrancier, P., and Bernard, J. M., 1985, Activation of alveolar macrophage tumoricidal activity and eradication of experimental metastases by freeze-dried liposomes containing a new lipophilic muramyl dipeptide derivative, *Cander Res.* **45:**125.

Polk, H. C., Galland, R. B., and Ausobsky, J. R., 1982, Nonspecific enhancement of resistance to bacterial infection. Evidence of an effect supplemental to antibiotics, *Ann. Surg.* **196:**436.

Pruul, H., Wetherall, B. L., and McDonald, P. J., 1981, Enhanced susceptibility of *Escherichia coli* to intracellular killing by human polymorphonuclear leukocytes after *in vitro* incubation with chloramphenicol, *Antimicrob. Agents Chemother.* **19:**945.

Root, R. K., 1981, Antibacterial properties of polymorphonuclear leukocytes and their interactions with antibiotics, in: *Microbiology 1981* (D. Schleissinger, ed.), pp. 176–180, American Society of Microbiology, Washington, DC.

Sackmann, W., and Dietrich, F. M., 1982, Experimental murine candidiasis: nonspecific resistance induced by synthetic compounds with immunostimulatory properties, in: *Current Chemotherapy and Immunotherapy* (P. Periti, and G. Grassi, eds.), pp. 1162–1165, American Society of Microbiology, Washington, DC.

Sandberg, T., Stenqvist, K., and Svanborg Eden, C., 1979, Effects of subminimal inhibitory concentrations of ampicillin, chloramphenicol and nitrofurantoin on the attachment of *Escherichia coli* to human uroepithelial cells *in vitro, Rev. Infect. Dis.* **1:**838.

Schwab, J. H., 1977, Modulation of the immune response by bacteria, in: *Microbiology 1977* (D. Schlessinger, ed.), pp. 366–373, American Society of Microbiology, Washington, DC.

Schwab, J. H., 1983, Bacterial interference with immunospecific defences, *Phil. Trans. R. Soc. Lond. B* **303:**123.

Scott, M. T., 1972, Biological effects of the adjuvant *Coryne-bacterium parvum*. II. Evidence for macrophage–T-cell interaction, *Cell. Immunol.* **5:**469.

Sharma, S. D., and Remington, J. S., 1980, The role of cell-mediated immunity in resistance to infection in the immunocompromised host, in: *Infections in the Immuno-compromised Host. Pathogenesis, Prevention and Therapy* (J. Verhoef, P. K. Peterson, and P. G. Quie, eds.), pp. 59–76, Elsevier, North Holland Biochemical Press, Amsterdam.

Smith, M. R., and Wood, W. B., 1956, An experimental analysis of the curative action of penicillin in acute bacterial infections. *II. The role of phagocytic cells in the process of recovery, J. Exp. Med.* **103:**499.

Sone, S., Mutsuura, S., Ogawara, M., Utsugi, T., and Tsubura, E., 1984, Activation by a new synthetic acyltripeptide and its analogs entrapped in liposomes of rat alveolar macrophages to the tumor cytotoxic state, *Cancer Immunol. Immunother.* **18:**169.

Staruch, M. J., and Wood, D. D., 1983, The adjuvanticity of interleukin-1 *in vivo, J. Immunol.* **130:**2191.

Stein, S. H., Plut, E. J., Shine, T. E., and Little, J. R., 1978, The importance of different murine cell types in the immunopotentiation produced by amphotericin methylester, *Cell. Immunol.* **40:**211.

Stinnett, J. D., Loose, L. D., Miskell, P., Tenney, C. L., Gonce, S. J., and Alexander, J. W., 1983, Synthetic immunomodulators for prevention of fatal infections in a burned guinea pig model, *Ann. Surg.* **198:**53.

Tanaka, A., Nagao, S., Nagao, R., Kotani, S., Shiba, T., and Kusumoto, S., 1979, Stimulation of the reticuloendothelial system of mice, by muramyl dipeptide, *Infect. Immun.* **24:**302.

Telzak, E., Conlon, Y., Wolff, S. M., Elkholy, A., Bahr, G. M., Morin, A., and Chedid, L., 1986, Clinical evaluation of the adjuvant-active derivative of MDP, Murabutide, associated with a tetanus toxoid vaccine. II. Study in non-previously vaccinated Egyptian healthy volunteers, *J. Infect. Dis.* **153**(3):628.

Thong, Y. H., and Ferrante, A., 1980, Effects of tetracycline treatment on immunological response in mice, *Clin. Exp. Immunol.* **39:**728.

Webb, D. R., Wieder, K. J., and Nowowiejski, I., 1982, Studies on the nonspecific regulation of immunocompetent cell function by prostaglandins, in: *Immunopharmacology and the Regulation of Leukocyte Function* (D. R. Webb, ed.), pp. 143–158, Marcel Dekker, New York.

Weinstein, L., and Dalton, A. C., 1968, Host determinants of response to antimicrobial agents, *N. Engl. J. Med.* **279:**467, 524, and 580.

Welch, W. D., Davies, D., and Thrupp, L. D., 1981, Effect of antimicrobial agents on human polymorphonuclear leukocyte microbicidal function, *Antimicrob. Agents Chemother.* **20:**15.

Yokota, Y., Mine, Y., Wakai, Y., Watanabe, Y., Nishida, N., Goto, S., and Kuwahara, S., 1983, Immunoactive peptides, FK-156 and FK-565. II. Restoration of host resistance to microbial infection in immunosuppressed mice, *J. Antibiotics (Japan)* **36:**1051.

Young, L. S., 1980, Immunoprophylaxis and immunotherapy of infection in the compromised host, in: *Infections in the Immunocompromised Host. Pathogenesis, Prevention and Therapy* (J. Verhoef, P. G. Quie, eds.), pp. 283–299, Elsevier, North Holland Biochemical Press, Amsterdam.

6

Immunity to Syphilitic Infection

RONALD F. SCHELL, BRET M. STEINER, and JEFFREY D. ALDER

1. INTRODUCTION

Despite centuries of interest in the clinical manifestations of syphilis, the nature of acquired resistance to infection by *Treponema pallidum* remains poorly understood. A number of factors, including to a great extent technical problems related to the nature of the organism, have been responsible for the slow progress in deciphering the immune responses. Although *T. pallidum* organisms have been grown *in vitro* with some success (Fieldsteel *et al.*, 1982; Norris, 1982), these studies have not progressed to accurate quantification of treponemes in tissue, for example, by colony counting. *T. pallidum* must be maintained in the laboratory by animal transfer and isolated from infected mammalian tissue. Any attempt to characterize its immunologic properties presents the problem of distinguishing treponemal antigens from those of the mammalian host. Although treponemal antigens have been isolated through recombinant DNA technology and improved biochemical techniques (Stamm and Bassford, 1982; Hanff *et al.*, 1983; Norgard and Miller, 1983; Van Embden *et al.*, 1983; Marchitto *et al.*, 1984; Hansen *et al.*, 1985), these antigens have not been shown to be important in the induction of acquired resistance. Finally, the hamster and rabbit appear to be the only nonprimates that can be regularly infected with *T. pallidum* organisms to produce sustained lesions (Turner and Hollander, 1957; Schell *et al.*, 1980b); however, the hamster and rabbit do not reflect the clinical manifestations of syphilis that are seen in humans.

2. ANTIBODY-MEDIATED IMMUNITY

The investigations that have been carried out in the face of these difficulties have suggested that antibody-mediated immunity plays only a modest role in

RONALD F. SCHELL • State Laboratory of Hygiene and Department of Medical Microbiology, University of Wisconsin, Madison, Wisconsin 53706. BRET M. STEINER • Department of Medical Technology, University of Wisconsin, Madison, Wisconsin 53706. JEFFREY D. ALDER • Department of Medical Microbiology, University of Wisconsin, Madison, Wisconsin 53706.

the host's response to infection with *T. pallidum* (Turner, 1939; Septejian *et al.*, 1973; Turner *et al.*, 1973; Perine *et al.*, 1973; Weiser *et al.*, 1976; Titus and Weiser, 1979; Bishop and Miller, 1983). Several factors indicating this are the following: (1) Primary and secondary syphilis progress despite the development of antibodies that immobilize and inactivate *T. pallidum in vitro* (Thompson *et al.*, 1950; Magnuson *et al.*, 1951; McLeod and Magnuson, 1953) or are highly specific, although their immune protective properties are largely unknown (Bishop and Miller, 1983). (2) Rabbits injected with killed *T. pallidum* (Waring and Fleming, 1951) or with sonicates of *T. pallidum* (Izzat *et al.*, 1971) developed Venereal Disease Research Laboratory, hemagglutination, and specific treponemal antibodies, yet are not protected against subsequent challenge with virulent treponemes. (3) Serum and immunoglobulin from rabbits resistant to syphilitic infection confer limited protection on recipients challenged with *T. pallidum* (Turner, 1939; Septejian *et al.*, 1973; Turner *et al.*, 1973; Perine *et al.*, 1973; Weiser *et al.*, 1976; Titus and Weiser, 1979) despite the fact that syphilitic rabbit serum has considerable treponemicidal activity *in vitro* (Bishop and Miller, 1983).

Attempts to elucidate the humoral mechanism by which animals respond to syphilitic infection have been hindered by the unavailability of suitable inbred animals. Although the rabbit is considered the animal model of choice for investigation of treponemal infections, inbred strains are not readily available. The inavailability of inbred animal strains that consistently produce symptomatic infection compromises immunologic studies involving the transfer of acquired resistance to syphilitic infection. The efficacy of syphilitic immune serum cannot be guaranteed in recipient outbred animals owing to allogeneic differences. This may account for the modest protection observed when serum from rabbits immune to infection with *T. pallidum* failed to confer complete protection on outbred recipients against challenge with the same strain. Inbred hamsters are readily available and have been shown to produce extensive chronic skin lesions after infection with *T. pallidum* subspecies *endemicum,* the causative agent of endemic syphilis (Schell *et al.*, 1980b).

T. pallidum subspecies *endemicum* (formerly *T. pallidum* (Bosnia A) was originally isolated from a dark-field-positive chancre on the shaft of the penis of a 35-year-old patient who resided in Bosnia, Yugoslavia (Turner and Hollander, 1957). When this strain was inoculated into inbred LSH/Ss LAK strain of hamster, it consistently produced extensive chronic skin lesions that persisted for 6–9 months (Fig. 1), even in the presence of peak antitreponemal antibody titers (Schell *et al.*, 1980b). The inguinal lymph nodes increased in weight and contained measurable number of treponemes. Most important, hamsters infected with *T. pallidum* subspecies *endemicum* developed substantial resistance to homologous reinfection and infection to other virulent treponemes (*T. pallidum* Nichols and *T. pallidum* subspecies *pertenue*) (Schell *et al.*, 1982). The ability to acquire resistance to reinfection, develop cross-immunity, and produce pathologic changes, which are amenable to quantitation, makes the hamster an ideal model to define the mechanism of resistance to syphilitic infection.

The most direct method to determine the mechanism of immunity would be to confer resistance on normal recipients by transferring serum or cells from

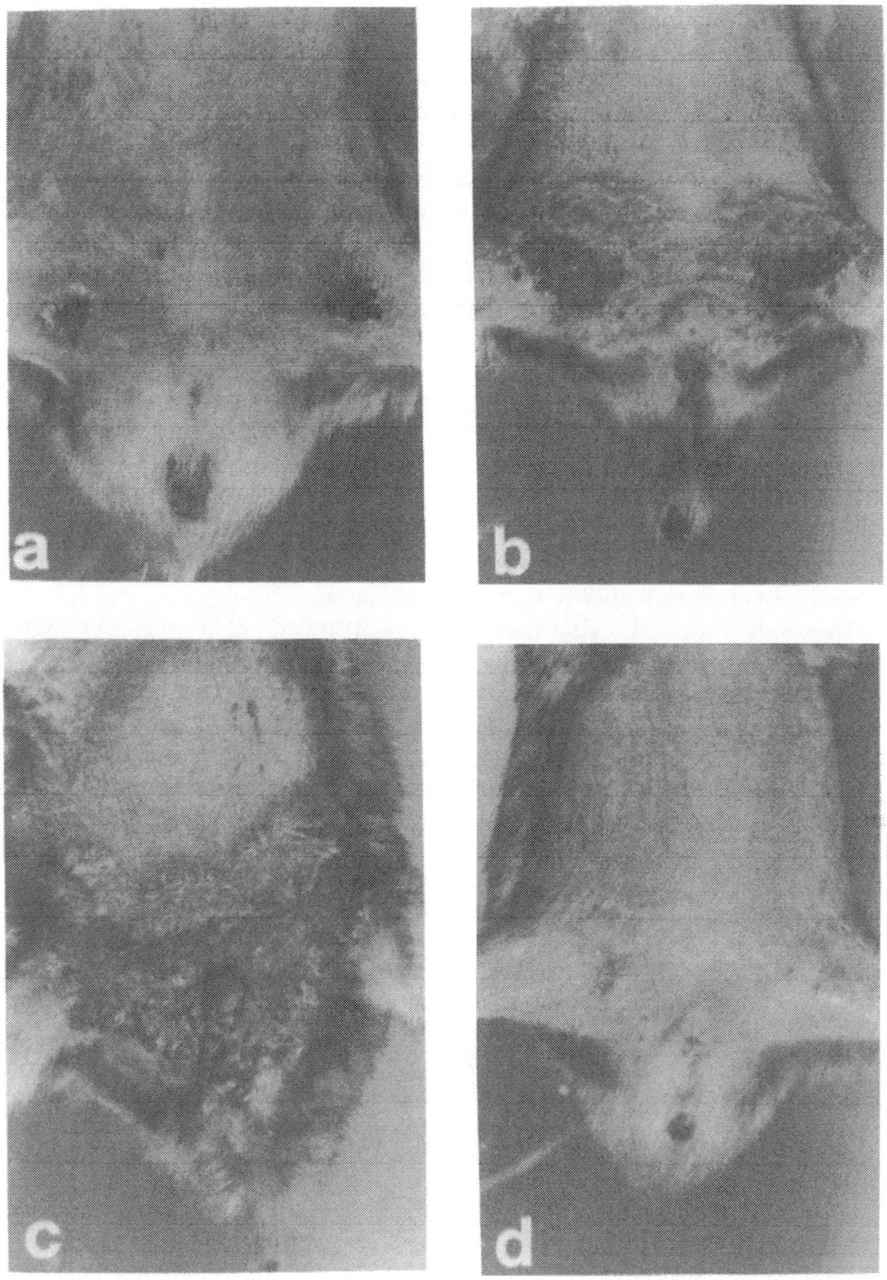

FIGURE 1. Development and regression of cutaneous lesions in representative LSH hamsters at (a) 4, (b) 8, (c) 16, and (d) 36 weeks after infection with *Treponema pallidum* subspecies *endemicum*.

animals immune to syphilitic infection. The ability of immune cells or serum to inhibit the development of syphilitic lesions would provide evidence as to how the host acquires resistance to infection with *T. pallidum*. The following parameters were used to measure the ability of immune serum and cells to confer resistance on recipient hamsters to syphilitic infection: inhibition of the development of cutaneous lesions, decreased number of treponemes, and weight of inguinal lymph nodes.

The purpose of this first experiment was to determine whether treatment of inbred hamsters with immune serum would protect animals from a subsequent infection with *T. pallidum* (Azadegan *et al.*, 1983). Seven groups of four hamsters each were injected intravenously with serum obtained from hamsters at various intervals after infection. An eighth group received normal serum. Subsequently all recipients were challenged.

Infection with *T. pallidum* produced lesions in most animals that had been adoptively immunized with serum from hamsters infected for less than 8 weeks (Table 1). These sera, however, contained sufficient antitreponemal activity to delay the onset of lesions and significantly decrease ($p \leq 0.01$) the weight and number of treponemes in the lymph nodes, compared with recipients of normal serum. In contrast, hamsters passively immunized with serum from hamsters infected for more than 8 weeks had no lesions. As a corollary, their lymph nodes weighed less and contained few or no treponemes.

These results suggest that the degree of protection conferred on recipients by immune serum correlates well with the progress of syphilitic disease in the donor hamsters at the time the immune serum is obtained.

Relatively high treponemicidal activity was found in serum of hamsters infected for 10 weeks or longer. Six groups of three or four hamsters each were injected intravenously with this serum, undiluted or in serial twofold dilutions,

TABLE 1. INCREASE IN PROTECTION CONFERRED BY TRANSFER OF IMMUNE SERUM FROM DONORS INFECTED FOR INCREASINGLY LONGER TIMES WITH *T. PALLIDUM* SUBSPECIES *ENDEMICUM*[a]

	Lesions		Lymph nodes	
Length of infection (weeks)	No./total sites	Development time (days)	Weight (mg)	Treponemes × 10^3
0 (normal)	8/8	13 ± 1	19 ± 2	1938 ± 1.2
1	8/8	17 ± 1	11 ± 1	39 ± 0.1
2	8/8	18 ± 1	11 ± 2	39 ± 0.1
4	6/8	23 ± 1	9 ± 1	21 ± 0.1
6	4/8	25 ± 1	8 ± 2	17 ± 0.1
8	0/8		5 ± 1	0.4 ± 0.1
10	0/8		6 ± 1	0
16	0/8		5 ± 2	0

[a] *T. pallidum* subspecies *endemicum* organisms were inoculated at two sites in each hamster. Animals were sacrificed 40 days after infection. All values in columns 3 through 5 are given as mean ± SE (N = 4 hamsters/group).

at 3-day intervals for 24 days. Three days after the first injection the hamsters were challenged with *T. pallidum*. Concomitantly a seventh group of three hamsters received normal serum and was infected with *T. pallidum* (Table 2). In all hamsters receiving immune serum or its dilutions (up to eightfold), lesions failed to develop, and the lymph node weights were significantly less ($p \leq 0.1$) than in controls. More important, no treponemes were detected in their lymph nodes, even after dark-field examination of extracts concentrated by centrifugation. Further dilutions of immune serum, however, decreased anti–*T. pallidum* activity. Lesions developed, and the weights and number of treponemes in the lymph nodes increased. All recipients of normal serum developed lesions, and their lymph nodes contained 10^5 treponemes.

Treatment of hamsters with cobra venum factor (CVF), a decomplementing agent, could prevent the passive transfer of protection with syphilitic immune serum on normal recipients challenged with *T. pallidum* (Azadegan *et al.*, 1984). Two groups of 10 hamsters each were injected intravenously with immune or normal serum (0.5 ml) at 3-day intervals for 4 weeks. Five hamsters from each group were also injected with CVF at weekly intervals for 4 weeks. Three days after the first injection of serum the hamsters were challenged intradermally with an inoculum of 2×10^5 *T. pallidum* organisms. Two groups of three hamsters each, with or without CVF treatment and receiving no serum, were also infected with *T. pallidum*. In all hamsters receiving immune serum but not CVF, lesions failed to develop, and the lymph node weights were significantly less ($p \leq 0.1$) than those of controls (Table 3). Few treponemes were detected in lymph nodes. By contrast, lesions developed after 14 ± 1 days in all other groups of hamsters. The weights and numbers of treponemes in the lymph nodes were significantly increased ($p \leq 0.01$) compared with those of hamsters receiving immune serum without CVF treatment. In addition, the progression of syphilitic lesions was accelerated significantly ($p \leq 0.01$) in all groups treated with CVF compared with controls (Table 3).

TABLE 2. EFFECT OF DILUTION OF SERUM ON PATHOLOGY OF RECIPIENT HAMSTERS CHALLENGED WITH *T. PALLIDUM* SUBSPECIES *ENDEMICUM*[a]

	Lesions		Lymph nodes	
Serum dilution	No./total sites	Development time (days)	Weight (mg)	Treponemes $\times 10^3$
Undiluted	0/8		5 ± 2	0
1:2	0/6		5 ± 2	0
1:4	0/6		5 ± 1	0
1:8	0/6		6 ± 4	0
1:16	4/6	24 ± 5	10 ± 4	4 ± 1
1:32	3/6	18 ± 4	18 ± 17	50 ± 4
Normal serum	6/6	15 ± 2	11 ± 1	96 ± 10

[a] *T. pallidum* subspecies *endemicum* organisms were inoculated at two sites in each hamster. Animals were sacrificed 10 days after the controls developed lesions. Lymph node values are given as mean ± SE (n = 3 or 4 hamsters/group).

TABLE 3. EFFECT OF COBRA VENOM FACTOR ON THE ABILITY OF SYPHILITIC IMMUNE SERUM TO CONFER PROTECTION ON HAMSTERS CHALLENGED WITH *T. PALLIDUM* SUBSPECIES *ENDEMICUM*[a]

Treatment groups	Lesions			Lymph nodes	
	No./total sites	Development time (days)	Diameter (mm)	Weight (mg)	Treponemes $\times 10^3$
Cobra venom factor					
Immune serum	9/10	16 ± 1	17 ± 1	33 ± 10	868 ± 68
Normal serum	10/10	15 ± 1	21 ± 1	42 ± 3	1148 ± 76
Control (no serum)	6/6	14 ± 1	20 ± 1	51 ± 2	1266 ± 112
No cobra venom factor					
Immune serum	0/10			12 ± 3	1 ± 1
Normal serum	10/10	16 ± 2	14 ± 1	24 ± 3	1219 ± 98
Control (no serum)	6/6	17 ± 2	13 ± 1	28 ± 4	2825 ± 177

[a] *T. pallidum* subspecies *endemicum* were inoculated at two sites in each hamster. Animals were sacrificed 10 days after the controls developed lesions. All values in columns 2 through 5 are given as mean ± SE (n = 3 or 5 hamsters/group). Cobra venom was administered once a week for 4 weeks.

These results showed that serum from hamsters immune to syphilitic infection can confer complete protection on recipient hamsters challenged with *T. pallidum*. Since cobra venom factor abrogates the passive transfer of resistance, these results indicate that complement influences the pathogenesis of syphilis. Bishop and Miller (1976, 1983) and Turner and Hollander (1957) showed that the treponemicidal activity of immune serum was dependent on heat-labile components.

3. CELL-MEDIATED IMMUNITY

The importance of cell-mediated immunity in the control of treponemal infection has been inferred from several clinical and experimental observations. (1) When skin-tested, patients with latent or tertiary syphilis were capable of eliciting delayed-type hypersensitivity responses to treponemal antigens (Marshall and Rothman, 1951; Csonka, 1950; Magnuson *et al.*, 1953). Delayed-type hypersensitivity is used frequently to access the development of cell-mediated immunity. (2) The gumma, a chronic lesion of late syphilis, is characterized by the massive infiltration of lymphocytes and macrophages (Turner, 1970), features commonly seen in other cell-mediated infections such as tuberculosis (Youmans, 1979) and brucellosis (Elberg, 1973). (3) Histologic studies have also shown that lymphocytes are depleted in the thymus-derived areas of the spleen in congenital syphilis in humans and in the lymph nodes of patients with late primary, secondary, and early latent syphilis (Turner and Wright, 1973). Similar observations have been made in neonatal rabbits infected with *T. pallidum* (Festenstein *et al.*, 1967). (4) Dysfunction of syphilitic T cells has been shown by the inability of infected rabbits to respond normally to stimulation with sheep erythrocytes, a T-dependent antigen (Baughn and Musher, 1978, 1979). (5) Syphilitic rabbits, however, are resistant to challenge with *Listeria monocytogenes*, an organism commonly used to monitor the development of acquired cellular resistance (Schell and Musher, 1974). This nonspecific resistance to *Listeria* could be conferred on recipient rabbits and abrogated by treatment of cells with antirabbit thymocyte serum and complement (Schell, 1975a,b). (6) Successful vaccination of rabbits to challenge with *T. pallidum* using viable irradiated treponemes has been reported (Miller, 1973). Classically, live vaccines have been more efficacious in inducing resistance against organisms that are associated with cell-mediated immunity (Smith *et al.*, 1968; Barclay, 1972).

Several *in vitro* methods for assessing lymphocyte responsiveness have also presented evidence for a role of cell-mediated immunity in syphilis. Patients with primary and secondary syphilis respond poorly to T-cell mitogens and treponemal and nontreponemal antigens such as *Candida*, *Trichophyton*, and tuberculin (Levene *et al.*, 1969; Musher *et al.*, 1974, 1975; Kantor, 1975; From *et al.*, 1976; Bey *et al.*, 1979). Decreased reactivity of lymphocytes from syphilitic rabbits to specific T-cell mitogens has been observed, while responses to classical B-cell mitogens have remained normal (Pavia *et al.*, 1976, 1977; Pavia and Folds, 1977). Inhibition of the migration of macrophages, an *in vivo* correlate of cell-mediated immunity, has also been reported to occur in patients with late or congenital syphilis (From *et al.*, 1976; Fulford and Brostoff, 1972) and in treponemal rabbits (Wicher and Wicher, 1975, 1977; Pavia *et al.*, 1975; Metzger *et al.*, 1977; Metzger, 1979).

These indirect approaches, however, have failed to demonstrate clearly whether cell-mediated responses are involved in protection from syphilitic infection. A significant effort was made by Metzger and Smoger (1975) to confer resistance on normal rabbits with immune cells from animals resistant to syphilitic infection. Adoptively immunized rabbits had fewer lesions and a significant delay in their appearance compared to rabbits that received normal lymphocytes or no cells. Although these studies were performed using outbred rabbits, the authors concluded that protection was due to cell-mediated immunity. By contrast, Baughn *et al.* (1977b) observed that chancre immunity could not be transferred to inbred rabbits with splenic cells from syngeneic rabbits resistant to rechallenge with *T. pallidum*. Recipients of immune cells developed syphilitic lesions that were indistinguishable from those in controls. The inbredness of these rabbits was verified by mixed-leukocyte reactions, and the transferability of cell-mediated responses was proven by the ability of these rabbits to elicit a delayed hypersensitivity response after receiving cells sensitized to tuberculin. These results suggested that cell-mediated mechanisms do not play a central role in immunity to syphilis.

Continued studies (Schell *et al.*, 1979, 1980a) have now shown that the hamster is especially suitable for studies of passive transfer of resistance with immune cells. We therefore sought direct evidence for T-cell involvement in syphilitic infection. Various populations of lymphocytes were separated without killing the T cells. Cell preparations were enriched for T cells by passage through glass and nylon–wool columns. This procedure has been shown to separate T from B cells in hamsters (Balsecki and Houston, 1977) and mice (Julius *et al.*, 1973).

Thirty-five hamsters were infected intradermally in the inguinal regions with 7×10^5 *T. pallidum*. Sixteen weeks after infection, the hamsters were treated with penicillin to terminate infection. Pooled immune cells (spleen and lymph node) were obtained from these hamsters 16 days after treatment with penicillin. Enriched suspensions of T cells were obtained from these and pooled normal cells by sequential filtration through glass and nylon–wool columns. Suspensions of fractionated or unfractionated immune or normal lymphoid cells were transferred intravenously into irradiated bone-marrow reconstituted hamsters. The recipient hamsters were subsequently infected intradermally with *T. pallidum*. All hamsters developed cutaneous lesions within 12 days after infection. In the two groups that had previously received immune cells, whether fractionated or unfractionated, the lesions then regressed, and in all but one animal they were completely healed by day 24 after infection (Table 4). In contrast, in the three groups that received bone marrow cells or fractionated or unfractionated suspensions of normal cells, the lesions enlarged and became necrotic. As a corollary, the lymph nodes of

TABLE 4. ABILITY OF CELL SUSPENSIONS TO CONFER RESISTANCE TO INFECTION WITH *T. PALLIDUM* IN IRRADIATED, BONE MARROW-RECONSTITUTED HAMSTERS[a]

Recipients of	Lesions/ total sites[b]	Lymph nodes: Weight (mg)	Lymph nodes: Treponemes × 10^3
Normal cells			
Unfractionated	8/8	29 ± 3	174
Fractionated	6/6	44 ± 7	146
Immune cells			
Unfractionated	0/8	17 ± 4	4
Fractionated	1/8[c]	15 ± 5	33
Control (received only bone marrow cells)	6/6	42 ± 10	123

[a]Results were obtained with hamsters infected 24 days previously with 7 × 10^5 *T. pallidum*. Fractionated cell suspensions were depleted of B cells and enriched in T cells. There were three to four hamsters per experimental group.
[b]Each hamster was inoculated at two sites.
[c]The lesion measured 1–2 mm, as compared with 9–12 mm in recipients of fractionated or unfractionated normal cells or bone marrow (control) cells.

animals that had received immune cells, whether fractionated or unfractionated, weighed less and had significantly few treponemes.

To verify that the fractionated lymphocyte suspensions were highly enriched in T cells, we incubated the unfractionated or fractionated immune cells with rabbit antihamster thymocyte serum (ATS) in the presence of complement, antihamster immunoglobulin, or classical T- or B-cell mitogens.

Fractionation reduced the proportion of cells bearing immunoglobulins from 36 ± 3% to only 3 ± 1% [mean ± standard error (SE)], a reduction of 94%. No fluorescent cells were detected after treatment with either control immunoglobulin (rabbit antihamster immunoglobulin and fluorescein-labeled goat antirabbit immunoglobulin) alone. In contrast, the proportion of cells susceptible to ATS plus complement increased from 37 ± 96% to 87 ± 4% (mean ± SE), an increase of 135%. The response of syphilitic lymphocytes to the four classical T- or B-cell mitogens, concanavalin A, phytohemagglutinin, lipopolysaccharide, and dextran sulfate, was decreased significantly by fractionation. The minimal decrease, however, was seen with the T-cell mitogen concanavalin A. As an additional measure of the proportion of B cells in the fractionated cells, immune hamsters were inoculated with sheep erythrocytes. After 5 days the number of direct and indirect plaque-forming cells in the fractionated and unfractionated cell populations was determined. Fractionation reduced the number of direct and indirect plaques by 92% and 88%, respectively.

The mechanism by which T cells confer protection on recipients against infection with virulent treponemes is not yet understood. Since T cells have not been shown to exert a direct treponemicidal effect, they may mediate destruction of treponemes via macrophages.

4. ROLE OF THE MACROPHAGE

Treponemal infection enhances the bactericidal activity of macrophages. Syphilitic infection of rabbits suppresses the growth of *L. monocytogenes* (Schell and Musher, 1974), an organism commonly used to monitor such activity (Mackaness, 1971). In addition, macrophages incubated with supernatants from lymphocytes cultured with *T. pallidum* develop resistance to challenge with *L. monocytogenes* (Lukehart, 1982).

Whether macrophages are ultimately responsible for the destruction of treponemes is in dispute. Harris and Thoen (1977) and Hardy *et al.* (1979) have shown that nonspecific activation of macrophages in rabbits inoculated with *Mycobacterium bovis* prevents the development or accelerates the resolution of treponemal lesions. In contrast, several investigators (Schell *et al.*, 1975b, 1983; Graves and Johnson, 1975; Baughn *et al.*, 1977a; Graves, 1980) have reported that neither *M. bovis* nor *Corynebacterium acnes* prevents or delays the appearance of syphilitic lesions, even in the presence of immune serum.

Interpretation of these studies has been hindered by an inability to confirm that macrophages have enhanced bactericidal activity at the site of syphilitic infection. We (Schell *et al.*, 1983) therefore vacinated hamsters intravenously and/or subcutaneously with *M. bovis* and determined their ability to resist local and systemic challenge with *L. monocytogenes* as an indication of macrophage activation. Other *M. bovis*-vaccinated hamsters were challenged with *T. pallidum*. The results of this study showed that subcutaneous and intravenous vaccination with *M. bovis* induced macrophage activation in hamsters, as determined by an enhanced ability to suppress the growth of *L. monocytogenes* in the livers, spleens, and inguinal lymph nodes. Hamsters challenged in the inguinal region with *T. pallidum* during periods of enhanced microbial resistance (3–8 weeks after *M. bovis* vaccination) developed lesions faster and with more necrosis. Increased numbers of treponemes were recovered from the regional lymph node of *M. bovis*-vaccinated hamsters than from nonvaccinated controls, although the differences were not statistically significant. These studies demonstrate that *M. bovis* vaccination influences the pathogenesis of syphilis without inducing macrophage-mediated treponemicidal activity.

One explanation for the discrepancy on the effect of *M. bovis*-activated macrophages may be that macrophages have to be activated in a specific sequence with other defense mechanisms (antibody). Since *M. bovis* vaccination in our study enhanced the production of specific antitreponemal antibody two- to fourfold, it is possible that *M. bovis* vaccination of rabbits prevented syphilitic lesions by the enhanced production of treponemicidal antibody or antibody that facilitated antibody-dependent phagocytosis of treponemes. Additional studies are needed to determine whether local production of treponemal antibody and induction of *M. bovis*-activated macrophages can destroy treponemes.

Another explanation may be that there is a selective reduction of macrophage aggressiveness toward *T. pallidum* despite retention of macrophage listericidal activity. We have shown that syphilitic hamster T cells exert suppression on macrophage C3b-mediated ingestion (Tabor *et al.*, 1984). In addition,

Lipopolysaccharide (LPS) -stimulated macrophages, which classically display an elevated level of activity, are also functionally inhibited from C3b-mediated ingestion by syphilitic hamster T cells. When macrophages from either LPS-treated or syphilitic hamsters are cultured in the absence of syphilitic T cells, they express significant C3b-mediated ingestion activity.

Impairment of macrophage C3b-mediated ingestion becomes most noticeable between 3 and 6 weeks after syphilitic infection and gradually wanes by the ninth week. The kinetics of this event parallel some of the circumstances leading to treponemal antibody production with subsequent elimination of treponemes in the hamster (Schell *et al.*, 1980b). Macrophages appear to be functioning early in response to treponemal challenge by promoting the emergence of antibody-producing cells since treponemal antibodies are detected shortly after infection (Schell *et al.*, 1980b; Sell *et al.*, 1982). As the disease progresses, effective treponemicidal antibody accumulates without destruction of these spirochetes (Azadegan *et al.*, 1983). During this period maximum suppression of macrophage C3b-mediated ingestion is detected. After 9 weeks, the number of treponemes in the tissues gradually declines (Schell *et al.*, 1980b) and the macrophage C3b-mediated ingestion resumes. Host utilization of the macrophage complement receptor has been shown to reflect an important defense mechanism against other infectious agents (Horwitz, 1980).

5. SUMMARY

Is there some way of putting together all these observations to formulate a unifying hypothesis? Infection of hamsters with *T. pallidum* produces extensive chronic skin lesions that persist for 6–9 months (Schell *et al.*, 1980b). The inability of hamsters to eliminate *T. pallidum* is a paradox, as their serum can kill treponemes *in vitro* and confer complete protection on nonimmune recipients. Furthermore, syphilitic immune T cells can confer protection on recipients challenged with *T. pallidum*.

Our studies support the hypothesis that syphilitic T cells render C3b receptors unavailable for processing antibody-complement-opsonized treponemes. It is possible that effective acquired immune responses are generated by syphilitic infection, but there are defects in macrophage processing initiated by T cells that prevent elimination of treponemes. However, other hypotheses may explain the protracted nature of syphilis despite the presence of effective immune responses. Baseman and colleagues suggest that the inability of humans or animals to eliminate the primary infection is due to the acquisition of host proteins that are associated with the outer envelope of *T. pallidum* (Alderete and Baseman, 1979; Baseman and Hayes, 1980; Baseman and Alderete, 1983). In support of this hypothesis, it is known that treponemes require "aging" before they react efficiently in the FTA–ABS test (Hardy and Nell, 1957). Treatment of treponemes with detergents is also required to achieve high reactivity in the indirect immunofluorescence assay (Penn and Rhodes, 1982). Zeigler *et al.* (1976) have shown that freshly isolated *T. pallidum* are coated with a polysaccharide "capsular" material which may prevent phagocytosis similar to the hyaluronic acid capsule of *Streptococcus pyogenes* (Ginsburg, 1972). Therefore, treponemes may evade direct cell killing, antibody-dependent phagocytosis, or complement-mediated lysis by coating themselves with protective "self" molecules that prevent immune recognition.

Sell and co-workers (1980; 1982) have argued that immunologically privileged sites, such as nerve tissue, may be responsible for the chronic course of syphilis. *T. pallidum* has been shown to be present in large numbers in nerves (Ovcinnikov and Delektorskij, 1975; Sell *et al.*, 1985). An intracellular location for small numbers of treponemes has also been suggested (Azar *et al.*, 1970; Sykes and Miller, 1971; Sykes *et al.*, 1974). This hypothesis is supported by the observation that syphilitic rabbits produce antibodies that block attachment of the treponemes to host cells in tissue culture (Fitzgerald *et al.*, 1977; Hayes *et al.*, 1977; Wong *et al.*, 1983; Rice and Fitzgerald, 1985). Presumably, syphilitic immune serum prevents penetration of treponemes into cells. A recent study showed attachment-blocking activity of immune serum correlated with resistance to infection (Wong *et al.*, 1983).

Considering the variety of mechanisms involved in *T. pallidum*'s evasion of host defenses, additional work is needed before an effective vaccine can be developed for this important disease.

ACKNOWLEDGMENTS. We thank Jegish P. Babu, John K. Chan, Ali A. Azadegan, and Dale R. Tabor for their contributions. This work was supported in part by Public Health Service grant no. A122199 from the National Institute of Allergy and Infectious Diseases and the World Health Organization.

REFERENCES

Alderete, J. F., and Baseman, J. B. 1979, Surface-associated host proteins on virulent *Treponema pallidum, Infect. Immun.* **26:**1048–1056.

Azadegan, A. A., Schell, R. F., and LeFrock, J. L., 1983, Immune serum confers protection against syphilitic infection on hamsters, *Infect. Immun.* **42:**42–47.

Azadegan, A. A., Tabor, D. R., Schell, R. F., and LeFrock, J. L., 1984, Cobra venom factor abrogates passive humoral resistance to syphilitic infection in hamsters, *Infect. Immun.* **44:**740–742.

Azar, H. A., Pham, T. D., and Kurban, A. K., 1970, An electron microscopic study of a syphilitic chancre, *Arch. Pathol.* **90:**143–150.

Balsecki, J. W., and Houston, K. J., 1977, Identification, functional characterization and partial purification of thymus-derived lymphocytes in inbred hamsters, *Immunology* **35:**1–11.

Barclay, W. R., 1972, BCG vaccination, in: *Immunization in Tuberculosis,* (E. C. Chamberlayne, ed.), pp. 245–247, Fogarty International Center Proc. no. 14, Washington, D.C.

Baseman, J. B., and J. F., Alderete, 1983, The parasitic strategies of *Treponema pallidum,* in: *Pathogenesis and Immunology of Treponemal Infection* (R. F. Schell and D. M. Musher, eds.), Marcel Dekker, New York.

Baseman, J. B., and Hayes, E. C., 1980, Molecular characterization of receptor binding proteins and immunogens of virulent *Treponema pallidum, J. Exp. Med.* **151:**573–586.

Baughn, R. E., and Musher, D. M., 1978, Altered immune responsiveness associated with experimental syphilis in the rabbit: Elevated IgM and depressed IgG responses to sheep erythrocytes, *J. Immunol.* **120:**1691–1695.

Baughn, R. E., and Musher, D. M., 1979, Aberrant secondary antibody responses to sheep erythrocytes in rabbits with experimental syphilis, *Infect. Immun.* **25**:133–138.

Baughn, R., Musher, D., and Knox, J., 1977a, Effect of sensitization with *Propionibacterium acnes* on the growth of *Listeria monocytogenes* and *Treponema pallidum* in rabbits, *J. Immunol.* **118**:109–113.

Baughn, R. E., Musher, D. M., and Simmons, C. B., 1977b, Inability of spleen cells from chancre-immune rabbits to confer immunity to challenge with *Treponema pallidum, Infect. Immun.* **17**:535–540.

Bey, R. F., Johnson, R. C., and Fitzgerald, T. J., 1979, Suppression of lymphocyte response to concanavalin A by mucopolysaccharide material from *Treponema pallidum* infected rabbits, *Infect. Immun.* **26**:64–69.

Bishop, N. H., and Miller, J. N., 1976, Humoral immunity in experimental syphilis. II. The relationship of neutralizing factors in immune serum to acquired resistance, *J. Immunol.* **117**:197–207.

Bishop, N. H., and Miller, J. N., 1983, Humoral immune mechanisms in acquired syphilis, in: *Pathogenesis and Immunology of Treponemal Infection* (R. F. Schell and D. M. Musher, eds.), pp. 241–269, Marcel Dekker, New York.

Csonka, G. W., 1950, "LuoTest": A preliminary evaluation in the diagnosis of late syphilis, *Med. Biol. Illus.* **4**:389–391.

Elberg, S. S., 1973, Immunity to *Brucella* infections, *Medicine* **52**:339–356.

Festenstein, H., Abraham, C., and Bokkenheusen, V., 1967, Runting syndrome in neonatal rabbits infected with *Treponema pallidum, Clin. Exp. Immunol.* **2**:311–320.

Fieldsteel, A. H., Cox, D. L., and Moeckli, R. A., 1982, Further studies on replication of virulent *Treponema pallidum* in tissue cultures of SflEp cells, *Infect. Immun.* **35**:449–455.

Fitzgerald, T. J., Johnson, R. C., Miller, J. N. and Sykes, J. A., 1977, Characterization of the attachment of *Treponema pallidum* (Nichols strain) to cultured mammalian cells and the potential relationship of attachment to pathogenicity, *Infect. Immun.* **18**:467–478.

From, E., Thestrup-Pederson, K., and Thulin, H., 1976, Reactivity of lymphocytes from patients with syphilis towards *T. pallidum* antigen in the leukocyte migration and lymphocyte transformation test, *Br. J. Vener. Dis.* **52**:224–229.

Fulford, K. W. M., and Brostoff, J., 1972, Leucocyte migration and cell-mediated immunity in syphilis, *Br. J. Vener. Dis.* **48**:483–488.

Ginsburg, I., 1972, Mechanism of cell and tissue injury induced by group A Streptococci: relationship to poststreptococcal sequelae, *J. Infect. Dis.* **126**:419–456.

Graves, S., 1980, Rate of clearance of virulent *Treponema pallidum* (Nichols) from the blood stream of normal, *Mycobacterium bovis* BCG-treated, and immune syphilitic rabbits, *Infect. Immun.* **27**:264–267.

Graves, S., and Johnson, R., 1975, Effect of pretreatment with *Mycobacterium bovis* (strain BCG) and immune syphilitic serum on rabbit resistance to *Treponema pallidum, Infect. Immun.* **12**:1029–1036.

Hanff, P. A., Miller, J. N., and Lovett, M. A., 1983, Molecular characterization of common treponemal antigens, *Infect. Immun.* **40**:825–828.

Hansen, E. B., Pedersen, P. E., Schouls, L. M., Severin, E., and van Embden, J. D. A., 1985, Genetic characterization and partial sequence determination of a *Treponema pallidum* operon expressing two immunogenic membrane proteins in *Escherichia coli, J. Bacteriol.* **162**:1227–1237.

Hardy, P. H., and Nell, E. E., 1957, Study of the antigenic structure of *Treponema pallidum* by specific agglutination, *Amer. J. Hyg.* **66**:160–172.

Hardy, P. H., Graham, D. J., Nell, E. E., and Dannenberg, A. M., 1979, Macrophages in immunity to syphilis: Suppressive effect of concurrent infection with *Mycobacterium bovis* (BCG) on the development of syphilitic lesions and growth of *Treponema pallidum* in tuberculin-positive rabbits, *Infect. Immun.* **26**:751–763.

Harris, D. L., and Thoen, C. O., 1977, Inhibition of skin lesions due to *Treponema pallidum* in BCG-infected rabbits, *Abstracts of the annual meeting of the American Society for Microbiology,* **E32**:86.

Hayes, N. S., Muse, K. E., Collier, A. M. and Baseman, J. B., 1977, Parasitism by virulent *Treponema pallidum* of host cell surfaces, *Infect. Immun.* **17**:174–186.

Horwitz, A. M., 1980, The roles of Fc and C3 receptors in the phagocytosis and killing of bacteria by human phagocytes, *J. Reticuloendothel. Soc.* **28**:175.

Izzat, N. N., Knox, J. M., Dacres, W. G., and Smith, E. B., 1971, Resistance and serological changes in rabbits immunized with virulent *Treponema pallidum* sonicate, *Acta Derm. Venereol.* **51**:157–160.

Julius, M. H., Simpson, E. and Herzenberg, L. A., 1973, A rapid method for the isolation of functional thymus-derived murine lymphocytes, *Eur. J. Immunol.* **3**:645–649.

Kantor, F. S., 1975, Infection, anergy and cell-mediated immunity, *N. Engl. J. Med.* **292**:629–634.

Levene, G. M., Turk, J. L., Wright, D. L., and Grimbel, A. G. S., 1969, Reduced lymphocyte transformation due to a plasma factor in patients with active syphilis, *Lancet* 2:246–247.

Lukehart, S. A., 1982, Activation of macrophages by products of lymphocytes from normal and syphilitic rabbits, *Infect. Immun.* **37**:64–69.

Mackaness, G., 1971, Resistance to intracellular infection, *J. Infect. Dis.* **123**:439–445.

Magnuson, H. J., Thompson, F. A., Jr., and McLeod, C. P., 1951, Relationship between treponemal immobilizing antibodies and acquired immunity in experimental syphilis, *J. Immunol.* **64**:41–48.

Magnuson, H. J., Thomas, E. W., Olansky, S., Kaplan, B. I., DeMello, L., and Culter, J. C., 1953, Inoculation syphilis in human volunteers, *Medicine* **35**:33–82.

Marchitto, K. S., Jones, S. A., Schell, R. F., Holmans, P. L., and Norgard, M. V., 1984, Monoclonal antibody analysis among pathogenic *Treponema pallidum* subspecies, *Infect. Immun.* **45**:660–666.

Marshall, L. C., and Rothman, S. 1951, Skin testing with a purified suspension of *Treponema pallidum, Am. J. Syph. Gonorr. Vener. Dis.* **35**:35–41.

McLeod, C. P., and Magnuson, H. J., 1953, Production of immobilizing antibodies unaccompanied by active immunity to *Treponema pallidum* as shown by injecting rabbits and mice with killed organisms, *Am. J. Syph.* **37**:9–22.

Metzger, M., 1979, Theme 2: Recent advances in the immunology of syphilis. Role of humoral versus cellular mechanisms of resistance in the pathogenesis of syphilis, *Br. J. Vener. Dis.* **55**:94–98.

Metzger, M., and Smoger, W. 1975, Passive transfer of immunity to experimental syphilis in rabbits by immune lymphocytes, *Arch. Immunol. Ther. Exp.* **23**:625–630.

Metzger, M., Podiwinska, J., and Smoger, W., 1977, Cell-mediated immunity in experimental syphilis in rabbits, *Arch. Immun. Ther. Exp.* **25**:25–34.

Miller, J. N., 1973, Immunity in experimental syphilis. VI. Successful vaccination of rabbits with *Treponema pallidum,* Nichols strain, attenuated by γ-irradiation, *J. Immunol.* **110**:1206–1215.

Musher, D. M., Schell, R. F., and Knox, J. M., 1974, *In vitro* lymphocytes response to *Treponema* refringens in human syphilis, *Infect. Immun.* **9**:654–657.

Musher, D. M., Schell, R. F., Jones, R. H., and Jones, A. M., 1975, Lymphocyte transformation in syphilis: An *in vitro* correlate of immune suppression *in vivo? Immunology* **11**:1261–1264.

Norgard, M. V., and Miller, J. N., 1983, Cloning and expression of *Treponema pallidum* (Nichols) antigens genes in *Escherichia coli, Infect. Immun.* **42**:435–445.

Norris, S. J., 1982, *In vitro* cultivation of *Treponema pallidum:* Independent confirmation, *Infect. Immun.* **36**:437–439.

Ovcinnikov, N. M., and Delektorskij, V. V., 1975, *Treponema pallidum* in nerve fibers, *Br. J. Vener. Dis.* **51**:10–18.

Pavia, C. S., and Folds, J. D., 1977, Selective response of lymphocytes from *Treponema pallidum* infected rabbits to mitogens and *Treponema reiters, Infect. Immun.* **15**:417–422.

Pavia, C. S., Baseman, J. B., and Folds, J. D., 1975, Development of macrophage migration inhibition in rabbits infected with virulent *Treponema pallidum., Infect. Immun.* **17**:651–654.

Pavia, C. S., Folds, J. D., and Baseman, J. B., 1976, Depression of lymphocyte response to concanavalin A in rabbits infected with *Treponema pallidum* (Nichols strain), *Infect. Immun.* **14**:320–322.

Pavia, C. S., Folds, J. D., and Baseman, J. B., 1977, Selective in vitro response of thymus-derived lymphocytes from *Treponema pallidum*-infected rabbits, *Infect. Immunn.* **18**:603–611.

Penn, C. W., and Rhodes, J. G., 1982, Surface-associated antigens of *Treponema pallidum* concealed by an inert outer layer, *Immunol.* **46**:9–16.

Perine, P. L., Weiser, R. S., and Klebanoff, S. J., 1973, Immunity to syphilis. I. Passive transfer in rabbits with hyper-immune serum, *Infect. Immun.* **8**:787–790.

Rice, M., and Fitzgerald, T. J., 1985, Detection and functional characterization of early appearing antibodies in rabbits with experimental syphilis, *Can. J. Microbiol.* **31**:62–67.

Schell, R. F., and Musher, D. M., 1974, Detection of non-specific resistance to *Listeria Monocytogenes* in infested rabbits with *Treponema pallidum, Infect. Immun.* **9**:658–662.

Schell, R. F., and Musher, D. M., 1975, Transfer of non-specific resistance to *Listeria monocytogenes* using spleen cells from syphilitic rabbits, *Proc. Soc. Exp. Biol. Med.* **148**:516–518.

Schell, R. F., Musher, D., Jacobson, K., Schwethelm, P., 1975a, Induction of acquired cellular resistance following transfer of thymus-dependent lymphocytes from syphilitic rabbits. *J. Immunol.* **114**:550–553.

Schell, R. F., Musher, D., Jacobson, K., Schwethelm, P., and Simmons, C., 1975b, Effect of macrophage activation on infection with *Treponema pallidum, Infect. Immun.* **12**:505–511.

Schell, R. F., Chan, J. K., and LeFrock, J. L., 1979, Endemic syphilis: Passive transfer of resistance with serum and cells in hamsters, *J. Infect. Dis.* **140**:378–383.

Schell, R. F., Chan, J. K., LeFrock, J. L., and Bagasra, O., 1980a, Endemic Syphilis: Transfer of resistance to *Treponema pallidum* strain Bosnia A in hamsters with a cell suspension enriched in thymus-derived cells, *J. Infect. Dis.* **141**:752–758.

Schell, R. F., LeFrock, J. L., Chan, J. K., and Bagasra, O., 1980b, LSH hamster model of syphilitic infection, *Infect. Immun.* **28**:909–913.

Schell, R. F., Azadegan, A. A., Nitskansky, S. G., and LeFrock, J. L., 1982, Acquired resistance of hamsters to challenge with homologous and heterologous virulent treponemes, *Infect. Immun.* **37**:617–621.

Schell, R. F., Azadegan, A. A., Nitskansky, S. G., and LeFrock, J. L., 1983, Effect of local and systemic macrophage activation in hamsters on infection with *Treponema pertenue* and *Treponema pallidum* Bosnia A, *J. Reticuloendothel. Soc.* **33**:231–238.

Sell, S., Baker-Zander, S., and Powell, H. C., 1982, Experimental syphilitic orchitis in rabbits. Ultrastructural appearance of *Treponema pallidum* during phagocytosis and dissolution by macrophages *in vivo, Lab Invest.* **46**:355–364.

Sell, S., Gamboa, D., Baker-Zander, S. A., Lukehart, S. A., and Miller, J. N., 1980, Host response to *Treponema pallidum* in intradermally infected rabbits: Evidence for persistance of infection at local and distant sites, *J. Invest. Dermatol.* **75**:470–475.

Sell, S., Salman, J., and Norris, S. J., 1985, Reinfection of chancre-immune rabbits with *Treponema pallidum.* 1. Light and Immunofluorescence studies, *Am. J. Pathol.* **118**:248–255.

Septejian, M., Salussola, D., and Thivolet, J., 1973, Attempt to protect rabbits against experimental syphilis by passive immunization, *Br. J. Vener. Dis.* **49**:335–337.

Smith, D. W., Grover, A. A., and Weigeshaus, E., 1968, Nonliving immunogenic substances of mycobacteria, *Adv. Tuberc. Res.* **16**:191–227.

Stamm, L. V., and Bassford, P. J., Jr., 1982, Cloning and expression of *Treponema pallidum* protein antigens in *Escherichia coli, DNA* **1**:329–333.

Sykes, J., and Miller, J. N., 1971, Intracellular location of *Treponema pallidum* (Nochols strain) in the rabbit testis, *Infect. Immun.* **4**:307–314.

Sykes, J. A., Miller, J. N., and Kapla, A. J., 1974, *Treponema pallidum* within cells of a primary chancre from a human female, *Br. J. Vener. Dis.* **50**:40–44.

Tabor, D. R., Azadegan, A. A., Schell, R. F., and LeFrock, J. L., 1984, Inhibition of macrophage C3b-mediated ingestion by syphilitic hamster T cell-enriched fractions, *J. Immunol.* **135**:2698–2705.

Thompson, F. A., Jr., Greenberg, B. G., and Magnuson, H. J., 1950, The relationship between immobilizing and spirocheticidal antibodies against *Treponema pallidum, J. Bacteriol.* **60**:473–480.

Titus, R. G., and Weiser, R. S., 1979, Experimental syphilis in the rabbit: Passive transfer of immunity with immunoglobulin G from immune serum, *J. Infect. Dis.* **140**:904–913.

Turner, D. R., and Wright, D. J. M., 1973, Lymphadenopathy in early syphilis, *J. Pathol.* **110**:305–308.

Turner, T. B., 1939, Protective antibodies in the serum of syphilitic rabbits, *J. Exp. Med.* **69**:867–890.

Turner, T. B., 1970, Syphilis and the Treponematoses, in: *Infectious Agents and Host Reactions* (S. Mudd, ed.), pp. 346–390, Saunders, Philadelphia.

Turner, T. B., and Hollander, D. H., 1957, Biology of the treponematoses, pp. 25–26, 33–66, 197–208, 214–233, W.H.O. Monogr. Ser. no. 35.

Turner, T. B., Hardy, P. H., Jr., Newman, B., and Nell, E. E., 1973, Effects of passive immunization on experimental syphilis in the rabbit, *Johns Hopkins Med. J.* **133**:241–251.

Van Embden, J. D., Van der Donk, H. J., Van Eijk, R. V., Van der Heide, H. G., DeJong, J. A., Van Olderen, M. F., Osterhaus, A. D., and Schouls, L. M., 1983, Molecular cloning and expression of *Treponema pallidum* DNA in *Escherichia coli* K-12, *Infect. Immun.* **42**:187–196.

Waring, G. W., and Fleming, W. L. 1951, Further attempts to immunize rabbits with killed *Treponema pallidum, Am. J. Syph.* **35**:568–572.

Weiser, R. S., Erickson, D., Perine, P. L., and Pearsall, N. N., 1976, Immunity to syphilis: Passive transfer in rabbits using serial doses of immune serum, *Infect. Immun.* **13**:1402–1407.

Wicher, V., and Wicher, K., 1975, Cell response in rabbits infected with *T. pallidum* as measured by the leucocyte migration inhibition test, *Br. J. Vener. Dis.* **51**:240–245.

Wicher, V., and Wicher, K., 1977, *In vitro* cell response of *Treponema pallidum* infected rabbits. I. Lymphocyte transformation, *Clin. Exp. Immunol.* **24**:480–486.

Wong, G. H. W., Steiner, B., and Graves, S., 1983, Effect of syphilitic rabbit sera taken at different periods after infection on treponemal motility, treponemal attachment to mammalian cells in vitro, and treponemal infection in rabbits, *Br. J. Vener. Dis.* **59**:220–224.

Youmans, G. P., 1979, *Tuberculosis,* pp. 209–224, Saunders, Philadelphia.

Zeigler, J. A., Jones, A. M., Jones, R. H., and Kubica, K. M., 1976, Demonstration of extracellular material at the surface of pathogenic *Treponema pallidum* cells, *Br. J. Vener. Dis.* **52**:1–8.

7

Immunology of Mycobacterial Infections

FRANK M. COLLINS

1. INTRODUCTION

Tuberculosis is still one of the most important and widespread of the life-threatening bacterial diseases occurring throughout the world today, affecting both human and domestic animal populations (Francis, 1958; Hershfield, 1979). There is clear archeologic evidence of tuberculous lesions in several of the early Egyptian mummies (Zimmerman, 1979), and pulmonary tuberculosis was widely recognized in Europe in the pre-Christian era. Such infections were probably sporadic at that time, but with the increasing density of population in Europe during the Middle Ages, the disease quickly became endemic, increasing at an alarming rate during the Industrial Revolution (Dubos and Dubos, 1952). A similar increase also occurred in the United States at this time, with mortality rates reaching as high as 800 per 100,000 per annum during the first half of the nineteenth century. Spread of the disease was undoubtedly promoted by the appalling living conditions, overcrowding, poor nutrition, alcoholism, ignorance, and superstition that occurred in most of the major industrial cities throughout the Western world. During this time, tuberculosis affected all segments of the population, reaching such proportions that it was frequently referred to as the "White Plague." Precise figures for the incidence of this disease in the United States do not extend back far beyond the turn of the twentieth century, but there is no reason to believe that the trend was any different from that seen in Britain and on the Continent prior to this time (Lowell, 1984; Styblo, 1980). Despite a general lack of scientific and medical knowledge about the cause, incidence, and distribution of tuberculosis (and leprosy) in Europe and the United States, a slow but striking shift in incidence occurred, beginning in the first half of the nineteenth century (Fine, 1984). The reasons for this worldwide change are still unclear but have been responsible for a gradual shift in the incidence and distribution of both diseases from Northern Europe and America

FRANK M. COLLINS • Trudeau Institute, Inc., Saranac Lake, New York 12983.

to the tropical countries of the Third World. Presumably, much of this shift reflects changing social rather than medical factors since they occurred prior to the impact of prophylactic or therapeutic agents on either disease (Anon, 1982c; Fine, 1981). It is a sobering thought that the landmark discoveries *Mycobacterium leprae* by Hansen in 1874 and *M. tuberculosis* by Koch in 1882 had little or no effect on this slow, steady decline in incidence in temperate and northern climates for more than 50 years (Fig. 1). The development of such diagnostic tools as x-rays and skin test antigens, the widespread introduction of milk pasteurization, and universal infant bacillus Calmette Guérin (BCG) vaccination all had some effect on this incidence rate, but it was only after the introduction of chemotherapy that a dramatic decline in mortality and morbidity was finally observed (Collins, 1984). The decline in pulmonary tuberculosis was so sharp that it was predicted that this disease would be totally eradicated from the Western world by the turn of this century (Styblo, 1980). Unfortunately, achievement of this goal now seems to be increasingly unlikely, even in the United States (Anon., 1986) although the mortality rate does continue to decline slowly (Comstock, 1982). The latter statistic provides little cause for comfort, however, and the incidence of these mycobacterioses in the United States has actually increased slightly over the past 5 years, owing to the influx of Southeast Asian refugees, many of whom were incubating the disease at the time of their entry to this country (Anon., 1985). Tuberculosis is still the most important life-threatening infectious disease reported in the United States today, with 22,000 new cases reported in 1984, and 1937 deaths directly attributable to this disease. Interestingly, the overall incidence of tuberculosis has once again begun to increase (Fig. 1). However, there is no corresponding change in the incidence of other mycobacterial and extrapulmonary forms of tuberculosis, especially in young children (Farer *et al.*, 1979).

The enormous economic and public health impact of these two major diseases is seen only when they are considered in their worldwide context where at

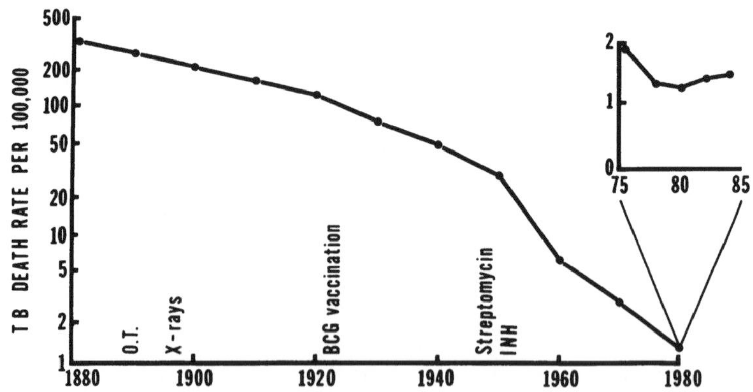

FIGURE 1. Annual mortality rates for pulmonary tuberculosis in the United States over the past century, in relation to major advances in tuberculosis research. The small increase around 1980 reflects the entry of large numbers of refugees (Anon., 1982c).

least 10 million persons are actively infected with the tubercle bacillus, resulting in over 200,000 potential new contacts occurring yearly (Hershfield, 1979). In 1977 there were an estimated 12 million cases of leprosy (Anon., 1977). There is also increasing evidence that drug-resistant strains are emerging in many parts of the world. These, together with the "atypical" mycobacteria, which are inherently drug resistant, represent an increasingly difficult management problem for the pulmonary physician (Anon., 1983; Davidson, 1979). Finally, there has been a striking emergence of *M. avium* complex infections associated with the acquired immune deficiency syndrome and it seems likely that these opportunistic lung pathogens will become even more of a problem with time (Collins, 1986).

Although tuberculosis seems to be largely under control in Europe and in the United States, in parts of Africa and Asia the incidence rate still runs as high as 200 per 100,000 individuals (Hershfield, 1979) and shows little sign of declining (Styblo, 1980). We have developed a highly effective live vaccine (Luelmo, 1982) and a number of effective chemotherapeutic agents (Raleigh, 1984) that should enable us to eradicate tuberculosis from the Western world; yet its incidence in many parts of the world is increasing (Glassroth *et al.*, 1980). Part of this problem stems from difficulties in locating and treating these patients. In the United States, many are poorly motivated, being members of the alcohol- and drug-abuse populations. The costs of locating and isolating these infectious index cases in the community is very high, although until this is done, these mycobacterial diseases will continue to be important public health problems for many decades to come.

The risk of tuberculin conversion in the United States and Canada has declined dramatically from 0.5–1.0% per annum in 1950 to less than 0.02% in 1980 (Fig. 2). However, the corresponding figures for many of the developing nations have not fallen appreciably, as can be seen from the data for Uganda and Kenya. The reasons for this difference are both diverse and complex, ranging from nutritional, economic, and social, to the availability of case-finding and health care personnel in many of the more remote rural areas (Lotte *et al.*, 1979;

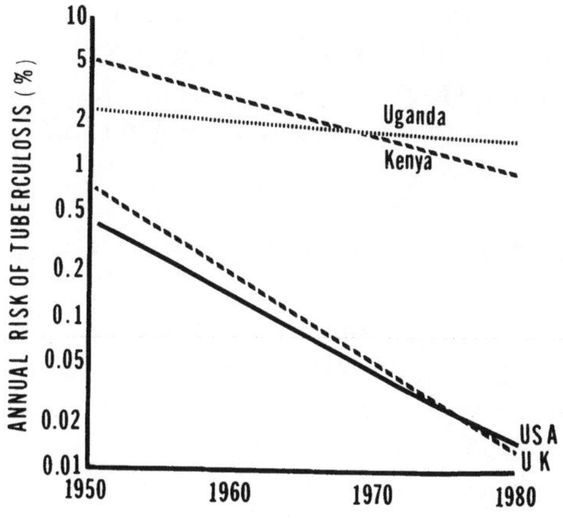

FIGURE 2. Decline in annual risk of contracting tuberculosis in the United States and Great Britain compared with two African countries over the past 30 years (Styblo, 1980).

Hock and Loy, 1981). In such communities, widespread childhood infection by *M. tuberculosis*, as well as by a variety of atypical mycobacteria, will have important secondary effects on overall adult tuberculosis rates later in life (Horwitz, 1973; Comstock, 1982). Several epidemiologic surveys indicate that these early childhood infections (many of which appear to be self-limiting) have important public health implications due to their endogenous reactivation in an increasing proportion of individuals late in life (Stead, 1967; Comstock and Edwards, 1975; Stead and Lofgren, 1983). At present, we badly need to use the appropriate markers (such as bacteriophage lytic patterns or drug susceptibility differences) to distinguish these strains of *M. tuberculosis* (Raleigh *et al.*, 1975; Raleigh, 1984), so that exogenous infection can be excluded in many of these patients (Sutherland *et al.*, 1982; Prabhakar *et al.*, 1987). There is excellent experimental (McCune *et al.*, 1966) as well as epidemiologic evidence (Horwitz *et al.*, 1973) that virulent tubercle bacilli can lie dormant within the tissues of apparently normal individuals for many years before being reactivated by some extrinsic factor. While the yearly risk of such reactivation is extremely small, it is also cumulative, so that eventually a substantial proportion of tuberculin-positive individuals will undergo a secondary attack of the disease (Comstock, 1982). One advantage of neonatal BCG vaccination is that it substantially reduces the incidence of tuberculosis during childhood and early adult life (Waaler and Rouillon, 1974; Sutherland, 1976). This effect probably also extends to many of the atypical mycobacteria, especially in schoolchildren (Romanus, 1983). Although there are questions raised about the overall efficacy of community-wide BCG vaccination programs, especially in tropical areas (Anon., 1980a; tenDam and Pio, 1982; Fine, 1984), it probably still represents the most cost-effective option for tuberculosis control in most developing countries (Holm, 1969; Lotte *et al.*, 1979; Anon., 1982a), where the annual tuberculin conversion rate is likely to be more than 0.5% (Styblo, 1980). In most European countries and in the United States, the incidence rate is too low to justify community-wide infant BCG vaccination programs (Rouillon and Waaler, 1976), although there may be some groups of individuals at special risk of exposure to this infection who should still be immunized (American Thoracic Society, 1975).

In recent years, there has been an increasing interest in the use of live BCG for the immunotherapy of human tumors (White, 1976). The use of such an immunologic modulator (Lagrange *et al.*, 1979) is largely beyond the scope of the present chapter, and the reader is referred to the excellent reviews by Mastrangelo *et al.* (1976) and Hersch *et al.* (1977) for further details.

2. MYCOBACTERIAL INFECTION AND DISEASE

Villemin first demonstrated the infectious nature of pulmonary tuberculosis by injecting rabbits with postmortem lung tissue taken from human patients. However, it was Koch who actually isolated the tubercle bacillus by culturing it on an inspissated serum medium (Collins, 1982). During these classic studies, Koch definitively established that this organism did cause the multitude of clinical forms of this disease seen in humans by taking material from a variety of tuberculous lesions (pulmonary, scrofulous, renal, meningeal) and showing that they all produced typical caseated tuberculous lesions in rabbits, guinea pigs, and mice (Koch, 1890). It was during these studies that he first proposed his famous postulates, thereby proving that *M. tuberculosis* was the causative agent of this important human disease (Chaparas, 1982b). During his studies of the tuberculous lesions in the guinea pig, Koch also noted an accelerated inflammatory response at the site of a secondary intradermal reinoculation with the same organism, the so-called "Koch phenomenon" (Lefford, 1981). This secondary inoculation site was associated with an accelerated mononuclear infiltrate in which the local lesion underwent complete regression, despite the fact that the primary lesions continued to develop. A delayed inflammatory response was also seen when heat-killed tubercle bacilli or an extract called old tuberculin was injected into the skin of the tuberculous individual. The diagnostic value of this tuberculin skin test was immediately recognized and was widely used clinically, although it would be nearly 50 years before the cellular nature of this response was fully appreciated (Chase, 1976).

Bovine tuberculosis was also widespread throughout the Western world, although it would be several years before it was shown to be caused by a different species of mycobacterium (*M. bovis*). This organism is excreted in large numbers in the milk of tuberculous cows, inducing cervical lymph node infections (scrofulosis), as well as severe pulmonary disease in children drinking unpasteurized milk (George and Payne, 1961). Many of these strains of *M. bovis* (Ravanel, Branch) are highly virulent for humans and may, in fact, be some of the most virulent isolates of the tubercle bacillus available for study in the laboratory today (Anon., 1980d). The resulting high mortality seen around the turn of the century in infants and young children due to this organism doubtless prompted Calmette and Guérin to use this species in their classical attenuation studies, which eventually resulted in development of the BCG vaccine (Guérin, 1957). Serial transfer of many strains of tubercle bacillus on artificial culture media for long periods of time had been widely recognized to result in a substantial loss of guinea pig virulence (Rich, 1951). Calmette cultured his *M. bovis* isolate repeatedly on a minimal potato–glycerol medium. Because of the intensely granular nature of the *M. bovis* growth on this culture medium, small quantities of ox bile were added to the medium to help disperse the growth and thus provide a smoother culture. Eventually the organism adapted to this new culture medium and it was transferred serially more than 200 times (despite the intervention of World War I) to produce a suitably attenuated organism. By 1920, attenuation had progressed to the point where the organism could no longer cause progressive disease in calves or guinea pigs (Guérin, 1957). After innumerable experimental protection studies, BCG was eventually fed to the newborn infant of a tuberculous mother in 1921. Twelve months later the infant was still free from tuberculosis, despite its heavy exposure to infection by virulent *M. tuberculosis*. Even more important, the infant showed no untoward side effects as a result of the oral vaccine. A number of other children, both in Europe and in the United States, were then vaccinated successfully and the new attenuated vaccine was finally pronounced a great success. Then came the Lübeck disaster (Rich, 1951), which rekindled all of the earlier concerns that the attenuated organism might revert to full virulence once it was reintroduced back into its natural environment. Careful assessment of the BCG vaccine in a variety of experimental animals, under all sorts of conditions, failed to show any evidence of such reversion to full virulence by the BCG, however. All of these tests indicated that the organism was stable, and it was concluded that accidental contamination of the vaccine with virulent tubercle bacilli had been the real cause of this disaster. However, its shadow greatly reduced enthusiasm for this vaccine, especially in the United States, although it was used quite extensively in Europe, Africa, and Asia over the next several decades (Rees, 1969a). However, clear epidemiologic evidence that the vaccine did have significant protective value against naturally acquired pulmonary tuberculosis became available only after World War II, when a number of extensive and carefully controlled field trials were carried out, both in Great Britain and in the United States (Comstock and Edwards, 1972; Anon., 1980g). These studies made proper allowance for many of the potential variables in the protection equation (tenDam *et al.*, 1976), such as the vaccine itself, its dosage and route of inoculation, the age and sex of the individuals included in both test and control groups, the level of tuberculin sensitivity, and potential levels of concurrent infection with other mycobacteria (Eickhoff, 1977; Rouillon and Waaler, 1976). Despite painstaking attention to these various factors, however, the level of protection achieved in nine major antituberculosis field trials has varied from as high as 80% to no increase in acquired resistance at all (Fig. 3). Much the same type of variability is true for the three major antileprosy trials held over this same period (Price, 1982). Although there is some evidence for increased antituberculous protection in infants vaccinated with live BCG (Tripathy, 1983; Romanus, 1983), there is still no real consensus regarding the underlying causes of the observed variability (Naganna, 1974; tenDam and Pio, 1982). Since many of these factors have been extensively

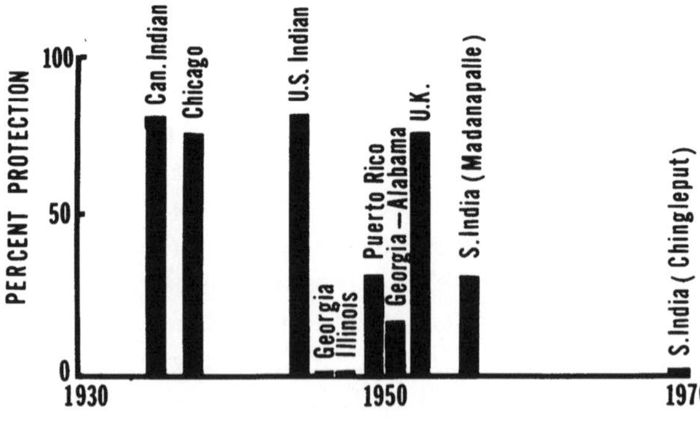

FIGURE 3. Protection by BCG vaccination determined in nine field trials over the past 50 years. Protection is expressed as the percent reduction in clinically or microbiologically proven pulmonary tuberculosis in the vaccinees, compared to controls receiving a placebo (Luelmo, 1982).

assessed in recent reviews (Anon., 1980b; Chaparas, 1982b; Rook and Stanford, 1981), they do not need further detailed discussion at this time.

2.1. *M. TUBERCULOSIS*

2.1.1. Human

Pulmonary tuberculosis is usually caused by *M. tuberculosis*, although some childhood infections may be due to *M. bovis* if unpasteurized milk is readily available (George and Payne, 1961). Direct pulmonary infections from tuberculous cows have also been reported in a few cases (Sigurdsson, 1945). Pulmonary disease can also be caused by some atypical mycobacteria (*M. kansasii, M. avium, M. intracellulare*) but these will be discussed in more detail later. Most strains of *M. tuberculosis* are highly virulent for rabbits, guinea pigs, mice, and monkeys (Collins, 1983), although drug-resistant isolates may exhibit some loss of mouse and guinea pig virulence (Collins and Miller, 1969; Collins, 1973). There is no evidence that their infectivity and virulence for humans is greatly affected, however (Brander *et al.*, 1968). A number of strains of *M. tuberculosis* occurring in Southeast Asia show substantially reduced virulence for guinea pigs (Mitchison *et al.*, 1963; Bhatia *et al.*, 1963; Prabhaker *et al.*, 1987). However, once again, there are no epidemiologic data to suggest that either their infectivity or virulence for humans is affected, and mortality and morbidity rates for the South Indian strains of *M. tuberculosis* differ very little from those reported for the corresponding European and North American isolates (Mitchison, 1964). Such variation takes on more than mere academic interest, however, since it has been suggested that the poor protective response shown by the Indian population at Chingleput (near Madras) to two European BCG vaccines may have been due to this difference in overall virulence by the *M. tuberculosis* strains (Anon., 1980a). Attempts to demonstrate this effect experimentally using aerogenically challenged, BCG-vaccinated guinea pigs have been quite unsuccessful, despite the substantial difference in virulence exhibited by the three challenge organisms (Hank *et al.*, 1981). However, it is possible that the Hartley strain of guinea pig is too susceptible to the tuberculous challenge to constitute the best experimental host for this type of comparative study (Collins, 1983). For this reason, the protection study was repeated using aerogenically challenged mice as a more relevant experimental model of the South Indian protection trial (Collins, 1985). The difference in guinea pig virulence shown by most South Indian strains of *M. tuberculosis* (compared with the standard American or British isolates, $H_{37}Rv$ or Erdman) was not readily detectable using this mouse infection model (Fig. 4). Both *M. tuberculosis* Erdman (TMC #107) and the South Indian (TMC #120) strain multiplied without an initial lag period in both the lungs and spleen (not shown) in aerogenically challenged mice. The growth rates did not differ substantially from each other (Fig. 4). On the other hand, the growth rate for the mouse attenuated strains (*M. avium* or BCG) were both markedly reduced al-

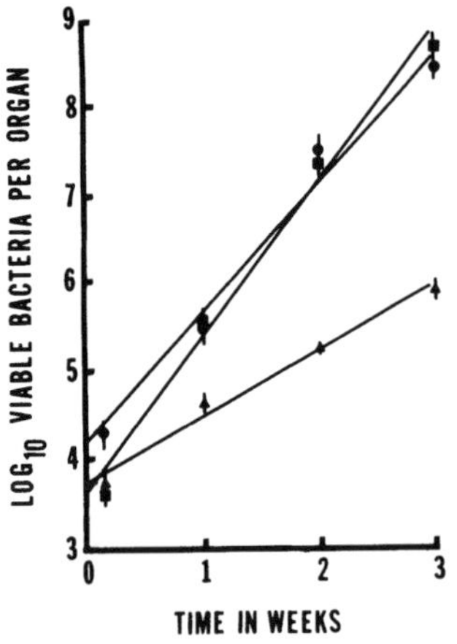

FIGURE 4. Growth of *M. tuberculosis* Erdman (■) or a south Indian strain, TMC 120 (●) in the lungs of aerogenically challenged mice. Growth of BCG Pasteur (▲) is included for comparative purposes. The vertical bars represent ± SEM for five determinations per time point. The regression coefficient R^2 for each of the test organisms was >0.98 ($p < 0.01$) (Collins, 1985).

though the lung growth curves were still exponential throughout the infection period (Collins, 1985).

2.1.2. Bovine

Cattle are highly susceptible to tuberculosis, developing life-threatening pulmonary disease that can be spread to other animals in the herd via droplet infection (Thorns and Morris, 1983). Heavily infected cows may also excrete large numbers of *M. bovis* in their milk, infecting their calves, as well as any children drinking the milk. Control of this disease depends on regular tuberculin testing of all milk herds and slaughter of the positive animals (Steele and Ranney, 1958). As a result, human infections due to this organism have now become quite rare, especially following the widespread introduction of pasteurization of bulk milk supplies in most Western countries (George and Payne, 1961). There has been no attempt to use BCG as a control measure for the bovine disease.

2.1.3. Murine

Although most rodents (mice, rats, hamsters, guinea pigs) can be readily infected with virulent *M. tuberculosis* in the laboratory, there are few reports that they suffer naturally from tuberculosis. Wells (1957) isolated an acid-fast organism from field voles in England which causes progressive, caseating lung infections in most rodents, although it appears to be quite nonpathogenic for humans. This organism (which he called *M. microti*) is almost indistinguishable from *M. tuberculosis* and has been used as an alternative to BCG in at least one British vaccine trial (Anon., 1972).

For many years, the rat was considered to be highly resistant to tuberculosis, and there was even some question whether it could develop tuberculin skin hypersensitivity (Hehre and Freund, 1939; Wessels, 1941). Subsequent studies made it clear that the rat was capable of developing both footpad and skin hypersensitivity responses (Volkman and Collins, 1971) as well as T-cell-mediated antituberculous immunity (Lefford *et al.*, 1973). Different inbred strains of rats and mice vary extensively in their ability to resist a tuberculous challenge, and there is evidence that this related to the ability of the host to develop delayed hypersensitivity (Gray, 1958; Plant and Glynn, 1974). Innate resistance may be a genetically determined trait expressed through the level of macrophage activation prior to the development of acquired antituberculous resistance (Skamene *et al.*, 1982; Orme *et al.*, 1986). However, such differences have proven to be very difficult to substantiate in macrophage cultures (Stokes *et al.*, 1986).

M. lepraemurium causes chronic skin lesions in rats and mice. The infection has been studied extensively in the laboratory since it resembles in some respects the skin lesions seen in human leprosy. For many years, the organism was considered an obligate rodent parasite that could not be cultured *in vitro*. However, in the past few years, it has been cultured successfully in Ogawa egg medium (Pattyn and Portaels, 1980) where it grows with a very long mean generation time (4–6 days). As a result, sparse growth begins to appear after 2–3 month's incubation in 5% CO_2-enriched air (Ogawa and Motomura, 1970). The cultured microorganism appears to retain its mouse virulence (Ogawa and Motomura, 1971). Its *in vitro* growth on this medium is too sparse to allow viable counts to be carried out on organ homogenates. As a result, growth can only be assessed in terms of an overall increase in total counts seen in stained homogenate smears (Collins *et al.*, 1980). This makes it difficult to detect antibacterial immunity in the organs of appropriated vaccinated mice (Closs, 1975), or even to estimate the mean generation time for this organism *in vivo*. *M. lepraemurium* gives rise to progressive infections in subcutaneously infected mice, but it is doubtful whether this infection pattern, which eventually results in substantial systemic involvement (Closs and Haugen, 1973), bears a direct relationship to human leprosy, especially since *M. leprae* cannot readily infect the normal rat, although it does produce substantial disease when the host is first immunosuppressed in some manner (Fieldsteel and Levy, 1976).

2.1.4. Avian

M. avium was first recognized as a poultry pathogen, but it can also produce a progressive, ultimately fatal lung infection in swine and foals (Francis, 1958). Human infections by *M. avium* were probably derived from animal sources originally, although some of them may be members of the *M. avium-intracellulare-scrofulaceum* complex (MAIS) with a water or soil origin. Most of the latter strains are of modest virulence for mice and guinea pigs, even when injected intravenously in large doses (Collins *et al.*, 1978; Collins, 1983). However, *M. avium* is an overt human pathogen and should always be handled in the laboratory under appropriate (P-3) containment conditions, particularly since these virulent organisms are drug-resistant. Interestingly, infections caused by *M. avium* are not uncommon in young children, where they usually produce a relatively mild localized lymphadenitis (Wickman, 1978). However, these same organisms can produce progressive, life-threatening lung infections in immunodepressed adults (Wolinsky, 1979) where their inherent drug resistance renders effective clinical management even more difficult (Davidson, 1979). In recent years, severe life-threatening infections due to members of the *M. avium* complex have been reported in heavily immunosuppressed cancer and transplant patients, as well as in AIDS patients (Anon., 1984; Fauci, 1984; Collins, 1986). Inapparent nasopharyngeal infections due to atypical mycobacteria of the MAIS complex are suprisingly common in many parts of the eastern and southern United States, as well as overseas (Wolinsky, 1979; Singer, 1965). In most cases these asymptomatic infections have been ascribed to the presence of dust- and water-derived organisms in infectious aerosols (Goslee and Wolinsky, 1976; Wendt *et al.*, 1980). Skin tests carried out with PPD-A, PPD-B, and PPD-G (prepared from the appropriate atypical microbacteria) are weakly positive in a surprisingly high proportion of apparently normal individuals, both in the United States and in Europe, suggesting widespread exposure to these environmental organisms (Snider, 1982). Most isolates appear to be *M. scrofulaceum*, but a variety of *M. avium*–complex (MAC) serotypes have been recovered from both lymphadenitis cases and asymptomatic individuals. Their presence has been blamed for the poor immunogenicity observed in several BCG vaccine trials (Edwards *et al.*, 1982; Stanford *et al.*, 1981). There is some experimental evidence that such interactions do occur (Palmer and Long, 1966; Collins, 1971), as well as epidemiologic data indicating extensive skin hypersensitivity to both PPD-A and PPD-B in test populations showing minimal immune responses to BCG vaccination (Edwards *et al.*, 1969; Anon., 1980c; Kumta and Sureka, 1980).

The possibility of interactions between BCG and environmental mycobacteria has been further emphasized by the recent finding that some protection was detected in the recent Chingleput trial if the test group was limited to infants and very young children who were less likely to have been exposed to atypical mycobacterial colonization (Tripathy, 1983). Localized mucosal colonization and lymphadenopathy due to *M. intracellulare* or *M. scrofulaceum*, whether animal or environmental in origin (Meissner and Anz, 1977; Falkinham *et al.*, 1980), may be responsible for the variable degree of cross-protection expressed against endemic tuberculosis in various parts of the world (Edwards *et al.*, 1982). Such cross-protection may be marginal, however, where malnutrition, parasitic infestations, intercurrent infections, and lack of medical treatment are widespread.

2.2. ATYPICAL MYCOBACTERIA

This is a heterogeneous group of acid-fast bacilli which include several overt respiratory pathogens (*M. kansasii*, *M. simiae*, and *M. szulgai*) as well as organisms that can cause severe, localized (skin) infections (*M. marinum* and *M. ulcerans*). Other soil and water strains of atypical mycobacteria can best be described as opportunistic human pathogens, causing disease only when potentiated by other factors (preexisting lung disease, silicosis, emphysema) or by severe immunosuppression (Jenkins, 1981; Abbott and Smith, 1981; Ortbals and Marr, 1978). Some members of this group (*M. simiae*, for instance) have been isolated from a surprisingly large number of sputum samples examined specifically for the presence of this technically demanding and opportunistic pathogen (Bell *et al.*, 1983; Lavy and Yoshpe-Parer, 1982). They may be responsible for progressive lung disease in some adults, but unless they are present in large numbers in multiple sputum or gastric washing samples, their presence is not usually associated with progressive infection or disease induction. The clinical significance of small numbers of nontuberculous mycobacteria in such specimens, especially from apparently normal individuals, remains unclear (Atwell and Pratt, 1960; Wolinsky, 1979). Atypical mycobacteria can colonize normal human nasopharyngeal membranes without going on to establish active disease or, at least, detectable pathology (Wolinsky, 1981). The underlying host–parasite interactions responsible for local tissue involvement, with spread to other host organs and ultimately the development of life-threatening disease, still remain largely unknown. In many cases, the existence of such infections can be confirmed by means of blood or bone marrow biopsy cultures (Cohen *et al.*, 1983), as was the case for *M. avium–intracellulare* complex infections in a surprising number of pre-AIDS patients (Collins, 1986).

2.3. JOHNE'S BACILLUS

This chronic intestinal infection occurs in cattle, goats, and sheep and is characterized by a massive infiltration of the intestinal mucosa by the acid-fast organism *M. paratuberculosis* (Merkal, 1984). The disease usually affects young animals most severely, causing a wasting syndrome associated with extensive thickening of the intestinal mucosae and maladsorption phenomena. Neonatal mice can be infected orally by large doses of this organism, but the resulting disease requires many months to reach clinically significant proportions. *M. paratuberculosis* requires special growth factors (usually provided as an alcoholic extract of *M. phlei*) when being cultured *in vitro*, where minute colonies are seen only after 4–6 weeks' incubation at 37°C on solid media. Reports of a similar type of localized infection with massive colonic involvement by acid-fast bacilli have been reported in some human patients suffering from Crohn's and Whipple's diseases (Elliott *et al.*, 1980; Strom and Gruninger, 1983). Significantly, extensive intestinal involvement also occurs in some AIDS patients (Sohn *et al.*, 1982; Zakowski *et al.*, 1982), although there is no suggestion that *M. paratuberculosis* is able to actively infect the intestinal tract of human patients.

2.4. *M. LEPRAE*

The leprosy bacillus was first described by Hansen in skin sections prepared from Norwegian lepers in 1868. Despite innumerable attempts, this organism has not been cultivated successfully in the laboratory, either in cell-free media or in tissue culture cells (Pattyn, 1973). Until relatively recently, there was not even a laboratory or experimental model of this human disease, which made it virtually impossible to study its immunology. Shepard (1960) reported the successful inoculation of nasal washings from lepromatous leprosy patients into the mouse footpad, followed by localized growth of the organisms over a 6–12-month period. Maximum yields were limited to about 10^6 AFB per footpad in normal, immunocompetent animals but this could be increased some 10- to 100-fold by cortisone treatment (Shepard and McRae, 1965) or by T-cell depletion (Rees, 1969a; Fieldsteel and Levy, 1976). Still greater yields of *M. leprae* have been obtained by inoculating *M. leprae* into the footpads of germ-free, congenitally athymic "nude" mice (Hastings *et al.*, 1981), resulting in up to 10^{10} AFB per gram of footpad tissue after about 18 months of infection. Analogous yields have been reported for intravenously infected nine-banded armadillos (Kirchheimer and Storrs, 1971; Kirchheimer and Sanchez, 1976) so that gram quantities of purified *M. leprae* cells have now been prepared from infected armadillo liver (Shepard *et al.*, 1980). However, the existence of a naturally acquired leprosylike disease in many of these animals (Binford *et al.*, 1977), combined with difficulties in breeding them under specific pathogen-free conditions in the laboratory (Job *et al.*, 1984), has raised questions regarding the widespread use of these experimental animals for leprosy research. However, until alternative experimental models are developed, armadillo-grown *M. leprae* will continue to be the most practical source of these bacteria for antigen fractionation and vaccine production for experimental and human protection studies (Shepard *et al.*, 1980; Convit *et al.*, 1979). Recent reports that a progressive leprosylike disease can be induced experimentally in the Mangabey monkey also offer hope for a better experimental model of this important human disease (Meyers *et al.*, 1981).

A number of other mycobacterial species have not been included in the present discussion, largely owing to space considerations. Further details about these mycobacteria can be found in recent reviews by Wolinsky (1979) and Chapman (1977).

3. PATHOGENESIS OF MYCOBACTERIAL INFECTIONS

The mycobacteria constitute some of the most successful microbial parasites known. As a group, they have adapted from a saprophytic existence in soil and water to an increasingly pathogenic role, ranging from the virtually nonvirulent opportunistic pathogens affecting both humans and animals, to strictly obligate intracellular parasites incapable of surviving outside the body (Barksdale and Kim, 1977). The distribution of these organisms in the community has changed extensively with time, and the factors responsible for these long-term shifts still remain unclear (Fine, 1984). For instance, 50 years ago tuberculosis was predominantly a disease of early childhood, with a decline in mortality rates around the time of puberty, followed by a secondary peak during early adult life. In most Western countries the incidence of this disease now peaks increasingly during late middle age, affecting men more than women, with a final increase during sensecence (Horwitz, 1973; Stead and Lofgren, 1983). It is still not clear whether this shift represents a change on the part of the host, the parasite, or both (Fine, 1984). Both *M. tuberculosis* and *M. leprae* are highly infectious for normal individuals, yet they give rise to active infection in only a small fraction of the exposed population (Rees and Meade, 1974). The complex host–parasite interactions that occur during the initial colonization phase results in the induction of local immunity which produces a subclinical infection and persistent systemic immunity in most of the exposed individuals (Collins, 1984; Godal and Negassi, 1973). It is important to clearly distinguish between the processes of infection and disease induction in this context (Collins, 1984; Wolinsky, 1981). Based on skin hypersensitivity conversion data, less than 5% of infected individuals go on to develop clinically active disease. The mechanism(s) responsible for this innate resistance are likely to be both complex and poorly defined and may vary depending on the size of the infectious inoculum, its route of entry into the tissues, the invasiveness of the test organism and the existence of secondary factors such as intercurrent viral infections or local tissue damage which may affect the ability of the normal cellular defenses to respond optimally to the mycobacterial sensitins (Fine, 1984; Collins, 1979). In fact, a spectrum of host responses will arise as the infection develops and later wanes, and this leads to variations in the overall effectiveness of the immunologic response within the developing lesion. Let us assume that a group of normal immunocompetent individuals have been exposed to a small aerogenic inoculum of highly virulent tubercle bacilli. A few lucky individuals will show no indication of any active infection whatsoever (Fig. 5). They will remain tuberculin negative and are presumably fully susceptible to any subsequent reexposure to the same pathogen. The minimum human infectious dose for a virulent tubercle bacillus is assumed to be only a few viable bacilli, since hospital cross-infections can occur under conditions involving only a few bacilli introduced as a result of aerial contamination (Riley *et al.*, 1962). Under these circumstances, a nonresponder can be assumed to have never been infected, the bacilli being adsorbed onto buccal or nasopharyngeal membranes and ultimately swallowed. Those that do reach the alveolar membranes may be removed mechanically by the resident alveolar macrophages up to the esophagus and eliminated by an essentially nonimmunologic mechanism (Green *et al.*, 1977), being swallowed and voided from the body (Kaltreider, 1976). During this process, however, some of the ingested tubercle bacilli may still enter the tissues via the Peyer's patches and the gut-associated lymphoid tissues. The chances of developing active infection by this oral route are several orders of magnitude lower than that for the aerogenic challenge, making it considerably less likely that active tuberculosis will develop (MacDonald and Carter, 1980; Collins and Stokes, 1987).

The majority of individuals exposed to aerogenic tubercle bacilli will develop a localized infection within the apical alveolar tissues. The normal resident alveolar macrophage present on the alveolar membranes is unable to kill virulent tubercle bacilli *in situ*, and these organisms then begin to multiply intracellularly within a developing lung granuloma (Mackaness, 1968, 1971b). As the orga-

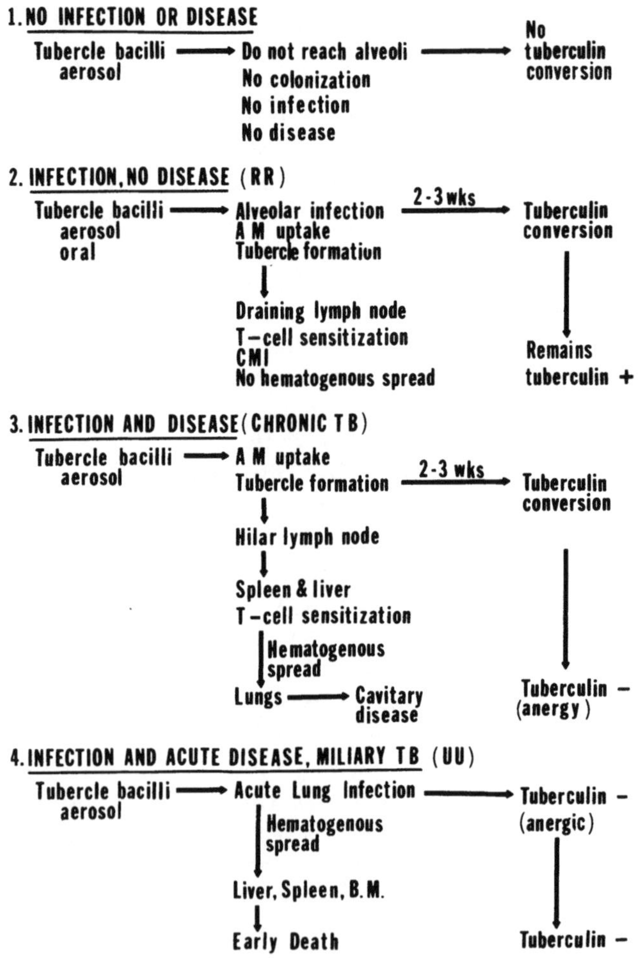

FIGURE 5. The spectrum of tuberculous disease seen in normal, immunocompetent individuals. AM, alveolar macrophage; CMI, cell-mediated immunity.

nisms multiply within the primary tubercle, they release antigenic products which serve as potent chemoattractants, drawing more and more blood monocytes into the lesion (North *et al.*, 1972). Some of these infected macrophages drain to the hilar and tracheobronchial lymph nodes where they establish secondary tubercles and begin to initiate the specific T-cell-mediated immune response (Mackaness, 1974). This response, occurring in perhaps 95% of exposed individuals, is often sufficient to limit the local growth of the pathogen before the infection can reach clinically significant proportions (Fig. 5). The only evidence that these innately resistant individuals have undergone infection is delayed cutaneous hypersensitivity to tuberculin developed some 2–4 weeks after exposure (Stead, 1978). Unless their normal cellular defenses have been reduced by severe malnutrition or immunosupressive drug therapy, they will not progress beyond this primary self-limiting infection (Collins, 1979). Individuals who go on to develop disease (Fig. 5) may have some genetically determined factor that potentiates the infection (Katz and Kunofsky, 1960; Fine, 1984). However, despite extensive investigation, using a wide variety of human populations, there is still no clear evidence that HLA antigens are in any way associated with enhanced susceptibility to tuberculosis (Chaparas, 1982a) or to leprosy (Fine *et al.*, 1979). However, there is experimental evidence that genetically determined factors are important, in both the mouse and rabbit models of human tuberculosis (Lurie *et al.*, 1952; Lagrange *et al.*, 1979; Skamene *et al.*, 1982). These genetic differences seem to be expressed through macrophage activation and possibly in the mobilization of the cellular hypersensitivity by some of the inbred mouse strains (Gros *et al.*, 1983; Plant and Glynn, 1974). However, it is still not clear whether similar genetic factors also occur in humans (Fine, 1984). There has been a long-standing debate regarding a possible role for such genetic factors in the predisposition of some individuals toward the polar lepromatous form of leprosy in humans (Thorsby *et al.*, 1973; Nakajima *et al.*, 1977). While some HLA antigens (HLA-D for instance) correlate with polar tuberculoid leprosy (Fine *et al.*, 1979) there is no such relationship with the polar lepromatous patient (although one has long been argued to exist, largely on the basis of immunologic dogma; Convit and Ulrich, 1978). Its apparent absence seems to suggest that complex predisposing factors may be involved in both human diseases (Stoner, 1978; Stoner *et al.*, 1982).

The relatively small percentage of exposed individuals who go on to develop active pulmonary disease (Fig. 5) present a spectrum of immunologic and clinical features varying from the immunologically reactive, chronically infected patient who expressed a clear cell-mediated response, to the totally unreactive form of the disease characterized as miliary tuberculosis (Lenzini *et al.*, 1977; Daniel *et al.*, 1981). A similar type of unresponsiveness to *M. avium* infections is seen in some AIDS patients (Fauci, 1984). Both infections can be correlated with histopathologic characteristics for the lesions seen across the disease spectrum (Ridley and Jopling, 1966). In leprosy, the polar lepromatous form is multibacillary in nature, with little or no evidence of any cell-mediated immunity (Bullock, 1981). On the other hand, the tuberculoid form of the disease exhibits strong skin (Mitsuda) reactivity, with few acid-fast bacilli present in the skin lesions. Both miliary tuberculosis (Daniel, 1980) and lepromatous leprosy (Bullock, 1978) are characterized by a persistent skin test anergy to the specific sensitin(s) with little or no evidence of any cell-mediated immune response (Bullock, 1975). Although the anergic host appears to be highly immunosuppressed, the same individual can produce extremely high levels of specific immunoglobulins directed against some of the mycobacterial cell-wall antigens (Harboe, 1982; Almeida, 1970). Many of these antibodies cross-react with antigens from other mycobacterial species (Brett *et al.*, 1983). There appears to be a direct correlation between this humoral antibody response and the type of disease pattern seen in the host (Touw *et al.*, 1982). High antibody titers correlate with persistent loss of skin hypersensitivity to this and other antigens of the pathogen (Miller *et al.*, 1973; Rooney *et al.*, 1976; Daniel *et al.*, 1981). Following successful chemotherapy, the elevated antibody titers seen in the anergic host decline quite sharply, although specific skin hypersensitivity may or may not return with time (Bullock, 1975; McMurray and Echeverri, 1978). This specific anergy has been ascribed to a persistent antigen overload within the lymphoreticular organs of the lepromatous leprosy patient, where enormous initial bacterial loads (10^9 AFB/g) can be demonstrated in many cases (Rook, 1975; Bullock, 1981). Some form of infectious tolerance (Gershon and Kondo, 1971) or immunomodulation involving supressor T cells, B cells, or macrophages has been invoked for these cases (Bloom and Mehra, 1984; Collins, 1982).

All these mycobacteria tend to be very slow growing. For instance, most virulent tubercle bacilli have a mean generation time within the lung of between 18 and 20 hr (Fine, 1984), while the leprosy bacillus grows even slower, with an estimated doubling time of about 12 days in the mouse footpad (Levy, 1976). These growth rates are not greatly increased by T-cell depletion (Morrison and Collins, 1975; Rees, 1969a) although the infection does spread to the bloodstream more rapidly and extensively in such animals, resulting in progressive involvement of the spleen, kidney, bone marrow, and lymph nodes, ultimately resulting in the death of many of the animals (North, 1973b; Collins *et al.*, 1975). This type of acute tuberculous pneumonia (once called "galloping consumption") is now relatively rare, although it can be seen in patients with severe combined immunodeficiency disease, some of whom have developed a fatal mycobacteriosis following BCG vaccination (Heilmann *et al.*, 1975). A similar sort of complication can occur in terminally ill cancer patients receiving massive doses of live BCG vaccine in an attempt at immunotherapy, usually following prolonged antitumor chemotherapy (Mathé, 1977). Interestingly, *M. leprae* will produce a systemic infection in T-cell-depleted mice, compared to the restricted footpad and popliteal lymph node involvement seen in immunocompetent rats and mice (Rees and Weddell, 1968). Careful counts of the number of solid staining *M. leprae* (Draper, 1983) present in the footpads of "nude" mice indicate that as many as 10^{10} acid-fast bacilli may be present per gram of tissue some 300 days after inoculation (Hastings *et al.*, 1981). Assuming that logarithmic growth occurs throughout this period, a mean generation time of 12 days can be estimated. However, this figure may be erroneously high, and mean generation times for solid-staining (putatively viable) bacilli only within the nude mouse footpad have been as low as 26 hr (Hastings and Morales, 1982). The majority of mycobacteria observed in the footpad lesions are probably nonviable, although there is no way of conclusively establishing this point at present. Recent advances in fluorescence staining technology offer the possibility that viable bacilli present in *M. leprae*–infected tissues may soon be identified more readily (Kvach *et al.*, 1984).

Both tuberculosis and leprosy have long incubation periods before clinical evidence of disease emerges; this may be 1 or 2 years for pulmonary tuberculosis and as long as 20 years for leprosy (Fine, 1984). The reason for the slow evolution of these mycobacterial diseases remains unclear but presumably reflects a combination of slow intracellular growth and efficient lymphoreticular organs draining the primary lesion, resulting in a slow spread by the infection to other tissues. In many cases, the primary lesion is too small to be readily identified and the primary route of infection is often in doubt (Fine, 1984). Lung tubercle may not be detectable radiologically and even at postmortem, the only indication of an earlier infection being a caseous or calcified tubercle in the draining tracheobronchial lymph nodes (Sutherland and Lindgren, 1979). Similarly, tracheobronchial lesions often constitute the only evidence of a reactivated tuberculous disease in elderly patients. It is possible that the real protective value of BCG vaccination is the development of an enhanced local lymph node response sufficient to prevent hematogenous spread (Fok *et al.*, 1976). The resulting pathology will be limited to little more than the primary tubercle, encased in a fibrous capsule, showing up on the x-ray negative as a residual scar. Such encapsulated lesion(s) will still contain some viable, virtually dormant bacteria which are protected against the bactericidal action of the cell-mediated defenses by the fibrous capsule. However, this steady "carrier" state is T-cell-mediated since active, progressive disease recurs at any time the normal cellular defenses are depressed as a result of immunosuppressive drug therapy, pregnancy, severe malnutrition, or age-related changes in T-cell reactivity (Collins, 1979; Gardner, 1980).

A small proportion of apparently normal, immunocompetent individuals (estimated to be between 2 and 5% of those exposed to the pathogen) will develop evidence of a primary cell-mediated immune response (tuberculin or lepromin skin reactivity), yet still go on to develop progressive, life-threatening disease (Fig. 5). Unlike most immunocompetent individuals, the pathogen spreads by lymphatic drainage to the bloodstream, producing progressive disease in the liver, spleen, kidneys, bone marrow, and central nervous system (Mackaness, 1970; Collins, 1979). This continued hematogenous spread, even in the face of an active cellular immune response, constitutes one of the most interesting enigmas associated with this infection. Two factors that undoubtedly play important roles during the development of active systemic disease are the primary route of infection and the size and virulence of the infecting inoculum (Fig. 6). Introduction of comparable numbers of *M. tuberculosis* into normal mice by the intravenous, footpad (subcutaneous), lung (aerogenic), or oral (intragastric) routes (Fig. 6) results in widely divergent growth patterns *in vivo*. Such differences become highly significant because the rate and level of cellular immunity seems to depend on the spread and persistence of the infecting inoculum within the spleen and, possibly, the draining lymph nodes (Collins *et al.*, 1977). The amount of growth by the pathogen within the reticuloendothelial organs largely determines the degree of pathogenicity. In the case of tubercle bacilli, this parameter appears to primarily involve the lungs (Fig. 7), the level of peripheral tuberculin hypersensitivity and systemic acquired cellular resistance induced by attenuated versus virulent organisms being almost entirely comparable. The heavily infected lungs will also constitute a greater infectious hazard for any susceptible contacts. In the case of multibacillary lepromatous leprosy patients, large numbers of bacilli are shed in the nasal secretions which seem to constitute the major infectious source of this disease within the community (Godal *et al.*, 1974; Huang, 1980). Tuberculoid and indeterminate leprosy pa-

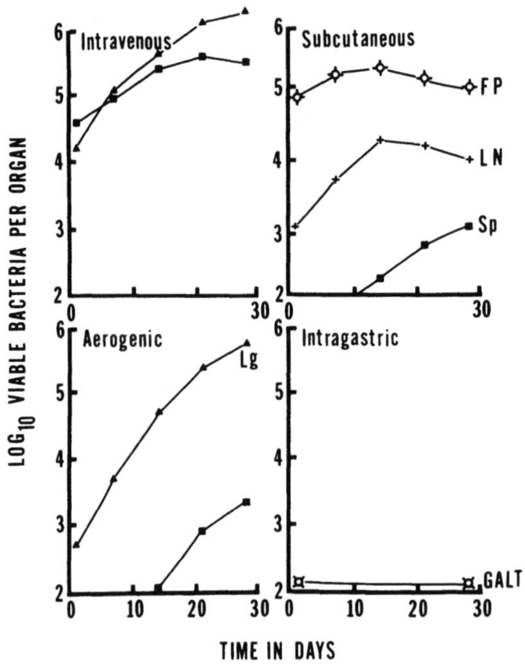

FIGURE 6. Growth of *M. tuberculosis* Erdman in B6D2 mice following intravenous, subcutaneous (footpad), aerogenic, or intragastric inoculation. Spleen (■), lungs (▲), popliteal lymph node (+), footpad (-○-), gut-associated lymphoid tissue (¤).

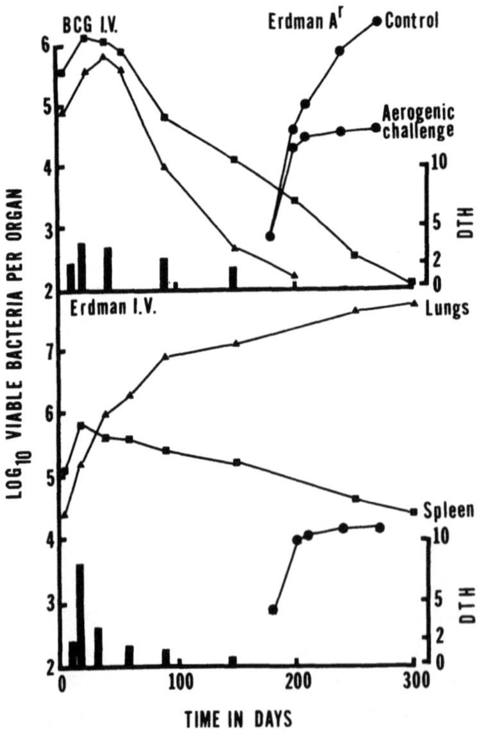

FIGURE 7. Growth curves for BCG Pasteur (top) and *M. tuberculosis* Erdman (bottom) following intravenous inoculation. Spleen (■), lungs (▲). Both groups of mice were reinfected aerogenically with an acriflavine-resistant strain of *M. tuberculosis* Erdman (A[r]) after 180 days and the lung counts were followed on acriflavine agar for up to 90 days (●). The histograms represent tuberculin hypersensitivity (0.2 mm or more increase in footpad thickness at 24 hr was statistically significant; $p = 0.02$). DTH, delayed-type hypersensitivity.

tients have much lower tissue loads of acid-fast bacilli and do not tend to shed large numbers of the bacilli into the environment and are thus not considered an infectious hazard (Bullock, 1981). Although there is still considerable debate regarding the usual route of infection in human leprosy (Huang, 1980), the known presence of leprosy bacilli in nasal secretions (Rees and Meade, 1974) makes a respiratory (or at least nasopharyngeal) route most likely. However, the intradermal route (possibly via an insect bite) would tend to involve the Langerhans cells as the primary antigen-presenting cells within the draining lymph node(s), and this may result in a larger proportion of tuberculoid leprosy cases compared to the nasal or alveolar-infected host, who may tend to develop the polar lepromatous form of the disease (Fine, 1984). Finally, ingestion of the same number of *M. leprae* may have little or no pathologic consequence for the host.

The same factors may hold true for human tuberculosis, too. Introduction of as few as 10 virulent *M. tuberculosis* or *M. bovis* directly into the human lung can result in a rapidly evolving, ultimately fatal disease. On the other hand, ingestion of a million times that number of mycobacteria in unpasteurized milk produces a slowly evolving, localized infection involving the cervical or mesenteric lymph nodes in which the resulting disease may be relatively self-limiting in most normal individuals. This seems to be particularly true for the opportunistic, nontuberculous mycobacteria such as the *M. avium* complex (Wolinsky, 1979). Reduced invasion by these atypical mycobacteria in orally infected adults may be responsible for the relative rarity of systemic infections in these individuals, despite evidence that these organisms repeatedly gain entry to the bronchus or gut-associated lymphoid tissues (Chapman, 1982). These atypical mycobacteria are found in many soil and natural water samples collected throughout the United States (Goslee and Wolinsky, 1976; Brooks *et al.*, 1984), and their presence in the tissues of apparently normal individuals may have important immunologic implications, which will be discussed later (Rook and Stanford, 1981).

Highly virulent mycobacteria will reach the local draining lymph nodes in large numbers as the primary infection develops. Eventually, some of these organisms will enter the bloodstream, where they will be removed by the liver and spleen (Fig. 7). Wherever they lodge, these organisms bring about the development of new tubercles in which the pathogen undergoes logarithmic growth, at least during the early stages of the infection (Collins, 1973). Subsequently, these secondary infections will be limited by the evolving acquired resistance, although it is not unusual for growth to continue in another specialized site such as the lungs, kidneys, or skin (Collins, 1982; Pejovic *et al.*, 1969). In many cases, the developing secondary lesions (tubercles) within the liver and spleen go on to produce further hematogenous spread back into previously uninvolved lobes of the lung (Fok *et al.*, 1976), the meningeal membranes, and the bone marrow (Lefford, 1981). As the lung tubercles in humans and rabbits (but not in rodents) undergo caseation, central liquefaction occurs as a result of hydrolytic enzymes released by the macrophages within the surrounding granuloma (Dannenberg and Sugimoto, 1976). Some of this tissue damage that ultimately contributes to cavity formation may be due to the ongoing delayed-type hypersensitivity characteristic of these lesions (Mackaness, 1968, 1974). Both tuberculin hypersensitivity and granuloma development (together with the concomitant cavitation) tend to be severely depressed in T-cell-depleted individuals, many of whom die from a rapidly fulminant tuberculous pneumonia characteristic of the miliary tuberculosis patient (Gatner and Anderson, 1982; North, 1973b). The lesions seen in such individuals are usually packed with large foamy macrophages, many of them containing viable tubercle bacilli that continue to multiply freely within the intracellular environment up to the time of death (Morrison and Collins, 1975).

In the chronically infected lung, the tubercle undergoes liquefaction and then cavitation and increases in size until eventually it erodes into a bronchiole. The entry of air into the cavity brings about increased growth of the tubercle bacilli, which can appear in large numbers in the sputum (Kaltreider, 1976). This can result in the release of large numbers of viable tubercle bacilli into the surroundings, and in a physically restricted environment (such as that occurring in a prison or nursing home or on shipboard) very high infection rates can be observed (Houk, 1980; Stead, 1978; Stead *et al.*, 1985). The open pulmonary tuberculosis patient constitutes a major health hazard wherever he or she is found and is the primary source of new cases of tuberculosis in any community. In the past, strenuous efforts have been made to identify, isolate, and treat such individuals (Styblo, 1980; Horwitz *et al.*, 1973), but such preventive measures are expensive and time consuming and in many communities have been judged to have only limited efficacy (Holm, 1969; Anon., 1982a). However, in some communities it may be the only cost-effective method for controlling the spread of this disease (Hock and Loy, 1981). Present indications are that even with combined case-finding, BCG vaccination, and chemotherapy programs, complete eradication of this disease is unlikely to occur in the foreseeable future, even in technically advanced nations (Rouillon and Waaler, 1976).

As cavitary disease develops, some tubercle bacilli may be carried mechanically from the infected area to other uninvolved lobes of the lung, especially during prolonged coughing spells (Dannenberg, 1982). This will further compound the severity and extent of the pulmonary involvement, which could help to explain why the infection spreads from the primary apical lesion to adjacent lung lobes despite the presence of an immunologically activated mononuclear defense that should be capable of limiting, even reversing, this growth behavior within the tissues (Collins, 1974). Despite clear evidence of a strong central immunity within the spleen and the draining lymph nodes, viable organisms paradoxically continue to multiply within these secondary lesions until eventually all functional lung activity is lost and the animal dies (Fig. 7). Initially, the infected host will exhibit a strong cellular hypersensitivity response to the specific mycobacterial sensitin (tuberculin), with erythema and induration of more than 20-mm diameter 48–72 hr after an intradermal injection of 5 TU of PPD (Daniel *et al.*, 1981). As the untreated disease progresses, however, a sharp decline is seen in this tuberculin skin reactivity (Fig. 7), the specific state of anergy being ascribed to antigen overload within the heavily infected tissues (Bullock, 1975; Rook, 1975). Successful chemotherapy will reduce the viable bacterial population to undetectable levels, usually resulting in a restoration of skin hypersensitivity, although in some very heavily infected individuals, unresponsiveness may persist for long periods of time (McMurray and Echeverri, 1978) and in lepromatous leprosy patients may be permanent (Bullock, 1975). Examination of lung biopsy or postmortem material taken from such treated patients reveals the presence of large numbers of silver-staining mycobacterial carcases months or years after chemotherapy (Nyka, 1963), making it likely that substantial amounts of mycobacterial antigen continue to persist almost indefinitely within the unresponsive host (Steenken, 1961; Bullock, 1981; Collins *et al.*, 1983).

4. ACQUIRED RESISTANCE TO MYCOBACTERIAL INFECTIONS

Open cavitary tuberculosis constitutes a continuous threat to public health at all levels in the community (Sutherland, 1976). However, the advent of combined chemotherapy using tuberculocidal drugs (Fox, 1979) allows us to render almost 100% of these sputum-positive patients noninfectious within a period of a few weeks (Mitchison, 1985). In most cases, some form of maintenance chemotherapy will be continued for an additional 6–12 months; the exact duration is still a matter of controversy. It must always be remembered, however, that regardless of the effectiveness of the treatment regimen selected, some residual viable organisms will persist almost indefinitely *in vivo* (in most cases, for the lifetime of the host). Evidence exists of relapses 20 or more years after successful chemotherapy, often resulting in the development of geriatric disease (Stead and Dutt, 1982). It seems likely that a residual nidus of infection persists within the fibrotic, calcified tubercle which protects the bacilli from the action of the circulating antibiotics; for technical reasons, this has been very difficult to prove experimentally (McCune *et al.*, 1966). The host–parasite interactions that lead to this lifelong "carrier" state are still largely unknown, as are the changes responsible for the endogenous reactivation of the tuberculous infection later in life when the host defenses have been ablated by malnutrition, alcoholism, pregnancy, or iatrogenic immunosuppression (Stead and Lofgren, 1983). Throughout the patient's lifetime, these residual organisms constitute a ticking time bomb for the host (Sutherland *et al.*, 1982), but they also perform the highly desirable function of maintaining a population of memory T cells to produce a protective anamnestic response against exogenous reinfection by the same organism (Collins, 1979; Orme and Collins, 1986b). Such resistance is never absolute and may be overwhelmed by a large enough inoculum of highly virulent tubercle bacilli introduced directly into the lung (Lefford, 1981; Orme and Collins, 1983). Furthermore, the immune host cannot prevent the establishment of this secondary pulmonary infection (Levy *et al.*, 1961), although it may mount an accelerated bacteriostatic response able to limit the continued growth and spread of the pathogen (compared to that seen in the nonsensitized control). This is usually sufficient to protect the host against the secondary challenge (Fig. 8). This recalled resistance is expressed by a population of memory T cells that exist

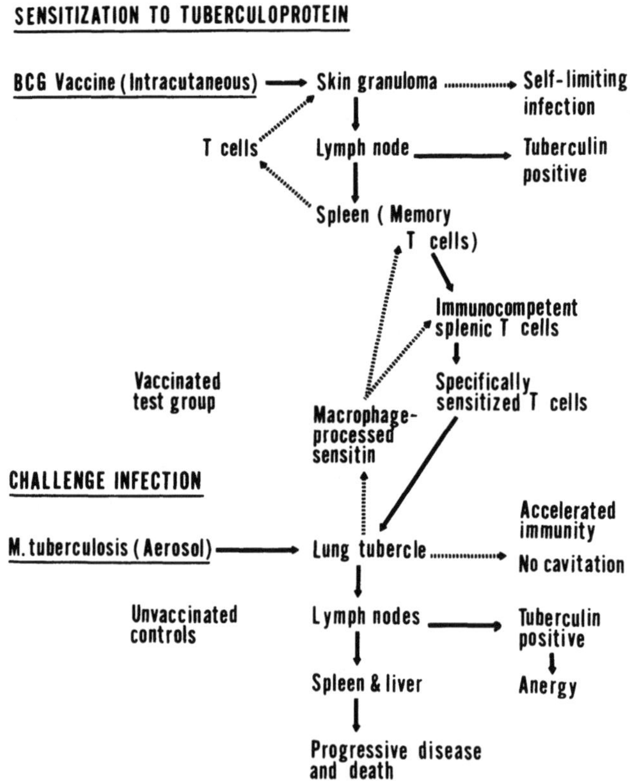

FIGURE 8. Interaction of cell-mediated responses by BCG-vaccinated mice subsequently challenged aerogenically with virulent *M. tuberculosis.*

within the lymphoreticular organs as a result of a primary BCG infection (which is usually introduced intradermally in humans). Experimentally, this virulent challenge infection is usually given about the time of peak resistance to the vaccinating infection (4–8 weeks) when the number of expressor T cells would be expected to be at a peak (Collins, 1972; Mackaness, 1971b). However, this idealized state of affairs is unlikely to occur in BCG-vaccinated human populations, so that the number of specific memory T cells induced by the different vaccines is a matter of considerable importance. Infection of the immune host with small numbers of virulent tubercle bacilli (Smith and Harding, 1977) will result in the stimulation of these memory T cells, which will, in turn, result in an accelerated expressor T-cell response that will limit the growth of the rapidly developing secondary lung infection (Koch reaction), compared to the nonvaccinated controls, which will develop progressive disease and death (Fig. 7). It is this ability to bring about an accelerated antituberculous response years later that constitutes the rationale for universal neonatal BCG vaccination in countries having an annual risk of exposure to tuberculosis above an arbitrarily selected level, which is usually around 0.2–0.5% (Styblo, 1980). Conversion rates in most Western countries has fallen below this point, and communitywide BCG vaccination for them is no longer considered to be cost-effective (Rouillon and Waaler, 1976). In addition, the number of complications arising from the BCG vaccination of all newborns may exceed the number of babies dying from the disease that the vaccine is meant to prevent (tenDam *et al.*, 1976). Although the actual level of protection afforded by BCG vaccination of infants in many Third World countries is still uncertain (tenDam and Hitze, 1980), it does reduce the incidence of miliary and meningeal tuberculosis in these children (Sutherland, 1976) and may also substantially reduce the number of cases of life-threatening pulmonary disease later in life (Sutherland and Lindgren, 1979). Such vaccination may also reduce the severity of some nontuberculous mycobacterial infections since small, but significant increases in the number of lymphadenopathies due to *M. avium–intracellulare* have been reported for Swedish children since the postnatal BCG vaccination program was ended in that country in 1975 (Romanus, 1983).

The other important factor that affects the level of acquired resistance is the size of the infectious inoculum itself (Godal *et al.*, 1974; Bullock, 1979b). This may be an important parameter even for an attenuated *Mycobacterium* such as BCG (Collins and Poulter, 1981). Thus, a large enough intravenous inoculum of modestly virulent mycobacteria may overwhelm the cellular defenses of the normal immunocompetent host, resulting in a rapidly disseminating systemic infection, progressive anergy, and death (Portelance *et al.*, 1976). However, the same number of viable bacilli given by the subcutaneous or intradermal routes tend to produce a chronic infection in which the plateau phase continues virtually indefinitely (Collins *et al.*, 1978). Such persistence may occur despite evidence of an activated cellular defense (Hepper and Collins, 1984a,b). On the other hand, a very small inoculum (10–100 CFU) of highly virulent tubercle bacilli introduced directly into the lung can induce a rapidly fulminating tuberculous pneumonia (Collins, 1985; Orme and Collins, 1986a). The minimum infectious dose for a highly virulent strain of *M. tuberculosis* (such as Erdman) may be as low as 1 or 2 viable units (Grover *et al.*, 1967), and the corresponding figure for humans may be only slightly higher, judging from figures reported for monkeys exposed experimentally to very small infectious aerosols of *M. tuberculosis* $H_{37}Rv$ (Janicki *et al.*, 1973). The normal lung offers very little resistance to the growth of these virulent bacilli (Orme and Collins, 1983), probably because the resident alveolar macrophage is totally unable to inactivate this organism *in situ* (Mackaness, 1971a). Even bacteriostasis must await the entry of immunologically sensitized T cells and emigrant blood monocytes into the developing lesion, which can then inactivate the virulent tubercle bacilli *in vivo* (Mackaness, 1971b; North *et al.*, 1972). During the interval between infection and the cessation of bacterial growth, a complex series of interactions occurs between the infected macrophages draining from the primary infectious focus within the lung to the hilar and tracheobronchial lymph nodes and the immunocompetent T cells present in the thymus-dependent areas of the lymph node. The infected macrophage must present specific mycobacterial sensitin(s) to those T cells which bear a specific recognition factor(s) on their surface that enables them to respond to the antigen (Howard, 1984). These T cells occur within the thymus-dependent areas within the medullary region of the node surrounding the postcapillary venules (Schwartz *et al.*, 1978; Cohen and

Livnat, 1976). In order to respond, the mycobacterial sensitin must be presented to the T cell in conjunction with an appropriate H-2 antigenic marker (Sprent *et al.*, 1980), after which the stimulated T cell undergoes blastogenic transformation producing an expanded clone of specifically sensitized T lymphocytes. The extent of this expansion varies depending on the number and immunogenicity (virulence) of the infecting organism, as well as on the route of inoculation (subcutaneous, aerogenic, intragastric) and the antigen-processing or -presenting cell. The presence of excess antigen within the lymph node may affect the cellular nature of this response, presumably by diverting an increasing proportion of the T cells into the humoral arm of the response (Mackaness *et al.*, 1973; Brown and Brown, 1982; Boyden, 1964). Histologically, this response can be visualized as a local influx of intensely pyroninophilic (rapidly dividing) lymphocytes which appear in the draining lymph node 10–12 days after immunization (Mackaness *et al.*, 1973; North *et al.*, 1972). Subsequently, a similar influx occurs at the periphery of the developing tubercle (Mackaness, 1974), where the emigrant T cells come in contact with the same mycobacterial sensitin being released by infected macrophages within the lesion. The antigenically stimulated T cell then releases a battery of lymphokines within the granuloma, which modulates the immune response to the infectious agent in various ways (Lafferty *et al.*, 1980; Watson, 1981). At least one of these agents, macrophage activation factor, arms the mononuclear phagocytes (Cohn, 1978) entering the granuloma from the bloodstream in such a way that the resulting enzymic and metabolic changes (North, 1978) enable them to kill (or at least to prevent further intracellular growth of) the pathogen within the phagolysosomes of the infected tissue macrophages (Rosenthal, 1980; Edelson, 1982). Experimentally, this response can be quantitated as there is a progressive reduction in the number of viable mycobacteria present in the lungs or the spleen with time (Mackaness, 1968). Adoptive transfer studies using spleen cells treated with anti-Thy 1.2 serum and complement have clearly demonstrated the pivotal role of T cells in this response (North, 1973b; Orme and Collins, 1983). However, it should not be overlooked that the substantial helper T-cell response also stimulates an increasing humoral response, especially after digestion of the killed organisms causes the release of large amounts of antigen into the bloodstream (Daniel *et al.*, 1981). Such antibodies play no active role in the expression of acquired antituberculous or antilepromatous immunity (Reggiardo and Middlebrook, 1974; Fine, 1984), although they may provide the means for developing discriminatory radioimmune or ELISA-based assays useful in the diagnosis of subclinical infections and early disease in humans (Daniel, 1980; Chapman, 1982; Hewitt *et al.*, 1982).

5. RELATIONSHIP BETWEEN TUBERCULIN DELAYED CUTANEOUS HYPERSENSITIVITY AND ANTITUBERCULOUS IMMUNITY

Naturally infected individuals usually develop delayed cutaneous hypersensitivity 3 or 4 weeks after exposure to the tubercle bacillus. Experimentally, this change coincides with a reduction in growth of the microorganisms within the liver and spleen, a T-lymphocyte-mediated response (North, 1974b; McGregor and Kostiala, 1976). Delayed cutaneous hypersensitivity and cell-mediated immunity appear to be related both temporally and functionally (Mackaness and Blanden, 1967; Collins and Mackaness, 1968, 1970). However, tuberculin hypersensitivity can be induced experimentally in the absence of antituberculous immunity, and *vice versa* (Rich, 1930). Mackaness (1967) pointed out the fallacy in the argument that they were therefore unrelated phenomena (Youmans, 1975) and proposed that both were mediated by a population of immunocompetent T lymphocytes present in the spleen and the lymph nodes following an active infection of the immunized host (Mackaness, 1968, 1971a, 1974). Such a conclusion seems consistent with recent reports of an accelerated antimicrobial response at the site of a heterologous delayed-type hypersensitivity response in the mouse footpad (Poulter and Lefford, 1968; Patel, 1981). However, not all investigators have observed this type of antimicrobial enhancement, and it seems likely that other factors, such as the mouse strain, the test organism, and the route of challenge, are involved in this reaction also (Løvik and Closs, 1982). The enhanced (or anamnestic) tuberculin response occurring within the lungs of the BCG-vaccinated individual following challenge with virulent tubercle bacilli probably involves a delayed type of inflammatory response which may be an important host defense mechanism against many intracellular pathogens (Mackaness, 1974). Metabolically active tubercle bacilli present within the developing tubercle release tuberculoproteins which induce a reaction equivalent to the "Koch" reaction originally seen in tuberculous guinea pig skin (Collins, 1984). The lymphokine-activated macrophages enhance elimination of the pathogen from the immune lung. It is these immunologically activated or "angry" macrophages (Mackaness, 1970) which produce the larger amounts of hydrogen peroxide and other activated oxygenated intermediaries responsible for their microbicidal activity (Lowrie, 1983) compared to the normal cell (Cohn, 1978). However, this is not the whole story, since many strains of *M. tuberculosis* are highly sensitive to the bactericidal action of hydrogen peroxide *in vitro*, yet are surprisingly resistant to immunologically activated macrophages (Walker and Lowrie, 1981). *M. tuberculosis* is resistant to most bactericidal agents, and so studies of intracellular killing by immunologically activated macrophages have been widely carried out with another facultative intracellular parasite, *Listeria monocytogenes* (Mackaness, 1969; North, 1981). This technical modification has demonstrated T-cell involvement in the bactericidal process using a variety of adoptive transfer systems (North, 1974a).

In recent years, studies with *Listeria* have shown antituberculous resistance in both adoptively immunized mice and rats (Lefford *et al.*, 1973; Lefford, 1975b; Orme and Collins, 1983). During these studies, both tuberculin hypersensitivity and antituberculous immunity can be transferred to naïve recipients using distinctive subsets of T cells differing in their Lyt-alloantigenic makeup (Orme and Collins, 1984; Kaufmann *et al.*, 1979). For unknown reasons, antituberculous immunity can be demonstrated in the lungs of adoptively immunized mice only with great difficulty, compared to that in the spleen (Lefford, 1975b). This may in part reflect differences in the cellular distribution patterns observed when immune lymphocytes are infused into normal syngeneic recipients. Radiolabeled splenic T cells injected intravenously into normal mice result in less than 1% of the label being recovered from the lungs 24 hr later, despite the fact that all of the infused cells must first traffic through that organ before lodging in the liver and spleen. Cells that were recovered from the lung were trapped within the capillary bed, and very few were recovered by bronchial lavage (Collins and Auclair, 1984). In the case of an intravenous inoculum, the tubercle bacilli and the sensitized T cells exist in relatively close physical apposition within the spleen compared to that occurring when the challenge inoculum was introduced directly into the lungs of the adoptively immunized host. In the latter, the sensitized T cells would have to be mobilized into the developing lung tubercle from the relatively distant splenic depot with a considerable delay in the onset of the antituberculous response. Transfer studies using memory T-cell suspensions from isoniazid-treated donors (Orme and Collins, 1986b) suggest that the T cells infused into the naïve recipient have a primarily instructional role with the recipient's spleen and that the lag before a detectable expressor T-cell population accumulates within the lung tubercle simply represents the time required to develop this T-cell immunity *de novo*. It is during this instructional period that the so-called "isogenic barrier" may be expressed by a cyclophosphamide-sensitive cell population present within the normal mouse spleen (Lefford, 1983; North *et al.*, 1982).

The number of sensitized T cells present in the spleens of intravenously challenged mice usually peaks several days after the bacterial growth curve slows from its early logarithmic rate to an arithmetic increase, before passing into a stationary growth phase (North, 1973a). This slowing of the early growth rate may be unrelated to the later immune T-cell activity and should be carefully differentiated from nonspecifically activated macrophage killing, which may also occur at this time (Hepper and Collins, 1984a). The only definitive way to achieve this conclusively is to use an adoptive transfer system, usually into sublethally irradiated or T-cell-depleted syngeneic recipients (Mackaness, 1969; Lefford, 1975b). Once the growth of the pathogen within the lungs and spleen has been brought under control, the specific expressor T-cell populations will be down-regulated to ensure that the activated cell-mediated defenses do not result in unnecessary tissue damage (Mackaness, 1974). Although the specific expressor T-cell population declines rapidly in number and reactivity *in vivo* (North and Spitalny, 1974), the convalescent host retains immunity for many months or even years against reinfection. This anamnestic response is mediated by the memory T cells induced during the primary infectious stimulus. The size and duration of the memory T-cell response within the lymphoreticular organs of the convalescent host constitute the most important parameter of any vaccination program. The long-lived memory T cells are relatively radiation and cyclophosphamide resistant and may persist within the spleen for long periods of time, even in the apparent absence of an ongoing antigenic stimulus (North and Deissler, 1975). The number of these cells will also depend on the dose, the route, the relative viability, the persistence, and the immunogenicity of the vaccine strain (Lagrange *et al.*, 1976; Gheorghia and Lagrange, 1983), as well as on the age and nutritional status of the test population (Comstock, 1982; Seth *et al.*, 1981).

6. ANTITUBERCULOUS IMMUNITY IN HUMANS

Pulmonary tuberculosis in the United States has declined to a yearly incidence of less than 10 per 100,000 (Comstock, 1982), most of which occurs in the lower socioeconomic groups where relatively poor housing, inadequate nutrition, and widespread alcohol and drug abuse contribute to its continued persistence. Although the overall incidence of pulmonary tuberculosis (especially in children) has declined sharply over the past two or three decades, the amount of extrapulmonary disease (much of it due to the drug-resistant atypical mycobacteria) has not declined correspondingly but still remains at about 2 per 100,000 per year (Farer *et al.*, 1979). The reason for this disparity is not immediately apparent but may have important implications in certain specialized subgroups (e.g., AIDS patients).

The overall incidence of tuberculosis in any given population is best estimated from the yearly tuberculin conversion rates (Styblo, 1980). However, these data merely reflect exposure to *M. tuberculosis* (or, in some cases, to atypical mycobacteria) over the past 12 months. Care must be taken in obtaining appropriate controls (Ferebee, 1967), having trained observers to measure the areas of induration seen 48–72 hr after the PPD injection, and in interpreting results. The resulting skin test data must be analyzed carefully (Naganna, 1974; Rook and Stanford, 1979). A positive skin test is generally accepted to be a minimum of 10 mm of induration at 48 hr following the intradermal injection of 0.1 μg (5 TU) of PPD (Reichman, 1979). Most epidemiologists agree that a response in the 5–9-mm range should be classed as "doubtful" and often is associated with infection by other mycobacteria. Such individuals should be retested with PPD-A, PPD-B, or PPD-G, which usually produces stronger skin responses in individuals infected with *M. avium, M. intracellulare,* or *M. scrofulaceum*. However, some apparently immunocompetent individuals remain skin-test-negative even following repeated exposures to BCG vaccine (Anon., 1955). The reason for this lack of responsiveness is still unclear, but it seems likely that these individuals are potentially of great interest immunologically, since there is clear epidemiologic evidence that they are unduly susceptible to tuberculosis later in life. Most BCG-vaccinated individuals develop strongly positive skin responses following a tuberculous challenge (20 mm or more of erythema and induration to 5 TU of PPD). In nonvaccinated individuals, such a level of skin reactivity is diagnostic of active disease (Palmer and Edwards, 1967). When untreated, such patients rapidly become completely unresponsive (anergic) to tuberculin (Nash and Douglas, 1980), a state that usually has serious prognostic implications for the patient (Kantor, 1975).

The immunologic basis for failure to develop delayed cutaneous hypersensitivity to an appropriate sensitin (tuberculin or lepromin) is still unclear (Chaparas, 1982a; Bullock, 1979a; Guinto, 1968). Originally, it was postulated that such individuals suffered from some kind of innate (or genetic) flaw that rendered them incapable of recognizing the specific mycobacterial antigen (sensitin). Such a host should also fail to develop a protective cell-mediated immune response against the infectious agent. However, whether this unresponsiveness is, in fact, genetically determined still remains controversial (Fine, 1984). There is substantial evidence for such genetically determined factors in mice (Orme *et al.*, 1986); however, in humans there seems little question that substantial age and sex differences exist with respect to skin hypersensitivity. Variations are also found between rural and urban populations with respect to the development of immune responses to a number of microbial parasites (Lagrange *et al.*, 1983). If a direct cellular relationship exists between delayed cutaneous hypersensitivity and acquired antimicrobial resistance (Plant and Glynn, 1974; Mackaness, 1971a; Hormaeche *et al.*, 1981), the observed differences in skin responsiveness seen in different individuals (Rook *et al.*, 1981) could also be expected to affect the response to the chronic infection produced by the leprosy and tuberculosis bacilli *in vivo* (Bullock, 1979a; Convit and Ulrich, 1978). A similar type of effect may influence the effectiveness of BCG vaccination against both these human pathogens (Price, 1982; Fine, 1984). Systemic diseases (syphilis, sarcoidosis, Hodgkin's disease, measles) can affect the level of tuberculin hypersensitivity present in tuberculous individuals (Kantor, 1975). The latter type of anergy seems to be the result of a nonspecific depletion of the helper T-cell population as a whole and need not imply the presence of a specific suppressor cell population induced by the intercurrent infection.

Persistent tuberculin or lepromin anergy both *in vivo* and *in vitro* in otherwise immunocompetent individuals (Closs *et al.*, 1982) suggested the absence of the specific T-cell clone(s) which normally recognize and respond to the specific sensitizing antigen(s) of these pathogens (Godal *et al.*, 1974). Because they are unable to recognize the specific sensitin, such individuals could be expected to develop the polar unresponsive form of the disease in which they are totally unable to control the growth of the virulent pathogen *in vivo* (Bullock, 1979b). Attempts to demonstrate a genetically determined defect in identical twins have provided, at best, equivocal data (Fine, 1981), and at the moment, the existence of such a defect must be considered moot, despite the fact that this theory has important implications for the development of immunization strategies against lepromatous leprosy (Convit and Ulrich, 1976, 1978) and other chronic pathogens (Bullock, 1979a). At present, there is no indication that susceptibility to infectious disease is in any way related to an inability to recognize the relevant microbial sensitins *in vivo* (Bloom and Mehra, 1984). Thus, the inability of 1 or 2% of the infected population to develop a protective cell-mediated immunity, while still able to produce a specific humoral response to one or more cell-wall antigens, constitutes one of the most intriguing enigmas of current immunology (Bullock, 1978; Rea and Levan, 1977; Bloom and Mehra, 1984).

Finally, the atypical mycobacterial infections (especially common in young children) can produce "doubtful" tuberculin reactions (5–9mm-diameter erythema and induration, 24–48 hr after 5 TU of PPD) in many normal, nontuberculous individuals (Snider, 1982). These self-limited lymph node infections may induce some cross-protection against the naturally acquired *M. tuberculosis* and *M. leprae*. However, such interactions may be a two-edged sword since it has been suggested that presence of endemic leprosy may adversely affect BCG vaccination responses and thus contribute to the apparent lack of increased antituberculosis protection reported in several carefully controlled BCG vaccination field trials (Anon., 1980c, 1982b; Stanford *et al.*, 1981). Such antigenic interactions are likely to be extraordinarily complex and may be largely irrelevant in actual practice since cases of active tuberculosis developing in leprosy patients (and *vice versa*) have been reported in the literature. In addition, leprosy patients can develop tuberculin hypersensitivity following BCG inoculation, despite the fact that these same individuals will remain totally unresponsive during lepromin skin testing (Lowe and McNulty, 1953).

7. IMMUNITY TO EXPERIMENTAL MYCOBACTERIAL INFECTIONS

Koch (1890) reported the accelerated immune reaction that still bears his name in reinfected tuberculous guinea pigs and was quick to realize the underlying protective implications from this response. A primary tuberculous skin lesion develops as a localized granuloma at the primary injection site with extensive involvement of the draining lymph node (Lefford, 1981). On the other hand, a secondary inoculation induces an acute indurative type of skin response peaking 2 or 3 days postchallenge with limited lymph node involvement, followed by an accelerated resolution of the skin lesion. This occurs at a time when the cellular response to the primary lesion was latent and the bacterial growth within the lesion was still virtually unrestricted (North *et al.*, 1972; Truitt and Mackaness, 1971). When the challenge was carried out in the mouse footpad, an influx of pyroniophilic T lymphocytes was seen in the draining lymph node(s), as well as in the primary lesion. The cellular nature of this response stems from the pioneering studies of Max Lurie (1928) in tuberculous rabbits. He quantitated the growth of mycobacteria in the lungs and spleens of normal and BCG-vaccinated animals by using a disgonic (minute colony) strain of *M. tuberculosis* as the challenge organism, thus allowing the *M. tuberculosis* colonies to be readily distinguished from the residual BCG colonies on the basis of their differing morphology (Lurie, 1929). He was thus able to show that the immunized host could limit the systemic growth of the virulent challenge organism (especially within the lung) thus providing protection against an otherwise lethal challenge. Lurie then demonstrated that mononuclear phagocytes harvested from immunized rabbits phagocytosed and killed virulent tubercle bacilli at a faster rate than similar cells taken from unvaccinated rabbits (Lurie, 1942). These early phagocytosis experiments carried out in the anterior chamber of the rabbit eye were confirmed and extended using peritoneal macrophage monolayer cultures harvested from normal and BCG-vaccinated animals (Mackaness, 1954). As a result, the idea of a purely "cell-mediated immunity" was conceived as a basis for acquired antituberculous resistance (Suter, 1956), which subsequently was expanded to include resistance to other intracellular parasites (Mackaness, 1964).

The immunocompetent T lymphocyte constitutes the cornerstone of this cell-mediated immune response (Mackaness, 1971a). However, the ultimate expressor cell for such acquired resistance is the activated mononuclear phagocyte

(Cohn, 1978). These tissue macrophages arise from emigrant blood monocytes which enter the developing lesions in large numbers. Within the granuloma, these cells are exposed to a variety of lymphokines, including macrophage activating factor released by the sensitin-stimulated T cells as they enter the developing granuloma (Watson, 1981; North, 1978; Nelson, 1981). Mouse peritoneal macrophages exposed *in vitro* to such lymphokines become immunologically armed and are capable of affecting the growth of even highly virulent tubercle bacilli (Zlotnick and Crowle, 1982). Unstimulated (control) macrophages can also phagocytose the virulent tubercle baccili, even in serum-free cultures, but the intracellular bacilli will not be inactivated to any real extent and may continue to multiply within these cells, protected from circulating bactericidal factors, antibodies, and antibiotics (Stokes *et al.*, 1986). Furthermore, the intracellular bacilli may be carried by the infected macrophage to other uninvolved organs, such as the meninges, kidney, liver, spleen, and bone marrow, to establish secondary tubercles (Lefford, 1981; Dannenberg, 1982).

The pivotal role of the immunocompetent T lymphocyte in the expression of antituberculous immunity has been clearly established by adoptive transfer assays (Lefford, 1975b). The importance of the T cell in this response was earlier inferred from infection studies carried out in T-cell-depleted animals (North, 1973b, 1974a; Collins *et al.*, 1975). Lack of antituberculous immunity was also demonstrated in BCG-infected congenitally athymic "nude" mice (Sher *et al.*, 1975; Ueda *et al.*, 1976). In the absence of immunocompetent T cells, both tuberculin hypersensitivity and antituberculous immunity are virtually ablated, but can be restored by a thymic graft (Gaugas *et al.*, 1971), by repeated injections of the thymic hormone, "thymosin" (Morrison and Collins, 1976), or by a direct infusion of the host with immune T cells (Collins and Morrison, 1979; Orme and Collins, 1983). Such adoptive transfers can be ablated by pretreatment of the spleen cells with anti-Thy 1.2 antiserum and complement (North, 1974a). Both tuberulin hypersensitivity and antituberculous immunity are mediated by specifically sensitized T cells, but the responsible cells may be distinguished from one another on the basis of their Lyt alloantigenic surface membrane markers (Orme and Collins, 1984). While it is generally agreed that delayed hypersensitivity is mediated by T cells bearing the Lyt $1^{+}23^{-}$ marker, there is less agreement regarding the cells responsible for the expression of cell-mediated immunity (Lefford, 1975a). All T cells express some Lyt1^{+} antigen on their membranes (Ledbetter *et al.*, 1980), but it is probable that many T cells bear insufficient amounts of this antigen on their surface membranes to allow their depletion from a suspension by means of anti-Lyt1^{+} antiserum treatment (Mage *et al.*, 1981). Thus, evidence that Lyt1^{+} cells are involved in the expression of cell-mediated immunity to infections by these facultative intracellular parasites still remains relatively tenuous (Kaufmann *et al.*, 1979). It is possible that the Lyt1^{+}(L3T4) cells that are responsible for the expression of delayed hypersensitivity response within an infected lesion may also help to potentiate the cell-mediated immunity expressed by a second T-cell population by intensifying macrophage activation within the developing tubercle (Hsu *et al.*, 1980; Patel, 1980; Mitsuyama *et al.*, 1982). There is increasing evidence that the two cellular reactivities (DTH and CMI) are expressed *in vivo* by alloantigenically distinctive lymphocyte subpopulations (Lefford, 1980; Orme and Collins, 1984). Recently, it was suggested that the T cell responsible for antilisterial immunity may bear the Lyt 1^+23^+ marker (Kaufman *et al.*, 1979), and these lymphocytes have also been successfully cloned *in vitro* (Kaufmann and Hahn, 1982). Cloned T lymphocytes are claimed to adoptively transfer both delayed hypersensitivity and cell-mediated immunity to naïve recipients (Fitch, 1981). However, the T lymphocytes used to develop these clones were harvested from animals immunized with heat-killed *L. monocytogenes* that provoke a humoral rather than a cellular response in the spleen. *In vivo* experiments indicate that a fully protective anti-*Listeria* immunity is induced only by live bacterial vaccine (Mackaness, 1971a; Collins, 1974). In BCG-vaccinated mice a fully protective antituberculous immunity is seen in T-lymphocyte-depleted mice infused with immune splenic T lymphocytes pretreated *in vitro* with anti-Lyt1^+ antiserum (Orme, unpublished data). This immune response to a small aerogenic tuberculosis challenge was ablated if the cells were pretreated with an anti-Lyt1^+ monoclonal antibody and complement (Orme and Collins, 1984).

8. VACCINATION AGAINST MYCOBACTERIAL DISEASES

Inactivated mycobacterial vaccines prepared from whole cells, cytoplasmic, and cell-wall components (Smith *et al.*, 1968; Daniel and Janicki, 1978; Ribi *et al.*, 1982) have been tested in mice, rats, guinea pigs, and monkeys. Nonetheless, these inactivated vaccines have required an oil-in-water adjuvant to induce delayed cutaneous hypersensitivity *and* antituberculous immunity. In addition, these preparations are too toxic for human use (Collins, 1974). Biodegradable preparations (such as Adjuvant-65 or alum) enhance humoral rather than cell-mediated responses (Myrvik, 1974). Heat-inactivated mycobacteria (up to 500 mg dry wt per ml of saline) without an oily adjuvant injected into normal mice and guinea pigs induce substantial levels of delayed hypersensitivity, and some antituberculous immunity (Shepard, 1975; Mehra and Bloom, 1979; Collins *et al.*, 1983). However, such dense suspensions produces persistent tissue granulomas (cold abcesses) at the injection site. This severe local inflammatory response also results in substantial tissue scarring, which limits its usefulness in humans. Furthermore, the persistence of the cellular immune response varies greatly depending on the strain of *Mycobacterium* used in the vaccine. For instance, *M. leprae* suspensions induce a persistent specific (lepromin) hypersensitivity response which is still readily detectable after more than 12 months (Mehra and Bloom, 1979; Collins *et al.*, 1983). Such animals may also express detectable levels of antilepromatous immunity up to this time (Shepard, 1975). With the availability of large amounts of purified, armadillo-grown *M. leprae* (Shepard *et al.*, 1980), a practical killed leprosy vaccine for human use may eventually become available (Bullock, 1979b). However, local tissue reactions from large inocula of killed leprosy bacilli are likely to be severe, and a more practical type of immunogen for human use may be a mixed vaccine consisting of live BCG and inactivated *M. leprae* which has been reported to serve as an effective therapeutic agent against this important human disease (Convit *et al.*, 1980).

At present, the only vaccine available for use against the human mycobacterioses is BCG (Chaparas, 1982b). In most communities, *M. tuberculosis* is still the most important pulmonary pathogen to be considered, with a much lower relative incidence (10%) of atypical mycobacterial infections, especially in normal immunocompetent adults (Jenkins, 1981; Rosenzweig, 1979). However, in some areas, the relative incidence of atypical lung infections seems to be on the increase as the number of *M. tuberculosis* cases steadily declines (Ahn *et al.*, 1979). Atypical mycobacterial infections due to the *M. avium* complex remains relatively common in young children (Lincoln and Gilbert, 1972), with increased numbers of such infections where BCG vaccination was not carried out during infancy (Wickman, 1978). Cross-protection between BCG and these atypical mycobacteria seems compatible with experimental data obtained in *M. avium*–challenged mice (Engbaek and Jespersen, 1966; Collins, 1971, 1985). Now with the recent outbreak of life-threatening, systemic *M. avium* complex infections in AIDS patients, the development of specific prophylactic agents for the immunization of high-risk groups in the community becomes increasingly desirable. However, even an attenuated live vaccine such as BCG constitutes an infectious hazard for immunosuppressed patients, since numerous reports of fatal mycobacterioses in congenitally and iatrogenically immunosuppressed individuals have been published (Heilman *et al.*, 1975; Dahlstrom and Sjogren, 1977; Ternani *et al.*, 1979).

Not all commercially available BCG vaccines are equally effective (Luelmo, 1982) despite the fact that they have all been derived from the same original strain of *M. bovis* developed by Calmette and Guérin at the Pasteur Institute over 60 years ago (Guérin, 1957). Serial subcultures of this strain over many decades resulted in a number of more or less minor cultural and biochemical changes which evolved as the organism was maintained in the different laboratories (Grange *et al.*, 1983). As a result, there are now at least 15 different substrains of BCG (Anon., 1980d), each with its own *in vitro* and *in vivo* growth characteristics (Dubos and Pierce, 1956a; Pierce *et al.*, 1956; Gheorghia and Lagrange, 1983). This is reflected in differences in immunogenicity when these strains are tested *in vivo* (Dubos and Pierce, 1956b; Mackaness *et al.*, 1973; Vallishayee *et al.*, 1974; Ladefoged *et al.*, 1976). Further variability of this type has now been largely obviated by the maintenance of type BCG strains as seed lots, either by lyophilization or by deep freezing (Levy *et al.*, 1968; Kim and Kubica, 1972). All commercial vaccine producers have now adopted this form of preservation for their BCG stocks (Osborne, 1983), but this still does not overcome the problem that BCG vaccines obtained from different commercial sources around the world are not of equivalent immunogenicity (Smith and Harding, 1977). This is especially important where the BCG vaccine is being used for its adjuvantive activity, which may vary extensively depending on the test conditions (Mackaness *et al.*, 1973; Sher *et al.*, 1973; Lagrange, 1978). Such variability probably derives in part from factors associated with the immunogenicity assays themselves, as well as from the way the resulting data have been interpreted by the different test laboratories (Wiegeshaus *et al.*, 1971). In particular, the experimental animal selected for the assay undoubtedly plays an important, probably crucial, role in determining the outcome of the protection test (Smith and Harding, 1977). However, there are almost certainly other contributory factors involved as well (Freudenstein *et al.*, 1979). With respect to the antituberculosis response, much of this variability may be more of academic than practical importance, since the specific challenge organism (*M. tuberculosis*) can be expected to extend and reinforce any residual BCG-induced immunity (Collins, 1984). The end result of the combined cellular response to the BCG and *M. tuberculosis* challenges is likely to be about the same regardless of the relative immunogenecity of the original BCG vaccine strain. Quantitative counts of *M. tuberculosis* present in aerogenically challenged mice indicate that substantial protection is expressed regardless of the immunogenicity of the BCG strain originally used to sensitize the host (Collins, 1984, 1985). However, the adjuvantive value of different BCG substrains using antigenically unrelated stimuli (red blood cells, tumor cells, contact sensitins) can be expected to vary far more widely. Under these conditions, BCG Pasteur was consistently more effective than most of the other vaccine substrains (Lagrange, 1978; Lagrange *et al.*, 1976; Gheorghia and Lagrange, 1983; Neveu, 1976).

Commercial BCG vaccines seem to be equally effective when tested as antituberculous immunogens in experimental animals (Ladefoged *et al.*, 1976; Smith *et al.*, 1979; Freudenstein *et al.*, 1979) or in humans (Vallishayee *et al.*, 1974). However, in the recent BCG field trial at Chingleput (South India), two widely proven European BCG vaccines (Pasteur and Copenhagen) were ineffective against naturally acquired human tuberculosis (Anon., 1979). The reason for this paradoxical loss of protective value in a carefully controlled community-wide trial has been the subject of intense debate and speculation (Anon., 1980a, 1981, 1982a). Decreasing protection seems to have been an ongoing trend over the last 20 years (Fig. 3), especially where the field trials were held in hot, dusty subtropical climates. One characteristic of this type of environment seems to be a widespread exposure to atypical mycobacterial infections (Fine, 1984). Unfortunately, this correlation does not seem to hold for the admittedly smaller number of BCG vaccine trials carried out against lepromatous leprosy in the same areas (Price, 1982). This lack of cross-protection by the environmental mycobacteria may be due to a wider antigenic disparity between *M. tuberculosis* and *M. leprae* (Bullock, 1979b) than between BCG and *M. tuberculosis*. Such divergency does not seem consistent with the recent suggestion (Anon., 1982b) that the presence of endemic leprosy within a test population may contribute to the ineffectiveness of the BCG vaccine against pulmonary tuberculosis. Whatever the real reason for the almost total failure of two proven European BCG vaccines in the Indian trial (Tripathy, 1983), an agonizing reappraisal of the general effectiveness of BCG vaccination has now been made, especially with respect to the Third World (Anon., 1982a; tenDam and Hitze, 1980; Clemens *et al.*, 1983). In the meantime, infant vaccination programs continue, often for no better reason than the lack of another alternative (tenDam and Pio, 1982; Waaler and Rouillon, 1974).

The first and most likely explanation for the lack of immunogenicity by the test vaccine(s) used at Chingleput is viability (Gheorghia and Lagrange, 1983). There is experimental evidence that different BCG preparations can vary widely in their relative viability (Mackaness *et al.*, 1973) as well as their allogenicity and immunogenicity, in both quantitative and qualitative terms. The presence of large numbers of dead organisms (or free antigen) in the vaccine will tend to induce strong humoral immune responses which may interfere with the cellular response (Brown *et al.*, 1979; Brown and Brown, 1982). The Chingleput trial was the first to use lyophilized BCG preparations, produced in Europe under carefully controlled conditions (Eickhoff, 1977). Both vaccines were quality controlled by means of immunogenicity tests carried out in mice and guinea pigs (tenDam *et al.*, 1976; Siebenmann and Barbara, 1974) and both were highly immunogenic. Unfortunately, tuberculin testing in BCG-vaccinated human volunteers was not attempted by the manufacturer, although this quality control had been recommended by several investigators for all lyophilized BCG preparations (Nyboe and Bunch-Christensen, 1966; Eickhoff, 1977). Comparative studies in Europe had indicated that freeze-dried vaccines are as immunogenic as freshly prepared suspensions (Springett and Sutherland, 1970; Hesselberg, 1970). However, lyophilized mycobacterial suspensions usually have lowered *relative* viabilities compared to either frozen or freshly prepared preparations (Lagrange *et al.*, 1976). Frequently, the growth behavior of the reconstituted suspension is less vigorous than that with the corresponding frozen suspensions, whether tested *in vivo* or *in vitro* (Dunbar *et al.*, 1968; Collins and Smith, 1969; Kim and Kubica, 1972). This decline in immunogenic vigor seems to be related to the increased mortality associated with the freeze-drying process, which cannot be completely overcome by the presence of protective additives (Lagrange *et al.*, 1976). The surviving lyophilized BCG also grows more slowly *in vitro* than nonlyophilized suspensions, often producing minute colonies (Mackaness *et al.*, 1973). Such colony forms are associated with slower growth *in vivo* and reduced adjuvantivity (Lagrange, 1978). In order to resolve these differences, carefully controlled comparative studies using freshly prepared versus lyophilized BCG preparations need to be carried out in a large group of tuberculin-negative (less than 5 mm induration 24–48 hr after skin testing with 250 TU of PPD) volunteers, with a protective assessment of the immunogenicity of such preparations over an extended period of time. Such studies would probably have to be carried out in Europe since it is clear that normal tuberculin-negative individuals are difficult to find in endemic areas such as Chingleput (Anon., 1979).

Low levels of tuberculin hypersensitivity to PPD-A and PPD-B were reported in both the test and control populations used in the Chingleput trial (Anon., 1979). This again raises the question of the role played by the so-called "doubtful" tuberculin reactors in determining background immunity levels in such populations (Anon., 1980c; Kumta and Sureka, 1980). Low levels of tuberculin reactivity due to widespread atypical mycobacterial infection have been reported, especially in children, where lymphadenitis due mainly to *M. scrofulaceum* is still relatively common (Hart *et al.*, 1962; Wolinsky, 1979; Stanford *et al.*, 1981). These infections have been blamed for poor protection in several earlier BCG trials in the southern United States (Palmer and Long, 1966; Springett, 1965; Hart, 1967). Most atypical mycobacteria cross-react antigenically with BCG and *M. tuberculosis* to some extent (Chaparas and Maloney, 1978; Closs *et al.*, 1980), with respect to both humoral and delayed hypersensitivity responses determined in appropriately immunized animals (Daniel, 1980). Such organisms should also induce substantial levels of cross-protection against a virulent *M. tuberculosis* challenge, and this was, in fact, established in both mice and guinea pigs (Palmer and Long, 1966; Pejovic *et al.*, 1969; Collins, 1971). The level of cell-mediated responsiveness induced by these opportunistic pathogens varies greatly depending on the ability of the organisms to establish progressive (or at least persistent) systemic infections within the host tissues (Fig. 9). Most *M. avium* and some *M. intracellulare* strains are able to establish persistent infections in the lungs and spleens of normal mice (Collins *et al.*, 1978). *M. avium* and *M. intracellulare* strains induce slowly progressive lung disease in heavily infected mice and this will eventually lead to the death of many of the chronically infected animals (Goto *et al.*, 1984). None of the *M. scrofulaceum* or *M. nonchromogenicum* strains tested produced stable infections in the livers and spleens of intravenously infected mice (Collins, 1983), which were eliminated by a nonimmunologically mediated response (Fig. 9). Preinfection of these animals with a second dose of the same organism produced no sign of an immunologically accelerated clearance (Collins and Niederbuhl, 1983; Collins and Montalbine, 1975). Such nonpersistent mycobacterial infections had no detectable effect (either promotional or suppressive in nature) with respect to the growth of a subsequent BCG inoculum *in vivo* or the

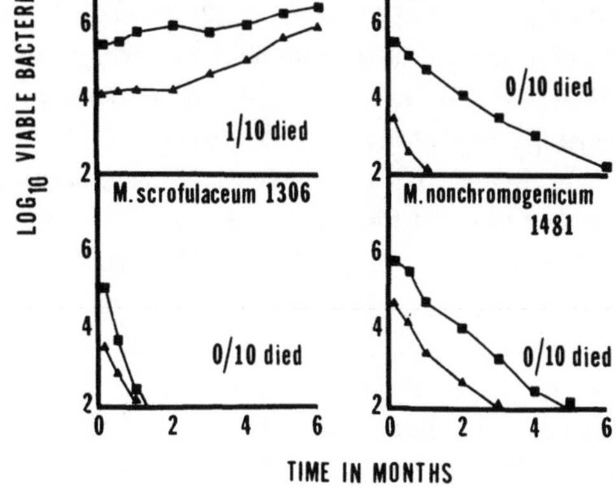

FIGURE 9. Growth curves for selected strains of *M. avium–intracellulare–scrofulaceum* complex following intravenous inoculation into normal mice. *M. nonchromogenicum* 1481 is also avirulent for mice. The figures on the graphs refer to the number of deaths/total number of mice infected 6 months earlier.

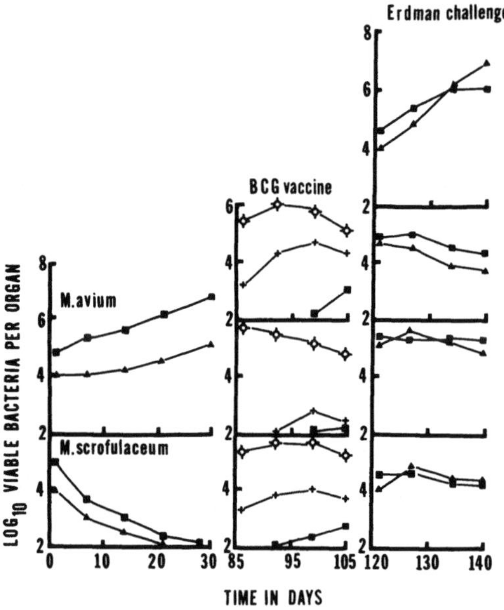

FIGURE 10. Effect of infection with *M. avium* (persister) or *M. scrofulaceum* (nonpersister) 3 months prior to BCG Pasteur vaccination. The level of antituberculous immunity was tested 1 month later. Control mice received BCG vaccine alone or saline alone. Spleen (■), lungs (▲), footpad (-◇-), popliteal lymph node (+). The difference in spleen and lung counts in the vaccinated versus saline-treated controls was statistically significant ($p < 0.001$).

level of antituberculous immunity (Fig. 10). Quantitation of the mixed populations within the lungs of the challenged mice suggested that any interactions between the atypical mycobacteria and the BCG vaccine were complementary rather than inhibitory, concerning the expression of an effective antituberculosis response (Collins, 1983). Although it may be true that certain atypical mycobacteria may interfere with the ability of the host to respond effectively to a subsequent tuberculous challenge (Weiszfeiler and Karasseva, 1981), the level of such interference appears to be too small to explain the failure to the Chingleput trial.

Atypical mycobacteria are mostly opportunistic human pathogens (Wolinsky, 1979), able to induce severe, systemic infections only in the immunosuppressed or immunologically immature individual. However, it is possible that some organisms in this group can produce persistent colonization of the mucosa-associated lymphoid tissues of the normal host (Collins and Stokes, 1987), and this may occur quite frequently in some areas of the country where such organisms are known to be widespread in the environment (Falkinham *et al.*, 1980). Sensitization to the corresponding mycobacterial sensitins appears to be equally widespread (Snider, 1982). If such infections occur before the cellular defenses are fully deployed, it is possible that specific immune tolerance could be developed which could affect the response to the same organisms later in life (Chapman, 1977; Codias and Reinhardt, 1979). Reinfection with the same organism might be expected to develop into significant bacterial populations involving the lung, the intestinal lymphoid organs, and the bone marrow, but because of the low toxicity of these organisms, relatively large bacterial loads may well develop without obvious signs of clinical disease. Such immune tolerance could also affect the ability of the host to respond to cross-reacting sensitins associated with the BCG vaccine *in vivo*, resulting in substantially lowered cellular responses by the vaccinated host (Rook and Stanford, 1981). The main difficulty with such an interference mechanism is the need to postulate the existence of a blocking effect against the BCG growth (to explain its lack of immunogenicity) without a similar blocking effect being expressed against the *M. tuberculosis* or *M. leprae* challenges (Axelson *et al.*, 1974; Chaparas *et al.*, 1978). Repeated experimental studies using *M. avium–* or *M. intracellulare–*infected mice and guinea pigs subjected to BCG vaccination (via either the subcutaneous or the intravenous route) have consistently failed to indicate any evidence of such an interference phenomenon, at least against a subsequent *M. tuberculosis* challenge (Collins, 1983); nor was it possible to demonstrate a different type of cellular response to the atypical mycobacteria (Rook *et al.*, 1981). Persistent strains (*M. avium* 706, 714, 724, and 801 and *M. intracellulare* D673) induced similar antituberculous resistance against the *M. tuberculosis* challenge, equivalent to that seen in the BCG-vaccinated controls (Collins, 1971, 1984). No evidence of a "*Listeria*-type" hypersensitivity response could be found in mice infected with nonpersister strains of *M. kansasii, M. intracellulare, M. scrofulaceum,* or *M. vaccae,* as had been predicted by Rook and Stanford (1979). Although it is possible that an aberrant type of cellular response does occur if atypical mycobacteria are introduced into the tissues via the oral or intranasal routes, such a response would still have to involve the cervical and/or mesenteric lymph nodes in some type of antigenic tolerance response (Rook and Stanford, 1981). It has proven to be surprisingly difficult to induce oral infections with these organisms in adult mice (Orme and Collins, unpublished data), and further studies using newborn or neonatal animals may be necessary.

Despite real concerns for the overall effectiveness of BCG vaccination in recent community-wide field trials, there are still cogent reasons for continuing the BCG vaccination of newborns, especially where case finding is difficult or impracticable and tuberculin conversion rates are above 0.5% per annum (Anon., 1980e, 1981). In Western Europe, North America, Japan, and Australia (where annual tuberculin conversion rates are low) BCG vaccination can be recommended only for those individuals likely to be at special risk from tuberculosis (American Thoracic Society, 1975; Anon., 1980f). Such individuals include nursing-home, health care, and laboratory personnel, as well as alcohol- and drug-abuse patients who are deemed unlikely to complete their antituberculosis drug therapy. In addition, migrants and other highly mobile individuals entering the community from high-risk areas should be actively considered as candidates for BCG vaccination.

Live attenuated vaccines may also be considered in relation to some other diseases in which specific vaccines have not yet been successfully developed (Mitchell, 1980; Collins and Campbell, 1982). Both *T. pallidum* and *M. leprae* have yet to be cultured serially in laboratory media or even in tissue culture (Pattyn, 1973; Fieldsteel *et al.*, 1977). The alternative is to grow the causative organisms in a suitable experimental animal and then to purify the resulting tissue-grown bacilli for use as BCG killed vaccines with an adjuvant (Convit *et al.*, 1982). *In vivo–*grown organisms also have the advantage that they may be more immunogenic than the corresponding *in vitro–*grown suspensions (Segal and Bloch, 1957). Much of this difference may be due to the physical dispersion of the bacilli, which affects the distribution of the suspension in the reticuloendothelial organs (Collins *et al.*, 1974; Collins and Montalbine, 1976). Sufficient antigenic material can now be harvested from actively infected experimental animals to provide a practical vaccine for human use. This approach has been used with armadillo-grown *M. leprae* (Shepard *et al.*, 1980), mosquito-grown *P. knowlesi* (Gwadz *et al.*, 1979), and rabbit-grown *T. pallidum* (Baseman *et al.*, 1974). Since the harvested organisms may still be viable and therefore virulent for humans, they must first be inactivated by as gentle a method as possible (irradiation seems to be the method of choice in most cases). Such killed vaccines will induce a predominantly humoral response, unless suspended in a suitable adjuvanting preparation (Sanchez *et al.*, 1980). Recent advances in liposome technology offer a number of exciting possibilities in this regard (van Rooijen and van Niewaemegen, 1980), which may permit the development of a predominantly cell-mediated response to the specific immunogen.

At present, the only adjuvant for this type of vaccine seems to be living BCG, although killed *C. parvum* and *B. pertussis* may also be effective under some circumstances (White, 1976). The BCG adjuvanting effect seems to be related to its ability to induce a strong T-cell response within the draining lymph node(s) which results in a concurrent cell-mediated response to any heterologous bacterial, parasitic, and tumor antigens included in the vaccinating preparation (Lagrange *et al.*, 1976). Such mixed vaccines have been proposed as a practical means for sensitizing lepromin-negative individuals and leprosy contacts to *M. leprae* antigens and thus increasing protection in areas where this disease is endemic (Convit and Ulrich, 1978). Live BCG vaccine has little or no therapeutic value against an established pulmonary infection (Collins, 1984), and it may even have a potentiating effect on individuals incubating the disease, a factor of considerable importance where chronic infections are widespread (Fine, 1984). Convit and his co-workers reported the presence of cross-reactive cell-wall antigens in BCG and *M. leprae* (Goihman-Yahr and Convit, 1972), but despite these common antigens, live BCG provides only modest local immunity in *M. leprae*–challenged animals (Shepard, 1975). However, repeated injections of live BCG vaccine can enhance this cellular resistance to an *M. leprae* challenge within the mouse footpad (Ganguly *et al.*, 1981). Extrapolating from this type of experimental study, Convit suggested the use of multiple doses of live BCG mixed with killed *M. leprae*, which were then injected intradermally into lepromatous and borderline leprosy patients. Some of these patients later showed enhanced clearance of *M. leprae* (Convit *et al.*, 1974). This was also associated with some restoration to lepromin sensitivity and a decline in the number of *M. leprae* in adjacent skin lepromas and their draining lymph nodes (Convit *et al.*, 1982). Such an immunotherapeutic approach offers the hope that a practical vaccine for human use may be developed in the not too distant future (Convit *et al.*, 1980). However, not all investigators have reported success using this type of approach (Smelt *et al.*, 1981), and much more work will be needed to finally resolve this matter (Rook, 1983). A similar vaccine might be helpful in intractable *M. ulcerans* skin infections (Rees, 1969b) and possibly in the life-threatening atypical mycobacterial infections reported in the severely immunosuppressed (AIDS) patients in this country (Fauci, 1984). Genetic engineering and DNA hybridation have produced new vaccine preparations for the prevention and therapy of some of these chronic mycobacterial infections (Young *et al.*, 1986; Collins, 1988). However, despite these novel immunologic approaches, both tuberculosis and leprosy are likely to be important public health problems in many parts of the world for decades to come.

ACKNOWLEDGMENTS. Work described in this chapter was supported by grants AI-14065 and HL-19774 from the National Institutes of Health, Bethesda, and a grant from the World Health Organization, Geneva.

REFERENCES

Abbott, M. R., and Smith, D. D., 1981, Mycobacterial infections in immunosuppressed patients, *Med. J. Aust.* **1**:351–353.

Ahn, C. M., Lowell, J. R., Onstad, G. D., Shurford, E. H., and Hurst, G. A., 1979, A demographic study of disease due to *M. kansasii* or *M. intracellulare–avium*, *Chest* **75**:120–125.

Almeida, J. O., 1970, Serology in leprosy, *Bull. World Health Organ.* **42**:673–702.

American Thoracic Society, 1975, BCG vaccines for tuberculosis, *Am. Rev. Respir. Dis.* **112**:478–480.

Anon., 1955, Experimental studies in vaccination, allergy and immunity in tuberculosis. V. Relation between degree of tuberculin sensitivity and resistance in the vaccinated individual, *Am. J. Hyg.* **62**:185–199.

Anon., 1972, BCG and vole bacillus vaccines in the prevention of tuberculosis in adolescence and early adult life, *Bull. World Health Organ.* **46**:371–385.

Anon., 1977, World Health Organization expert committee on leprosy. Fifth report, *World Health Organ. Tech. Rep.* #607:7–48.

Anon., 1979, Trial of BCG vaccines in South India as tuberculosis prevention: First report, *Bull. World Health Organ.* **57**:667–682.

Anon., 1980a, Vaccination against tuberculosis, *World Health Organ. Tech. Rep. #651:7–23.*

Anon., 1980b, BCG vaccination policies. Report of a WHO Study Group, *World Health Organ. Tech. Rep.* **652**:5–17.

Anon., 1980c, Trial of BCG vaccines in south India for tuberculosis prevention: First report, *Bull. Intern. Union Tuberc.* **55**:14–22.

Anon., 1980d, *Mycobacterial Culture Collection Catalogue*, NIH Publication no. 80-289.

Anon., 1980e, BCG vaccination in the newborn, *Br. Med. J.* **281**:1445–1446.

Anon., 1980f, BCG vaccination: What should Australia do?, *Med. J. Aust.* **2**:534–535.

Anon., 1980g, Effectiveness of BCG vaccination in Great Britain in 1978, *Br. J. Dis. Chest* **74**:215–227.

Anon., 1981, BCG vaccination after the Madras study, *Lancet* **1**:309–310.

Anon., 1982a, Tuberculosis control, *Tubercle* **63**:157–169.

Anon., 1982b, BCG: Sibling rivalry, *Lancet* **1**:206.

Anon., 1982c, Centennial: Koch's discovery of tubercle bacillus, Morbid. Mortal. Wkly. Rep. **31**:121–123.

Anon., 1983, Interstate outbreak of drug-resistant tuberculosis involving children, *Morbid. Mortal. Wkly. Rep.* **32**:516–518.

Anon., 1984, Acquired immunodeficiency syndrome—an assessment of the present situation in the world: Memorandum from a WHO meeting, *Bull. World Health Organ.* **62**:419–432.

Anon., 1985, Tuberculosis—United States 1984, *Morbid. Mortal. Wkly. Rep.* **34**:299–307.

Anon., 1986, Tuberculosis and acquired immunodeficiency syndrome—Florida, *Morbid. Mortal. Wkly. Rep.* **36**:587–590.

Atwell, R. J., and Pratt, P. I., 1960, Unclassified mycobacteria in the gastric contents of healthy personnel and patients of a tuberculosis hospital, *Am. Rev. Respir. Dis.* **81**:888–892.

Axelson, N. H., Harboe, M., Closs, O., and Godal, T., 1974, BCG antibody profiles in tuberculoid and lepromatous leprosy, *Infect. Immun.* **9**:952–958.

Barksdale, L., and Kim, K-S., 1977, Mycobacterium, *Bacteriol. Rev.* **41**:217–372.

Baseman, J. B., Nichols, J. C., Rumpp, J. W., and Hayes, N. S., 1974, Purification of *T. pallidum* from infected rabbit tissue: Resolution into two treponemal populations, *Infec. Immun.* **10**:1062–1067.

Bell, R. C., Higuchi, J. H., Donovan, W. N., Krasnow, I., and Johanson, W. G., 1983, *Mycobacterium simiae*. Clinical features and follow-up of twenty four patients, *Am. Rev. Respir. Dis.* **127**:35–38.

Bhatia, A. L., Jacob, C. V., Hitze, K. L., Ramachandran, K., and Selkon, J. B., 1963, A comparison of the virulence in the guinea pig of tubercle bacilli from Thai and British patients, *Bull. World Health Organ.* **29**:483–490.

Binford, C. H., Meyers, W. M., Walsh, G. P., Storrs, E. E., and Brown, H. L., 1977, Naturally acquired leprosy-like disease in the nine-banded armadillo (*Dasypus novencinctus*). Histopathologic and microbiologic studies of tissues, *J. Reticuloendothel. Soc.* **22**:377–388.

Bloom, B. R., and Mehra, V., 1984, Immunological unresponsiveness in leprosy, *Immunol. Rev.* **80**:5–28.

Boyden, S. V., 1964, The immune response to antigens of the tubercle bacillus in relation to specific acquired immunity, *Acta Tuberc. Scand.* **58**:19–26.

Brander, E., Aho, K., and Patiala, J., 1968, Spread of drug-resistant tubercle bacilli, *Am. Rev. Respir. Dis.* **98**:407–415.

Brett, S. J., Draper, P., Payne, S. N., and Rees, R. J. W., 1983, Serological activity of a characteristic phenolic glycolipid from *M. leprae* in sera from patients with leprosy and tuberculosis, *Clin. Exp. Immunol.* **52**:271–279.

Brooks, R. W., Parker, B. C., Gruft, H., and Falkinham, J. O., 1984, Epidemiology of infection by non-tuberculous mycobacteria. V. Numbers in Eastern United States soils and correlation with soil characteristics, *Am. Rev. Respir. Dis.* **130**:630–633.

Brown, C. A., and Brown, I. N., 1982, *Mycobacterium bovis* BCG, modulation of murine antibody responses; influence of dose and degree of aggregation of live or dead organisms, *Br. J. Exp. Pathol.* **63**:133–143.

Brown, C. A., Brown, I. N., and Sljivic, V. S., 1979, Suppressed or enhanced antibody responses *in vitro* after BCG treatment of mice: Importance of BCG viability, *Immunology* **38**:481–488.

Bullock, W. E., 1975, Anergy and infection, *Adv. Intern. Med.* **21**:149–173.

Bullock, W. E., 1978, Leprosy: A model of immunological perterbation in chronic infection, *J. Infect. Dis.* **137**:341–354.

Bullock, W. E., 1979a, Mechanisms of anergy in infectious diseases, in: *Immunological Aspects of Infectious Diseases* (G. Dick, ed), pp. 269–294, University Press, Baltimore.

Bullock, W. E., 1979b, Immunology and the therapeutics of leprosy, *Ann. Intern. Med.* **91**:482–484.

Bullock, W. E., 1981, Immunobiology of leprosy, *Compr. Immunol.* **8**:369–390.

Chaparas, S. D., 1982a, The immunology of mycobacterial infections, *CRC Crit. Rev. Microbiol.* **9**:139–197.

Chaparas, S. D., 1982b, Immunity to tuberculosis, *Bull. World Health Organ.* **60**:447–462.

Chaparas, S. D., and Maloney, C. J., 1978, An analysis of cross-reactions among Mycobacteria by *in vivo* and *in vitro* assays of cellular hypersensitivity, *Am. Rev. Respir. Dis.* **117**:897–902.

Chaparas, S. D., Brown, T. M., and Hyman, I. S., 1978, Antigenic relationships of various mycobacterial species with *M. tuberculosis*, *Am. Rev. Respir. Dis.* **117**:1091–1098.

Chapman, J. S., 1977, *The Atypical Mycobacteria and Human Mycobacteriosis*, Plenum Press, New York.

Chapman, J. S., 1982, The atypical mycobacteria, *Am. Rev. Respir. Dis.* **125**:S119–124.

Chase, M. W., 1976, Developments in delayed-type hypersensitivities: 1950–1975, *J. Invest. Dermatol.* **67**:136–148.

Clemens, J. D., Chuong, J. H., and Feinstein, A. R., 1983, The BCG controversy, *JAMA* **249**:2362–2369.

Closs, O., 1975, Experimental murine leprosy: growth of *M. lepraemurium* in C3H and C57BL mice after footpad inoculation, *Infect. Immun.* **12**:480–489.

Closs, O., and Haugen, O. A., 1973, Experimental murine leprosy. 1. Clinical and histological evidence for varying susceptibility of mice to infection with *M. lepraemurium*, *Acta Pathol. Microbiol. Scand.* **81A**:401–410.

Closs, O., Harboe, M., Axelson, N. H., Bunch-Christensen, K., and Magnusson, M., 1980, The antigens of *Mycobacterium bovis* strain BCG studied by cross-immunoelectrophoresis: a reference system, *Scand. J. Immunol.* **12**:249–263.

Closs, O., Reitan, L. J., Negassi, K., Harboe, M., and Belehu, A., 1982, *In vitro* stimulation of lymphocytes in leprosy patients, healthy contacts of leprosy patients and subjects not exposed to leprosy, *Scand. J. Immunol.* **16**:103–115.

Codias, E. K., and Reinhardt, D. J., 1979, Distribution of serotypes of the *Mycobacterium-avium-intracellulare-scrofulaceum* complex in Georgia, *Am. Rev. Respir. Dis.* **119**:965–970.

Cohen, I. R., and Livnat, S., 1976, The cell-mediated immune response: interactions of initiator and recruited T lymphocytes, *Transplant. Rev.* **29**:24–58.

Cohen, R. J., Samozuk, M. K., Busch, D., and Lagios. M., 1983, Occult infections with *M. intracellulare* in bone marrow biopsy specimens from patients with AIDS, *N. Engl. J. Med.* **308**:1475–1476.

Cohn, Z. A., 1978, The activation of mononuclear phagocytes: Fact, fancy and future, *J. Immunol.* **121**:813–816.

Collins, F. M., 1971, Immunogenicity of various mycobacteria and the corresponding levels of cross-protection developed between species, *Infect. Immun.* **4**:688–696.

Collins, F. M., 1972, Acquired resistance to mycobacterial infection, *Adv. Tuberc. Res.* **18**:1–30.

Collins, F. M., 1973, The relative immunogenicity of virulent and attenuated strains of tubercle bacilli, *Am. Rev. Respir. Dis.* **107**:1030–1040.

Collins, F. M., 1974, Vaccines and cell-mediated immunity, *Bacteriol. Rev.* **38**:371–402.

Collins, F. M., 1979, Cellular antimicrobial immunity, *CRC Crit. Rev. Microbiol.* **7**:27–91.

Collins, F. M., 1982, The immunology of tuberculosis, *Am. Rev. Respir. Dis.* **125**:S42–49.

Collins, F. M., 1983, Kinetics of the delayed-type hypersensitivity response in tuberculous guinea pigs and mice tested with several mycobacterial antigens, *Am. Rev. Respir. Dis.* **127**:599–604.

Collins, F. M., 1984, Tuberculosis, in: *Bacterial Vaccines* (R. Germanier, ed.), pp. 373–418, Academic Press, New York.

Collins, F. M., 1985, Protection afforded by BCG vaccines against an aerogenic challenge by three mycobacteria of decreasing virulence, *Tubercle* **66**:267–276.

Collins, F. M., 1986, *M. avium*–complex infections and development of the acquired immunodeficiency syndrome: Casual opportunist or causal cofactor? *Int. J. Leprosy* **54**:458–474.

Collins, F. M., 1988, Cell-mediated and humoral responses to mycobacterial disease, *Clin. Immunol. Newsletter* (in press).

Collins, F. M., and Auclair, L. K., 1984, Effect of *M. bovis* (BCG) infection on the kinetics of mononuclear cell response within the lung, *J. Leukocyte Biol.* **35**:321–332.

Collins, F. M., and Campbell, S. G., 1982, Immunity to intracellular bacteria, *Vet. Immunol. Immunother.* **3**:5–66.

Collins, F, M., and Mackaness, G. B., 1968, Delayed hypersensitivity and Arthus reactivity in relation to host resistance in *Salmonella*-infected mice, *J. Immunol.* **101**:830–845.

Collins, F. M. and Mackaness, G. B., 1970, The relationship of delayed hypersensitivity to acquired antituberculous immunity. 1. Tuberculin sensitivity and resistance to reinfection in BCG-vaccinated mice, *Cell. Immunol.* **1**:253–265.

Collins, F. M., and Miller, T. E., 1969, Growth of drug-resistant strains of *M. bovis* (BCG) in normal and immunized mice, *J. Infect. Dis.* **120**:517–533.

Collins, F. M., and Montalbine, V., 1975, Relative immunogenicity of streptomycin-susceptible and resistant strains of BCG. II. Effect of the route of inoculation on growth and immunogenicity, *Am. Rev. Respir. Dis.* **111**:43–51.

Collins, F. M., and Montalbine, V., 1976, Distribution of Mycobacteria grown *in vivo* in the organs of intravenously infected mice, *Am. Rev. Respir. Dis.* **113**:281–286.

Collins, F. M., and Morrison, N. E., 1979, Restoration of T-cell responsiveness by thymosin: Expression of antituberculous immunity in mouse lungs, *Infect. Immun.* **23**:330–335.

Collins, F. M., and Niederbuhl, C. J., 1983, Mechanisms of defense against intestinal pathogens: A review, in: *Experimental Bacterial and Parasitic Infections* (G. Keusch and T. W. Wadstrom, eds.), pp. 181–194, Elsevier Press, New York.

Collins, F. M., and Poulter, L. W., 1981, Effector and escape mechanisms in tuberculosis and leprosy, in: *Immunological Aspects of Leprosy, Tuberculosis, and Leishmaniasis* (D. P. Humber, ed.), pp. 97–113, Excerpta Medica, Amsterdam.

Collins, F. M., and Smith, M. M., 1969, A comparative study of the virulence of *Mycobacterium tuberculosis* measured in mice and guinea pigs, *Am. Rev. Respir. Dis.* **100:**631–639.

Collins, F. M., and Stokes, R. W., 1987, *M. avium*–complex infections in normal and immunodeficient mice, *Tubercle* **68:**127–136.

Collins, F. M., Wayne, L. G., and Montalbine, V., 1974, The effect of cultural conditions on the distribution of *M. tuberculosis* in the spleens and lungs of specific pathogen-free mice, *Am. Rev. Respir. Dis.* **110:**147–156.

Collins, F. M., Congdon, C. C., and Morrison, N. E., 1975, Growth of *Mycobacterium bovis* (BCG) in T-lymphocyte depleted mice, *Infect. Immun.* **11:**57–64.

Collins, F. M., Auclair, L. K., and Mackaness, G. B., 1977, *Mycobacterium bovis* (BCG) infection of the lymph nodes of normal, immune and cortisone-treated guinea pigs, *J. Natl. Cancer Inst.* **59:**1527–1535.

Collins, F. M., Morrison, N. E., and Montalbine, V., 1978, Immune response to persistent mycobacterial infection in mice, *Infect. Immun.* **20:**430–438.

Collins, F. M., Morrison, N. E., Dhople, A. M., and Watson, S. R., 1980, Microscopic counts carried out on *M. leprae* and *M. tuberculosis* suspensions, *Int. J. Leprosy* **48:**402–407.

Collins, F. M., Morrison, N. E., and Watson, S. R., 1983, Fernandez and Mitsuda reactivity in guinea pigs sensitized with heat-killed *M. leprae:* Persistence and specificity of skin reactivity to soluble and particulate antigens, *Int. J. Leprosy* **51:**481–489.

Comstock, G. W., 1982, Epidemiology of tuberculosis, *Am. Rev. Respir. Dis.* **125:**S8–15.

Comstock, G. W., and Edwards, P. Q., 1972, An American view of BCG vaccination illustrated by results of a controlled trial in Puerto Rico, *Scand. J. Respir. Dis.* **53:**207–217.

Comstock, G. W., and Edwards, P. Q., 1975, The competing risks of tuberculosis and hepatitis for adult tuberculin reactors, *Am. Rev. Respir. Dis.* **111:**573–577.

Convit, J., and Ulrich, M., 1976, Recent advances in the immunology of leprosy, *Int. J. Dermatol.* **15:**157–170.

Convit, J., and Ulrich, M., 1978, General ideas concerning a vaccine against leprosy: A basis for discussion during the eleventh International Leprosy Congress, *Int. J. Leprosy* **46:**61–63.

Convit, J., Pinardi, M. E., Rodriguez-Ochoa, G., Ulrich, M., Kavila, J. L. and Goihman-Yahr, M., 1974, Elimination of *M. leprae* subsequent to local *"in vivo"* activation of macrophages in lepromatous leprosy by other Mycobacteria, *Clin. Exp. Immunol.* **17:**261–265.

Convit, J., Aranzazu, N., Pinardi, M., and Ulrich, M., 1979, Immunological changes observed in indeterminate and lepromatous leprosy patients and Mitsuda-negative contacts after inoculation of a mixture of *M. leprae* and BCG, *Clin. Exp. Immunol.* **36:**214–220.

Convit, J., Ulrich, M., and Aranzazu, N., 1980, Vaccination in leprosy—Observations and interpretations, *Int. J. Leprosy.* **48:**62–65.

Convit, J., Aranzazu, N., Ulrich, M., Pinardi, M., Reyes, O., and Alvarado, J., 1982, Immunotherapy with a mixture of *M. leprae* and BCG in different forms of leprosy and in Mitsuda-negative contacts, *Int. J. Leprosy.* **50:**415–424.

Dalhstrom, G., and Sjogren, I., 1977, Side effects of BCG vaccination, *J. Biol. Standard* **5:**147–148.

Daniel, T. M., 1980, The immunology of tuberculosis, *Clin. Chest Med.* **1:**189–201.

Daniel, T. M., and Janicki, B. W., 1978, Mycobacterial antigens: A review of their isolation, chemistry and immunological properties, *Microbiol. Rev.* **42:**84–113.

Daniel, T. M., Oxtoby, M. J., Pinto, E., and Moreno, E. J., 1981, The immune spectrum in patients with pulmonary tuberculosis, *Am. Rev. Respir. Dis.* **123:**556–559.

Dannenberg, A.M., 1982, Pathogenesis of pulmonary tuberculosis, *Am. Rev. Respir. Dis.* **125:**S25–29.

Dannenberg, A.M. and Sugimoto, M., 1976, Liquefaction of caseous foci in tuberculosis, *Am. Rev. Respir. Dis.* **113:**257–259.

Davidson, P. T., 1979, The management of disease with atypical mycobacteria, *Clin. Notes Resp. Dis.* **18**:3–13.

Draper, P., 1983, The bacteriology of *M. leprae*, *Tubercle* **64**:43–56.

Dubos, R. J., and Dubos, J., 1952, *The White Plague: Tuberculosis, Man and Society*, Little, Brown, Boston.

Dubos, R. J., and Pierce, C. H., 1956a, Differential characteristics *in vitro* and *in vivo* of substrains of BCG. I. Multiplication and survival *in vitro*, *Am. Rev. Tuberc.* **74**:655–666.

Dubos, R. J. and Pierce, C. H., 1956b, Differential characteristics *in vitro* and *in vivo* of substrains of BCG. IV. Immunizing effectiveness, *Am. Rev. Tuberc.* **74**:655–666.

Dunbar, F. P., Pejovic, J., Cacciatore, R., Peric-Golia, L., and Runyon, E. H., 1968, *Mycobacterium intracellulare*. Maintenance of pathogenicity in relation to lyophilization and colony form, *Scand. J. Respir. Dis.* **49**:153–162.

Edelson, P. J., 1982, Intracellular parasites and phagocytic cells: Cell biology and pathophysiology, *Rev. Infect. Dis.* **4**:124–135.

Edwards, L. B., Acquaviva, F. A., Liversay, V. T., Cross, F. W., and Palmer, C. E., 1969, An atlas of sensitivity to tuberculin, PPD-B and histoplasmin in the United States, *Am. Rev. Respir. Dis.* **99**:51–132.

Edwards, M. L., Goodrich, J. M., Muller, D., Pollack, A., Ziegler, J. E., and Smith, D. W., 1982, Infection with *M. avium–intracellulare* and the protective effects of bacille Calmette-Guérin, *J. Infect. Dis.* **145**:733–741.

Eickhoff, T. C., 1977, The current status of BCG immunization against tuberculosis, *Annu. Rev. Med.* **28**:411–428.

Elliott, P. R., Buraham, W. R., White, S., and Stanford, J. L., 1980, Further data on skin testing with mycobacterial antigens in inflammatory bowel disease, *Lancet* **2**:483–484.

Engbaek, H. C., and Jespersen, A., 1966, Effect of BCG vaccination on experimental infection with *M. avium*, *Acta Path. Microbiol. Scand.* **67**:503–513.

Falkinham, J. O., Parker, B. C., and Gruft, H., 1980, Epidemiology of infection by nontuberculous mycobacteria. 1. Geographic distribution in the Eastern United States, *Am. Rev. Respir. Dis.* **121**:931–937.

Farer, L. S., Lowell, A. M., and Meador, M. P., 1979, Extrapulmonary tuberculosis in the United States, *Am. J. Epidemiol.* **109**:205–217.

Fauci, A. S., 1984, Acquired immunodeficiency syndrome: Epidemiologic, clinical, immunologic and therapeutic considerations, *Ann. Intern. Med.* **100**:92–106.

Ferebee, S. H., 1967, An epidemiological model of tuberculosis in the United States, *Natl. Tuberc. Assoc. Bull.* **53**:5186–5189.

Fieldsteel, A. H., Becker, F. A., and Stout, J. G., 1977, Prolonged survival of virulent *Treponema pallidum* (Nichols strain) in cell-free and tissue culture systems, *Infect. Immun.* **18**:173–182.

Fieldsteel, H., and Levy, L., 1976, Neonatally thymectomized Lewis rats infected with *Mycobacterium leprae:* Response to primary infection, secondary challenge and large inocula, *Infect. Immun.* **14**:736–741.

Fine, P. E. M., 1981, Immunogenetics of susceptibility to leprosy, tuberculosis, and leishmaniasis. An epidemiological perspective, *Int. J. Leprosy* 49:437–454.

Fine, P., 1984, Leprosy and tuberculosis—An epidemiological comparison, *Tubercle* **65**:137–153.

Fine, P. E. M., Wolf, E., Prichard, J., Watson, B., Bradley, D., Festenstein, H., and Chacks, C. S. G., 1979, HLA-linked genes and leprosy: A family study in Karigiri, South India, *J. Infect. Dis.* **140**:152–161.

Fitch, F. W., 1981, T-lymphocyte clones having defined immunological functions, *Transplantation* **32**:171–176.

Fok, J. S., Ho, R. S., Arora, P. K., Harding, G .E., and Smith, D. W., 1976, Host-parasite relationships in experimental airborne tuberculosis. IV. Lack of hematogenous dissemination of *Mycobacterium tuberculosis* to the lungs in animals vaccinated with bacille Calmette–Guérin, *J. Infect. Dis.* **133**:137–144.

Fox, W., 1979, The chemotherapy of pulmonary tuberculosis: A review, *Chest* **76**:S785–796.

Francis, J., 1958, *Tuberculosis in Animals and Man*, Cassell, London.

Freudenstein, H., Pranter, W., and Schweinsberg, H., 1979, Assessment of several BCG vaccines in different animal test systems, *J. Biol. Stand.* **7**:203–212.

Ganguly, N. K., Kumar, B., Kaur, S., Sharma, S., Kaur, M., and Chukravanti, R. K., 1981, BCG induced immunity to *M. leprae* in mice, *Leprosy in India* **53**:57–62.

Gardner, I. D., 1980, The effect of aging on susceptibility to infection, *Rev. Infect. Dis.* **2**:801–810.

Gatner, E. M. S., and Anderson, R., 1982, Immune responses and immunostimulation in tuberculous therapy, *S. Afr. Med. J.* **62**:707–710.

Gaugas, J. M., Rees, R. J. W., Weddell, A. G., and Palmer, E., 1971, Reversal effect of thymus grafts on lepromatous leprosy in thymectomized-irradiated mice, *Int. J. Leprosy* **39**:388–395.

George, J. T. A., and Payne, D. J. H., 1961, Tuberculosis from tuberculin tested milk with a note on the frequency of *Brucella abortus* in consumer milk, *Monthly Bull. Ministry Health* **20**:49–102.

Gershon, R. K., and Kondo, K., 1971, Infectious immunological tolerance, *Immunology* **21**:903–914.

Gheorghia, M., and Lagrange, P. H., 1983, Viability, heat stability and immunogenicity of 4 BCG vaccines prepared from 4 different BCG strains, *Ann. Immunol.* **134C**:125–148.

Glassroth, J., Robins, A. G., and Snider, D. E., 1980, Tuberculosis in the 1980's, *N. Engl. J. Med.* **302**:1441–1450.

Godal, T., and Negassi, K., 1973, Subclinical infection in leprosy, *Br. Med. J.* **2**:557–559.

Godal, T., Myrvang, B., Stanford, J. L. and Samuel, D. R., 1974, Recent advances in the immunology of leprosy with special reference to new approaches in immunoprophylaxis, *Bull. Inst. Past.* **72**:273–310.

Goihman-Yahr, M., and Convit, J., 1972, Cross-reactivity of *M. leprae* and BCG, *Int. J. Leprosy* **40**:62–67.

Goslee, S., and Wolinsky, E., 1976, Water as a source of potentially pathogenic mycobacteria, *Am. Rev. Respir. Dis.* **113**:287–292.

Goto, Y., Nakamura, R. M., Takahashi, H., and Tokunaga, T., 1984, Genetic control of resistance to *M. intracellulare* infection in mice, *Infect. Immun.* **46**:135–140.

Grange, J. M., Gibson, J., Osborn, T. W., Collins, C. H., and Yates, M. D., 1983, What is BCG? *Tubercle* **64**:129–139.

Gray, D. F., 1958, Immunity, natural anergy and artificial desensitization in experimental tuberculosis, *Am. Rev. Tuberc.* **78**:235–250.

Green, G. M., Jakab, G. J., Low, R. B., and Davis, G. S., 1977, Defense mechanisms of the respiratory membrane, *Am. Rev. Respir. Dis.* **115**:479–514.

Gros, P., Skamene, E., and Forget, A., 1983, Cellular mechanisms of genetically controlled host resistance to *M. bovis* (BCG), *J. Immunol.* **131**:1966–1972.

Grover, A. A., Kim, H. K., Wiegeshaus, E. H., and Smith, D. W., 1967, Host-parasite relationships in experimental airbourne tuberculosis. II. Reproducible infection by means of an inoculum preserved at −70°C, *J. Bacteriol.* **94**:832–835.

Guérin, C., 1957, The early history of BCG, in: *BCG Vaccination against Tuberculosis* (S. R. Rosenthal, ed.), pp. 48–53, Little, Brown, Boston.

Guinto, R. S., 1968, Skin tests in leprosy, *Ann. NY Acad. Sci.* **154**:149–156.

Gwadz, R. W., Cochrane, A. H., Nussenzweig, V., and Nussenzweig, R. S., 1979, Preliminary studies on the vaccination of Rhesus monkeys with irradiated sporozoites of *Plasmodium knowlesi* and characterization of surface antigens of these parasites, *Bull. World Health Organ.* **51**-(Suppl):165–173.

Hank, J. A., Chan, J. K., Edwards, M. L., Muller, D., and Smith, D. W., 1981, Influence of the virulence of *M. tuberculosis* on protection by bacille Calmette–Guérin in guinea pigs, *J. Infect. Dis.* **143**:734–738.

Harboe, M., 1982, Significance of antibody studies in leprosy and experimental models of the disease, *Int. J. Leprosy* **50**:342–350.

Hart, P. D'A., 1967, Efficacy and applicability of mass BCG vaccination in tuberculosis control, *Br. Med. J.* **1**:587–592.

Hart, P. D'A., Miller, C. L., and Sutherland, I., 1962, Sensitivity to avian and human Old Tuberculin in man in Great Britain, *Tubercle* **43**:268–286.

Hastings, R. C., and Morales, M. J., 1982, Observations, calculations and speculations on the growth and death of *M. leprae in vivo*, *Int. J. Leprosy* **50**:579–582.

Hastings, R. C., Chehl, S. K., Black, G., Morales, M. J., Shannon, E. J., and Kirchheimer, W. F., 1981, Growth of *M. leprae* in nude (athymic) mice, in: *Immunological Aspects of Leprosy, Tuberculosis, and Leishmaniasis* (D. P. Humber, ed.), pp. 166–176, Excerpta Medica, Amsterdam.

Hehre, E., and Freund, J., 1939, Sensitization, antibody formation and lesions produced by tubercle bacilli in the albino rat, *Arch. Pathol.* **27**:289–306.

Heilmann, K., Brehmer, W., and Schraegle, H., 1975, Fatal course of generalized BCG histocytosis in congenital immune deficiency, *Deutsche Med. Wochenschr.* **100**:2604–2610.

Hepper, K. P., and Collins, F. M., 1984a, Immune responsiveness in mice heavily infected with *Mycobacterium kansasii*, *Immunology* **53**:357–364.

Hepper, K. P., and Collins, F. M., 1984b, Adoptive transfer of acquired resistance to *M. kansasii* infections by splenic T-cells harvested from chronically infected mice, *Immunology* **53**, 819–825.

Hersch, E. M., Gutterman, J. V., and Mavligit, G. M., 1977, BCG as adjuvant immunotherapy for neoplasia, *Annu. Rev. Med.* **28**:489–516.

Hershfield, E. S., 1979, Tuberculosis in the world, *Chest* **76**:S805–811.

Hesselberg, I., 1970, Standardization of freeze-dried BCG vaccine by means of vaccination tests, *Scand. J. Respir. Dis.* **72**:S53–58.

Hewitt, J., Coates, A. R. M., Mitchison, D. A., and Ivanyi, J., 1982, The use of murine monoclonal antibodies without purification of antigen in the serodiagnosis of tuberculosis, *J. Immunol. Methods* **55**:205–211.

Hock, G. E., and Loy, F. K., 1981, A dynamic model of tuberculosis epidemiology for Singapore, *Ann. Acad. Med. Sing.* **10**:40–49.

Holm, J., 1969, Tuberculosis in the world today, *Bull. Int. Union Tuberc.* **43**:26–33.

Hormaeche, C. E., Fuhrenbrog, M. C., Pettifor, R. A., and Brock, J., 1981, Acquired immunity to *S. typhimurium* and delayed (footpad) hypersensitivity in BALB/c mice, *Immunology* **43**:547–554.

Horwitz, O., 1973, Disease, cure and death: Epidemiologic and clinical parameters for chronic diseases illustrated by a model–Tuberculosis, *Am. J. Epidemiol.* **97**:148–159.

Horwitz, O., Edwards, P. Q., and Lowell, A. M., 1973, National tuberculosis control program in Denmark and the United States, *Public Health Rep.* **88**:493–498.

Houk, V. N., 1980, Spread of tuberculosis via recirculated air in a naval vessel: The Byrd study, *Ann. NY Acad. Sci.* **353**:10–24.

Howard, J. C., 1984, The T-cell antigen receptor—Paradigm regained, *Immunol. Today.* **5**:188–189.

Hsu, H. S., Miller, K. B., and Nakoneczua, I., 1980, The role of delayed hypersensitivity in the enhancement of host resistance to infection, *Can. J. Microbiol.* **26**:1438–1442.

Huang, C. L-H., 1980, The transmission of leprosy in man, *Int. J. Leprosy.* **48**:309–318.

Janicki, B. W., Good, R. C., Minden, P., Affronti, L. F., and Hymes, W. F., 1973, Immune responses in Rhesus monkeys after bacillus Calmette–Guérin vaccination and aerosol challenge with *M. tuberculosis*, *Am. Rev. Respir. Dis.* **107**:359–366.

Jenkins, P. A., 1981, Nontuberculous mycobacteria and disease, *Eur. J. Respir. Dis.* **62**:69–71.

Job, C. K., Sanchez, R. M., Kirchheimer, W. F., and Hastings, R. C., 1984, Attempts to breed the nine-banded armadillo (*Dasypus novemcinctus*) in captivity—A preliminary report, *Int. J. Leprosy* **52**:362–364.

Kaltreider, H. B., 1976, Expression of immune mechanisms in the lung, *Am. Rev. Respir. Dis.* **113**:347–379.

Kantor, F. S., 1975, Infection, anergy and cell-mediated immunity, *N. Engl. J. Med.* **292**:629–634.

Katz, J. and Kunofsky, S., 1960, Environmental versus constitutional factors in the development of tuberculosis among Negroes, *Am. Rev. Respir. Dis.* **81**:17–25.

Kaufmann, S. H. E., and Hahn, H., 1982, Biological functions of T-cell lines with specificity for the intracellular bacterium *L. monocytogenes in vitro* and *in vivo*, *J. Exp. Med.* **155**:1754–1764.

Kaufmann, S. H. E., Simon, M. M., and Hahn, H., 1979, Specific Lyt 123 T-cells are involved in protection against *L. monocytogenes* and in delayed type hypersensitivity to listerial antigens, *J. Exp. Med.* **150**:1033–1038.

Kim, T. H., and Kubica, G. P., 1972, Long-term preservation and storage of mycobacteria, *Appl. Microbiol.* **24**:311–317.

Kirchheimer, W. F., and Sanchez, R. M., 1976, Quantitation aspects of leprosy in armadillos, *Int. J. Leprosy* **44**:84–87.

Kirchheimer, W. F., and Storrs, E. E., 1971, Attempts to establish the armadillo (*Dasypus novemcinctus Linn*) as a model for the study of leprosy, *Int. J. Leprosy* **39:**693–702.

Koch, R., 1890, Weitere Mittheilung uber ein Heilmittel gegen Tuberkulose, *Deutsch Med. Wochenschr.* **16:**1029–1032.

Kumta, N. B., and Sureka, P. R., 1980, Current status of BCG vaccination, *Indian J. Pediatr.* **47:**449–455.

Kvach, J. T., Mungnia, G., and Strand, S. H., 1984, Staining tissue-derived *M. leprae* with fluorescein diacetate and ethidium bromide, *Int. J. Leprosy* **52:**176–182.

Ladefoged, A., Bunch-Christensen, K., and Guld, J., 1976, Tuberculin hypersensitivity in guinea pigs after vaccination with varying doses of BCG of 12 different strains, *Bull. World Health Organ.* **53:**435–444.

Lafferty, K. J., Andras, L., and Prowse, S. J., 1980, Role of lymphokine and antigen in control of specific T-cell responses, *Immunol. Rev.* **51:**279–314.

Lagrange, P. H., 1978, Comparative studies of different strains of BCG vaccine in mice: T-cell dependent immune responses, *Dev. Biol. Standard* **38:**223–229.

Lagrange, P. H., Miller, T. E., and Mackaness, G. B., 1976, Parameters conditioning the potentiating effect of BCG as the immune response, in: *BCG in Cancer Immunotherapy* (G. Lamoureux, R. Turcotte, and V. Portelance, eds.), pp. 23–36, Grune & Stratton, New York.

Lagrange, P. H., Hurtrel, B., and Thickstun, P. M., 1979, Immunological behavior after mycobacterial infection in selected lines of mice with high and low antibody responses, *Infect. Immun.* **25:**39–47.

Lagrange, P. H., Hurtrel, B., Brandely, M., and Thickstun, P. M., 1983, Experimental evaluation of factors conditioning the protective immunity in mycobacterial infections and related consequences for vaccination in humans, in: *Experimental Bacterial and Parasitic Infections* (G. Keusch and T. Wadstrom, eds.), pp. 417–430, Elsevier Press, New York.

Lavy, A., and Yoshpe-Parer, Y., 1982, Isolation of *M. simiae* from clinical specimens in Israel, *Tubercle* **63:**279–285.

Ledbetter, J. A., Rouse, R. V., Micklem, H. S., and Herzenberg, L. A., 1980, T-cell subsets defined by expression of Lyt 123 and Thy-1 antigens. Two parameter immunofluorescence and cytotoxicity analysis with monoclonal antibodies modified current views, *J. Exp. Med.* **152:**280–295.

Lefford, M. J., 1975a, Delayed hypersensitivity and immunity in tuberculosis, *Am. Rev. Respir. Dis.* **111:**243–246.

Lefford, M. J., 1975b, Transfer of adoptive immunity to tuberculosis in mice, *Infect. Immun.* **11:**1174–1181.

Lefford, M. J., 1980, Properties of peritoneal exudate lymphocytes that mediate tuberculin delayed-type hypersensitivity and anti-tuberculous immunity. II. Velocity sedimentation analysis, *Immunology* **41:**634–651.

Lefford, M. J., 1981, Immunology of *Mycobacterium tuberculosis, Comprehens. Immunol.* **8:**345–368.

Lefford, M. J., 1983, Suppression of adoptive antituberculous immunity by normal recipient animals, *Infect. Immun.* **41:**257–263.

Lefford, M. J., McGregor, D. D., and Mackaness, G. B.. 1973, Immune response to *M. tuberculosis* in rats, *Infect. Immun.* **8:**182–189.

Lenzini, L., Rottoli, P., and Rottoli, L., 1977, The spectrum of human tuberculosis, *Clin. Exp. Immunol.* **27:**230–237.

Levy, L., 1976, Studies of the mouse footpad technique for cultivation of *M. leprae*. 3. Doubling time during logarithmic multiplication, *Leprosy Rev.* **47:**103–108.

Levy, F. M., Conge, G. A., Pasquier, J. F., Mauss, H., Dubos, R. J., and Schaedler, R. N., 1961, The effect of BCG vaccination on the fate of virulent tubercle bacilli in mice, *Am. Rev. Respir. Dis.* **84:**28–36.

Levy, F. M., Mande, R., Conge, G., Fillastre, C., and Orssand, E., 1968, Perspectives for BCG standardization, *Adv. Tuberc. Res.* **16:**63–176.

Lincoln, E. M., and Gilbert, L. A., 1972, Disease in children due to mycobacteria other than *M. tuberculosis, Am. Rev. Respir. Dis.* **105:**683–714.

Lotte, A., Wasz-Hochhert, O., Lert, F., Dumitrescu, N., and Poisson, N., 1979, Balance of risks and balance of costs regarding tuberculosis and antituberculous vaccination, *Dev. Biol. Stand.* **43:**111–119.

Løvik, M., and Closs, O., 1982, Repeated delayed-type hypersensitivity reactions against *M. lepraemurrium* antigens at the infection site do not affect bacillary multiplication in C3H mice, *Infect. Immun.* **36**:768–774.

Lowe, J., and McNulty, E., 1953, The effect of BCG in lepromatous cases of leprosy, *Int. J. Leprosy.* **21**:173–177.

Lowell, A. M., 1984, Tuberculosis: its social and economic impact and some thoughts on epidemiology, in: *The Mycobacteria—a Source Book* (G. P. Kubica and L. G. Wayne, eds.), pp. 1021–1055, Marcel Dekker, New York.

Lowrie, D. B., 1983, How macrophages kill tubercle bacilli, *J. Med. Microbiol.* **16**:1–12.

Luelmo, F., 1982, BCG vaccination, *Am. Rev. Respir. Dis.* **125**:S70–72.

Lurie, M. B., 1928, The fate of human and bovine tubercle bacilli in various organs of the rabbit, *J. Exp. Med.* **48**: 155–182.

Lurie, M. B., 1929, The fate of tubercle bacilli in the organs of reinfected rabbits, *J. Exp. Med.* **50**:747–765.

Lurie, M. B., 1942, Studies on the mechanism of immunity in tuberculosis: The fate of tubercle bacilli ingested by mononuclear phagocytes derived from normal and immunized animals, *J. Exp. Med.* **75**:247–267.

Lurie, M. B., Abramson, S., and Heppleston, A. G., 1952, On the response of genetically resistant and susceptible rabbits to the quantitative inhalation of human type tubercle bacilli and the nature of resistance to tuberculosis, *J. Exp. Med.* **95**:119–134.

MacDonald, T. T., and Carter, P. B., 1980, Cell-mediated immunity to intestinal infection, *Infect. Immun.* **28**:516–523.

Mackaness, G. B., 1954, The growth of tubercle bacilli in monocytes from normal and vaccinated rabbits, *Am. Rev. Tuberc.* **69**:495–504.

Mackaness, G. B., 1964, The immunological basis of acquired cellular resistance, *J. Exp. Med.* **120**:105–120.

Mackaness, G. B., 1967, The relationship of delayed hypersensitivity to acquired cellular resistance, *Br. Med. Bull.* **23**:52–54.

Mackaness, G. B., 1968, The immunology of antituberculous immunity, *Am. Rev. Respir. Dis.* **97**:337–344.

Mackaness, G. B., 1969, The influence of immunologically committed lymphoid cells on macrophage activity *in vivo*, *J. Exp. Med.* **129**:973–992.

Mackaness, G. B., 1970, The monocyte in cellular immunity, *Semin. Hematol.* **7**:172–184.

Mackaness, G. B., 1971a, Resistance to intracellular infection, *J. Infect. Dis.* **123**:439–445.

Mackaness, G. B., 1971b, The induction and expression of cell-mediated hypersensitivity in the lung, *Am. Rev. Respir. Dis.* **104**:813–828.

Mackaness, G. B., 1974, Delayed hypersensitivity in lung diseases, *Ann. NY Acad. Sci.* **221**:312–316.

Mackaness, G. B., and Blanden, R. V., 1967, Cellular immunity, *Prog. Allergy* **11**:89–140.

Mackaness, G. B., Auclair, D. J. and Lagrange, P. H., 1973, Immunopotentiation with BCG. 1. Immune response to different strains and preparations, *J. Natl. Cancer Inst.* **51**:1655–1667.

Mage, M., Mathieson, B., Sharrow, S., McHugh, L., and Hammerling, V., 1981, Preparative nonlytic separation of Lyt2^{+} and Lyt2^{-} T-lymphocytes, functional analyses of the separated cells and demonstration of synergy in graft-vs-host reaction of Lyt2^{+} and Lyt2^{-} cells, *Eur. J. Immunol.* **11**:228–235.

Mastrangelo, M. J., Berd, D., and Bellet, R. E., 1976, Critical review of previously reported clinical trials of cancer immunotherapy with non-specific immunostimulants, *Ann. NY Acad. Sci.* **227**:94–123.

Mathé, G., 1977, Interspersion of chemotherapy and active systemic immunotherapy. Experimental and rational bases, *Cancer Immunol. Immunother.* **3**:1–3.

McCune, R. M., Feldman, F. M., Lambert, H. P., and McDermott, W., 1966, Microbial persistence. 1. The capacity of tubercle bacilli to survive sterilization in mouse tissues, *J. Exp. Med.* **123**:455–468.

McGregor, D. D., and Kostiala, A. A. I., 1976, Role of lymphocytes in cellular resistance to infection, *Contemp. Topics Immunobiol.* **5**:237–266.

McMurray, D. N., and Echeverri, A., 1978, Cell-mediated immunity in anergic patients with pulmonary tuberculosis, *Am. Rev. Respir. Dis.* **118**:827–834.

Mehra, V., and Bloom, B. R., 1979, Induction of cell-mediated immunity to *Mycobacterium leprae* in guinea pigs, *Infect. Immun.* **23**:787–794.

Meissner, G., and Anz, W., 1977, Sources of *M. avium* complex infection resulting in human disease, *Am. Rev. Respir. Dis.* **116**:1057–1064.

Merkal, R. S., 1984, Paratuberculosis, in: *The Mycobacteria—a Source Book* (G. P. Kubica and L. G. Wayne, eds.), pp. 1237–1249, Marcel Dekker, New York.

Meyers, W. M., Walsh, G. P., Brown, H. L., Burford, C. H., Gerone, P. J., Wolf, R. H., Gormus, B. J., and Martin L., 1981, Leprosy in the Mangabey monkey, *Int. J. Leprosy* **49**:500–502.

Miller, T. E., Mackaness, G. B., and Lagrange, P. H., 1973, Immunopotentiating with BCG. II. Modulation of the response to sheep red blood cells. *J. Natl. Cancer Inst.* **51**:1669–1676.

Mitchell, G. F., 1980, Responses to infection with metazoan and protozoan parasites in man, *Adv. Immunol.* **28**:451–511.

Mitchison, D. A., 1964, The virulence of tubercle bacilli from patients with pulmonary tuberculosis in India and other countries, *Bull. Intern. Union Tuberc.* **35**:287–305.

Mitchison, D. A., 1985, The action of antituberculosis drugs in short course chemotherapy, *Tubercle* **66**:219–225.

Mitchison, D. A., Selkon, J. B., and Lloyd, J., 1963, Virulence in the guinea pig, susceptibility to hydrogen peroxide and catalase activity of isoniazid sensitive tubercle bacilli from South Indian and British patients, *J. Pathol. Bact.* **86**:377–386.

Mitsuyama, M., Nomoto, K., and Takeya, K., 1982, Direct correlation between delayed footpad reaction and resistance to local bacterial infection, *Infect. Immun.* **36**:72–79.

Morrison, N. E., and Collins, F. M., 1975, Immunogenicity of an aerogenic BCG vaccine in T-cell depleted and normal mice, *Infect. Immun.* **11**:1110–1121.

Morrison, N. E., and Collins, F. M., 1976, Restoration of T-cell responsiveness by thymosin: Development of antituberculous resistance in BCG-infected animals, *Infect. Immun.* **13**:554–563.

Myrvik, Q. N., 1974, Adjuvants, *Ann. NY Acad. Sci.* **221**:324–330.

Naganna, K., 1974, Some of the BCG trials and certain aspects involved in them, *Am. Rev. Respir. Dis.* **109**:497–499.

Nakajima, S., Kobayashi, S., Nohara, M., and Sato, S., 1977, HLA-antigen and susceptibility to leprosy, *Int. J. Leprosy.* **45**:273–277.

Nash, D. R., and Douglas, J. E., 1980. Anergy in active pulmonary tuberculosis. A comparison between positive and negative reactors and an evaluation of 5 TU and 25 TU skin test doses, *Chest* **77**:32–37.

Nelson, D. S., 1981, Macrophages: progress and problems, *Clin. Exp. Immunol.* **45**:225–233.

Neveu, P. J., 1976, The effects of BCG on both cellular and humoral immunity during the early response to a hapten carrier complex, *Clin. Exp. Immunol.* **26**:169–172.

North, R. J., 1973a, Cellular mediators of anti-Listeria immunity as an enlarged population of short-lived, replicating T-cells, *J. Exp. Med.* **138**:342–355.

North, R. J., 1973b, Importance of thymus-derived lymphocytes in cell-mediated immunity to infection, *Cell. Immunol.* **7**:166–176.

North, R. J., 1974a, T-cell dependence of macrophage activation and mobilization during infection with *M. tuberculosis*, *Infect. Immun.* **10**:66–71.

North, R. J., 1974b, Cell-mediated immunity and the response to infection, in: *Mechanisms of Cell Mediated Immunity* (R. T. McLuskey and S. Cohen, eds.), pp. 185–200, Wiley, New York.

North, R. J., 1978, The concept of the activated macrophage, *J. Immunol.* **121**:806–808.

North, R. J., 1981, Immunity to *Listeria monocytogenes*, *Comprehens. Immunol.* **8**:201–219.

North, R. J., and Deissler, J. F., 1975, Nature of "memory" in T-cell-mediated antibacterial immunity: cellular parameters that distinguish between active immune response and a state of "memory," *Infect. Immun.* **12**:761–767.

North, R. J., and Spitalny, G., 1974, Inflammatory lymphocyte in cell-mediated antibacterial immunity: Factors governing the accumulation of mediator T-cells in peritoneal exudates, *Infect. Immun.* **10**:489–498.

North, R. J., Mackaness, G. B., and Elliott, R. W., 1972, The histogenesis of immunologically committed lymphocytes, *Cell. Immunol.* **3**:680–694.

North, R. J., Dye, E. S., Mills, C. D., and Chandler, J. P., 1982, Modulation of antitumor immunity-immunobiologic approaches, *Springer Sem. Immunopathol.* **5**:193–220.

Nyboe, J., and Bunch-Christensen, K., 1966, Assay in man of different BCG products, *Bull. World Health Organ.* **35**:645–650.
Nyka, W., 1963, Studies on *M. tuberculosis* in lesions in the human lung, *Am. Rev. Respir. Dis.* **88**:670–679.
Ogawa, T., and Motomura, K., 1970, Studies on immune leprosy bacillus. 1. Attempt to cultivate *in vitro* the Hawaiian strain of *M. lepramurium, Kitasato Arch. Exp. Med.* **43**:21–36.
Ogawa, T., and Motomura, K., 1971, Studies on murine leprosy bacillus. IV. Further report on primary *in vitro* isolation, subcultivation, reproduction of the disease in mice of slow growing acid-fast organisms supposedly murine leprosy bacillus, *Kitasato Arch. Exp. Med.* **44**:167–183.
Orme, I. M., and Collins, F. M., 1983, Protection against *M. tuberculosis* infection by adoptive immunotherapy, *J. Exp. Med.* **158**:74–83.
Orme, I. M., and Collins, F. M., 1984, Adoptive protection of the *M. tuberculosis*-infected lung. Dissociation between cells that passively transfer protective immunity and those that transfer delayed-type hypersensitivity to tuberculin, *Cell. Immunol.* **84**:113–120.
Orme, I. M., and Collins, F. M., 1986a, Aerogenic vaccination with *M. bovis* BCG, *Tubercle* **67**:133–140.
Orme, I. M., and Collins, F. M., 1986b, Cross-protection against nontuberculous mycobacterial infections by *M. tuberculosis* memory immune T lymphocytes, *J. Exp. Med.* **163**:203–208.
Orme, I. M., Stokes, R. W., and Collins, F. M., 1986, Genetic control of natural resistance to nontuberculous mycobacterial infections in mice, *Infect. Immun.* **54**:56–62.
Ortbals, D., and Marr, J., 1978, A comparative study of tuberculosis and other mycobacterial infections and their association with malignancy, *Am. Rev. Respir. Dis.* **117**:39–45.
Osborne, T. W., 1983, Changes in BCG strains, *Tubercle* **64**:1–13.
Palmer, C. E., and Edwards, L. B., 1967, The tuberculin test in retrospect and prospect, *Arch. Environ. Health* **15**:792–808.
Palmer, C. E., and Long, M. W., 1966, Effect of infection with atypical Mycobacteria on BCG vaccination and tuberculosis, *Am. Rev. Respir. Dis.* **94**:554–568.
Patel, P. J., 1980, Expression of antibacterial resistance at the site of a delayed hypersensitivity reaction, *Infect. Immun.* **29**:59–65.
Patel, P. J., 1981, Antibacterial resistance in mice infected with *M. lepraemurium, Clin. Exp. Immunol.* **45**:654–661.
Pattyn, S. R., 1973, The problem of cultivation of *Mycobacterium leprae, Bull. World Health Organ.* **49**:403–410.
Pattyn, S. R., and Portaels, F., 1980, *In vitro* cultivation and characterization of *M. lepraemurium, Int. J. Leprosy* **48**:7–14.
Pejovic, J., Dunbar, F. P., Cacciatore, R., Valdez, H., and Sanders, E. 1969, Immunity to *Mycobacterium tuberculosis* in mice. Effect of heterologous mycobacteria in relation to quantitative and qualitative differences in their organ localization, *Am. Rev. Respir. Dis.* **99**:786–789.
Pierce, C. H., Dubos, R. J., and Schaefer, W. B., 1956, Differential characteristics *in vitro* and *in vivo* of several substrains of BCG. III. Multiplication and survival *in vivo, Am. Rev. Tuberc.* **74**:683–698.
Plant, J., and Glynn, A. A., 1974, Natural resistance to Salmonella infection, delayed hypersensitivity and Ir genes in different strains of mice, *Nature* **248**:345–347.
Portelance, V., Quevillon, M., and Page, Y.. 1976, Toxic effects of massive doses of BCG in mice, guinea pigs and monkeys, *Cancer Immunol. Immunother.* **1**:63–68.
Poulter, L. W., and Lefford, M. J., 1978, The relationship between delayed type hypersensitivity and the progression of *M. lepraemurium* infection in mice, *Infect. Immun.* **20**:530–540.
Prabhakar, R., Venkataraman, P., Vallishayee, R. S., *et al.*, 1987, Virulence for guinea pigs of tuberculin bacilli isolated from the sputum of participants in the BCG trial—Chingleput district South India, *Tubercle* **68**:3–17.
Price, J. E., 1982, BCG vaccination in leprosy, *Int. J. Leprosy.* **50**:205–212.
Raleigh, J. W., 1984, Chemotherapy of tuberculosis, in: *The Mycobacteria—A Source Book* (G. P. Kubica and L. G. Wayne, eds.), pp. 1007–1020, Marcel Dekker, New York.
Raleigh, J. W., Wichelhausen, R. H., Rado, T. M., and Bates, J. H., 1975, Evidence for infection by two distinct strains of *M. tuberculosis* in pulmonary tuberculosis: Report of 9 cases, *Am. Rev. Respir. Dis.* **112**:497–503.

Rea, T. H., and Levan, N. E., 1977, Current concepts in the immunology of leprosy, *Arch. Dermatol.* **113**:345–352.

Rees, R. J. W., 1969a, New prospects for the study of leprosy in the laboratory, *Bull. World Health Organ.* **40**:785–800.

Rees, R. J. W., 1969b, BCG vaccination in Mycobacterial infections, *Br. Med. Bull.* **25**:183–188.

Rees, R. J. W., and Meade, T. W., 1974, Comparison of the modes of spread and the incidence of tuberculosis and leprosy, *Lancet* **1**:47–48.

Rees, R. J. W., and Weddell, A. G. M., 1968, Experimental models for studying leprosy, *Ann. NY Acad. Sci.* **154**:214–236.

Reggiardo, Z., and Middlebrook, G., 1974, Failure of passive serum transfer of immunity against aerogenic tuberculosis in guinea pigs, *Proc. Soc. Exp. Biol. Med.* **145**:173–175.

Reichman, L. B., 1979, Tuberculin skin testing. The state of the art, *Chest* **76**:S764–770.

Ribi, E., Granger, D. L., Milner, K. C., Yamamoto, K., Strain, S. M., Parker, R., Smith, R. W., Brehmer, W., and Azuman, I., 1982, Induction of resistance to tuberculosis in mice with defined components of mycobacteria and with some unrelated materials, *Immunology* **46**:297–305.

Rich, A. R., 1930, Observations on the relation of allergy to immunity, *Bull. Johns Hopkins Hosp.* **47**:189–214.

Rich, A. R., 1951, *The Pathogenesis of Tuberculosis*, 2nd ed., Charles C. Thomas, Springfield, IL.

Ridley, D. S., and Jopling, W. H., 1966, Classification of leprosy according to immunity, *Int. J. Leprosy* **34**:255–273.

Riley, R. L., Mills, C. D., O'Grady, F., Sultan, L. V., Wittstadt, F., and Shivpurs, D. N., 1962, Infections of air from a tuberculosis ward, *Am. Rev. Respir. Dis.* **85**:511–525.

Romanus, V., 1983, Childhood tuberculosis in Sweden, *Tubercle* 64:101–110.

Rook, G. A. W., 1975, The immunological consequences of antigen overload in experimental mycobacterial infections in mice, *Clin. Exp. Immunol.* **19**:167–178.

Rook, G. A. W., 1983, The immunology of leprosy, *Tubercle* **64**:297–312.

Rook. G. A. W., and Stanford, J. L., 1979, The relevance to protection of three forms of delayed skin-test response evoked by *M. leprae* and other Mycobacteria in mice. Correlation with the classical work in the guinea pig, *Parasite Immunol.* **1**:111–123.

Rook, G. A. W., and Stanford, J. L., 1981, The heterogenicity of the immune response to mycobacteria and the relevance of the common antigens to their pathogenicity, *Ann. Immunol.* **132D**:155–164.

Rook, G. A. W., Bahr, G. M., and Stanford, J. L., 1981, The effect of two distinct forms of cell-mediated response to Mycobacteria on the protective efficacy of BCG, *Tubercle* **62**:63–68.

Rooney, J. J., Croceo, J. A., and Kramer, S., 1976, Further observations on tuberculin reactions in active tuberculosis, *Am. J. Med.* **60**:517–522.

Rosenthal, A. S., 1980, Regulation of immune response—Role of the macrophage, *N. Engl. J. Med.* **303**:1153–1156.

Rouillon, A., and Waaler, H.. 1976, BCG vaccination and epidemiological situation: A decision making approach to the use of BCG, *Adv. Tuberc. Res.* **19**:64–126.

Rozenzweig, D. Y., 1979, Pulmonary mycobacterial infections due to *M. intracellulare–avium* complex, *Chest* **75**:115–119.

Sanchez, Y., Ionescu-Matin, I., Dreesman, G. R., Kramp, N., Six, H. R., Hollinger, F. B., and Melinsk, J. L., 1980, Humoral and cellular immunity to hepatitis B virus-derived antigens: comparative activity of Freund's complete adjuvant, alum and liposomes, *Infect. Immun.* **30**:728–733.

Schwartz, R. H., Yano, A., and Paul, W. E., 1978, Interaction between antigen-presenting cells and primed T-lymphocytes, *Immunol. Rev.* **40**:153–180.

Segal, W., and Bloch, H., 1957, Pathogenic and immunogenic differentiation of *M. tuberculosis* grown *in vitro* and *in vivo*, *Am. Rev. Tuberc.* **75**:495–500.

Seth, V., Kukreja, N., Sundaram, K. R., and Malaviya, A. N., 1981, Delayed hypersensitivity after BCG in preschool children in relation to their nutritional status, *Indian J. Med. Res.* **74**:392–398.

Shepard, C. C., 1960, The experimental disease that follows the injection of human leprosy bacilli into footpads of mice, *J. Exp. Med.* **112**:445–454.

Shepard, C. C., 1975, Vaccination of mice against *M. leprae* infection, *Int. J. Leprosy.* **44**:222–226.

Shepard, C. C., and McRae, D. H., 1965, *M. leprae* in mice: Minimal infectious dose, relationship between staining quality and infectivity and the effect of cortisone, *J. Bacteriol.* **89**:365–372.

Shepard, C. C., Draper, P., Rees, R. J. W., and Lowe, C., 1980, Effect of purification steps on the immunogenicity of *M. leprae, Br. J. Exp. Pathol.* **61:**376–379.

Sher, N. A., Chaparas, S. D., Pearson, J., and Chirigos, M., 1973, Virulence of six strains of *Mycobacterium bovis* (BCG) in mice, *Infect. Immun.* **8:**736–742.

Sher, N. A., Chaparas, S. D., Greenberg, L. E., Merchant, E. B., and Vickers, J. H., 1975, Response of congenitally athymic (nude) mice to infection with *Mycobacterium bovis* (strain BCG), *J. Natl. Cancer Inst.* **54:**1419–1426.

Siebenmann, C. O., and Barbara, C., 1974, Quantitative evaluation of the effectiveness of Connaught freeze-dried BCG vaccine in mice and guinea pigs, *Bull. World Health Organ.* **51:**283–290.

Sigurdsson, S., 1945, Studies on the risk of infection with bovine tuberculosis to the rural population, *Acta Tuberc. Scand.* **15:**S147–154.

Singer, E., 1965, Nonspecific sensitization to Old Tuberculin: asymptomatic infection with mycobacteria, *Tubercle* **46:**270–272.

Skamene, E., Gros, P., Forget, A., Kongshavn, A. L., St. Charles, C., and Taylor, B. A., 1982, Genetic regulation of resistance to intracellular pathogens, *Nature* **297:**506–509.

Smelt, A. H. M., Rees, R. J. W., and Liew, F. Y., 1981, Failure to induce delayed-type hypersensitivity to *M. leprae* in long-term treated lepromatous leprosy patients, *Clin. Exp. Immunol.* **44:**507–511.

Smith, D. W., and Harding, G. E., 1977, Approaches to the validation of animal test systems for assay of protective potency of BCG vaccines, *J. Biol. Standard.* **5:**131–138.

Smith, D. W., Grover, A. A., and Wiegeshaus, E. H., 1968, Non-living immunogenic substances of mycobacteria, *Adv. Tuberc. Res.* **16:**191–227.

Smith, D. W., Harding, G., Chan, J., Edwards, M., Hank, J., Muller, D., and Sobhi, F., 1979, Potency of 10 BCG vaccines as evaluated by their influence on the bacillemic phase of experimental airborne tuberculosis in guinea pigs, *J. Biol. Stand.* **7:**179–198.

Snider, D. E., 1982, The tuberculin skin test, *Am. Rev. Respir. Dis.* **125:**S108–118.

Sohn, C. C., Schroff, R. W., Kliewer, K. E., Lebel, D. M., and Fligiel, S., 1982, Disseminated *M. avium–intracellulare* infection in homosexual men with acquired immunodeficiency: A histologic and immunologic study of two cases, *Am. J. Clin. Pathol.* **79:**247–252.

Sprent, J., Korngold, R., and Molnar-Kimber, K., 1980, T-cell recognition of antigen *in vivo:* role of the H-2 complex, *Springer Sem. Immunopathol.* **3:**213–246.

Springett, V. K., 1965, The value of BCG vaccination, *Tubercle* **46:**76–84.

Springett, V. H., and Sutherland, I., 1970, Comparison of the efficacy of liquid and freeze-dried strains of BCG vaccine in preventing tuberculosis, *Br. Med. J.* **11:**184–150.

Stanford, J. L., Shield, M. J., and Rook, G. A. W., 1981, How environmental mycobacteria may predetermine the protective efficacy of BCG, *Tubercle* **62:**55–62.

Stead, W. W., 1967, Pathogenesis of the first episode of chronic tuberculosis in man: Recrudescence of residuals of the primary infection or exogenous reinfection, *Am. Rev. Respir. Dis.* **95:**729–745.

Stead, W. W., 1978, Undetected tuberculosis in prison, *JAMA* **240:**2544–2547.

Stead, W. W., and Dutt, A. K., 1982, Chemotherapy for tuberculosis today, *Am. Rev. Respir. Dis.* **125:**S94–101.

Stead, W. W., and Lofgren, J. P., 1983, Does the risk of tuberculosis increase in old age? *J. Infect. Dis.* **147:**951–955.

Stead, W. W., Lofgren, J. D., Warren, E., and Thomas, C., 1985, Tuberculosis as an endemic and nosocomial infection among the elderly in nursing homes, *N. Engl. J. Med.* **312:**1483–1487.

Steele, J. H., and Ranney, A. F., 1958, Animal tuberculosis, *Am. Rev. Tuberc.* **77:**908–922.

Steenken, W., 1961, The persistence of tubercle bacilli in caseous lesions in the experimental animal (guinea pig), *Am. Rev. Respir. Dis.* **83:**550–554.

Stokes, R. W., Orme, I. M., and Collins, F. M., 1986, The role of phagocytes in the expression of resistance and susceptibility to *M. avium* in mice, *Infect. Immun.* **54:**811–819.

Stoner, G. L., 1978, Ir genes and leprosy, *Int. J. Leprosy.* **46:**217–220.

Stoner, G. L., Mshana, R. N., Touw, J., and Belehu, A., 1982, Studies on the defect in cell-mediated immunity in lepromatous leprosy using HLA-D identical siblings. Absence of circulating suppressor cells and evidence that the defect is in the T-lymphocyte rather than in the monocyte population, *Scand. J. Immunol.* **15:**33–48.

Strom. R. L., and Gruninger, R. P., 1983, AIDS with *M. avium-intracellulare* lesions resembling those of Whipple's disease, *N. Engl. J. Med.* **309:**1323–1324.

Styblo, K., 1980, Recent advances in epidemiological research in tuberculosis, *Adv. Tuberc. Res.* **20:**1–63.

Suter, E., 1956, Interaction between phagocytes and pathogenic micro-organisms, *Bacteriol. Rev.* **20:**94–132.

Sutherland, I., 1976, Recent studies in the epidemiology of tuberculosis, based on the risk of being infected with tuberculosis, *Adv. Tuberc. Res.* **19:**1–63.

Sutherland, I. and Lindgren, I., 1979, The protective effect of BCG vaccination as indicated by autopsy studies, *Tubercle* **60:**225–231.

Sutherland, I., Svandovor, E., and Radhakrishma, S., 1982, The development of clinical tuberculosis following infection with tubercle bacilli, *Tubercle* **63:**255–268.

tenDam, H. G., and Hitze, K. L., 1980, Does BCG vaccination protect the newborn and young infants? *Bull. World Health Organ.* **58:**37–41.

tenDam, H. G., and Pio, A., 1982, Pathogenesis of tuberculosis and effectiveness of BCG vaccination, *Tubercle* **63:**225–233.

tenDam, M. G., Toman, K., Hitze, K. L., and Guld, J., 1976, Present knowledge of immunization against tuberculosis, *Bull. World Health Organ.* **54:**255–269.

Ternani, R., Zimmerman, A., and Moroll, A., 1979, BCG septacemia as fatal complication of BCG vaccination, *Schweizer Med. Wochenschr.* **109:**708–713.

Thorns, C. J., and Morris, J. A., 1983, The immune spectrum of *M. bovis* infections in some mammalian species: a review, *Vet. Bull.* **53:**543–550.

Thorsby, E., Godal, T., and Myrvang, B., 1973, HLA antigens and susceptibility to disease. II. Leprosy, Tissue antigens **3:**373–377.

Touw, T., Lengenditk, E. M. T., Stoner, G. L. and Belehu, A., 1982, Humoral immunity in leprosy: immunoglobulin G and M antibody responses to *M. leprae* in relation to various disease patterns, *Infect. Immun.* **36:**885–892.

Tripathy, S. P., 1983, The case for BCG, *Ann. Natl. Acad. Med. Sci. (India)* **10:**11–20.

Truitt, G. L., and Mackaness, G. B., 1971, Cell-mediated resistance to aerogenic infection in the lung, *Am. Rev. Respir. Dis.* **104:**829–843.

Ueda, K., Yamazaki, S., and Someya, S., 1976, Experimental mycobacterial infection in congenitally athymic "nude" mice, *J. Reticuloendothel. Soc.* **19:**77–90.

Vallishayee, R. S., Shashidhara, A. N., and Bunch-Christensen, K., 1974, Tuberculin sensitivity and skin lesions in children after vaccination with 11 different BCG strains, *Bull. World Health Organ.* **51:**489–494.

van Rooijen, N. and van Nieuaemegen, R., 1980, Liposomes in immunology: Evidence that their adjuvant effect results from surface exposition of the antigens, *Cell. Immunol.* **49:**402–407.

Volkman, A., and Collins, F. M., 1971, The restorative effect of peritoneal macrophages on delayed hypersensitivity following ionizing radiation, *Cell. Immunol.* **2:**552–566.

Waaler, H. T., and Rouillon, A., 1974, BCG-vaccination policies according to the epidemiological situation, *Bull. Int. Union Tuberc.* **49:**166–189.

Walker, L. and Lowrie, D. B., 1981, Killing of *M. microti* by immunologically activated macrophages, *Nature* **293:**69–70.

Watson, J. D., 1981, Lymphokines and the induction of immune responses, *Transplantation* **31:**313–317.

Weiszfeiler, J. G., and Karasseva, V., 1981, Mixed mycobacterial infections, *Rev. Infect. Dis.* **3:**1081–1083.

Wells, A. Q., 1957, The vole bacillus vaccine, *Adv. Tuberc. Res.* **8:**61–78.

Wendt, S. L., George, K. L., Parker, B. C., Gruft, H., and Falkinham, J. O., 1980, Epidemiology of infection by nontuberculous mycobacteria. III. Isolation of potentiallly pathogenic mycobacteria from aerosols, *Am. Rev. Respir. Dis.* **122:**259–263.

Wessels, C. C., 1941, Tuberculosis in the rat. II. The fate of tubercle bacilli in the various organs of the rat, *Am. Rev. Tuberc.* **43:**459–474.

White, R. G., 1976, The adjuvant effect of microbial products on the immune response, *Annu. Rev. Microbiol.* **30:**579–600.

Wickman, K., 1978, BCG vaccination and infection by atypical mycobacteria, *Labartidningen.* **75**:113–114.

Wiegeshaus, E. H., Harding, G. E., McMurray, D., Grover, A. A., and Smith, D. W., 1971, A cooperative evaluation of test systems used to assay tuberculosis vaccines, *Bull. World Health Organ.* **45**:543–550.

Wolinsky, E., 1979, Non-tuberculous mycobacteria and associated diseases, *Am. Rev. Respir. Dis.* **119**:107–159.

Wolinsky, E., 1981, When is an infection disease? *Rev. Infect. Dis.* **3**:1025–1027.

Youmans, G. P., 1975, Relation between delayed hypersensitivity and immunity in tuberculosis, *Am. Rev. Respir. Dis.* **111**:109–118.

Young, D. B., Kent, L., Rees, A. D. M., Lamb, J., and Ivanyi, D., 1986, Immunological activity of a 38 kilodalton protein purified from *M. tuberculois, Infect. Immun.* **54**:177–183.

Zakowski, P., Gligid, S., Berlin, G. W., and Johnson, B. L., 1982, Disseminated *M. avium–intracellulare* infection in homosexual men dying of acquired immunodeficiency, *JAMA* **248**:2980–2982.

Zimmerman, M. R., 1979, Pulmonary and osseous tuberculosis in an Egyptian mummy, *Bull. NY Acad. Med.* **55**:604–608.

Zlotnik, A., and Crowle, A. J., 1982, Lymphokine-induced mycobacteriostatic activity in mouse pleural macrophages, *Infect. Immun.* **37**:786–793.

8

Interactions between Macrophages and Fungi

H. JEAN SHADOMY and ROBERT A. FROMTLING

1. INTRODUCTION

Macrophages play a key role in two important host defense mechanisms: (1) ingestion and inactivation of invading organisms, and (2) macrophage interactions with other immune cells, i.e., T lymphocytes and B lymphocytes, that promote antigen recognition and processing. Among the many functions of macrophages in host defense, attachment to and engulfment of microorganisms are vitally important. After phagocytosis, most organisms are degraded, stored, and transferred by macrophages. Degraded or processed antigen then may be presented to other lymphoid cells on the surface of the macrophage. These factors demonstrate the significant contribution that macrophages make to host immune competence (Fauci, 1978).

Most research in the area of macrophage/fungi interactions has been done with three organisms that are agents of the life-threatening mycoses *Histoplasma capsulatum, Cryptococcus neoformans,* and *Candida albicans.* Some investigators have begun to study a few other mycotic agents, but it is readily apparent that this area of study is a relatively new and blossoming field. The tenacious character of many of the fungi following interactions with macrophages or monocytes warrants further study in this important area of host defense against disease.

The diseases produced by fungi in humans generally are placed into three categories: dermatophytic, systemic, and opportunistic (Emmons *et al.*, 1977; Rippon, 1982). Dermatophytes generally produce superficial infections, invading the skin, mucous membranes, nails, and hair, and seldom invade deeper tissues; they are not included in this discussion of macrophage/fungi interactions.

H. JEAN SHADOMY • Department of Microbiology and Immunology, Medical College of Virginia, Virginia Commonwealth University, Richmond, Virginia 23298. ROBERT A. FROMTLING • Department of Basic Microbiology, Merck Institute for Therapeutic Research, Rahway, New Jersey 07065.

In systemic fungal infections, cell-mediated immunity is generally a more significant factor than humoral immunity, and activated macrophages appear to be the major line of defense (Fromtling and Shadomy, 1986). The activated macrophages sensitize T lymphocytes, while at the same time they stimulate B lymphocytes to produce antibodies; B lymphocytes also can be stimulated indirectly by T lymphocytes. Although these antibodies apparently have little or no role in killing the fungi, their detection by various serologic and immunologic techniques has become a useful diagnostic/prognostic tool. In the majority of individuals, the appearance of cellular immunity to the causal organism leads to control and resolution of the disease process and may confer immunity. The interaction of macrophages and etiologic agents of blastomycosis, coccidioidomycosis, and histoplasmosis is discussed in this section. In opportunistic fungal infections, i.e., infections produced by ubiquitous organisms with a very low virulence, the major line of defense is the neutrophil. Establishment of disease depends on the lowered resistance of the host, either due to underlying local or systemic disease or to administration of immunosuppressive or antimicrobial agents. In recent years, various opportunistic fungal infections have increased in number because of extensive therapies for malignancies and immunologic disorders (Hotchi *et al.*, 1980). Recovery from infection due to opportunistic fungi generally does not confer specific resistance to the causal organism(s). Discussion of interaction of macrophages and the etiologic agents of aspergillosis, zygomycosis, cryptococcosis, and candidiasis is included in this section.

2. INTERACTION OF MACROPHAGES AND SYSTEMIC FUNGI

2.1. BLASTOMYCOSIS

Blastomycosis, historically known as North American blastomycosis, is a chronic granulomatous and suppurative disease with a clinical spectrum ranging from primary pulmonary to disseminated multisystem involvement. The dimorphic fungus *Blastomyces dermatitidis* is the sole etiologic agent. The major endemic region is the Mississippi River Valley basin in the United States (Rippon, 1982). Three clinical forms of blastomycosis are recognized: pulmonary, disseminated, and cutaneous. The microconidia, which are less than 5 μm in diameter, are produced by the mycelial phase of the fungus and have been shown to readily undergo transformation to the *in vivo* yeastlike form (Garrison and Boyd, 1978). This favors the view that the microconidia are the infectious particles and that primary infection is pulmonary following inhalation of the conidia.

Sixby *et al.* (1979) studied the *in vitro* interaction between human granulocytes and the yeast phase of *B. dermatitidis*. They observed that granulocytes readily phagocytized the yeast cells, but only 29% of the inoculum was killed. They suggested that the difference between phagocytosis and intracellular killing may be due to fungal resistance to the antimicrobial mechanisms of the granulocytes. In addition, they speculated that this resistance to killing may be related to the "relentless progression" of untreated systemic blastomycosis. Brummer *et al.* (1985) noted that polymorphonuclear cells from naïve mice did not elicit an oxidative burst when exposed to cells of *B. dermatitidis* and the yeast cells were not killed. However, cells from mice immunized with *B. dermatitidis* elicited a large oxidative burst when exposed to cells of *B. dermatitidis*, which was correlated with a significant reduction in colony-forming units of the yeast. A critical role for hydrogen peroxide was demonstrated in this study. Drutz and Frey (1985) also have demonstrated that both conidia and yeast cells are killed by predominantly oxidative polymorphonuclear mechanisms. However, the role of monocyte-derived macrophages also was noted; these macrophages killed 90% of conidia and 40% of yeast cells. Miyaji and Nishimura (1983) studied granuloma formation and killing functions of granuloma in congenitally athymic nude mice infected with *B. dermatitidis*. They noted that although mononuclear cells had migrated to the foci of infection and formed a granulomatous response, yeast cells generally were not killed and continued to multiply within the granuloma.

Brummer *et al.* (1980) studied the *in vitro* interaction of macrophages with yeast cells of *B. dermatitidis*. Compared with control cultures, yeast cell growth was significantly inhibited in the presence of concanavalin A (Con A)–activated macrophages and macrophages from mice that had recovered from nonlethal subcutaneous *Blastomyces* infection. In contrast to resident macrophages activated *in vivo* by *B. dermatitidis* infection, the authors noted that macrophages from oil-induced exudate cells from infected mice lacked activity in the assay on initial isolation. Additional studies by Brummer *et al.* (1983) demonstrated that the activation of peritoneal macrophages for fungicidal activity by Con A treatment was dose dependent and transient. Their data showed that murine peritoneal macrophages were capable of killing *B. dermatitidis in vitro*, but that the addition of specific antibody or complement or both did not enhance the killing of the yeast cells.

Macrophages have been shown to be important cells in host defense against blastomycosis (Morozumi *et al.*, 1981a,b; Brummer *et al.*, 1981; Brummer and Stevens, 1982). Susceptibility differences of inbred strains of mice to blastomycosis have been noted (Morozumi *et al.*, 1981a,b). Two strains of mice (C3H/HeJ and A/HeJ), which are known to have macrophage cytotoxicity defects, were shown to be significantly more susceptible to pulmonary blastomycosis following infection when compared to strains of mice that did not have a macrophage cytotoxicity defect. Additionally, Brummer *et al.* (1981) have reported a correlation of virulence of strains of *B. dermatitidis in vivo* with ability of the yeast cells to escape from macrophage inhibition of yeast replication *in vitro*. The authors noted that in 24-hr cocultures the replication of three different strains (attenuated, avirulent, and virulent) of *B. dermatitidis* was inhibited by macrophages from (1) untreated mice, (2) mice that had recovered from nonlethal blastomycosis, (3) concanavalin A–treated mice, or (4) peritoneal exudates elicited by thioglycollate. Significantly, after 72 hr of incubation of the first three types of cocultures, the attenuated and avirulent strains of *B. dermatitidis* remained inhibited, but the virulent strain was not inhibited.

Recently, Brummer and Stevens (1982) reported opposing effects of human monocytes, macrophages, and polymorphonuclear neutrophils on the replication of *B. dermatitidis in vitro.* They noted that peripheral blood monocyte monolayers significantly inhibited the replication of a virulent and avirulent strain of *B. dermatitidis* in 24- and 72-hr cultures. Macrophage monolayers, derived from monocytes by *in vitro* culturing for 9 days, also inhibited the replication of both strains in 24- and 72-hr cultures. Polymorphonuclear neutrophils, however, stimulated the replication of both strains of yeast under identical conditions. The authors suggested that the phagocytic cells tested may have similar opposing effects on the replication of *B. dermatitidis in vivo.* McDaniel and Cozad (1983) demonstrated by scanning electron microscopy that sensitized mouse macrophages readily phagocytized yeast cells of *B. dermatitidis.* In addition, there appeared to be increased metabolic activity as measured by chemiluminescence. Sensitized macrophages controlled replication of *B. dermatitidis* yeast cells *in vitro* better than nonsensitized cells. Although the sensitized macrophages did not completely kill the phagocytized yeast cells, the inhibition of growth suggests a role for the sensitized macrophage in host defense against this disease. Bradsher *et al.* (1985) studied the intracellular growth and phagocytosis of *B. dermatitidis* by monocyte-derived macrophages from previously infected and normal human subjects and noted that yeast cells were more readily phagocytized and killed by macrophages from the previously infected donors when compared to normal donors. This greater efficiency in phagocytosis and killing reflects the role of this mechanism of human cellular immunity to this fungal infection.

2.2. COCCIDIOIDOMYCOSIS

Coccidioidomycosis is a benign, subclinical upper-respiratory infection that may spontaneously resolve or progress into a severe systemic disease that may prove fatal. Disseminated disease may include fungal colonization of any visceral organ or the meninges, bones, joints, or subcutaneous or cutaneous tissues (Rippon, 1982).

Coccidioides immitis is the only etiologic agent of this disease. The organism is a dimorphic soil saprophyte which has a geographical distribution limited to Lower Sonoran biomes, areas that are hot and semiarid. The saprophyte mycelial phase produces small (4×6 μm), thick-walled, barrel-shaped cells, termed arthroconidia, which are specialized segments of hyphae. Inhalation of a single arthroconidium can initiate infection. Following inhalation, the arthroconidia convert into the *in vivo* spherule or sporangium phase. This morphologic form of the fungus is the phase that interacts with cells of the host defense (Emmons *et al.,* 1977; Rippon, 1982).

The immune response to coccidioidomycosis that leads to protection has not been characterized adequately although new information is learned monthly. Drutz and Catanzaro (1978) noted that the sequence of events presumed to follow contact with *C. immitis* is (1) the fungus is phagocytized and processed by macrophages, (2) antigens are presented which sensitize lymphocytes, (3) cell-mediated immunity and delayed-type hypersensitivity to the fungus develop, (4) macrophages become activated, (5) specific antibodies are synthesized, and (6) the immune response becomes regulated by the activity of T lymphocytes with either helper or suppressor functions.

Evidence for the development of a phagocytic response by polymorphonuclear leukocytes was first reported by Wegner *et al.* (1972). Polymorphonuclear leukocytes recovered from the venous blood of dogs that had been inoculated with live arthroconidia of *C. immitis* or vaccinated with an irradiated spherule vaccine showed significantly greater *in vitro* phagocytic activity than leukocytes obtained from control dogs. The authors were able to show a significant correlation between the complement fixation antibody titer and the phagocytic index in inoculated dogs. Additionally, humans who reacted to a coccidioidin skin test demonstrated a larger phagocytic index than nonreactive persons. The use of coccidioidin in *in vitro* studies using rabbit macrophages suggested a potential toxicity of some coccidioidin preparations for the immune cells (Wallraff and van Liew, 1967). Antigen preparations for such studies must be controlled carefully.

Savage and Madin (1968) observed phagocytized arthroconidia develop into spherules within macrophages from infected mice. Fewer spherules were observed in alvelolar macrophages from lung washings or immunized than from nonimmunized mice after intranasal challenge with arthroconidia, suggesting that macrophages from immunized animals may kill arthroconidia and/or immature spherules of *C. immitis*. How arthroconidia of *C. immitis* survive in macrophages of normal animals is not known; however, Beaman and Holmberg (1980) reported that in experiments performed in nonimmune monkeys, both morphologic forms of the fungus inhibited phagosome–lysosome fusion. The surviving arthroconidia developed into spherules. Additional studies by Galgiani *et al.* (1982) determined that arthroconidia became septated spherules in the presence of human leukocytes, but grew as mycelia when viable leukocytes were absent *in vitro*. In a newly developed assay, they were able to show that human neurophils, lymphocytes, monocytes, and rabbit alveolar macrophages inhibited spherules from forming germ tubes. Their studies indicated that leukocytes influenced coccidioidal dimorphism toward the spherule phase, and that this effect may be important to host defense against this disease. Further studies of phagocytosis by Drutz and Huppert (1983) revealed that macrophages can ingest arthroconidia and endospores, but appear to be unable to kill them unless lymphokines stimulate phagolysosomal fusion. It is still unclear whether spherules can be killed by macrophages. Clemons *et al.* (1985) investigated the role of macrophages in host resistance to experimental infection with *C. immitis* in mice with selectively enhanced or impaired macrophage function. They observed that nonspecific resistance to coccidioidomycosis could not be obtained by bacille Calmette Guérin (BCG) induction, but that pretreatment with formalin-killed mycelia of *C. immitis* enhanced protection in the mice. Collectively, their data demonstrated that macrophages play a key role in host resistance to coccidioidomycosis in specifically immunized mice and that a compartmentalization of site-specific resistance against infection by *C. immitis* may occur.

Kashkin *et al.* (1977) reported that peritoneal macrophages from guinea pigs, when infected in the presence of immune lymphocytes, phagocytized arthroconidia of *C. immitis.* They suggested that the immune lymphocytes were producing fungicidal factors that decreased the viability of the phagocytized arthroconidia. Beaman *et al.* (1977) showed that T lymphocytes from immune mice were required for the transfer of resistance to a lethal infection to an immunologically naive host. Beaman *et al.* (1981) noted that peritoneal macrophages from normal mice did phagocytize arthroconidia and endospores of *C. immitis,* but did not kill the spores after 4 hr of incubation. When macrophages were infected in the presence of lymphocytes from immune mice, the viability of the phagocytized conidia was markedly reduced. The authors confirmed earlier observations that fusion of phagosomes and lysosomes was inhibited in normal macrophages that had ingested conidia of *C. immitis.* Significantly, they reported that phagosome–lysosome fusion did occur in macrophages that had been infected in the presence of immune lymphocytes (Beaman *et al.,* 1981). Another study by Beaman *et al.* (1983) demonstrated that activation of macrophages can be effected after incubation with lymphokines. The authors reported that $Ly1^{+}2^{-}$T lymphocytes produced soluble factor(s) that enhanced the antifungal action of macrophages, leading to increased fusion of phagosomes containing *C. immitis* with lysosomes and increased killing of the fugus. They further demonstrated that this activation of macrophages by the lymphokine could occur across H-2 histocompatibility barriers. In a more recent study, Beaman and Pappagianis (1985) demonstrated that monocytes from any human source were capable of killing yeast cells of *Candida albicans;* however, monocytes killed arthroconidia of *C. immitis* only in the presence of lymphocytes from immune individuals. Additional studies are needed to determine how *C. immitis* is killed and to define the mechanism(s) whereby lymphocytes activate macrophages against *C. immitis.*

2.3. HISTOPLASMOSIS

Histoplasmosis may be an acute, subacute, chronic pulmonary or systemic mycosis. The organism has a predilection for the reticuloendothelial system and may involve any organ including the central nervous system. The dimorphic soil saprophyte *Histoplasma capsulatum* is the etiologic agent. The organism has been isolated worldwide from soils enriched by the excreta of bats, chickens, starlings, and other birds. Microconidia are 2–3 μm in diameter and are produced by growth of the saprophytic phase in soils. The microconidia are believed to be the infectious particles, and infection is initiated by inhalation of them (Emmons *et al.,* 1977; Rippon, 1982). Acute pulmonary histoplasmosis is the most common form of the disease (Rippon, 1982).

The organism is an intracellular parasite, and parasitization of the host mononuclear phagocyte by the yeast phase of *H. capsulatum* is the central event in disseminated histoplasmosis. Phagocytosis of the yeast cells is common, but intracellular killing does not seem to occur. Some have proposed that the susceptible macrophage not only provides a protective environment for the yeast, but may support intracellular growth of the organism and act as a vehicle of dissemination (Goodwin and Des Prez, 1978; Caldwell and Sprouse, 1982). The mechanism of this defect in cellular immunity is unknown at present.

The first report of macrophage interactions with yeast cells of *H. capsulatum* was an electron microscopic study by Edwards *et al.* (1959). They noted that ultrastructural modifications of splenic cells, specifically macrophages, that had phagocytized yeast cells suggested that the macrophages attempted to degrade the yeast cells, but were unsuccessful. Hill and Marcus (1960) demonstrated increased intracellular digestion in immune macrophages independent of mediation by circulating or passive antibody. Wu and Marcus (1963) further reported that such immune phagocytes were much more active in digestion than normal macrophages. Preliminary exposure of yeast cells of *H. capsulatum* to specific antibody and complement was shown not to affect intracellular proliferation of the fungus *in vitro* (Howard, 1965). However, mononuclear phagocytes from immunized animals restricted intracellular growth of *H. capsulatum* (Howard *et al.*, 1971; Howard, 1981b).

An *in vivo* study of the ultrastructural changes during phagocytosis of *H. capsulatum* by unstimulated, unsensitized hamster peritoneal macrophages at intervals varying from 5 min to 48 hr after injection revealed that fungi were present in two types of phagosomes, a "tight" and a "loose" type (Dumont and Robert, 1970). Neither structure appeared to be effective in killing internalized yeast cells. Although phagolysosomes formed, no progressive alteration of fungi was observed. The authors concluded that few fungi show signs of degeneration following phagocytosis and most of the yeasts multiplied within the different vacuoles of macrophages. They suggested that the host response may be characterized by successive phagocytosis of the same fungus by different macrophages, thus promoting chronicity of the disease.

Howard (1973) studied the interaction between yeast cells of *H. capsulatum* and macrophages from mice immunized by sublethal infection. He noted that autoradiographic analysis determined that protein synthesis by the fungus *in situ* was impaired, but that although the growth of the yeast cells within the immune macrophages was inhibited, the fungus was not killed by the encounter. A study by DeSanchez and Carbonell (1975) using macrophages obtained from normal and sensitized rabbits noted that yeast cells of *H. capsulatum* were readily phagocytized by both macrophage populations, but protoplasts (cell wall free) of the organism were not engulfed. They reported that the addition of immune sera activated phagocytosis of both forms of the fungus, particularly immune sera against cell wall components; however, intracellular killing of the fungus was not detected. These observations were later verified by Schlitzer *et al.* (1981).

Howard and Otto (1977) determined that immune lymphocytes mediate the activity of peritoneal exudate cells from mice immunized with *H. capsulatum*. Supernatants from cultures of immune lymphocytes or from cultures of mitogen-stimulated lymphocytes were partially able to arm macrophages from immunized animals, thus promoting phagocytosis and inhibition of intracellular growth of yeast cells of *H. capsulatum*. In a related study, Wu-Hsieh *et al.* (1984) reported that supernatants from a T-cell hybridoma clone also activated macrophages to suppress intracellular growth of *H. capsulatum*, and that this activity was similar to that found in supernatants from immune spleen cell cultures. Tewari *et al.* (1980) varied the role of immune lymphocytes in host defense against histoplasmosis and determined that immunity to histoplasmosis is mediated by T lymphocytes and that replication of the sensitized T cells is essential for expression of protective immunity. A study by Nickerson *et al.* (1981) demonstrated potent suppressor cell activity during the course of experimental histoplasmosis in C3H/Anf mice. Spleen cells from infected mice suppressed antibody response *in vitro* of normal syngeneic spleen cells to both T-dependent and T-independent antigens. One population was identified as T cells that were nonadherent to nylon wool columns. The second suppressor population had macrophagelike characteristics; thus, macrophages or macrophagelike cells may play a role in the immunoregulation of histoplasmosis and perhaps other mycoses. Other investigators have devised studies that also demonstrate that a strong interaction of several cell types may occur in host defense and immune regulation in histoplasmosis (Deepe *et al.*, 1984; Taylor *et al.*, 1984; Ruiz and Carvajal, 1986).

In another study poly-2-vinylpyridine-*N*-oxide, a compound known to prevent or reduce the toxic, fibrogenic, and immunosuppressive effects of silica, was shown to enhance the resistance of mice to an experimental infection of *H. capsulatum*, thus supporting the hypothesis that macrophages play a key role in host defense against histoplasmosis (Von Behren *et al.*, 1983b). These investigators also studied the effect of silica on the susceptibility of mice to experimental histoplasmosis and further verified that macrophages play an essential role in natural immunity to this fungal disease (Von Behren *et al.*, 1983a). Tiku *et al.* (1985) studied macrophage-related fibrinolysis in experimental disseminated histoplasmosis in a CBA/J mouse model and determined that spleen and peritoneal macrophages from *H. capsulatum*–infected mice exhibited increased fibrinolysis, which is indicative of macrophage activation. The activation resulted from the interaction of T cells and macrophages and from the action of *Histoplasma* products on the macrophages.

Kimberlin *et al.* (1981) studied the interactions between *H. capsulatum* and macrophages from normal and *Mycobacterium bovis* BCG-treated mice. This study is unique in the sense that the authors examined the interaction between macrophages and both the yeast and mycelial phase of the fungus. Their study showed that alveolar macrophages phagocytized small mycelial-phase particles, i.e., microconidia (3–5 μm) and small hyphal fragments (3–10 μm), at a rate similar to that of yeast-phase cells. Peritoneal and alevolar macrophages from all groups of mice were observed to surround and attach to macroconidia and long hyphal fragments that were too large to be phagocytized. Only the BCG-treated macrophages were able to kill the mycelial-phase particles. Fungal cells (mycelial-phase and yeast-phase) in all other macrophage preparations survived phagocytosis and eventually killed the macrophage, thus confirming the observations of Howard *et al.* (1971).

Calderone and Peterson (1979) extracted a lysosomal-rich fraction from normal rabbit alveolar macrophages and examined its effect on protein synthesis of

H. capsulatum. The incorporation of all amino acids tested, including leucine, proline, and glutamic acid, was inhibited by the extract. The inhibition of amino acid uptake was determined to be specific (2-deoxy-D-glucose, uridine, and rubidium uptake was not affected) and resulted in decreased protein synthesis, an observation earlier reported by Howard (1973). Viability of *H. capuslatum* was reduced significantly when yeast cells were incubated in the lysosomal extract for at least 20 hr. Electron microscopic examination of treated cells revealed the yeasts to be highly vacuolated with loss of cytoplasmic organization. An interesting series of studies by Howard (1981a, 1983) demonstrated a variability in the sensitivity of strains of *H. capsulatum* to hydrogen peroxide. Howard noted large, consistent differences in the levels of catalase among strains of the fungus. Resistance to hydrogen peroxide did not correspond to levels of catalase, and there was no obvious correlation of peroxide sensitivity and virulence of the strains. These studies lend support to the hypothesis that some isolates of *H. capsulatum* may not be adversely affected by this phase of host cell defense.

Immunization protocols developed by Sharma *et al.* (1980) and Williams *et al.* (1981a) have helped elucidate the role of macrophages and other immune cells in host defense against histoplasmosis. Sharma *et al.* (1980) immunized mice with ribosomes and ribosome proteins from yeast cells of *H. capuslatum.* The immunity could be transferred to syngeneic recipients with lymphocytes or peritoneal exudate cells, suggesting a role for macrophages in the immune responses. Additional studies by Tewari *et al.* (1977, 1978) demonstrated that peritoneal exudate cells are more efficient than spleen cells in adoptive transfer of immunity and that the immune cell that conferred protection was the T lymphocyte. A recent study by Williams *et al.* (1981a) confirms the finding of Howard and Otto (1977), Tewari *et al.* (1977, 1978), and Sharma *et al.* (1980) concerning the role of T lymphocytes and macrophages in host defense against histoplasmosis and further clarifies that adoptive transfer of immunity to *H. capsulatum* is multifactorial and depends on both sensitized T lymphocytes and activated macrophages.

3. INTERACTION OF MACROPHAGES AND OPPORTUNISTIC FUNGI

3.1. ASPERGILLOSIS

Aspergillosis is an opportunistic mycosis caused by members of the ubiquitous, large genus *Aspergillus*. A majority of these organisms are soil and air saprophytes. The most common etiologic agents of human infections are *Aspergillus fumigatus, A. flavus*, and *A. niger*, although other species have been implicated as well. Aspergillosis most commonly is divided into allergic bronchopulmonary, fungus ball, or invasive disease. Aspergillosis is an opportunistic infection primarily affecting debilitated, immunologically altered patients. Pulmonary aspergillosis is the most common form of the disease. The infectious particles are the phialoconidia, which measure 4–8 μm in diameter.

Experimental studies involving conidia of *Aspergillus* and macrophages have not been extensive. Epstein *et al.* (1967) studied the early stages in the pathogenesis of experimental pulmonary aspergillosis in cortisone-treated mice. Normal and cortisone-treated mice were exposed to aerosols of *A. flavus* conidia, and the ability of alveolar macrophages to kill conidia *in vitro* was assessed. The authors reported a similar degree of early phagocytosis of conidia by macrophages in both groups. However, the *in vitro* studies showed that a larger number of viable conidia remained after incubation with experimental macrophages than after incubation with control macrophages. The findings indicated that the alveolar macrophages of cortisone-treated animals were not as competent as control macrophages in arresting growth and proliferation, as well as in killing conidia of *A. flavus.*

Merkow *et al.* (1968, 1971) performed similar studies noting that conidia within alveolar macrophages of control mice either remained dormant or became fragmented, while conidia within macrophages of steroid-treated mice showed evidence of pregermination and germination after 2–4 hr (Merkow *et al.*, 1971). Electron microscopic examination of the cells revealed that the lysosomes of control macrophages showed extensive fusion with the phagocytic membranes surrounding spores, but the lysosomes of macrophages from animals treated with cortisone revealed little interaction with the membranes (Merkow *et al.*, 1968). This diminished lysosomal response in forming phagocytic vacuoles may be important in the subsequent development of the hyphal bronchopneumonia which frequently occurred in the steriod-treated mice exposed to spores of *A. flavus.* This study was supported further by similar observations of the importance of lysosomal mechanisms of killing using yeast cells of *Saccharomyces cerevisiae* (Hart, 1981; Hart and Young, 1975) and *C. immitis* (Beaman and Holmberg, 1980). A study by Schaffner *et al.* (1982), however, demonstrated that a host can rely on two independent phagocytic cell lines to mount a defense against *Aspergillus* species. They observed that macrophages prevented germination and killed conidia of *Aspergillus fumigatus in vitro* and rapidly eradicated conidia *in vivo* even in neutropenic and athymic mice. In addition, polymorphonuclear cells were shown to kill mycelia of this fungus *in vitro.* Interestingly, these two lines of defense functioned in the absence of a specific immune response. In a subsequent study, Schaffner *et al.* (1983) proposed that the anatomic source of macrophages was important in the ability of the macrophages to kill conidia of *Aspergillus.* These researchers examined alveolar and peritoneal macrophages from rabbits and noted that alveolar macrophages were efficient in phagocytizing and killing 82–90% of the *Aspergillus* conidia while peritoneal macrophages were unable to inhibit the germination of the *Aspergillus* conidia. In addition, the alveolar macrophages were capable of killing *Aspergillus* conidia in the absence of T-lymphocyte modulation or generation of oxygen intermediates.

Lehrer and Jan (1970) studied the ability of peripheral blood neutrophils and monocytes to kill ingested conidia of *A. fumigatus.* After 3 hr of incubation, the phagocytized conidia were not killed. The authors noted an apparent ineffectiveness of chloride in the system and suggested that this factor could underlie the observed inability of the cells tested to kill ingested conidia of *A. fumigatus* despite their competence to kill the yeast *Candida albicans*.

Diamond *et al.* (1978) established that human neutrophils could damage and probably kill hyphae of *A. fumigatus* and *Rhizopus oryzae in vitro* and kill conidia of *A. fumigatus* (Levitz and Diamond, 1985) primarily by oxygen-dependent mechanisms active at the cell surface. Further study by Diamond and Clark (1982) determined that the hyphae of these two fungi could be damaged by oxidative and nonoxidative microbicidal products *in vitro*. This observation subsequently was confirmed by Diamond *et al.* (1983), who studied cells from patients with hereditary leukocytic dysfunctions and reported that hyphal damage sometimes may be mediated by myeloperoxidase-independent and even totally nonoxidative mechanisms. Waldorf *et al.* (1984) studied the ability of bronchoalveolar macrophages from normal, diabetic, and cortisone-treated mice to inhibit conidia germination and kill conidia of *A. fumigatus* and *Rhizopus oryzae in vivo*. The authors determined that alveolar macrophages control invasive fungal diseases by different mechanisms. Prevention of zygomycosis (*R. oryzae*) was attained by inhibition of conidia germination by alveolar macrophages; prevention of aspergillosis was attained by actual killing of conidia by the macrophages. In reality, it is probable that human neutrophils, monocytes, and macrophages may interact *in vivo* to determine the outcome of potentially invasive aspergillosis.

Several additional studies have demonstrated the ability of rabbit alveolar macrophages to phagocytize conidia of *A. fumigatus* (Lundborg and Holma, 1972, 1974; Richard and Thurston, 1975). Normal rabbit alveolar macrophages were shown to be very effective in phagocytizing conidia of *A. fumigatus* in comparison to other fungal conidia and polystyrene particles (Lundborg and Holma, 1972, 1974). Intracellular fate of the conidia was not assayed. Richard and Thurston (1975) noted that alveolar macrophages from rabbits given varying doses of aflatoxin B_1 had reduced phagocytic acitvity when compared to macrophages from control rabbits. Incorporation of serum from the aflatoxin-treated rabbits in the *in vitro* culture system resulted in less stimulation of phagocytosis by macrophages from control rabbits when compared with the same system incorporating control serum. These authors noted that stimulation of phagocytosis by macrophages from control rabbits when compared with the same system incorporating control serum. These authors noted that stimulation of phagocytosis by macrophages was least when both serum and macrophages from aflatoxin-treated rabbits were used in the *in vitro* system. The ability of the macrophage populations to kill the conidia was not investigated.

Williams *et al.* (19G1b) examined the immunity to infection with *A. fumigatus* conidia in the athymic nude (nu/nu) mouse. Three routes of infection were studied: intratracheal, intravenous, and intraperitoneal. Although nu/nu mice were not significantly more susceptible than their heterozygous (nu/+) littermates, only nu/+ animals could be protected by prior immunization. Protection was transferred to nu/nu recipients by immune glass-adherent cells. Additionally, gold sodium thiomalate, which interferes with macrophage and monocyte function, greatly increased the susceptibility of mice to infection with conidia of *A. fumigatus.* Thus, the data suggested a role for both T cells and macrophages in host defense against aspergillosis.

3.2. ZYGOMYCOSIS

Zygomycosis is an acute, usually fulminating infection which most often afflicts immunologically or physiologically compromised patients (Rippon, 1982; Kitz *et al.*, 1981). A variety of filamentous fungi in the class Mucorales are the etiologic agents in the disease. Genera most often encountered are *Rhizopus, Absidia, Rhizomucor,* and *Mucor.* Synonyms of zygomycosis include mucormycosis and phycomycosis. The invading fungi have a predilection for blood vessels and often lead to embolism and necrosis of adjacent tissues.

Little information is available on the interaction of the etiologic agents of zygomycosis and macrophages. Diamond *et al.* (1982) studied the antihyphal activity of human monocytes against *Rhizopus oryzae.* The hyphae of this fungus are too large to be completely phagocytized and digested. *Diamond et al.* (1982) observed that monocytes attached to hyphae and appeared to destroy them in the absence of serum. The presence of 10% human serum did not augment attachment or damage, but monocyte-mediated damage was abrogated by inhibitors of oxidative metabolism, suggesting that monocytes are capable of damaging and perhaps killing hyphae of this fungus via oxidative mechanisms. Diamond (1983) later showed that monocyte-mediated damage to hyphae of *R. oryzae* is inhibited by steroid hormones, although the effect of 10 μM hydrocortisone against *R. oryzae* was less marked than the inhibition of monocyte activity against hyphae of *A. fumigatus* and *C. albicans.*

Waldorf *et al.* (1984) demonstrated that bronchoalveolar macrophages inhibited conidia germination of *R. oryzae* in an experimental model of murine zygomycosis resulting in reduced pathogenicity of the fungal agent. Bronchoalveolar macrophages from diabetic and cortisone-treated mice were not capable of inhibiting conidia germination and permitted infection with the *Rhizopus.* This study demonstrated that macrophages participate in host defense against this opportunistic fungal pathogen.

Much remains to be learned concerning the interaction of macrophages and the several etiologic agents of zygomycosis. This group of pathogens are an emerging threat; the numbers of compromised patients who may be susceptible to this disease are increasing rapidly as successful cancer treatments and organ transplants, as well as cases of acquired immune deficiency syndrome, rise. Greater attention must be directed to understanding specific host defense, i.e., macrophage interaction, against these potentially life-treatening filamentous fungi.

3.3. CRYPTOCOCCOSIS

Cryptococcosis is a chronic, subacute, or acute mycosis that can be pulmonary, meningeal or systemic. The etiologic agent is an encapsulated yeast, *Cryp-*

tococcus neoformans, a saprophyte that has been isolated worldwide, most frequently from soils enriched with pigeon excreta (Rippon, 1982). The infectious particle is believed to be a small yeast cell that enters the body by inhalation (Neilson *et al.*, 1977). A primary pulmonary infection follows inhalation of yeast cells with subsequent localization or fatal dissemination. The organism has a marked predilection for the central nervous system, particularly the meninges (Emmons *et al.*, 1977; Rippon, 1982).

A large proportion of research concerned with the interaction between fungi and macrophages has been performed using *C. neoformans*. Reviews by Gadebusch (1979) and Fromtling and Shadomy (1982) describe the historical and scientific developments in the area of immunity and cryptococcosis.

There is disagreement among researchers as to whether monocytes and/or macrophages can kill cells of *C. neoformans*. Generally, neutrophils appear to be capable of killing many species of fungi, primarily via the peroxidase systems inherent in these cells (Lehrer, 1969; Diamond, 1981a). Specifically in cryptococcosis, human neutrophilis have been shown to ingest and kill "small" capsule forms of *C. neoformans* (Diamond *et al.*, 1972). Additionally, neutrophils have been shown to mediate extracellular killing of cryptococci in the presence of anticryptococcal antibodies (Diamond and Allison, 1976, Miller and Kohl, 1983). A recent study by Diamond and Erickson (1982) showed that chemotaxis of human neutrophils and monocytes was stimulated by sera activated with cells of *C. neoformans*. Concentrations of cryptococcal capsular polysaccharide less than 1 mg/ml did not generate chemotactic factors from sera.

Cline and Lehrer (1968) demonstrated that human monocytes were capable of phagocytizing cells of *C. neoformans*. They noted that serum or serum factors were required for phagocytosis. Intracellular killing of the yeast cells was not examined in this study. Rinehart *et al.* (1974) studied the effect of corticosteroids *in vitro* on human monocyte function and determined that hydrocortisone phosphate and cortisone acetate at concentrations of 120 μg/ml reduced monocyte phagocytosis of *C. neoformans* yeast cells by 20%. The authors suggested that corticosteroid-induced defects in monocyte function may contribute to reduced cellular defense during corticosteroid therapy (Rinehart *et al.*, 1974, 1975), a fact now well established and recognized. Diamond (1974, 1977) and Diamond and Allison (1976) presented evidence that monocytes may be capable of killing cells of *C. neoformans* (by a nonphagocytic mechanism) in the presence of specific antibody. The specific cell or cells mediating and/or responsible for the nonphagocytic killing were not characterized, although the evidence that monocytes played a role was substantial.

Schneerson-Porat *et al.* (1965) reported the formation of histiocyte rings in response to infection by *C. neoformans*. This reaction, which was observed in both peritoneal exudates and the omentum of intraperitoneally inoculated mice, occurred only when the yeast was heavily encapsulated and previously exposed to immune serum. Although the authors reported that the yeast cells appeared to be degenerated, they were unable to verify killing by cultural methods. Kletter *et al.* (1980) were able to induce *in vitro* the formation of giant cells from macrophages surrounding cells of *C. neoformans*. Larger numbers of giant cells were induced when lymphocytes were added to the cultures with the addition of calcium ions or cyclic GMP.

Tissue reactions to *C. neoformans* appear to be dependent on the degree of encapsulation of the invading yeast cells. Naturally occurring capsule-deficient forms of *C. neoformans* have been reported in humans (Farmer and Komorowski, 1973; Gutierrez *et al.*, 1975). In both cases host response was significant and was characterized by marked histiocytic and fibroblastic reactions, including the formation of giant cells (Gutierrez *et al.*, 1975). Experimental murine models also have demonstrated strong host responses to *C. neoformans*. Fromtling *et al.* (1982a) observed peritoneal granulomas in mice that had been inoculated with chemically induced acapsular mutants of *C. neoformans*. The granulomas were characterized by a large, central mass consisting of intact, degenerating, and necrotic yeast cells surrounded by concentric layers of a broad band of histiocytes, a narrow band of fibroblasts, and, around the periphery, a mass of lymphocytes and plasma cells. Miyaji and Nishimura (1981) verified that a granulomatous response against *C. neoformans* infection occurs in the organs that are rich in sessile macrophages. In their study, granulomas were formed easily in the organs rich in macrophages and lymphocytes, such as the liver, spleen, lymph nodes, thymus, lungs, small intestine, and fat tissue. On the contrary, in organs poor in macrophages, such as the brain, heart, pancreas, kidneys, adrenal glands, and testes, they noted that the chief histopathologic feature was a gelantinous cyst containing numerous yeast cells but few host immune cells. Similar histopathologic observations in experimental murine cryptococcosis have been reported by Shadomy and Lurie (1971).

Kalina *et al.* (1974) studied the interaction between polymorphonuclear cells and monocytes and reported that several hydrolases, such as acid phosphatase, β-glucuronidase, and nonspecific esterase, were released from the phagocytic cells into the enclosed yeast. Disintegration of the capsule was observed as well as phagocytosis of its material. Papadimitriou *et al.* (1978) examined the ultrastructural interaction between macrophages and cells of *C. neoformans*. Long, cytoplasmic processes from the macrophages penetrated the cryptococcal capsule. The authors reported obvious degradative changes within the enclosed yeast cells. The cell body of the yeast was disrupted, the capsule thinned, and, less often, the cell wall fragmented or fibrillated. The cell wall proved to be a resistant structure and remained relatively intact in most cases.

In spite of ultrastructural studies which suggest that cells of *C. neoformans* can be degraded by macrophages, a number of investigators have presented evidence that *C. neoformans* is not readily killed following phagocytosis, and that the polysaccharide capsule inhibits macrophage attachment to and phagocytosis of the yeast cells. Bulmer and Sans (1967, 1968) reported that human leukocytes phagocytized nonencapsulated cells more readily than encapsulated cells of *C. neoformans*. Yeast cells that are engulfed by human leukocytes, however, are killed by mechanisms dependent on the generation of hydrogen peroxide (Tacker *et al.*, 1972; Diamond *et al.*, 1972). In macrophage *in vitro* models, kinetic studies showed that inhibition of phagocytosis is directly related to the presence of cryptococcal polysaccharide at the yeast surface rather than to an indirect effect by the polysaccharide on serum components necessary for phagocytosis (Kozel and Mastroianni, 1976; Kozel, 1977). Cryptococcal polysaccharide has been shown to mask immunoglobulin G (McGaw and Kozel, 1979), alter rates of phagocytosis (Breen *et al.*, 1982; Mathai and Radhakrishnan, 1983), and passively inhibit phagocytosis (Kozel and Gotschlich, 1982; Kozel, 1983). Opsonizing IgG plays a major role during the attachment phase of cryptococci to macrophages (Kozel and McGaw, 1979; Kozel and Follette, 1981); thus interference of phagocytosis by capsular material is a major problem in effective host response to cryptococcosis. A study by Griffin (1981) suggests that a lymphokine that enhances macrophage complement receptor function may, in part, overcome the antiphagocytic effect of the capsular material. He noted that murine peritoneal macrophages avidly phagocytized C3-coated, encapsulated cryptococci in the presence of the lymphokine. Morgan *et al.* (1983), however, have reported a heat-stable, trypsin-sensitive, allospecific lymphokine derived from the supernatant of cultured lymphocytes taken from *C. neoformans*–infected mice which nonspecifically suppressed phagocytosis by macrophages. It is apparent that much remains to be learned concerning the interaction of lymphokines and macrophages in cryptococcosis.

Although polymorphonuclear leukocytes clearly have been shown to kill phagocytized cells of *C. neoformans*, macrophages appear to be less effective killers. Bulmer and Tacker (1975) demonstrated that guinea pig alveolar macrophages could ingest nonencapsulated cells of C. neoformans, but were unable to kill the engulfed cells during the first few hours following exposure. This observation was supported by Karaoui *et al.* (1977a,b) using alveolar and peritoneal macrophages from sensitized mice, by Diamond (1977) using guinea pig macrophages, and by Diamond and Bennett (1973), who demonstrated growth of *C. neoformans* within human macrophages that had been activated by *C. neoformans*. Mitchell and Friedman (1972), however, examined the intracellular fate of *C. neoformans* in rat peritoneal exudate cells and noted that many of the yeast cells did not multiply and were presumably dead. Sethi and Pelster (1974) reported that murine peritoneal macrophages could actively inhibit the intracellular replication of thinly encapsulated cells of *C. neoformans*, although cells having medium to large capsules were capable of multiplication intracellularly. With the few exceptions reported above, the apparent inability of macrophages to kill engulfed cells of *C. neoformans in vitro* may be attributed to (1) a deficiency of peroxidase-based killing mechanism, (2) inability of primary and secondary lysosomes to fuse, or (3) delayed activation of macrophage oxidative metabolism (Gadebusch, 1979).

Macrophages can be activated by a variety of procedures (Evans, 1974). Some studies have demonstrated a nonspecific protective role of macrophages *in vivo*. Gentry and Remington (1971) noted that persistent infection with intracellular bacteria and protozoa conferred resistance against infection with *C. neoformans*. The activated macrophages in their model resisted destruction following phagocytosis of *C. neoformans*. Regelson *et al.* (1970, 1974) and Fromtling *et al.* (1982b, 1984) used pyran copolymer to activate murine macrophages and demonstrated that pretreatment of mice resulted in the conversion of *C. neofor-*

mans–infected animals to a healthy carrier state for prolonged periods. Monga (1981), using intravenous silica injections to destroy macrophages in a murine model of cryptococcosis, presented evidence that macrophages may play a role in natural or nonspecific cell-mediated immunity to murine cryptococcosis. Anderson and Dykstra (1984) observed increased resistance to experimental murine cryptococcosis following vaccination with formalin-killed cells of *C. neoformans* administered either intranasally, intravenously, or intraperitoneally. Increased phagocytic activity of alveolar macrophages against cells of *C. neoformans*, however, was observed only in macrophages derived from intranasally vaccinated mice, suggesting a possible interaction of alveolar macrophages and immune lymphocytes in host response to this disease. In another study, Kitz *et al.* (1984) determined that bone marrow–derived macrophage cell lines from a variety of experimental mouse strains were capable of inhibiting the growth of a lightly encapsulated isolate of *C. neoformans*. Data from this study supported the hypothesis of complement fraction C3b as an important modulator of interactions between macrophages and the encapsulated yeast *C. neoformans*, a hypothesis recently examined by Kozel and Pfrommer (1986), whose data also suggest that iC3b is the ligand that most likely interacts with phagocyte C3 receptors involved in the phagocytosis of *C. neoformans*.

Amid the conflicting data, it is apparent that macrophages do play a key role in host defense against cryptococcosis. The specific role of these specialized immune cells, particularly their ability to phagocytize and kill engulfed yeast cells of *C. neoformans*, remains unknown.

3.4. CANDIDIASIS

Candidiasis may be a primary or secondary infection caused by yeasts of the genus *Candida*. The most common etiologic agent is *Candida albicans*, although numerous other *Candida* spp. may be involved. *Candida spp.* are endogenous as normal flora on the mucocutaneous regions and alimentary tracts of humans and animals (Emmons *et al.*, 1977; Rippon, 1982). The ubiquity of *C. albicans* makes it difficult to distinguish between the mucosal, commensal, and etiologic agent of a disease process. *C. albicans* is a saprophytic organism under normal conditions and a limited pathogen in large numbers, but is distinctly pathogenic in a host with depressed or otherwise compromised defense mechanisms (Rippon, 1982). Predisposing factors include long-term chemotherapy, indwelling catheters, diabetes, drug addiction, underlying disease processes, surgical procedures, or a number of factors that alter the host's immunocompetency. Candidiasis may be manifested as a superficial skin eruption, chronic infection of nails, disease of the mouth, throat, or vagina, or a frequently fatal systemic disease which may involve the lungs, heart, gastrointestinal tract, and other organs (Emmons *et al.*, 1977; Rippon, 1982).

A majority of the studies concerned with the interaction between phagocytic cells and fungi have been directed to *C. albicans*. As with *C. neoformans*, there is much disagreement among researchers as to how effective phagocytic cells are in killing yeastlike or hyphallike cells of *C. albicans*. Louria and Brayton (1964) reported that human polymorphonuclear leukocytes readily phagocytized yeast cells of *C. albicans*. In spite of the rapid phagocytosis, the fungus was able to survive within the phagocytic cells and eventually kill the phagocyte by formation of germ tubes and hyphae. However, subsequent evidence by other investigators suggested that leukocytes are capable of killing cells of *C. albicans*. Lehrer and Cline (1969a) and Patterson-Delafield and Lehrer (1977) described a simple quantitative method for measuring the ability of humans neutrophils to kill ingested cells of *C. albicans*. It was determined that neutrophils were efficient killers of *C. albicans* and that myeloperoxidase activity was the primary candidacidal factor (Lehrer and Cline, 1969b; Diamond and Krzesicki, 1978). A recent report by Bistoni *et al.* (1982) described a radiolabeled chromium-release microassay for quantifying intracellular killing of *C. albicans*. The technique described permits easy-to-perform, simultaneous testing of a variety of phagocytes even when only small quantities are available.

Additional studies clarified the importance of the third component of complement (C3) in intracellular killing of *C. albicans*. Morelli and Rosenberg (1971) reported that phagocytosis of the yeast by mouse blood leukocytes was enhanced by specific antiserum and by a complete hemolytic complement system. The extent of intracellular killing of *Candida* in murine peritoneal exudate cells, however, was not influenced by the presence of antibody or a complete complement system. Yamamura and Valdimarrson (1977) demonstrated that *C. albicans* opsonized by C3-deficient serum was ingested, but not killed *in vitro* by human polymorphonuclear leukocytes. Killing could be induced by adding purified C3 to the C3-deficient serum. A recent study by Morrison and Cutler (1981) verified the importance of complement in the phagocytosis and intracellular killing of *C. albicans* by neutrophils and macrophages. A recent report by Diamond (1981b) presents a review of host mechanisms of resistance to *C. albicans* with special emphasis on the role of neutrophils in combatting this disease.

Specific interactions between monocytes and cells of *C. albicans* also have been studied. Lehrer (1970) presented data suggesting that human monocytes via a myeloperoxidase-linked system were fungicidal for *C. albicans*. Monocytes were found to contain cationic proteins, other than myeloperoxidase, that killed cells of *Candida parapsilosis in vitro*. These components, which differed substantially from the fungicidal proteins of the neutrophil, were proposed as a second, nonmyeloperoxidase-based mechanism of intracellular killing of *Candida* possessed by monocytes (Lehrer, 1975) and macrophages (Lehrer *et al.*, 1980). Diamond and Haudenschild (1981) recently confirmed the importance of the myeloperoxidase-linked mechanism of killing cells of *C. albicans*. In an interesting "turnabout," Danley and Polakoff (1986) reported that yeast cells of *C. albicans* are capable of generating significant quantities of hydrogen peroxide and may, under certain conditions, damage monocytes that have phagocytized them. Membrane damage to monocytes that had phagocytized yeast cells of *C. albicans* were detected using an ethidium bromide technique. The role of this observation *in vivo* has not been determined as yet, but the ability of some isolates of this fungus to damage phagocytizing monocytes may play a role in the pathogenesis of this organism.

Patterson-Delafield *et al.* (1980, 1981) reported that rabbit macrophages contain two highly cationic microbicidal proteins, which were shown to be distinct from histones. These macrophage proteins were active against *C. albicans* and *C. parapsilosis*, as well as several bacterial isolates, and had not been demonstrated previously in any mammalian macrophage. The two microbicidal proteins were designated MCP-1 (rich in arginine and half-cystine) and MCP-2 (less arginine and half-cystine than MCP-1). MCP-1 was shown to kill 99.6% of *C. albicans* in 20 min at 1.0 μg/ml; MCP-2 was shown to kill ca. 80% under similar experimental conditions. Both proteins were shown to suppress oxygen consumption and induce a loss of radiolabeled rubidium (Patterson-Delafield *et al.*, 1981). Further experimentation is required to clarify the role of these microbicidal cationic proteins in host protection against fungal infection.

Leijh *et al.* (1977) studied the kinetics of phagocytosis and intracellular killing of *C. albicans* by human granulocytes and monocytes. In the presence of serum, 96% of the yeast cells were phagocytized; granulocytes and monocytes killed 58 and 50% of the ingested *C. albicans*, respectively. An additional kinetics study by Schuit (1979) suggested that monocytes play an important role in host defenses against fungal diseases, and that the relative virulence of a pathogenic yeast in human disease may be related to the ability of a fungus to survive after being ingested by a circulating phagocyte. More specifically, mononuclear phagocytes may be important components of host defense mechanisms against invasive candidiasis by contributing to prevention of dissemination or by limiting the extent of mucocutaneous lesions.

The ability of macrophages to kill cells of *Candida* species has been studied extensively. Stanley and Hurley (1969) reported that mouse macrophage cultures were destroyed within 24 hr after culture with cells of *C. albicans* and *C. tropicalis*, both of which formed mycelium and pseudomycelium. Macrophage cultures were destroyed within 72 hr after culture with *C. krusei*, *C. pseudotropicalis*, and *C. parapsilosis*. The yeast cells produced germ tubes and pseudymycelium, which ruptured and killed the phagocytizing macrophages. Ozato and Uesaka (1974) reported that murine peritoneal macrophages were unaffected 2 hr after ingesting cells of *C. albicans* and that the macrophages suppressed metabolism and growth of the phagocytized yeast cells. However, by 3 hr the metabolic activity of the macrophages decreased markedly and growth of the ingested yeast proceeded, with subsequent rupture of the macrophage. Arai *et al.* (1977) reported similar data using rabbit alveolar macrophages.

More recent evidence by several investigators suggests that macrophages are capable of killing cells of *C. albicans*. Meister *et al.* (1977) induced hepatic granulomas in mice by injection of yeast cells, cell walls, and glucan of *C. albicans*. Fungal units appeared in liver macrophages and were detectable in granulomas and Kupffer cells for 20 days postinjection. Degradation of phagocytized cells, reduction of cytoplasm, and cell wall deformation and collapse were evidence of enzymatic breakdown of *Candida* cell wall mannan by macrophages. Natural cell-mediated cytotoxicity studies by Baccarini *et al.* (1983) and Bistoni *et al.* (1983), silica pretreatment of athymic and euthymic mice analyses by Lee and Balish (1983), and macrophage–lymphocyte interaction studies by Hadfield and Marcus (1982) all have provided strong experimental evidence that the macrophage plays a key role in host defense against infection and disease with *C. albicans*. Bistoni *et al.* (1986) provide evidence of macrophage-mediated protection against lethal infection with *C. albicans*. Using an experimental mouse model, these investigators primed mice by infection with a low-virulence isolate of *C. albicans*. Following this priming, the mice were challenged with a virulent isolate of *C. albicans*, and a nonspecific increase in resistance was observed with a concomitant increase in granulocyte and macrophagelike cells. The authors stated that these cell lines were crucial to host protection against candidiasis.

Peterson and Calderone (1978) reported inhibition of specific amino acid uptake in *C. albicans* by lysosomal extracts from rabbit alevelor macrophages. Viability of *Candida* cells, as measured by colony-forming ability, decreased after 16 hr of incubation with the extract. A study of the relationship between *C. albicans* and rabbit alveolar macrophages determined that germ tube formation by cells of *C. albicans* that had been phagocytized was inhibited in contrast to extracellular (nonphagocytized) *C. albicans*. A 71–93% inhibition of macromolecular synthesis in the fungus was obtained when the organism was incubated with alveolar macrophages (Peterson and Calderone, 1977).

Rogers and Balish (1977) reported that *in vivo* activation of mouse macrophages by vaccination with *Mycobacterium bovis* (BCG strain) did not protect the animals against renal candidiasis. Maiti *et al.* (1980), however, reported that activated macrophages obtained from either BCG-vaccinated mice or by *in vitro* exposure of normal macrophages to phytohemagglutinin-induced lymphokines exhibited higher phagocytic and candidacidal activities than did normal macrophages. The incorporation of immune mouse serum into the culture medium enhanced both the phagocytic and candidacidal activity of the macrophages. Evron (1980) reported increased phagocytosis and intracellular killing of *C. albicans* in mouse peritoneal macrophages that had been sensitized to *C. albicans*. *Corynebacterium parvum* vaccination of mice significantly increased the number of leukocytes adherent to hepatic vessels (Sawyer *et al.*, 1981). Perfused *C. parvum*–treated livers killed significantly more *C. albicans* than did livers not treated with *C. parvum*. This effect was reversed by the macrophage inhibitors silica, iodoacetate, and phenylbutazone (Sawyer *et al.*, 1981). Sasada and Johnston (1980) reported that macrophages elicited by injection of lipopolysaccharide or obtained from animals infected with BCG killed cells of *C. albicans* and *C. parapsilosis* more effectively than did resident cells. The killing of the yeasts was inhibited by scavengers of oxygen radicals, which suggested that oxygen metabolites may play an important role in macrophage candidacidal activity. In a recent study in which the candidacidal activity of macrophages from normal and thymus-deficient (nude) mice was assayed, macrophages from normal mice were candidastatic while macrophages from nude mice were candidacidal (Cutler and Poor, 1981; Poor and Cutler, 1981). Results of gradient fractionations of peritoneal exudates indicated that more than one cell type may be responsible for candidacidal activity by nude mouse macrophage-rich cells. Additionally, hyphal-phase cells appeared to be more resistant than yeast-phase cells to killing by normal and nude mouse phagocytic cells.

In spite of the large amount of research investigating the role of immune cells, macrophages in particular, in host defense against candidiasis, the conflicting data suggest that additional study of the interaction between macrophages and *C. albicans* is needed.

4. CONCLUSIONS

Recent increases in the number of reported mycoses appear to be related not only to improved education and diagnosis, but also to the rise in the number of immunocompromised hosts, i.e., organ transplant recipients, cancer patients undergoing prolonged chemotherapy, patients suffering from acquired immunodeficiency syndrome. These compromised patients are prime candidates for development of the opportunistic mycoses. We are beginning to realize how much conflicting data exists and how little is known about host defenses against mycotic agents. If the major advances that have been made in immunology, molecular biology, and medicine are aggressively applied to the study of the mycoses, then much valuable information can be learned and applied in the near future to alleviate morbidity and mortality due to the mycoses.

REFERENCES

Anderson, D. M., and Dykstra, M. A., 1984, Resistance to challenge and macrophage activity in mice previously vaccinated with formalin-killed *Cryptococcus neoformans, Mycopathologia* **68:**169–177.

Arai, T., Mikami, Y., and Yokoyama, K., 1977, Phagocytosis of *Candida albicans* by rabbit alvelolar macrophages and guinea pig neutrophils, *Sabouraudia* **15:**171–178.

Baccarini, M., Bistoni, F., Puccetti, P., and Garaci, E., 1983, Natural cell-mediated cytotoxicity against *Candida albicans* induced by cyclophosphamide: Nature of the *in vitro* cytotoxic effector, *Infect. Immun.* **42:**1–9.

Beaman, L., and Holmberg, C. A., 1980, *In vitro* response of alveolar macrophages to infection with *Coccidioides immitis, Infect. Immun.* **28:**594–600.

Beaman, L., and Pappagianis, D., 1985, Fate of *Coccidioides immitis* arthroconidia in human peripheral blood monocyte cultures *in vitro,* in: *Coccidioidomycosis, Proceedings of the Fourth International Congress* (H. E. Einstein and A. Catanzaro, eds.), pp. 170–180, National Foundation for Infectious Diseases, Washington, DC.

Beaman, L., Pappagianis, D., and Benjamini, E., 1977, Significance of T cells in resistance to experimental murine coccidioidomycosis, *Infect. Immun.* **17:**580–585.

Beaman, L., Benjamini, E., and Pappagianis, D., 1981, Role of lymphocytes in macrophage induced killing of *Coccidioides immitis in vitro, Infect. Immun.* **34:**347–353.

Beaman, L., Benjamini, E., and Pappagianis, D., 1983, Activation of macrophages by lymphokines: Enhancement of phagosome–lysosome fusion and killing of *Coccidioides immitis, Infect. Immun.* **39:**1201–1207.

Bistoni, F., Baccarini, M., Blasi, E., Puccetti, P., and Marconi, P., 1982, A radiolabel release microassay for phagocytic killing of *Candida albicans, J. Immunol. Meth.* **52:**369–377.

Bistoni, F., Baccarini, M., Blasi, E., Marconi, P., Puccetti, P., and Garaci, E., 1983, Correlation between *in vivo* and *in vitro* studies of modulation of resistance to experimental *Candida albicans* infection by cyclophosphamide in mice, *Infect. Immun.* **40:**46–55.

Bistoni, F., Vecchiarelli, A., Cenci, E., Puccetti, P., Marconi, P., and Cassone, A., 1986, Evidence for macrophage-mediated protection against lethal *Candida albicans* infection, *Infect. Immun.* **51**:668–674.

Bradsher, R. W., Ulmer, W. C., Marmer, D. J., Townsend, J. W., and Jacobs, R. J., 1985, Intracellular growth and phagocytosis of *Blastomyces dermatitidis* by monocyte-derived macrophages from previously infected and normal subjects, *J. Infect. Dis.* **151**:57–64.

Breen, J. F., Lee, I. C., Vogel, F. R., and Friedman, H., 1982, Cryptococcal capsular polysaccharide-induced modulation of murine immune responses, *Infect. Immun.* **36**:47–51.

Brummer, E., and Stevens, D. A., 1982, Opposite effects of human monocytes, macrophages, and polymorphonuclear neutrophils on replication of *Blastomyces dermatitidis in vitro, Infect. Immun.* **36**:297–303.

Brummer, E., Morozumi, P. A., and Stevens, D. A., 1980, Macrophages and fungi; *in vitro* effects of method and macrophage induction, activation by different stimuli, and soluble factors on *Blastomyces, J. Reticuloendoth. Soc.* **28**:507–518.

Brummer, E., Morozumi, P. A., Philpott, D. E., and Stevens, D. A., 1981, Virulence of fungi: Correlation of virulence of *Blastomyces dermatitidis in vivo* with escape from macrophage inhibition of replication *in vitro, Infect. Immun.* **32**:864–871.

Brummer, E., Sugar, A. M., and Stevens, D. A., 1983, Activation of peritoneal macrophages by concanavalin A or *Mycobacterium bovis* for fungicidal activity against *Blastomyces dermatitidis* and effect of specific antibody and complement, *Infect. Immun.* **39**:817–822.

Brummer, E., Sugar, A. M., and Stevens, D. A., 1985, Enhanced oxidative burst in immunologically activated but not elicited polymorphonuclear leukocytes correlates with fungicidal activity, *Infect. Immun.* **49**:396–401.

Bulmer, G. S., and Sans, M. D., 1967, *Cryptococcus neoformans*. II. Phagocytosis by human leukocytes, *J. Bacteriol.* **94**:1480–1483.

Bulmer, G. S., and Sans, M. D., 1968, *Cryptococcus neoformans*. III. Inhibition of phagocytosis, *J. Bacteriol.* **95**:5–8.

Bulmer, G. S., and Tacker, J. R., 1975, Phagocytosis of *Cryptococcus neoformans* by alveolar macrophages, *Infect. Immun.* **11**:73–79.

Calderone, R. A., and Peterson, E., 1979, Inhibition of amino acid uptake and incorporation into *Histoplasma capsulatum* by a lysosomal extract from rabbit alveolar macrophages, *J. Reticuloendothel. Soc.* **26**:11–19.

Caldwell, C. W., and Sprouse, R. F., 1982, Iron and host resistance in histoplasmosis, *Am. Rev. Respir. Dis.* **125**:674–677.

Clemons, K. V., Leathers, C. R., and Lee, K. W., 1985, Role of activated macrophages in resistance to experimental *Coccidioides immitis* infections, in: *Coccidioidomycosis, Proceedings of the Fourth International Congress* (H. E. Einstein and A. Catanzaro, eds.), pp. 149–159, National Foundation for Infectious Diseases, Washington, DC.

Cline, J. J., and Lehrer, R. I., 1968, Phagocytosis by human monocytes, *Blood* **32**:423–435.

Cutler, J. E., and Poor, A. H., 1981, Effect of mouse phagocytes on *Candida albicans* in *in vivo* chambers, *Infect. Immun.* **31**:1110–1116.

Danley, D. L., and Polakoff, J., 1986, Rapid killing of monocytes *in vitro* by *Candida albicans* yeast cell, *Infect. Immun.* **51**:307–313.

Deepe, G. S., Jr., Watson, S. R., and Bullock, 1984, Cellular origins and target cells of immunoregulatory factors in mice with disseminated histoplasmosis, *J. Immunol.* **132**:2064–2071.

DeSanchez, S. B., and Carbonell, L. M., 1975, Immunological studies on *Histoplasma capsulatum, Infect. Immun.* **11**:387–394.

Diamond, R. D., 1974, Antibody dependent killing of *Cryptococcus neoformans* by human peripheral blood mononuclear cells, *Nature (London)* **247**:148–150.

Diamond, R. D., 1977, Effects of stimulation and suppression of cell-mediated immunity on experimental cryptococcosis, *Infect. Immun.* **17**:187–194.

Diamond, R. D., 1981a, Antifungal properties of granulocytes, in: *Microbiology—1981* (D. Schlessinger, ed.), pp. 184–187, American Society for Microbiology, Washington, DC.

Diamond, R. D., 1981b, Mechanisms of host resistance to *Candida albicans*, in: *Microbiology—1981* (D. Schlessinger, ed.), pp. 200–204, American Society for Microbiology, Washington, DC.

Diamond, R. D., 1983, Inhibition of monocyte-mediated damage to fungal hyphae by steroid hormones, *J. Infect. Dis.* **147**:160.

Diamond, R. D., and Allison, A. C., 1976, Nature of the effector cells responsible for antibody dependent cell-mediated killing of *Cryptococcus neoformans, Infect. Immun.* **14**:716–720.

Diamond, R. D., and Bennett, J. E., 1973, Growth of *Cryptococcus neoformans* within human macrophages *in vitro, Infect. Immun.* **7**:231–236.

Diamond, R. D., and Clark, R. A., 1982, Damage to *Aspergillus fumigatus* and *Rhizopus oryzae* hyphae by oxidative and nonxoidative microbicidal products of human neutrophils *in vitro, Infect. Immun.* **38**:487–495.

Diamond, R. D., and Erickson, III, N. F., 1982, Chemotaxis of human neutrophils and monocytes induced by *Cryptococcus neoformans, Infect. Immun.* **38**:380–382.

Diamond, R. D., and Haudenschild, C. C., 1981, Monocyte-mediated serum-independent damage to hyphal and pseudohyphal forms of *Candida albicans in vitro, J. Clin. Invest.* **67**:173–182.

Diamond, R. D., and Krzesicki, R., 1978, Mechanisms of attachment of neutrophils to *Candida albicans* pseudohyphae in the absence of serum, and of subsequent damage to pseudohyphae by microbicidal processes of neutrophils *in vitro, J. Clin. Invest.* **61**:360–369.

Diamond, R. D., Root, R. K., and Bennett, J. E., 1972, Factors influencing killing of *Cyrptococcus neoformans* by human leukocytes *in vitro, J. Infect. Dis.* **125**:367–376.

Diamond, R. D., Krzesicki, R., Epstein, B., and Jao, W., 1978, Damage to hyphal forms of fungi by human leukocytes *in vitro:* A possible host defense mechanism in aspergillosis and mucormycosis, *Am. J. Pathol.* **91**:313–328.

Diamond, R. D., Haudenschild, C. C., and Erickson, N. F., III, 1982, Monocyte-mediated damage to *Rhizopus oryzae* hyphae *in vitro, Infect. Immun.* **38**:292–297.

Diamond, R. D., Huber, E., and Haudenschild, C. C., 1983, Mechanisms of destruction of *Aspergillus fumigatus* hyphae mediated by human monocytes, *J. Infect. Dis.* **147**:474–483.

Drutz, D. J., and Catanzaro, A., 1978, Coccidioidomycosis. Part 1, *Am. Rev. Respir. Dis.* **117**:559–585.

Drutz, D. J., and Frey, C. L., 1985, Intracellular and extracellular defenses of human phagocytes against *Blastomyces dermatitidis* conidia and yeasts, *J. Lab. Clin. Med.* **105**:737–750.

Drutz, D. J., and Huppert, M., 1983, Coccidioidomycosis: factors affecting the host-parasite interaction, *J. Infect. Dis.* **147**:372–390.

Dumont, A., and Robert, A., 1970, Electron microscope study of phagocytosis of *Histoplasma capsulatum* by hamster peritoneal macrophages, *Lab. Invest.* **23**:278–286.

Edwards, G. A., Edwards, M. R., and Hazen, F. L., 1959, Electron microscope study of *Histoplasma* in mouse spleen, *J. Bacteriol.* **77**:429–438.

Emmons, C. W., Binford, C. H., Utz, J. P., and Kwon-Chung, K. J., 1977, *Medical Mycology,* 3rd ed., Lea & Febiger, Philadelphia.

Epstein, S. M., Verney, E., Miale, T. D., and Sidransky, H., 1967, Studies on the pathogenesis of experimental pulmonary aspergillosis, *Am. J. Pathol.* **51**:769–788.

Evans, R., 1974, Specific and nonspecific activation of macrophages, in: *Activation of Macrophages. Proccedings of the Second International Workshop Conference Hoechst,* Volume 2 (W. H. Wagner and H. Hahn, eds.), pp. 305–315, Excerpta Medica, Amsterdam.

Evron, R., 1980, *In vitro* phagocytosis of *Candida albicans* by peritoneal mouse macrophages, *Infect. Immun.* **28**:963–971.

Farmer, S. G., and Komorowski, R. A., 1973, Histologic response to capsule-deficient *Cryptococcus neoformans, Arch. Pathol.* **96**:383–387.

Fauci, A. S., 1978, *Host-Defense Mechanisms against Disease,* Upjohn, Kalamazoo, MI.

Fromtling, R. A., and Shadomy, H. J., 1982, Immunity in cryptococcosis: An overview, *Mycopathologia* **77**:183–190.

Fromtling, R. A., and Shadomy, H. J., 1986: An overview of macrophage–fungal interactions, *Mycopathologia* **93**:77–93.

Fromtling, R. A., Shadomy, H. J., and Jacobson, E. S., 1982a, Decreased virulence in stable, acapsular mutants of *Cryptococcus neoformans, Mycopathologia* **79**:23–29.

Fromtling, R. A., Shadomy, H. J., and Kaplan, A. M., 1982b, *Cryptococcus neoformans: In vivo* protection of mice by pretreatment with pyran copolymer, *Abstr. Proc. XIII Int. Congress Microbiol.,* Royal Society of London, Manchester.

Fromtling, R. A., Shadomy, H. J., and Kaplan, A. M., 1984. *Cryptococcus neoformans: In vivo* protection of mice by pretreatment with pyran copolymer, *Mycopathologia* **85**:3–7.

Gadebusch, H. H., 1979. Native and acquired resistance to infection with *Cryptococcus neoformans*, in: *Phagocytes and Cellular Immunity* (H. H. Gadebusch, ed.), pp. 137–157, CRC Press, Boca Raton, FL.

Galgiani, J. N., Hayden, R., and Payne, C. M., 1982, Leukocyte effects on the dimorphism of *Coccidioides immitis*, *J. Infect. Dis.* **146**:56–63.

Garrison, R. G., and Boyd, K. S., 1978, Role of the conidium in dimorphism of *Blastomyces dermatitidis*, *Mycopathologia* **64**:29–33.

Gentry, L. O., and Remington, J. S., 1971, Resistance against *Cryptococcus* conferred by intracellular bacteria and protozoa, *J. Infect. Dis.* **123**:22–31.

Goodwin, R. A., and Des Prez, R. M., 1978, Histoplasmosis, *Am. Rev. Respir. Dis.* **117**:929–955.

Griffin, F. M., Jr., 1981, Roles of macrophage Fc and C3b receptors in phagocytosis of immunologically coated *Cryptococcus neoformans*, *Proc. Natl. Acad. Sci. USA* **78**:3853–3857.

Gutierrez, F., Fu, Y. S., and Lurie, H. I., 1975, Cryptococcosis histologically resembling histoplasmosis, *Arch. Pathol.* **99**:347–352.

Hadfield, T. L., and Marcus, S., 1982, Macrophage and lymphocyte contributions in resistance to *Candida albicans* infections, *Immunol. Commun.* **11**:201–216.

Hart, P. D'Arcy, 1981, Macrophage antimicrobial activity: Evidence for participation by lysosomes in the killing of *Saccharomyces cerevisiae* by normal resident macrophages, *Infect. Immun.* **31**:828–830.

Hart, P. D'Arcy, and Young, M. R., 1975, Interference with normal phagosome-lysosome fusion in macrophages using ingested yeast cells and suramin, *Nature (London)* **256**:47–49.

Hill, G. A., and Marcus, S., 1960, Study of cellular mechanisms in resistance to systemic *Histoplasma capsulatum*, *J. Immunol.* **85**:6–13.

Hotchi, M., Okada, M., and Nasu, T., 1980, A statistical survey of mycoses in the total autopsy cases in Japan during the period from 1966 to 1975, in: *Human and Animal Mycology, Proceeding of the VII Congress of ISHAM* (E. S. Kuttin, and G. L. Baum, eds.), pp. 26–29, Excerpta Medica, New York.

Howard, D. H., 1965, Intracellular growth of *Histoplasma capsulatum*, *J. Bacteriol.* **89**:518–523.

Howard, D. H., 1973, Further studies on the inhibition of *Histoplasma capsulatum* within macrophages from immunzied animals, *Infect. Immun.* **8**:577–581.

Howard, D. H., 1981a, Comparative sensitivity of *Histoplasma capsulatum* conidiospores and blastospores to oxidative antifungal systems, *Infect. Immun.* **32**:381–387.

Howard, D. H., 1981b, Mechanisms of resistance in the systemic mycoses, in: *Immunology of Human Infection* (A. J. Nahmias and R. J. O'Reilly, eds.), pp. 475–494, Plenum Press, New York.

Howard, D. H., 1983, Studies on the catalase of *Histoplasma capsulatum*, *Infect. Immun.* **39**:1161–1166.

Howard, D. H., and Otto, V., 1977, Experiments on lymphocyte mediated cellular immunity in murine histoplasmosis, *Infect. Immun.* **16**:226–231.

Howard, D. H., Otto, V., and Gupta, R. K., 1971, Lymphocyte-mediated cellular immunity in histoplasmosis, *Infect. Immun.* **4**:605–610.

Kalina, M., Kletter, Y., and Aronson, M., 1974, The interaction of phagocytes and the large-sized parasite *Cryptococcus neoformans: Cytochemical and ultrastructural study*, *Cell. Tissue Res.* **152**:165–174.

Karaoui, R. M., Hall, N. K., and Larsh, H. W., 1977a, Role of macrophages in immunity and pathogenesis of experimental cryptococcosis induced by the airborne route. Part I: Pathogenesis and acquired immunity of *Cryptococcus neoformans*, *Mykosen* **20**:380–388.

Karaoui, R. J., Hall, N. K., and Larsh, H. W., 1977b, Role of macrophages in immunity and pathogenesis of experimental cryptococcosis induced by the airborne route. Part II: Phagocytosis and intracellular fate of *Cryptococcus neoformans*, *Mykosen* **20**:409–422.

Kashkin, K. P., Likholetov, S. M., and Lipnitsky, A. V., 1977, Studies on mediators of cellular immunity in experimental coccidioidomycosis, *Sabouraudia* **15**:59–68.

Kimberlin, C. L., Hariri, A. R., Hempel, H. O., and Goodman, N. L., 1981, Interactions between *Histoplasma capsulatum* and macrophages from normal and treated mice: Comparison of the mycelial and yeast phases of alveolar and peritoneal macrophages, *Infect. Immun.* **34**:6–10.

Kitz, D. J., Embree, R. M., and Cazin, Jr., J., 1981, Comparative virulence of *Absidia corymbifera* strains in mice, *Infect. Immun.* **33**:395:400.

Kitz, D. J., Johnson, C. R., Kobayashi, G. S., Medoff, G., and Little, J. R., 1984, Growth inhibition of *Cryptococus neoformans* by cloned cultured murine macrophages, *Cell. Immunol.* **88**:489–500.

Kletter, Y., Paran, N., and Aronson, M., 1980, *In vitro* formation of giant cells from macrophages surrounding *Cryptococcus neoformans* cells, in: *Human and Animal Mycology, Proceedings of the VII Congress of ISHAM* (E. S. Kuttin, and G. L. Baum, eds.), pp. 102–105, Excerpta Medica, New York.

Kozel, T. R., 1977, Non-encapsulated variant of *Cryptococcus neoformans*. II. Surface receptors for cryptococcol polysaccharide and their role in inhibition of phagocytosis by polysaccharide, *Infect. Immun.* **16**:99–106.

Kozel, T. R., 1983, Dissociation of a hydrophobic surface from phagocytosis of encapsulated and non-encapsulated *Cryptococcus neoformans, Infect. Immun.* **39**:1214–1219.

Kozel, T. R., and Follette, J. L., 1981, Opsonization of encapsulated *Cryptococcus neoformans* by specific anticapsular antibody, *Infect. Immun.* **31**:978–984.

Kozel, T. R., and Gotschlich, E. C., 1982, The capsule of *Cryptococcus neoformans* passively inhibits phagocytosis of the yeast by macrophages, *J. Immunol.* **129**:1675–1680.

Kozel, T. R., and Mastroianni, R. P., 1976, Inhibition of phagocytosis by cryptococcal polysaccharide: dissociation of the attachment and ingestion phases of phagocytosis, *Infect. Immun.* **14**:62–67.

Kozel, T. R., and McGaw, T. G., 1979, Opsonization of *Cryptococcus neoformans* by human immunoglobulin G: Role of immunoglobulin G in phagocytosis by macrophages, *Infect. Immun.* **25**:255–261.

Kozel, T. R., and Pfrommer, G. S. T., 1986, Activation of the complement system by *Cryptococcus neoformans* leads to binding of iC3b to the yeast, *Infect. Immun.* **52**:1–5.

Lee, K. W., and Balish, E., 1983, Systemic candidosis in silica-treated athymic and euthymic mice, *Infect. Immun.* **41**:902–907.

Lehrer, R. I., 1969, Antifungal effects of peroxidase systems, *J. Bacteriol.* **99**:361–365.

Lehrer, R. I., 1970, The fungicidal activity of human monocytes: A myeloperoxidase linked mechanism, *Clin. Res.* **18**:408.

Lehrer, R. I., and Cline, M. J., 1969a, Interaction of *Candida albicans* with human leukocytes and serum, *J. Bacteriol.* **98**:996–1004.

Lehrer, R. I., and Cline, M. J., 1969b, Leukocyte myeloperoxidase deficiency and disseminated candidiasis: the role of myleoperoxidase in resistance to *Candida* infection, *J. Clin. Invest.* **48**:1478–1488.

Lehrer, R. I., and Jan, R. G., 1970, Interaction of *Aspergillus fumigatus* spores with human leukocytes and serum, *Infect. Immun.* **1**:345–350.

Lehrer, R. I., 1975, The fungicidal mechanisms of human monocytes. I. Evidence for myeloperoxidase–linked and myeloperoxidase–independent candidacidal mechanisms, *J. Clin. Invest.* **55**:338–346.

Lehrer, R. I., Ferrari, L. G., Patterson-Delafield, J., and Sorrell, T., 1980, Fungicidal activity of rabbit alveolar and peritoneal macrophages against *Candida albicans, Infect. Immun.* **28**:1001–1008.

Leijh, P. C. J., van den Barselaar, M. T., and van Furth, R., 1977, Kinetics of phagocytosis and intracellular killing of *Candida albicans* by human granulocytes and monocytes, *Infect. Immun.* **17**:313–318.

Levitz, S. M., and Diamond, R. D., 1985, Mechanisms of resistance of *Asperquillus fumigatus* conidia to killing by neutrophils *in vitro, J. Infect. Dis.* **152**:33–42.

Louria, D. B., and Brayton, R. G., 1964, Behavior of *Candida* cells within leukocytes, *Proc. Soc. Exp. Biol. Med.* **115**:93–98.

Lundborg, M., and Holma, B., 1972, *In vitro* phagocytosis of fungal spores by rabbit lung macrophages, *Sabouraudia* **10**:152–156.

Lundborg, M., and Holma, B., 1974, The influence of *Aspergillus fumigatus* spores and polystyrene particles on the number of lung macrophages in rabbits, *Sabouraudia* **12**:105–111.

Maiti, P. K., Kumar, R., and Mohapatra, L. N., 1980, Candidacidal activity of mouse macrophages *in vitro, Infect. Immun.* **29**:477–482.

Mathai, A., and Radhakrishnan, V. V., 1983, Role of hyaluronidase in *in vitro* phagocytosis of encapsulated yeasts of *Cryptococcus neoformans, Indian J. Exp. Biol.* **21**:285–288.

McDaniel, L. S., and Cozad, G. C., 1983, Immunomodulation by *Blastomyces dermatitidis:* Functional activity of murine peritoneal macrophages, *Infect. Immun.* **40**:733–740.

McGaw, T. G., and Kozel, T. R., 1979, Opsonization of *Cryptococcus neoformans* by human immunoglobulin G: Masking of immunoglobulin G by cryptococcal polysaccharide, *Infect. Immun.* **25**:262–267.

Meister, H., Heymer, B., Schafer, H., and Haferkamp, O., 1977, Role of *Candida albicans* in granulomatous tissue reactions. II. *In vivo* degradation of *C. albicans* in hepatic macrophages of mice, *J. Infect. Dis.* **135**:235–242.

Merkow, L., Pardo, M., Epstein, S. M., Verney, E., and Sidransky, H., 1968, Lysosomal stability during phagocytosis of *Aspergillus flavus* spores by alveolar macrophages of cortisone-treated mice, *Science* **160**:79–81.

Merkow, L. P., Epstein, S. M., Sidransky, H., Verney, E., and Pardol, M., 1971, The pathogenesis of experimental pulmonary aspergillosis, *Am. J. Pathol.* **62**:57–74.

Miller, G. P., and Kohl, S., 1983, Antibody-dependent leukocyte killing of *Cryptococcus neoformans*, *J. Immunol.* **131**:1455–1459.

Mitchell, T. G., and Friedman, L., 1972, *In vitro* phagocytosis of intracellular fate of variously encapsulated strains of *Cryptococcus neoformans*, *Infect. Immun.* **5**:491–498.

Miyaji, M., and Nishimura, K., 1981, Studies on organ specificity in experimental murine cryptococcosis, *Mycopathologia* **76**:145–154.

Miyaji, M., and Nishimura, K., 1983, Granuloma formation and killing functions of granuloma in congenitally athymic nude mice infected with *Blastomyces dermatitidis* and *Paracoccidioides brasiliensis*, *Mycopathologia* **82**:129–141.

Monga, D. P., 1981, Role of macrophages in resistance of mice to experimental cryptococcosis, *Infect. Immun.* **32**:975–978.

Morelli, R., and Rosenberg, L. T., 1971, The role of complement in the phagocytosis of *Candida albicans* by mouse peripheral blood leukocytes, *J. Immunol.* **107**:476–482.

Morgan, M. A., Blackstock, R. A., Bulmer, G. S., and Hall, N. K., 1983, Modification of macrophage phagocytosis in murine cryptococcosis, *Infect. Immun.* **40**:493–500.

Morozumi, P. A., Brummer, E., and Stevens, D. A., 1981a, Strain differences in resistance to infection reversed by route of challenge: Studies in blastomycosis, *Infect. Immun.* **34**:623–625.

Morozumi, P. A., Halpern, J. W., and Stevens, D. A., 1981b, Susceptibility differences of inbred strains of mice to blastomycosis, Infect. Immun. **32**:160–168.

Morrison, R. P., and Cutler, J. E., 1981, Interaction of murine phagocytic cells with *Candida albicans: In vitro*, *J. Reticuloendothel. Soc.* **29**:23–34.

Neilson, J. B., Fromtling, R. A., and Bulmer, G. S., 1977, *Cryptococcus neoformans:* Size range of infectious particles from aerosolized soil, *Infect. Immun.* **17**:634–638.

Nickerson, D. A., Havens, R. A., and Bullock, W. E., 1981, Immunoregulation in disseminated histoplasmosis: Characterization of splenic suppressor cell populations, *Cell. Immunol.* **60**:287–297.

Ozato, K., and Uesaka, I., 1974, The role of macrophages in *Candida albicans* infection *in vitro*, *Jpn. J. Microbiol.* **18**:29–35.

Papadimitriou, J. M., Robertson, T. A., Kletter, Y., Aronson, M., and Walters, M. N-I., 1978. An ultrastructural examination of the interaction between macrophages and *Cryptococcus neoformans*, *J. Pathol.* **124**:103–109.

Patterson-Delafield, J., and Lehrer, R. I., 1977, A simple microscopic method for identifying and quantitating phagocytic cells *in vitro*, *J. Immunol. Methods* **18**:377–379.

Patterson-Delafield, J., Martinez, R. J., and Lehrer, R. I., 1980, Microbicidal cationic proteins in rabbit alveolar macrophages: A potential host defense mechanism, *Infect. Immun.* **30**:180–132.

Patterson-Delafield, J., Szklareki, D., Martinez, R. J., and Lehrer, R. I., 1981, Microbicidal cationic proteins of rabbit alveolar macrophages: Amino acid composition and functional attributes, *Infect. Immun.* **31**:723–731.

Peterson, E. M., and Calderone, R. A., 1977, Growth inhibition of *Candida albicans* by rabbit alveolar macrophages, *Infect. Immun.* **15**:910–915.

Peterson, E. M., and Calderone, R. A., 1978, Inhibition of specific amino acid uptake in *Candida albicans* by lysosomal extracts from rabbit alveolar macrophages, *Infect. Immun.* **21**:506–513.

Poor, A. H., and Cutler, J. E., 1981, Analysis of an *in vivo* model to study the interaction of host factors with *Candida albicans*, *Infect. Immun.* **31**:1104–1109.

Regelson, W., Munson, A., and Wooles, W., 1970, The biologic activity of synthetic polymers: interferon inducers and reticuloendothelial response; polyanionic phylaxis against microorganisms and alterations of drug metabolism, *Symp. Series Immunobiol. Standard.* **14**:227–236.

Regelson, W., Morahan, P., Kaplan, A. M., Baird, L. G., and Munson, J. A., 1974. Synthetic polyanions: Molecular weight, macrophage activation and immunologic response, in: *Activation of Macrophages. Proceedings of the Second International Workshop Conference Hoechst,* Volume 2 (W. H. Wagner and H. Hahn, eds.), pp. 97–110, Excerpta Medica, Amsterdam.

Richard, J. L., and Thurston, J. R., 1975, Effect of aflatoxin an phagocytosis of *Aspergillus fumigatus* spores by rabbit alveolar macrophages, *Appl. Microbiol.* **30**:44–47.

Rinehart, J. J., Balcerzak, S. P., Sagone, A. L., and LoBuglio, A. F., 1974, Effects of corticosteriods on human monocyte function, *J. Clin. Invest.* **54**:1337–1343.

Rinehart, J. J., Sagone, A. L., Balcerzak, S. P., Ackerman, G. A., and LoBuglio, A. F., 1975. Effects of corticosteroid therapy on human monocyte function, *N. Engl. J. Med.* **292**:236–241.

Rippon, J. W., 1982, *Medical Mycology. The Pathogenic Fungi and the Pathogenic Actinomycetes,* 2nd ed., Saunders, Philadelphia.

Rogers, T. J., and Balish, E., 1977. The role of activated macrophages in resistance to experimental renal candidiasis, *J. Reticuloendothel. Soc.* **22**:309–318.

Ruiz, B. H., and Carvajal, R. E., 1986. Immunomodulation by *Histoplasma capsulatum* products; polyclonal activation and mitogenic effects, *Mycopathologia* **93**:113–119.

Sasada, M., and Johnston, R. B., 1980. Macrophage microbicidal activity. Correlation between phagocytosis-associated oxidative metabolism and the killing of candida by macrophages, *J. Exp. Med.* **152**:85–98.

Savage, D. C., and Madin, S. H., 1968. Cellular responses in lungs of immunized mice to intranasal infection with *Coccidioides immitis, Sabouraudia* **66**:94–102.

Sawyer, R. T., Moon, R. J., and Beneke, E. S., 1981. Trapping and killing of *Candida albicans* by *Corynebacterium parvum*–activated livers, *Infect. Immun.* **32**:945–950.

Schaffner, A., Douglas, H., and Braude, A., 1982. Selective protection against conidia by mononuclear and against mycelia by polymorphonuclear phagocytes in resistance to aspergillus, *J. Clin. Invest.* **69**:617–631.

Schaffner, A., Douglas, H., Braude, A. I., and Davis, C. E., 1983. Killing of *Aspergillus* spores depends on the anatomical source of the macrophage, *Infect. Immun.* **42**:1109–1115.

Schlitzer, R. L., Chandler, F. W., and Larsh, H. W., 1981. Primary acute histoplasmosis in guinea pigs exposed to aerosolized *Histoplasma capsulatum, Infect. Immun.* **33**:575–582.

Schneerson-Porat, S., Shahar, A., and Aronson, M., 1965, Formation of histiocyte rings in response to *Cryptococcus neoformans* infection, *J. Reticuloendoth. Soc.* **2**:249–255.

Schuit, K. E., 1979, Phagocytosis and intracellular killing of pathogenic yeasts by human monocytes and neutrophils, *Infect. Immun.* **24**:932–938.

Sethi, K. K., and Pelster, B., 1974, Destruction of *Cryptococcus neoformans* by nonspecifically activated macrophages in: *Activation of Macrophages. Proceedings of the Second International Workshop Conference Hoechst,* Volume 2 (W. H. Wagner and H. Hahn, eds.), pp. 269–279, Excerpta Medica, Amsterdam.

Shadomy, H. J., and Lurie, H. I., 1971, Histoplathological observations in experimental cryptococcosis caused by a hypha-producing strain of *Cryptococcus neoformans* (Coward strain) in mice, *Sabouraudia* **9**:6–9.

Sharma, D., Chaudhary, S., Burt, W. R., and Tewari, R. P., 1980. Characteristics of delayed hypersensitivity and macrophage migration inhibition factor in mice immunized with ribosomes or live yeast cells of *Histoplasm capsulatum,* in: *Human and Animal Mycology, Proceedings of the VII Congress of ISHAM* (E. S. Kuttin, and G. L. Baum, eds.), pp. 89–92, Excerpta Medica, New York.

Sixby, J. W., Fields, B. T., Sun, C. N., Clark, R. C., and Nolan, C. M., 1979, Interactions between human granulocytes and *Blastomyces dermatitidis, Infect. Immun.* **23**:41–44.

Stanley, V. D., and Nurley, R., 1969, The growth of *Candida* species in cultures of mouse peritoneal macrophages, J. Pathol. **97**:356–366.

Tacker, J. R., Farhi, F., and Bulmer, G. S., 1972, Intracellular fate of *Cryptococcus neoformans, Infect. Immun.* **6**:162–167.

Taylor, M. L., Diaz, S., Gonzalez, P. A., Sosa, A. C., and Toriello, C., 1984, Relationship between pathogenesis and immune regulation mechanisms in histoplasmosis: A hypothetical approach, *Rev. Infect. Dis.* **6**:775–782.

Tewari, R. P., Sharma, D., Solotorowsky, M., Lafemina, R., and Blaint, J., 1977, Adoptive transfer of immunity from mice immunized with ribosomes or live cells of *Histoplasma capsulatum, Infect. Immun.* **15**:789–795.

Tewari, R. P., Sharma, D. K., and Mathur, A., 1978, Significance of thymus-derived lymphocytes in immunity elicited by immunization with ribosomes or live yeast cells of *Histoplasma capsulatum, J. Infect. Dis.* **138**:605–613.

Tewari, R. P., Sharma, D., and Mathur, A., 1980, Importance of T lymphocytes in acquired immunity to histoplasmosis, in: *Human and Animal Mycology, Proceedings of the VII Congress of ISHAM* (E. S. Kuttin, and G. L. Baum, eds.), pp. 93–96, Excerpta Medica, New York.

Tiku, M. L., McNabb, P. C., and Tomasi, Jr., T. B., 1985, Macrophage-related fibrinolysis in experimental disseminated histoplasmosis, *Infect. Immun.* **49**:641–646.

Von Behren, L. A., Chaudhary, S., Khardori, N., Rabinovich, S., Shu, M., and Tewari, R. P., 1983a, Effect of silica on the susceptibility of mice to experimental histoplasmosis, *J. Reticuloendothel. Soc.* **34**:99–111.

Von Behren, L. A., Chaudhary, S., Rabinovich, S., Shu, M-E., and Tewari, R. P., 1983b, Protective effect of poly-2-vinylpyridine-*N*-oxide on susceptibility of silica-treated mice to experimental histoplasmosis, *Infect. Immun.* **42**:818–823.

Waldorf, A. R., Levitz, S. M., and Diamond, R. D., 1984, *In vivo* bronchoalveolar macrophage defense against *Rhizopus oryzae* and *Aspergillus fumigatus, J. Infect. Dis.* **150**:752.760.

Wallraff, E. B., and van Liew, R. M., 1967. Rabbit alveolar macrophage response to *in vivo* addition of coccidioidin, in: *Coccidioidomycosis, Proceedings II Coccidioidomycosis Symposium* (L. Ajello, ed.), pp. 179–181, University of Arizona Press, Tucson.

Wegner, T. N., Reed, R. E., Trautman, R. J., and Beavers, C. D., 1972, Some evidence for the development of a phagocytic response by polymorphonuclear leukocytes recovered from the venus blood of dogs inoculated with *Coccidioides immitis* or vaccinated with an irradiated spherule vaccine, *Am. Rev. Respir. Dis.* **105**:845–849.

Williams, D. M., Graybill, J. R., and Drutz, D. J., 1981a, Adoptive transfer of immunity to *Histoplasma capsulatum* in athymic nude mice, *Sabouraudia* **19**:39–48.

Williams, D. M., Weiner, M. H., and Drutz, D. J., 1981b, Immunologic studies of disseminated infection with *Aspergillus fumigatus* in the nude mouse, *J. Infect. Dis.* **143**:726–733.

Wu, W. G., and Marcus, S., 1963. Humoral factors in cellular resistance. I. The effects of heated and unheated homologous and heterologous sera on phagocytosis and cytopedis by normal and immune macrophages, *J. Immunol.* **91**:313–322.

Wu-Hsieh, B., Zlotnik, A., and Howard, D. H., 1984, T-cell hybridoma-produced lymphokine that activates macrophages to suppress intracellular growth of *Histoplasma capsulatum, Infect. Immun.* **43**:380–385.

Yamamura, M., and Valdimarsson, H., 1977, Participation of C_3 in intracellular killing of *Candida albicans, Scand. J. Immunol.* **6**:591–594.

9

Macrophages and Natural Resistance to Virus Infections

SØREN C. MOGENSEN

1. INTRODUCTION

More than two decades ago Mims (1964) stated in his classical review of aspects of the pathogenesis of virus diseases: "Macrophages are literally in a position to control the susceptibility of animals to virus infections, since they monitor the main body compartments and may control the entry of viruses to target organs like the liver." Since then, the position of macrophages as important elements in the host defense armamentarium against a variety of virus infections has been further substantiated and the interplay between macrophages and other defense mechanisms elucidated (Gresser and Lang, 1966; Allison, 1974; Mogensen, 1979; Morahan and Morse, 1979; Virelizier, 1981; Morahan, 1984; Mogensen and Virelizier, 1987).

The body has at its disposal several sequential lines of defense against virus infections, and macrophages figure prominently in many of these. In the uninfected host there is a series of natural, nonspecific barriers to virus penetration, e.g., cutaneous epithelium, ciliated respiratory epithelium, nonspecific virucidal substances in secretions, and the fixed macrophages in all organs and tissues. Among these, the barrier function of macrophages has been most widely studied (Mims, 1964; Mogensen, 1979). Soon after initiation of infection additional lines of nonspecific antiviral defenses are induced. Most studied have been interferon (Baron, 1973; Haller, 1981; Virelizier, 1981), natural killer cells (Welsh, 1981), and, again, macrophages, which may become nonspecifically activated (Morahan, 1984). Taken together, these preexisting or induced nonspecific resistance mechanisms may have important implications for the course and final outcome of a virus infection, because they may interfere with the establishment and early dissemination of the virus in the host. Although they probably work in concert in the early phases of the infection, their relative roles may vary depending on the virus, the target organ, and the phase of a particular infection.

SØREN C. MOGENSEN • Institute of Medical Microbiology, University of Aarhus, DK-8000 Aarhus C, Denmark.

Final recovery from a fully established virus infection is mediated through the specific immunologic response. Both the humoral and cell-mediated immune responses are dependent on a functioning macrophage system. Thus, macrophages are necessary as accessory antigen-presenting cells in the inductive phase of immunity; they play a role as modulators of the response through monokine production; and finally, they can act as ultimate antiviral effectors after having been recruited into infectious foci as part of the immune inflammatory process (Blanden, 1974; Oehler *et al.*, 1978; Unanue, 1981).

Thus macrophages may be pivotal cells in both resistance to, and recovery from, virus infection. The present chapter focuses on the role they play in the natural, nonspecific defense against virus infections, with special emphasis on their key role as a first line of defense at the portal of entry of the virus or in target organs of crucial importance for the outcome of the infection.

2. VIRUS–MACROPHAGE INTERACTIONS

Cells of the mononuclear phagocyte system (monocytes and macrophages) are well suited to function as an important element in the defense against the invasion of viruses (Mims, 1964). They are long-lived and strategically placed at the portal of entry (e.g., the alveolar macrophages of the lung) and are widely distributed in most organs and tissues in close contact with the circulating blood (e.g., Kupffer cells of the liver). Thus, they are in a position for viral encounter early in the infectious process.

From a functional point of view, the ability of macrophages to destroy virus particles and to interfere with their replication may prevent the access of virus to susceptible cells and reduce the amount of virus produced in these cells. Furthermore, the ability of the mononuclear phagocyte system to provide an influx of new cells from the bone marrow and blood monocyte compartments into foci of infection and the potential of these newly arrived and activated cells to exert various antiviral functions may also be important in determining the outcome of a virus infection.

There are two kinds of *in vitro* interaction between viruses and macrophages: *intrinsic* and *extrinsic* (Morahan and Morse, 1979).

2.1. INTRINSIC VIRUS–MACROPHAGE INTERACTION

Intrinsic interaction refers to the outcome of virus replication in the macrophage per se. Macrophages may possess various degrees of intrinsic susceptibility (permissiveness) or resistance (nonpermissiveness) to virus replication. In the nonpermissive situation the virus is internalized in the macrophage by endocytosis and destroyed by lysosomal enzymes, or the macrophage is infected by the virus, but virus replication is aborted within the cell. In both instances the macrophage represents a "blind alley" for propagation of the infection. In the permissive situation, on the other hand, the virus replicates productively in the macrophage, and progeny particles are released to surrounding permissive cells. The infection may thus proceed with destruction of organs and tissues. Furthermore, infected monocytes in the circulation may, because of their migration through the body, serve as a vehicle for dissemination (Mims, 1964; McKendall *et al.*, 1981).

A number of experimental variables may influence the outcome of virus–macrophage interactions. First, the source of macrophages must be taken into account. Because of their accessibility, peritoneal macrophages have been most widely used, and results obtained with these cells have been extrapolated to predict the potentials of tissue macrophages. However, macrophages from different anatomic compartments may differ in permissiveness for a virus. Thus, Rodgers and Mims (1981) found a considerable difference between permissiveness of mouse alveolar and peritoneal macrophages for influenza virus antigen production. Alveolar macrophages, which may be important in the pathogenesis of the infection, were completely permissive, in contrast to peritoneal macrophages, which were much more restrictive. Conversely, Plaeger-Marshall *et al.* (1982) found rabbit peritoneal macrophages more permissive than alveolar macrophages for herpes simplex virus type 1. Generally, blood monocytes are more restrictive than tissue macrophages, but may gain variable degrees of permissiveness by being allowed to differentiate to macrophages in culture (Daniels *et al.*, 1978; Linnavuori and Hovi, 1981).

A second variable is the widespread use of nonspecific irritants such as proteose-peptone and thioglycollate to increase macrophage yields from the peritoneal cavity. Thus, Lopez and Dudas (1979) found peritoneal macrophages elicited with thioglycollate much more susceptible to replication of herpes simplex virus type 1 than normal or proteose–peptone-stimulated macrophages. Similar results have been reported by others (Morahan, 1984). Third, culture conditions are also important, as illustrated in the study by Lopez and Dudas (1979), in which precultivation of mouse macrophages for 3–7 days increased their permissiveness. Also, blood monocytes from adult humans become partially permissive upon culture for several days (Daniels *et al.*, 1978; Linnavuori and Hovi, 1981). This effect has been related to differentiation of the monocytes to macrophages in culture. Thus, infection of monocytes with a high multiplicity of infection was found to inhibit this differentiation, leaving the cells restrictive to virus replication (Linnavuori and Hovi, 1981, 1983). Other culture conditions known to influence macrophage permissiveness are the type and concentration of serum used (Lavelle and Bang, 1971). Finally, antiviral antibody has in some instances enhanced virus replication in macrophage cultures (Halstead and O'Rourke, 1977), presumably because of Fc receptor–mediated virus uptake into the cells (Schlesinger and Brandriss, 1981).

Macrophages are generally more restrictive for virus replication than most other cell types (Mogensen, 1979). Therefore, it is often necessary to use indirect measures such as the infectious center assay to detect virus replication. This kind of assay only indicates the number of macrophages in a culture that support virus replication, but gives no information on the amount of virus produced in each cell. With this assay, it has been shown that even though the majority of macrophages may be infected (as detected by immunofluorescence), only a minor fraction of the cells are releasing infectious virus. It is currently of major interest to relate permissiveness to any known particular heterogeneity of macrophages such as morphology, stage of cell cycle, stage of differentiation, or surface antigen expression (Morahan 1984).

Some virus–macrophage systems yield enough virus to allow a direct titration of the extracellular or intracellular virus produced. For instance, mouse hepatitis virus type 2 (Shif and Bang, 1970a) and avian influenza A virus (Lindenmann *et al.*, 1978) replicate to high titers in mouse macrophage cultures. However, in many of these instances the virus has been adapted to grow in macrophages, or special culture procedures favoring productive replication have been used, thereby making their relevance to the natural infection more questionable.

The mechanisms that render macrophages more restrictive than most other cell types to a number of viruses are still an enigma. The data of Stevens and Cook (1971) indicate that herpes simplex virus undergoes an abortive infection in macrophages from adult mice. Both viral DNA and the major virus-specified proteins were synthesized, but only few virus particles with central dense cores and very few enveloped capsids were seen by electron microscopy. This suggests that virus-specified macromolecules are either structurally defective, or produced in insufficient amounts, or that errors exist in the assembly of the virus particles.

Since the discovery of interferon in 1957 (Isaacs and Lindenmann, 1957), several attempts have been made to relate interferon to macrophage permissiveness for viruses (Mogensen and Virelizier, 1987). In models of age-related and genetically determined resistance to herpes simplex virus in the mouse it has been found that interferon production by macrophages correlates with restriction of virus replication in these cells (Hirsch *et al.*, 1970; Brücher *et al.*, 1984; Ellermann–Eriksen *et al.*, 1986). The nonpermissiveness of human monocytes for herpes simplex virus type 1 replication at a high multiplicity of infection has also been linked to interferon production in the cultures (Linnavuori and Hovi, 1981, 1983). With other viruses interferon production by macrophages has not reflected virus replication. Thus, Hanson *et al.* (1969) found that mouse macrophages with a differential ability to replicate flaviviruses produced equal amounts of interferon in response to these viruses. However, in the flavivirus–macrophage system, restrictive macrophages from resistant animals were more sensitive to the protective action of interferon than were more permissive macrophages from susceptible mice. Such a differential sensitivity to interferon between permissive and nonpermissive macrophages has also been described for myxoviruses (Haller *et al.*, 1979a) and mouse hepatitis virus type 3 (Virelizier, 1981). In a recent study from Gresser's group (Belardelli *et al.*, 1984) on macrophage intrinsic restriction of vesicular stomatitis and encephalomyocarditis viruses, it was suggested that low, undetectable levels of endogenous interferon maintain an antiviral state in macrophages, since antiinterferon treatment of normal mice *before* macrophage harvest rendered these cells permissive to a subsequent viral challenge. A similar observation has been made with influenza virus (Haller *et al.*, 1979a).

2.2. EXTRINSIC VIRUS–MACROPHAGE INTERACTION

Extrinsic interaction refers to macrophage influence on extracellular virus and virus replication in surrounding cells (Morahan and Morse, 1979). Thus, in experimental *in vitro* systems macrophages have exhibited a potent extrinsic inhibitory effect on the replication of several viruses in permissive cells. Generally, this is measured by adding macrophages to monolayers of infected permissive cells and observing reduced cytopathic effect, plaque reduction, or reduced amounts of intracellular or extracellular virus produced.

Morahan and Morse (1979) have outlined the various ways in which macrophages might theoretically interfere with all the steps in the replication of a virus in a permissive cell. The vulnerable stage of the infection will depend on the specific virus–cell interaction and, therefore, requires individual assessment with each virus–cell system. Special interest has focused on whether interferon production by macrophages or direct macrophage cytotoxicity to infected cells is involved. In the herpes simplex system, which has been studied in detail, these mechanisms do not seem to be important (Morse and Morahan, 1981). Although macrophage–cell contact was required for optimal activity, no substantial cell lysis was detected. An interferon-mediated mechanism was also discarded, since culture supernatants did not contain detectable levels of interferon, and the antiviral activity was not affected by antiinterferon. Moreover, mouse macrophages were active in xenogeneic Vero cells. It was suggested that the effect was related to cytostasis through disturbance of host cell macromolecular synthesis. A similar conclusion was reached by Hayashi *et al.* (1980). Alternatively, it has been suggested that arginase produced by activated macrophages might be responsible for the extrinsic inhibition of herpes simplex and vaccinia virus multiplication by depleting the medium of arginine, which is essential for replication of the viruses (Wildy *et al.*, 1982). This effect does not require macrophage–cell contact.

Extrinsic antiviral activity is generally not a prominent feature of normal, unstimulated macrophages (Morahan and Morse, 1979; Hayashi *et al.*, 1980; Koff *et al.*, 1983), although it has been reported to occur in some instances (Schlabach *et al.*, 1979; Stohlman *et al.*, 1982). Therefore, extrinsic restriction of virus replication can probably not be considered a "ready-to-function" barrier to the establishment of virus infections in the sense that has been described for intrinsic restriction. However, since macrophage activation is an early event in many virus infections (Morahan and Morse, 1979), extrinsic activity of activated macrophages accumulating in established foci of infection might be important in diminishing further spread of the infection.

It is currently not known whether intrinsic and extrinsic restriction of viruses by macrophages are related phenomena. The fact that a given macrophage population may show a high activity in one and not the other character may indicate that this is not the case. For instance, resident macrophages are generally more intrinsically resistant to herpes simplex virus replication than thioglycollate-induced macrophages, whereas only the latter show high extrinsic restriction (Morahan, 1984). Likewise, Stohlman *et al.* (1982) were unable to correlate extrinsic and intrinsic macrophage antiviral activity toward mouse hepatitis virus. Preliminary information from the study of Wildy *et al.* (1982) on the role of macrophage arginase in extrinsic restriction of viruses indicates that self-starvation of arginine may not be the effector mechanism in intrinsic restriction. However, a recent study has shown that γ-interferon-activated macrophages express intrinsic restriction and that this is possibly mediated by arginase production, since addition of increasing doses of arginine to the macrophage cultures abrogated the induced restriction (Sethi, 1983). Probably, the effector mechanisms in both systems, and thus their relationship, may depend on the particular virus–macrophage pair studied and on the experimental protocol used.

3. INDICATIONS FOR A ROLE OF MACROPHAGES IN NATURAL RESISTANCE TO VIRUS INFECTIONS

No model exists of complete macrophage deficiency comparable to the complete lack of elements of the lymphoid system in certain human immunodeficiency diseases or experimental animals (e.g., the T-cell-deficient nude mouse). Therefore, evidence for a primary role of macrophages in natural resistance to virus infections has accumulated from a multitude of observations from experimental studies, in which manipulations of the macrophage system in an upward or downward direction have influenced the outcome of virus infections. In other instances, naturally occurring models with differences in virus–macrophage interactions have been exploited to correlate the outcome of this interaction with the overall resistance or susceptibility of the host (Mogensen, 1979). Examples are provided below of studies that have led to the concept that the early virus–macrophage interaction has vital importance for the final outcome of a virus infection.

3.1. STUDIES ON MACROPHAGE TRANSFER

This approach has been successfully used in experimental infections with a number of viruses, including Semliki Forest and encephalomyocarditis (Glasgow, 1970), herpes simplex type 1 (Hirsch *et al.*, 1970) and type 2 (Mogensen, 1978), murine cytomegalovirus (Selgrade and Osborn, 1974), Coxsackie B-3 (Rager-Zisman and Allison, 1973), rabies (Turner and Ballard, 1976), and mouse hepatitis viruses (Stohlman *et al.*, 1980).

Most studies on the effect of macrophage transfer have been performed in inbred strains of mice, in which macrophages from virus-resistant adults have been transferred to usually more susceptible young or suckling mice. Protection of the recipient against a subsequent virus infection has then been ascribed to the transferred cells. The effect of macrophage transfer is generally monitored by studying overall mortality, mean survival time, or the kinetics of virus growth and lesion development in selected target organs. Of these, the latter are more sensitive markers for macrophage protection and also disclose at which stage of the infection the transferred cells are exerting their effect. As an example, death from encephalitis of herpes simplex virus–infected suckling mice was significantly reduced only when 6×10^6 stimulated macrophages were transferred (Hirsch *et al.*, 1970), whereas we have observed markedly reduced early virus growth in the liver of young mice after intravenous transfer of 2×10^6 resident peritoneal macrophages from resistant adult mice (Mogensen, 1978). Probably, macrophages are less important in the central nervous system than in the organs of the reticuloendothelial system. Furthermore, the intravenous route of transferring cells may be especially suited to protect the liver from subsequent infection, since approximately 60% of peritoneal macrophages injected by this route are localized in the liver within a few hours after inoculation (Roser, 1965).

3.2. STUDIES ON MACROPHAGE BLOCKADE

As mentioned, no good model of naturally occurring macrophage deficiency exists. However, a number of studies have tried to evaluate the role of macrophages in host resistance to virus infections by treatment of animals with antimacrophage serum or chemical agents more or less selectively toxic to macrophages (Mogensen, 1979).

The first attempts to modify the course of virus infections by antimacrophage serum were made in mice by Panijel and Cayeux (1968), with equivocal results: increasing susceptibility to yellow fever virus and apparent protection against encephalomyocarditis virus infection. Later studies have shown increase in susceptibility to vesicular stomatitis (Hirsch *et al.*, 1969), herpes simplex (Zisman *et al.*, 1970), and rabies (Turner and Ballard, 1976) viruses. These studies were done with crude xenogeneic antisera with inherent specificity problems. The use of monoclonal antibodies reacting specifically with macrophages should allow more precise interpretations in future studies.

Among the macrophage toxic agents, silica has been the most widely used (Mogensen, 1979) and treatment with this compound can increase the severity of many experimental virus infections (reviewed by Mogensen, 1979). It was originally described by Allison and co-workers (1966) to be selectively toxic to macrophages. However, in recent years some silica preparations have been found also to affect adversely the function of other cell types, including lymphocytes (Wirth *et al.*, 1980). On the other hand, silica treatment may be followed by an enhanced antibody response to viruses, either because of an increased antigenic stimulus due to the unchecked viral growth, or because macrophages are involved in suppression of antibody-producing cells (Knoblich *et al.*, 1983). Therefore, when using silica, it is important to look for an effect during the early phases of the infection and to use standardized silica preparations, for instance the original Dörentrup silica described by Allison (Allison *et al.*, 1966).

The complex interaction of silica with virus infections of macrophages was also illustrated in the study of duBuy (1975) on lactic dehydrogenase virus infection of mice. This virus replicates mainly, or exclusively, in monocytes/

macrophages. Administration of silica i.p. actually increased virus replication in the peritoneal cavity, presumably because of a silica-induced rapid influx of monocytes (and thus new targets for the virus) into the peritoneal cavity. Conversely, Schindler *et al.* (1984) found that i.p., but not i.v., silica treatment conferred early protection to mice against i.p. infection with mouse hepatitis virus type 3. The protection was correlated with a temporary elimination by silica of peritoneal macrophages, and the effect was ascribed to a depletion of this early target cell for virus replication.

3.3. STUDIES ON MACROPHAGE ACTIVATION

Induced or naturally occurring nonspecific activation of macrophages is often accompanied by an enhanced resistance to virus infection, an indication for a role of macrophages in the natural defense against the infections (Mogensen, 1979).

Mice treated with immunomodulators such as pyran copolymer and bacterial adjuvants such as BCG or *Propionibacterium acnes* (formerly known as *Corynebacterium parvum*) exhibit increased resistance to a number of viruses together with augmented extrinsic and intrinsic antiviral activity of their macrophages (Morahan and Morse, 1979). However, the sole role of macrophages in this phenomenon may be questioned. Immunomodulators are known to activate other potential early defense systems, for instance interferon, which might exert an antiviral activity in its own right or through activation of natural killer cells (Morahan and Morse, 1979; Kirchner, 1982). Pyran and *P. acnes* still induce resistance in ^{89}Sr-marrow-ectomized mice (Morahan *et al.*, 1982). These mice are severely monocytopenic and show low natural killer cell activity without impairment of tissue macrophage functions (Volkman *et al.*, 1983). This indicates that the immunomodulators do not act through activation of a marrow-dependent cell, but may exert a direct action on tissue macrophages. However, this assumption is challenged by the finding that resistance induced by pyran and *P. acnes* is not reversed by treatment with silica (Morahan *et al.*, 1982).

Further evidence for the antiviral potential of nonspecifically activated macrophages has emerged from studies in congenitally athymic nude mice and mice with a graft-versus-host reaction. Common features of these animals are severe immunodeficiency together with hyperactivity of the mononuclear phagocyte system (Mogensen, 1979). They are generally considered highly susceptible to a number of naturally occurring infectious diseases because of their inefficiency in mounting a specific immune reaction. However, in several studies it has been documented that they actually have an unexpected enhanced primary resistance to experimental infections with a number of pathogens, including bacteria, yeasts, and viruses (Mogensen, 1979). Thus, nude mice are considerably more resistant than their normal littermates to the induction of focal necrotizing hepatitis by herpes simplex virus type 2, with a pronounced ability to restrain virus multiplication in the liver during the first days of the infection (Mogensen and Andersen, 1978). Macrophage involvement in this early resistance was indicated by the findings that silica treatment abolished the augmented resistance and that macrophages from nude mice exhibited enhanced intrinsic restriction of the virus. Likewise, Rao *et al.* (1977) have demonstrated restriction of vaccinia virus replication in macrophages from nude mice. Probably, an increased intestinal bacterial flora of nude mice is responsible for a sustained nonspecific macrophage activation, since macrophages from nude mice, raised under germ-free conditions (Rao *et al.*, 1977) or treated with antibiotics (Nickol and Bonventre, 1977), do not have activated macrophages.

3.4. STUDIES ON VIRUS STRAIN- OR TYPE-RELATED RESISTANCE

Closely related virus strains or virus types differing in pathogenicity or virulence for a particular host have been used as valuable means of defining host defense mechanisms. In several cases, a difference in the ability of a pair of such viruses to replicate productively in macrophages from an animal has been correlated to a differential ability of the viruses to initiate infection in the intact animal. This was first shown by Bang and Warwick (1957) with a virulent and an avirulent strain of Newcastle disease virus. Later, a similar correlation was shown with virulent and avirulent strains of ectromelia (Roberts, 1964), vaccinia (Mims, 1964), mouse hepatitis type 2 (Shif and Bang, 1970b), and lymphocytic choriomeningitis (Tosolini and Mims, 1971) viruses. In other instances such a correlation has not been found. Thus, virulent and avirulent strains of mouse cytomegalovirus (Selgrade and Osborn, 1974; Brautigam *et al.*. 1979) and influenza virus (Rodgers and Mims, 1981) did not differ in their ability to replicate in macrophages, which indicates that this property is only one among several virulence determinants of viruses.

Also, inherent differences in pathogenicity for different types of related viruses have been correlated with differences in virus replication in macrophages. Mouse hepatitis virus exists in several closely related serotypes with various degrees of pathogenicity in mice. The nonpathogenic type 1 does not multiply in mouse macrophages (Allison, 1974), whereas the pathogenic types 2 and 3, which after i.p. inoculation produce fatal hepatitis, readily infect mouse peritoneal macrophages and grow to high titers with distinct cytopathic effect (Bang and Warwick, 1960; Virelizier, 1981). The marked difference in mouse hepatotropism between the two types of herpes simplex virus (type 2 readily induces focal necrotic hepatitis in mice, in contrast to type 1) was correlated with a far better restriction of herpes simplex virus type 1 than of type 2 in macrophages from mice (Mogensen, 1977a). Furthermore, blockade of the macrophage function of the animals by silica almost completely eliminated the pathogenic distinction between liver infections with the two types, allowing also type 1 to induce liver necrosis (Mogensen and Andersen, 1977).

3.5. STUDIES ON AGE-RELATED RESISTANCE

The ability of many viruses to replicate productively in macrophages dramatically decreases with increasing age of the macrophage donor (Mogensen, 1979). Likewise, extrinsic antiviral activity is age-related (Schlabach *et al.*, 1979;

Hayashi *et al.*, 1980). This maturation of macrophage antiviral potencies occurs concomitantly with the well-known age-related increase in resistance to many experimental virus infections, which is particularly obvious from the neonatal to the weaning period. In most instances this phenomenon is observed only after peripheral inoculation of the virus; adult animals remain highly susceptible to intracerebral challenge. This led very early to the concept of the development of "barriers" to the spread of virus from the periphery to highly susceptible target organs (Sabin and Olitsky, 1937). Early macrophage restriction of virus replication has been considered such a barrier in a number of infections (reviewed by Mogensen, 1979).

The cellular basis for age-dependent resistance of mice to herpes simplex virus encephalitis was first demonstrated by Johnson (1964). He showed by immunofluorescence that the macrophage represents the cellular level at which the infection is arrested in adult mice. Later studies with macrophage blockade (Zisman *et al.*, 1970) and macrophage transfer (Hirsch *et al.*, 1970) corroborated this notion. The age-dependent development of resistance to visceral infection with herpes simplex virus has also been related to maturation of the antiviral potency of macrophages (Mogensen, 1978). The human counterpart to these animal model infections is neonatal infection with herpes simplex virus (Nahmias *et al.*, 1975). This life-threatening infection is characterized by a widespread dissemination of the virus with lesions in many organs, including the central nervous system and the liver. A disproportionate number of cases are seen in premature infants (1:1.4 against expected 1:8 to 1:10), suggesting that immature host defenses (for example, macrophage function) are responsible for the ease in contracting the infection.

3.6. STUDIES ON GENETICALLY DETERMINED RESISTANCE

The earliest reports on a genetic basis for resistance to animal virus infections were published half a century ago (Gowen and Schott, 1933; Lynch and Hughes, 1936; Webster, 1937). Since then, the inheritability of resistance to virus infections has been demonstrated in several animal models and the genetics explored (Bang, 1978). In some instances a single dominant gene has a major impact on resistance. Classical examples of this in mice are resistance to flaviviruses and susceptibility to mouse hepatitis virus type 2, both of which are inherited as monogenic autosomal dominant characters. Some examples are discussed below of virus infections in which macrophages express at the cellular level the genetically determined resistance seen *in vivo*.

3.6.1. Flaviviruses

During the thirties, Webster (1937) developed, by selective breeding, strains of mice that differed greatly in resistance to flaviviruses. Genetic analysis indicated that resistance was dominant and unifactorial. In the fifties, Sabin (1952) extended this work on nonselected PRI and C_3H strains of mice and found PRI mice resistant and C_3H mice susceptible to a number of flaviviruses. The work of Sabin also indicated that resistance operated at the level of individual cells.

In extensive studies with West Nile virus, Goodmann and Koprowski (1962) concluded that cells of the reticuloendothelial system expressed the gene for flavivirus resistance. Thus, cultures of peritoneal and splenic macrophages from resistant and susceptible mice differed greatly in permissiveness for the virus, and this trait segregated as expected in macrophage cultures from backcross mice. Also macrophages from congenic resistant C_3HRV mice, differing from susceptible C_3H mice only at the gene for flavivirus resistance, expressed the resistance phenotype *in vitro*.

The depressed flavivirus replication in cells from resistant mice was found in the original reports to involve macrophages selectively. Later studies revealed that other cell types, for instance, brain cells and fibroblasts, also express flavivirus resistance (Darnell and Koprowski, 1974; Brinton, 1981). However, this finding does not rule out a role for the macrophage barrier in resistance, but only implies that the resistance gene also operates, if the barrier has been broken.

The mechanism of action of the flavivirus resistance gene is not fully understood. Neither differential interferon production nor interferon sensitivity seems to play any particular role (Brinton, 1981). Studies by Darnell and Koprowski (1974) have indicated that the production of defective interfering particles by resistant cells may be important. However, this may not constitute the only, or even major, mechanism by which virus replication is blocked in resistant cells.

3.6.2. Coronaviruses

Infection of mice with mouse hepatitis virus type 2 provides the best example of susceptibility inherited as a dominant, monogenic trait. Studies over two decades by Bang and co-workers (reviewed by Bang, 1978) on mouse resistance to infection with this virus have yielded evidence that macrophages represent a cellular barrier to virus access to the liver. Genetic analyses of several strains of mice have shown a complete agreement between mouse and macrophage susceptibility to the virus. It is interesting to notice that the same two strains of mice (PRI and C_3H) have been used in studies of the genetics of macrophage resistance to mouse hepatitis virus and flaviviruses and that the strain susceptible to mouse hepatitis virus is resistant to flaviviruses and vice versa. This indicates that virus–macrophage interactions require individual assessment with each virus–host system.

Infection with mouse hepatitis virus type 3, another member of the coronavirus family, also differs in various mouse strains (Virelizier, 1981). Most strains undergo a lethal, systemic infection with this virus, whereas the A mouse is fully resistant. C_3H mice show an intermediate susceptibility in that they resist the acute phase of the infection, but develop a chronic progressive disease. In all cases there is a precise correlation between the ability of the virus to replicate productively in macrophages with formation of multinucleated giant cells and the *in vivo* course of the infection (Virelizier, 1981). With the neurotropic mouse hepatitis virus strain JHM, resistance was also found to be correlated to intrinsic restriction of virus replication in macrophages, whereas macrophages from resistant and susceptible mice showed the same degree of extrinsic antiviral activity (Stohlman *et al.*, 1982).

Several attempts have been made to elucidate the mechanism of macrophage restriction of mouse hepatitis viruses (Bang, 1978; Virelizier, 1981). Resistance is clearly expressed at the level of individual cells, and interferon production by macrophages does not seem important. Although various manipulations of macrophage cultures (transfer of extracts from susceptible to resistant cultures, cortisone treatment, and addition of supernatants from allogeneic mixed lymphocyte cultures or Con A–stimulated lymphocytes) have been found to modify the virus–macrophage interaction, these studies have not clarified the situation. On the basis of studies with antiinterferon treatment, Virelizier (1981) has suggested that endogenous interferon, produced in response to the virus infection in question and also to past exposure to other infections may be essential for the acquisition of the ability of macrophages to restrict virus replication, and that the efficiency of interferon to do this varies in different mouse strains.

3.6.3. Herpesviruses

Studies of inbred mice have revealed a marked difference in resistance among various strains to both herpes simplex virus type 1 and type 2. Thus, Lopez (1975) found resistance to herpes simplex virus type 1–induced encephalitis to be dominant and governed by at least two independent, non-H-2-linked genes, whereas we found resistance to hepatitis induction by herpes simplex virus type 2 to be influenced by one major X-linked dominant gene (Mogensen, 1977b; Pedersen *et al.*, 1983).

With both types of the virus, macrophages from resistant strains of mice exert higher intrinsic restriction of virus replication than macrophages from susceptible strains, a difference that is not seen in fibroblast cultures. In the case of herpes simplex virus type 2, macrophage restriction of virus replication correlated with resistance in the F_1 generation. This has been confirmed by Armerding *et al.* (1981). Furthermore, high and low restriction segregated close to a 1 : 1 ratio in macrophage cultures prepared from individual mice in the BC_1 generation, indicating that a single gene is at work. With the type 1 virus, Lopez and Dudas (1979) were not able to relate the genetics of macrophage restriction of virus replication to resistance. Although F_1 mice were resistant to the infection, macrophages from these mice replicated the virus as well as did macrophages from the susceptible parent. One reason for the different results obtained with the two virus types might well be found in the experimental approaches used. The results with type 2 were obtained with normal peritoneal macrophages infected *in vivo* or the day after plating in culture, whereas the results with type 1 were derived from thioglycollate-induced or 4-day-cultured macrophages, which have lost much of the native restriction capability against the virus.

Interferon is probably involved in the genetically determined antiviral action of macrophages against herpes simplex virus. Thus, we have found that resistance to herpes simplex virus type 2 in mice, macrophage restriction of virus replication, and the very early interferon response *in vivo* and by macrophages *in vitro* all show the same genetics, with the influence of an X-linked locus (Mogensen, 1977b; Pedersen *et al.*, 1983; Ellermann–Eriksen *et al.*, 1986). The finding of an identical, and, in the context of natural resistance to virus infections, unusual mode of inheritance of two defense mechanisms clearly indicates that they may be interdependent. Recently we have also provided evidence that macrophages from virus–resistant mice exhibit a genetically determined higher interferon sensitivity than do macrophages from virus susceptible mice (Ellermann-Eriksen and Mogensen, submitted for publication).

3.6.4. Myxoviruses

Mice of the A2G strain express a pronounced and selective resistance to various orthomyxoviruses (reviewed by Haller, 1981). Resistance has been attributed to a single dominant gene designated *Mx*. For many years the phenotypic expression of the *Mx* gene was an enigma. In 1978 it was shown that peritoneal macrophages and Kuppfer cells from A2G and F_1 mice, as opposed to macrophages from a number of susceptible mouse strains, resisted replication of influenza virus (Lindenmann *et al.*, 1978). Furthermore, resistance and susceptibility of individual mice and their macrophages cosegregated in a ratio close to 1 : 1 in backcrosses between resistant F_1 mice and susceptible mice, indicating that resistance *in vivo* and macrophage restriction of virus replication were related. However, in elegant studies with radiation chimeras it was later found that lethally irradiated Mx^+ mice substituted with macrophage precursors from susceptible Mx^- mice, and vice versa, expressed the resistance phenotype of the recipient, in spite of the fact that their macrophages expressed the phenotype of the donor (Haller *et al.*, 1979b). Thus, macrophage resistance and resistance of the intact animal seemed to not be causally related. Instead, it is now clear that the *Mx* gene determines the sensitivity of individual cells, including macrophages but also parenchymal cells, to an influenza virus–specific antiviral action of interferon (Haller *et al.*, 1980). However, only macrophages show a constitutive expression of the *Mx* gene (the Mx protein), indicating that these cells respond more efficiently to low levels of endogenous interferon (Dreiding *et al.*, 1985; Staeheli and Haller, 1987).

4. CONCLUDING REMARKS

Macrophages represent an important early barrier to the establishment of virus infections through their ability to exert intrinsic and extrinsic restriction of virus replication. Of these two antiviral mechanisms, intrinsic restriction is probably the most important during initiation of the infection, since it represents a constitutional part of the body's defense, which is immediately available. Extrinsic restriction, on the other hand, may become more important during the course of the infection, since it is exclusively or most effectively carried out by activated macrophages. Other natural defense systems like interferon and natural killer cells are only briefly discussed. This does not reflect an underestimation of their obvious (interferon) or potential (natural killer cells) importance in resistance; probably they work in concert during virus infection (Mogensen and Virelizier, 1987). Future work should assess their relative importance and cooperation in the early defense network against virus infections. The basic nature of the antiviral action of macrophages and the possibilities for positive modulation of the antiviral properties of these cells should also receive more attention.

REFERENCES

Allison, A. C., 1974, Interactions of antibodies, complement components and various cell types in immunity against viruses and pyogenic bacteria, *Transplant. Rev.* **19:**3–55.

Allison, A. C., Harington, J. S., and Birbeck, M., 1966, An examination of the cytotoxic effects of silica on macrophages, *J. Exp. Med.* **124:**141–153.

Armerding, D., Mayer, P., Scriba, M., Hren, A., and Rossiter, H., 1981, *In-vivo* modulation of macrophage functions by herpes simplex virus type 2 in resistant and sensitive inbred mouse strains, *Immunobiology* **160:**217–227.

Bang, F. B., 1978, Genetics of resistance of animals to viruses: I. Introduction and studies in mice, *Adv. Virus Res.* **23:**270–348.

Bang, F. B., and Warwick, A., 1957, The effect of an avirulent and a virulent strain of Newcastle virus (*Myxovirus multiforme*) on cells in tissue culture, *J. Pathol. Bacteriol.* **73:**321–330.

Bang, F. B., and Warwick, A., 1960, Mouse macrophages as host cells for the mouse hepatitis virus and the genetic basis of their susceptibility, *Proc. Natl. Acad. Sci. USA* **46:**1065–1075.

Baron, S., 1973, The defensive and biological roles of the interferon system, in: *Interferons and Interferon Inducers* (N. B. Finter, ed.), pp. 267–293, American Elsevier, New York.

Belardelli, F., Vignaux, F., Proietti, E., and Gresser, I., 1984, Injection of mice with antibody to interferon renders peritoneal macrophages permissive for vesicular stomatitis virus and encephalomyocarditis virus, *Proc. Natl. Acad. Sci. USA* **81:**602–606.

Blanden, R. V., 1974, T cell response to viral and bacterial infection, *Transplant. Rev.* **19:**56–88.

Brautigam, A. R., Dutko, F. J., Olding, L. B., and Oldstone, M. B. A., 1979, Pathogenesis of murine cytomegalovirus infection: the macrophage as a permissive cell for cytomegalovirus infection, replication and latency, *J. Gen. Virol.* **44:**349–359.

Brinton, M. A., 1981, Genetically controlled resistance to flavivirus and lactate-dehydrogenase-elevating virus-induced disease, *Curr. Top. Microbiol. Immunol.* **92:**1–14.

Brücher, J., Domke, I., Schröder, C. H., and Kirchner, H., 1984, Experimental infection of inbred mice with herpes simplex virus. VI. Effect of interferon on *in vitro* virus replication in macrophages, *Arch. Virol.* **82:**83–93.

Daniels, C. A., Kleinerman, E. S., and Snyderman, R., 1978, Abortive and productive infections of human mononuclear phagocytes by type 1 herpes simplex virus, *Am. J. Pathol.* **91:**119–136.

Darnell, M. B., and Koprowski, H., 1974, Genetically determined resistance to infection with group B arboviruses. II. Increased production of interfering particles in cell cultures from resistant mice, *J. Infect. Dis.* **129:**248–256.

Dreiding, P., Staeheli, P., and Haller, O., 1985, Interferon–induced protein Mx accumulates in nuclei of mouse cells expressing resistance to influenza virus, *Virology* **140:**192–196.

duBuy, H., 1975, Effect of silica on virus infections in mice and mouse tissue culture, *Infect. Immun.* **11:**996–1002.

Ellermann-Eriksen, S., Liberto, M. C., Iannelo, D., and Mogensen, S. C., 1986, X–Linkage of the early *in vitro* α/β interferon response of mouse peritoneal macrophages to herpes simplex virus type 2, *J. Gen. Virol.* **67:**1025–1033.

Glasgow, 1970, Transfer of interferon-producing macrophages: new approach to viral chemotherapy, *Science* **170:**854–856.

Goodmann, G. T., and Koprowski, H., 1962, Study on the mechanism of innate resistance to virus infection, *J. Cell. Comp. Physiol.* **59**:333–373.

Gowen, J. W., and Schott, R. H., 1933, Genetic constitution in mice as differentiated by two diseases, pseudorabies and mouse typhoid, *Am. J. Hyg.* **18**:674–687.

Gresser, I., and Lang, D. J., 1966, Relationships between viruses and leukocytes, *Prog. Med. Virol.* **8**:62–130.

Haller, O., 1981, Inborn resistance of mice to myxoviruses, *Curr. Top. Microbiol. Immunol.* **92**:25–52.

Haller, O., Arnheiter, H., Gresser, I., and Lindenmann, J., 1979, Genetically determined, interferon-dependent resistance to influenza virus in mice, *J. Exp. Med.* **149**:601–612.

Haller, O., Arnheiter, H., Lindenmann, J., 1979b, Natural, genetically determined resistance toward influenza virus in hemopoietic mouse chimeras. Role of mononuclear phagocytes, *J. Exp. Med.* **150**:117–126.

Haller, O., Arnheiter, H., Lindenmann, J., and Gresser, I., 1980, Host gene influences sensitivity to interferon action selectively for influenza virus, *Nature* **283**:660–662.

Halstead, S. B., and O'Rourke, E. J., 1977, Dengue viruses and mononuclear phagocytes. I. Infection enhancement by non-neutralizing antibody, *J. Exp. Med.* **146**:201–217.

Hanson, B., Koprowski, H., Baron, S., and Buckler, C. E., 1969, Interferon-mediated natural resistance of mice to arbo B virus infection, *Microbios* **1B**:51–68.

Hayashi, K., Kurata, T., Morishima, T., and Nassery, T., 1980, Analysis of the inhibitory effect of peritoneal macrophages on the spread of herpes simplex virus, *Infect. Immun.* **28**:350–358.

Hirsch, M. S., Gary, G. N., and Murphy, F. A., 1969, *In vitro* and *in vivo* properties of anti-macrophage sera, *J. Immunol.* **102**:656–661.

Hirsch, M. S., Zisman, B., and Allison, A. C., 1970, Macrophages and age-dependent resistance to herpes simplex virus in mice, *J. Immunol.* **104**:1160–1165.

Isaacs, A., and Lindenmann, J., 1957, Virus interference. I. The interferon, *Proc. R. Soc. London Sec. B.* **147**:258–267.

Johnson, R. T., 1964, The pathogenesis of herpes virus encephalitis. II. A cellular basis for the development of resistance with age, *J. Exp. Med.* **120**:359–374.

Kirchner, H., 1982, *Immunobiology of Infection with Herpes Simplex Virus*, Karger, Basel.

Knoblich, A., Görtz, J., Härle-Grupp, U., and Falke, D., 1983, Kinetics and genetics of herpes simplex virus–induced antibody formation in mice, *Infect. Immun.* **39**:15–23.

Koff, W. C., Showalter, S. D., Seniff, D. A., and Hampar, B., 1983, Lysis of herpesvirus-infected cells by macrophages activated with free or liposome-encapsulated lymphokine produced by a murine T cell hybridoma, *Infect. Immun.* **42**:1067–1072.

Lavelle, G. C., and Bang, F. B., 1971, Influence of type and concentration of sera in vitro on susceptibility of genetically resistant cells to mouse hepatitis virus, *J. Gen. Virol.* **12**:233–238.

Lindenmann, J., Deuel, E., Fanconi, S.. and Haller, O., 1978, Inborn resistance of mice to myxoviruses: Macrophages express phenotype *in vitro*, *J. Exp. Med.* **147**:531–540.

Linnavuori, K., and Hovi, T., 1981, Herpes simplex virus infection in human monocyte cultures: Dose dependent inhibition of monocyte differentiation resulting in abortive infection, *J. Gen. Virol.* **52**:381–385.

Linnavuori, K., and Hovi, T., 1983, Restricted replication of herpes simplex virus in human monocyte cultures: Role of interferon, *Virology* **130**:1–9.

Lopez, C., 1975, Genetics of natural resistance to herpesvirus infection in mice, *Nature* **258**:152–153.

Lopez, C., and Dudas, G., 1979, Replication of herpes simplex virus type 1 in macrophages from resistant and susceptible mice, *Infect. Immun.* **23**:432–437.

Lynch, C. J., and Hughes, T. P., 1936, The inheritance of susceptibility to yellow fever encephalitis in mice, *Genetics* **21**:104–112.

McKendall, R. R., Carrigan, D. R., and Johnson, K. P., 1981, Lymphoid cell infection by measles virus in newborn hamsters. Role for monocytes in virus spread to distant sites, *J. Neuroimmunol.* **1**:261–274.

Mims, C. A., 1964, Aspects of the pathogenesis of virus diseases, *Bacteriol. Rev.* **28**:30–71.

Mogensen, S. C., 1977a, Role of macrophages in hepatitis induced by herpes simplex virus types 1 and 2 in mice, *Infect. Immun.* **15**:686–691.

Mogensen, S. C., 1977b, Genetics of macrophage-controlled resistance to hepatitis induced by herpes simplex virus type 2 in mice, *Infect. Immun.* **17**:268–273.

Mogensen, S. C., 1978, Macrophages and age-dependent resistance to hepatitis induced by herpes simplex virus type 2 in mice, *Infect. Immun.* **19**:46–50.

Mogensen, S. C., 1979, Role of macrophages in natural resistance against virus infections, *Microbiol. Rev.* **43**:1–26.

Mogensen, S. C., and Andersen, H. K., 1977, Effect of silica on the pathogenic distinction between herpes simplex virus type 1 and 2 hepatitis in mice, *Infect. Immun.* **17**:274–277.

Mogensen, S. C., and Andersen, H. K., 1978, Role of activated macrophages in resistance of congenitally athymic nude mice to hepatitis induced by herpes simplex virus type 2, *Infect. Immun.* **19**:792–798.

Mogensen, S. C., and Virelizier, J. L., 1987, The interferon–macrophage alliance, in: *Interferon 8* (I. Gresser, ed.) pp. 55–84, Academic Press, London.

Morahan, P. S., 1984, Interactions of herpesviruses with mononuclear phagocytes, in: *Immunobiology of Herpes Simplex Virus Infections* (B. T. Rouse and C. Lopez, eds.), pp. 71–89, CRC Press, Boca Raton, Fl.

Morahan, P. S., and Morse, S. S., 1979, Macrophage-virus interactions, in: *Virus–Lymphocyte Interactions: Implications for Disease* (M. R. Profitt, ed.), pp. 17–35, Elsevier/North Holland, New York.

Morahan, P. S., Coleman, P. H., Morse, S. S., Volkman, A., 1982, Resistance to infections in mice with defects in the activities of mononuclear phagocytes and natural killer cells: Effects of immunomodulators in beige mice and ^{89}Sr-treated mice, *Infect. Immun.* **37**:1079–1085.

Morse, S. S., and Morahan, P. S., 1981, Activated macrophages mediate interferon-independent inhibition of herpes simplex virus, *Cell. Immunol.* **58**:72–84.

Nahmias, A. J., Visintine, A. M., Reimer, C. B., Del Buono, I., Shore, S. L., and Starr, S. E., 1975, Herpes simplex virus infection of the fetus and newborn, *Prog. Clin. Biol. Res.* **3**:63–67.

Nickol, A. D., and Bonventre, P. F., 1977, Anomalous high native resistance of athymic mice to bacterial pathogens, *Infect. Immun.* **18**:636–645.

Oehler, J. R., Herberman, R. B., and Holden, H. T., 1978, Modulation of immunity by macrophages, *Pharmacol. Ther.* **A2**:551–593.

Panijel, J., and Cayeux, P., 1968, Immunosuppressive effects of macrophage antiserum, *Immunology* **14**:769–780.

Pedersen, E. B., Haahr, S., and Mogensen, S. C., 1983, X-linked resistance of mice to high doses of herpes simplex virus type 2 correlates with early interferon production, *Infect. Immun.* **42**:740–746.

Plaeger-Marshall, S., Wilson, L. A., and Smith, J. W., 1982, Permissiveness of rabbit monocytes and macrophages for herpes simplex virus type 1, *Infect. Immun.* **35**:151–156.

Rager-Zisman, B., and Allison, A. C., 1973, The role of antibody and host cells in the resistance of mice against infection by Coxsackie B-3 virus, *J. Gen. Virol.* **19**:329–338.

Rao, G. R., Rawls, W. E., Perey, D. Y. E., and Tompkins, W. A. F., 1977, Macrophage activation in congenitally athymic mice raised under conventional and germ-free conditions, *Res. J. Reticuloendothel. Soc.* **21**:13–20.

Roberts, J. A., 1964, Growth of virulent and attenuated ectromelia virus in cultured macrophages from normal and ectromelia-immune mice, *J. Immunol.* **92**:837–842.

Rodgers, B., and Mims, C. A., 1981, Interactions of influenza virus with mouse macrophages, *Infect. Immun.* **31**:751–757.

Roser, B., 1965, The distribution of intravenously injected peritoneal macrophages in the mouse, *Aust. J. Exp. Biol. Med. Sci.* **43**:553–562.

Sabin, A. B., 1952, Nature of inherited resistance to viruses affecting the nervous system, *Proc. Natl. Acad. Sci. USA* **38**:540–546.

Sabin, A. B., and Olitsky, P. K., 1937, Influence of host factors on neuroinvasiveness of vesicular stomatitis virus. I. Effect of age on the invasion of the brain by virus instilled in the nose, *J. Exp. Med.* **66**:15–34.

Schindler, L., Brücher, J., and Kirchner, H., 1984, Protection of mice against infection with mouse hepatitis virus type 3 by injection of silica, *Immunobiology* **166**:62–71.

Schlabach, A. J., Martinez, D., Field, A. K., and Tytell, A. A., 1979, Resistance of C58 mice to primary systemic herpes simplex virus infection: Macrophage dependence and T-cell independence, *Infect. Immun.* **26:**615–620.

Schlesinger, J. J., and Brandriss, M. W., 1981, Growth of 17 D yellow fever virus in a macrophage-like cell line, U937: role of Fc and viral receptors in antibody-mediated infection, *J. Immunol.* **127:**659–665.

Selgrade, M. K., and Osborn, J. E., 1974, Role of macrophages in resistance to murine cytomegalovirus, *Infect. Immun.* **10:**1383–1390.

Sethi, K. K., 1983, Contribution of macrophage arginase in the intrinsic restriction of herpes simplex virus replication in permissive macrophage cultures induced by gamma-interferon containing products of activated spleen cells, *Immunobiology* **165:**459–474.

Shif, I., and Bang, F. B., 1970a, *In vitro* interaction of mouse hepatitis virus and macrophages from genetically resistant mice. I. Adsorption of virus and growth curves, *J. Exp. Med.* **131:**843–850.

Shif, I., and Bang, F. B., 1970b, *In vitro* interaction of mouse hepatitis virus and macrophages from genetically resistant mice. II. Biological characterization of a variant virus MHV(C_3H) isolated from stocks of MHV(PRI), *J. Exp. Med.* **131:**851–862.

Staeheli, P., and Haller, O., 1987, Interferon–induced Mx protein: a mediator of cellular resistance to influenza virus, in: *Interferon 8* (I. Gresser, ed.) pp. 1–23, Academic Press, London.

Stevens, J. G., and Cook, M. L., 1971, Restriction of herpes simplex virus by macrophages. An analysis of the cell–virus interaction, *J. Exp. Med.* **133:**19–38.

Stohlman, S. A., Frelinger, J. A., and Weiner, L. P., 1980, Resistance to fatal central nervous system disease by mouse hepatitis virus, strain JHM. II. Adherent cell-mediated protection, *J. Immunol.* **124:**1733–1739.

Stohlman, S. A., Woodward, J. G., and Frelinger, J. A., 1982, Macrophage antiviral activity: Extrinsic versus intrinsic activity, *Infect. Immun.* **36:**672–677.

Tosolini, F. A., and Mims, C. A., 1971, Effect of murine strain and viral strain on the pathogenesis of hymphocytic choriomeningitis infection and a study of footpad responses, *J. Infect. Dis.* **123:**134–144.

Turner, G. S., and Ballard, R., 1976, Interaction of mouse peritoneal macrophages with fixed rabies virus *in vivo* and *in vitro*, *J. Gen. Virol.* **30:**223–231.

Unanue, E. R., 1981, The regulatory role of macrophages in antigenic stimulation. Part two: Symbiotic relationship between lymphocytes and macrophages, *Adv. Immunol.* **31:**1–136.

Virelizier, JL., 1981, Role of macrophages and interferon in natural resistance to mouse hepatitis virus infection, *Curr. Top. Microbiol. Immunol.* **92:**53–64.

Volkman, A., Chang, N. C., Strausbauch, P. H., and Morahan, P. S., 1983, Differential effects of chronic monocyte depletion on macrophage populations, *Lab. Invest.* **49:**291–298.

Webster, L. T., 1937, Inheritance of resistance of mice to enteric bacterial and neurotropic virus infections, *J. Exp. Med.* **65:**261–286.

Welsh, R. M., 1981, Natural cell-mediated immunity during viral infections, *Curr. Top. Microbiol. Immunol.* **92:**83–106.

Wildy, P., Gell, P. G. H., Rhodes, J., and Newton, A., 1982, Inhibition of herpes simplex virus multiplication by activated macrophages: A role for arginase, *Infect. Immun.* **37:**40–45.

Wirth, J. J., Carney, W. P., and Weelock, E. F., 1980, The effect of particle size on the immunosuppressive properties of silica, *J. Immunol. Meth.* **32:**357–373.

Zisman, B., Hirsch, M. S., and Allison, A. C., 1970, Selective effects of anti-macrophage serum, silica and anti-lymphocyte serum on pathogenesis of herpes virus infection of young adult mice, *J. Immunol.* **104:**1155–1159.

10

Virus-Induced Modulation of Reticuloendothelial Function

STEVEN SPECTER

1. INTRODUCTION

The ability of viruses to modify the functions of cells of the lymphoreticular system has been extensively investigated and reviewed (Woodruff and Woodruff, 1975; Specter and Friedman, 1978; Nash, 1985; Friedman *et al.*, 1986; Rouse and Horohov, 1986). Frequently this information is amassed in such a manner that the importance of the interactions of viruses and macrophages receives little attention. In the present chapter these interactions are the focus of the discussion and review. The exception to inclusion of viruses in this chapter are the RNA tumor viruses, which are dealt with in Chapter 13 in this volume.

The interactions of viruses and the immune system are bidirectional, so that the immune cells can limit virus spread while the viruses might modify lymphoreticular cell function. In the sections that follow I examine these interactions and explain the mechanisms by which they occur as well as the ultimate outcome of infection on the macrophage and the host in general.

2. MACROPHAGES IN HOST DEFENSES AGAINST VIRUSES

The protection of the host against virus infection involves several interactions with macrophages (Mogensen, 1986). These host–virus encounters may involve nonspecific functions, such as phagocytosis and digestion of the virus particle, or may involve specific immune phenomena that require the stimulation of macrophages by products released from lymphocytes during immune sensitization. Ingestion and degradation of the virion results in antigenic components being presented on the macrophage surface, which is important for the triggering of lymphocytes. This process is referred to as accessory cell function.

STEVEN SPECTER • Department of Medical Microbiology and Immunology, University of South Florida College of Medicine, Tampa, Florida 33612.

Another function that results in activation of lymphocytes is the production of interleukin 1 (IL-1). This monokine is most noted for the activation of T lymphocytes but has been demonstrated to have other important activities in inflammatory and immune responses (Dinarello, 1984). In addition, macrophages may be activated by lymphokines or interaction with T lymphocytes and then become capable of killing virus-infected cells directly (Cook and Lewis, 1984) or participating in antibody-dependent killing of virus-infected cells (Kohl *et al.*, 1977). Cook and Lewis have noted that macrophages activated by the mycobacterial vaccine BCG readily kill adenovirus 2–infected hamster cells but are less effective at killing adenovirus type 12–infected cells. This has been related to the oncogenic potential of type 12 and related to expression of viral T antigen on the surface of these cells. Expression of this antigen is associated with viral transformation of cells and may be a protective mechanism that prevents macrophage killing.

The most basic function of macrophages in resisting virus infections, i.e., phagocytosis and digestion, has been demonstrated to be determined by both genetics and the age of the host (Lopez, 1975; Stiehm, 1980; Mogensen, 1986). The genetics will determine whether the macrophage is resistant (nonpermissive) or susceptible (permissive) to virus replication. The nonpermissive macrophage readily destroys the virus and prevents or limits infection. Permissive macrophages may not only replicate virus, but also act as a vehicle to transmit virus to distant sites in the host (Mims, 1964). Culturing nonpermissive macrophages *in vitro* frequently results in their conversion to a permissive state (Lopez and Dudas, 1979; Morahan and Morse, 1979). Thus the intrinsic genetic restriction of replication can be overcome under culture conditions and possibly during immune suppression in the host.

Finally, macrophages may be important as regulatory cells for immune responses in virus infections. Suppressor macrophages have been demonstrated from infected hosts that can modify a variety of immune responses (Varesio, 1983). It is noted here that this may not always be detrimental to the host since inhibition of immunopathogenesis may limit damage of vital organs that are infected by viruses.

3. ALTERATION OF MACROPHAGE FUNCTIONS BY VIRUSES

The list of viruses capable of infecting monocytes/macrophages *in vivo* and/or *in vitro* is extensive (Table 1). The interactions of these viruses and macrophages often determine whether disease will ensue or the macrophage will be able to abort infection. The consequences of interaction between macrophage and virus may be without a general effect on the host or they may result in mild to severe manifestations affecting host immunity (Tables 2 and 3).

3.1. ADENOVIRUSES

There is no evidence that adenoviruses are capable of infecting macrophages. However, there are indications that macrophages are important in ade-

TABLE 1. VIRUSES CAPABLE OF INFECTING MONOCYTE/MACROPHAGES

Arenaviruses	Paramyxoviruses
Lymphocytic choriomeningitis virus	Measles
Junin	Mumps
Pichinde	Parainfluenza viruses (Sendai virus)
	Respiratory syncytial virus
Coronaviruses	Poxviruses
Mouse hepatitis virus	Ectromelia
Enteroviruses	Vaccinia
Poliovirus	Reoviruses
Herpesviruses	Retroviruses (nononcogenic only)
Bovine herpesvirus	Caprine arthritis encephalitis virus
Cytomegalovirus	Equine infectious anemia
Human	Human immunodeficiency virus
Mouse	Visna virus
Guinea pig herpes like virus	Rhabdoviruses
Herpes simplex virus	Vesicular stomatis virus
Infectious laryngotracheitis virus	Togaviruses
Pseudorabies virus	Dengue
Varicella zoster virus	Japanese encephalitis virus
Influenza	Lactate dehydrogenase virus
Papovaviruses	Rubella
	West Nile
	Yellow fever

novirus-induced immunosuppression. Berensci and co-workers (1982, 1985) related suppressed antibody production against sheep erythrocytes to macrophages in adenovirus-infected mice. This was based on the observation that silica gel, which is selectively toxic for macrophages, when given to mice infected with adenovirus type 6 prevents the immunosuppression. Furthermore, although adenovirus does not replicate in macrophages, it can infect primate lymphocytes. Andiman and Miller (1982) suggest that macrophages from cord blood are responsible for the maintenance of persistent infection. The mechanism by which the macrophage is capable of this was not elaborated.

TABLE 2. TYPES OF INTERACTIONS BETWEEN MACROPHAGES AND VIRUSES

Type of infection	Alteration of cell function	Viruses involved
Attachment and penetration	None identified	Most viruses
Abortive (non-permissive)	None to slight impairment	HSV, influenza, measles
Productive (permissive)	Moderate to severe impairment	HIV, measles, polio, togaviruses
Persistence	None to moderate impairment	LDHV, MCMV, CAEV, parainfluenza

TABLE 3. MACROPHAGE FUNCTIONS REPORTED TO BE ALTERED BY VIRUS INFECTION

Accessory cell function (antigen processing and/or presentation)
Adherence to glass or plastic surfaces
Antibody-dependent cellular cytotoxicity
Degradation of viral particles, antigens
Direct cytotoxic/cytocidal activity
Microbicidal activity
Phagocytosis
Production of monokines/inflammatory molecules (e.g., IL-1, interferon-α, prostaglandins)
Suppressor cell activity
Suppressor/cytotoxic factor production

3.2. ARENAVIRUSES

Although these viruses are seldom the cause of infections in humans (at least in the "developed" nations), they can cause severe, life-threatening diseases. Macrophages are a major target for infection by arenaviruses, along with lymphocytes (Murphy *et al.*, 1976, 1977; Doyle and Oldstone, 1978; Laguens *et al.*, 1983a,b). The most frequently studied model of arenavirus infection is lymphocytic choriomeningitis (LCM) virus in mice. LCM has a unique relationship with the immune system, in that T lymphocytes mediate the acute symptoms of the disease in the central nervous system (Buchmeier *et al.*, 1980).

In vitro infection of murine macrophages by LCM virus minimally affects their function. Phagocytosis is normal and the level of hydrolytic enzymes in these cells is unaltered (Mims and Wainwright, 1968; Oldstone *et al.*, 1973; Schwartz *et al.*, 1970). Some reports do indicate that enzyme activity and dye uptake are altered (Allison, 1967; Yarborough *et al.*, 1967). These studies suggest that infection of macrophages by LCM causes only minimal changes. However, *in vitro* studies in mice suggest a more important role for macrophages in LCM infection. Gledhill *et al.* report a decrease in phagocytosis of colloidal carbon following infection (1965). Jacobs and Cole further support the importance of a macrophage defect by showing that depressed T- and B-lymphocyte blastogenic responses to mitogens can be reversed by addition of macrophages from normal mice (1976). It has been suggested that the macrophage function in these studies may only be enhancing viability of the lymphocytes (Bro-Jorgensen, 1978). If this is so, then the defect could be the inhibition of IL-1 production or some other monokine by virus-infected macrophages.

Two other arenaviruses, Junin virus and Pichinde virus, have been reported to alter mouse macrophage function (Gonzalez *et al.*, 1982; Laguens *et al.*, 1983b; Friedlander *et al.*, 1984). Pichinde has been noted to inhibit the ability of mouse macrophages to respond to macrophage growth factor, resulting in a failure of these cells to proliferate (Friedlander *et al.*, 1984). The ability of Junin virus to replicate in macrophages *in vitro* is directly related to virulence, with virulent strains replicating and attenuated strains being inhibited (Laguens *et al.*, 1983b).

This suggests that virulence may be related to the ability of the virus to inhibit macrophage function. However, the only macrophage function tested was phagocytosis, and this was unaltered when compared to uninfected macrophages (Gonzalez *et al.*, 1982).

3.3. AVIAN INFECTIOUS BURSAL DISEASE

Avian infectious bursal disease (IBDV) is a double-stranded RNA virus containing 32 capsomeres in a naked nucleocapsid. Thus, this is reminiscent of the Reoviridae. Experimentation using IBDV is performed in chickens that exhibit a profound immunodepression of B- and T-lymphocyte responses following infection. This disease progression appears to be age related, as chicks infected *in ovo* or at birth show greater immunodepressive effects than birds infected at 3 weeks of age (Sivanandan and Maheswaran, 1980). Immunodepressive effects related to phytohemagglutinin-induced lymphoblast transformation have been attributed to a "macrophagelike" suppressor cell found in the spleen of infected chicks (Sharma and Lee, 1983). These suppressor cells were not altered by anti-T- or anti-B-lymphocyte antisera but could be depleted from spleen cell suspensions by plastic adherence or ingestion of carbonyl iron.

There are no reports on the infectivity of chicken macrophages by IBDV, suggesting that the effect of the virus on these cells may be indirect. In addition, phagocytosis of *Staphylococcus aureus* is normal in circulating phagocytes from IBDV-infected animals (Santivatr *et al.*, 1981).

3.4. MOUSE HEPATITIS VIRUS (CORONAVIRUS)

Mouse hepatitis virus (MHV) readily replicates in macrophages (peritoneal and Kupffer cells) of susceptible mice. The ability to replicate in macrophages is genetically determined and is responsible for susceptibility or resistance of the host (Bang and Warwick, 1960; Taguchi *et al.*, 1976). Two or more recessive genes are involved in control of susceptibility, at least one of which is H-2 linked (Levy-LeBlond *et al.*, 1979). This resistance/susceptibility of mouse spleen cell cultures is readily reversed by *in vitro* manipulation. Resistance can be conferred on susceptible cells by stimulation with concanavalin A (Con A) (Weiser and Bang, 1977) while resistant cells can be rendered susceptible following treatment with silica, cortisone, or lymphokines (Weiser and Bang, 1976; Taguchi *et al.*, 1980; Taylor *et al.*, 1981).

MVH strains can cause acute or chronic infection which is related to persistent infection of macrophages. Chronically infected mice respond to antigenic challenge with a depressed antibody response to sheep erythrocytes and *Escherichia coli* lipopolysaccharide, presumably because of macrophage dysfunction in antigen processing or presentation (Leray *et al.*, 1982). However, antibody responses to MVH, per se, were detectable in both acute and chronic infection. These responses were limited to the IgM class (Virelizier *et al.*, 1975). Although macrophage accessory cell function seems to be altered, phagocytic activity was normal when measured by yeast phagocytosis *in vitro* and sheep erythrocyte uptake *in vivo* (Krzystyniak and Dupuy, 1983).

3.5. HEPATITIS B VIRUS

Hepatitis B virus (HBV) has not been detected within macrophages, including attempts to measure HBV DNA in such cells (Korba *et al.*, 1986). The only activity related to macrophages that has been tested during HBV infection is interferon-α (IFN-α) production, and this was normal (Davis *et al.*, 1984). Nevertheless, it is not possible to entirely rule out a role for macrophages in HBV infection since virus particles persist in the liver, where they are in close contact with Kupffer cells. Furthermore, there is an important immunopathologic aspect to disease progression which could readily involve macrophages as well as lymphocytes.

3.6. HERPESVIRUSES

A number of herpesviruses of humans and animals have been reported to alter immune responses of the host. Most of these viruses, as described in the following sections, can interact with macrophages and have an effect on immune responsiveness. A few, however, including herpes sylvilagus in rabbits (Kramp *et al.*, 1985) and mouse thymic virus in mice (Cohen *et al.*, 1975), do not appear to infect macrophages.

3.6.1. Human Cytomegalovirus

Monocytes have been described as the primary target site for human cytomegalovirus (HCMV), although this may often be an abortive replication (Einhorn and Ost, 1984; Rice *et al.*, 1984). Nevertheless, infectious HCMV has been isolated from mononucleosis patients (Carney and Hirsch, 1981), and infected monocytes are capable of inhibiting autologous lymphocytes from responding to Con A (Rinaldo *et al.*, 1977, 1980). The mechanism of the depression of the Con A response by monocytes has not been ascertained, but recently Rodgers *et al.* (1985) demonstrated a 95-kd protein in human monocyte cultures infected with CMV strain AD 169, which specifically inhibits IL-1.

3.6.2. Epstein–Barr Virus

Monocytes do not become infected with Epstein–Barr virus (EBV) but are often in the microenvironment of infected B lymphocytes. Junker *et al.* (1986) have demonstrated that monocyte numbers remain normal in infectious mononucleosis due to EBV. In fact, macrophages seem to be activated during such infections in hematopoietic tissues, and often small granulomas are seen in the bone marrow (Rothwell, 1975). Activated macrophages often produce increased levels of prostaglandins, which in turn can depress T-lymphocyte function. Thus, macrophages appear to be involved in EBV-induced immunosuppression, although they are not directly infected.

3.6.3. Herpes Simplex Virus

Herpes simplex virus (HSV) has been shown to readily infect macrophages of both humans and experimental animals. HSV can infect macrophages *in vivo,* but this is generally an abortive infection; however, *in vitro* they may replicate and cause destruction of the infected macrophages (Morahan *et al.*, 1985). These authors also report that macrophage surface receptors, phagocytosis, oxidative metabolism, phagosome–lysosome fusion, and microbial killing are all altered by HSV infection. Decreases in natural killer cell activity, ADCC, and Fc receptor expression have also been reported (Rhodes, 1985). These decreases in immune function can be related to increased PGE_2 production by macrophages and resulting increases in intracellular cAMP levels. In addition, susceptible monocyte–macrophage cultures treated with PGE yield higher titers of HSV1. Nick *et al.* reported the suppression of antibody responses to HSV1 or HSV2 when mice or rats were given HSV2 intraperitoneally (Nick *et al.*, 1986). HSV2-induced suppression could be abrogated if the animals were treated with silica. This implicates the macrophages as the cell population responsible for suppression, but a mechanism was not investigated. The studies just cited suggest that this may be due to increased cAMP and/or prostaglandins.

3.6.4. Varicella–Zoster Virus

Macrophages have been reported to be susceptible to infection by varicella–zoster virus (VZV). Macrophages infected with VZV are capable of inhibiting mixed lymphocyte reactions *in vitro* (Twomey *et al.*, 1974). Arneborn and Biberfeld also report suppressive activity of monocytes on T-cell responses during varicella (1983). In contrast, removal of phagocytic cells did not result in a loss of suppression, suggesting that either monocytes/macrophages were not responsible for suppression or the damage to T cells by the infected macrophages was irreversible.

3.6.5. Bovine Herpesvirus

Bovine herpesvirus type 1 (BHV-1) is noted for causing infectious bovine rhinotracheitis. In this infection model, defects in macrophage function are readily observed between 4 and 7 days postinfection. In addition, polymorphonuclear leukocyte and lymphocyte dysfunctions are observed. Macrophages infected *in vitro* by BHV-1 express very few Fc receptors and have a reduced uptake of erythrocytes or yeast (Forman and Babiuk, 1982). Mitogen-induced proliferation of lymphocytes is reduced in the presence of BHV-1-infected macrophages, even when these infected cells comprise only 1% of the cells in the reaction (Bendixen *et al.*, 1981). Suppression of lymphocytes is most notable 3–5 days postinfection and has been demonstrated to be due to a soluble factor. Several macrophage-produced soluble products have been linked to the possible reduction in lymphocyte function. These include IFN-α or PGE_2, as well as toxic oxygen radicals like H_2O_2 or superoxide, all of which are increased in BHV-1 infection (Babiuk *et al.*, 1987). Indeed, plasma from BHV-1-infected cattle suppressed lymphocyte blast transformation by 80%. The authors speculate that PGE_2 may work via inhibition of IL-1, and that toxic oxygen radicals may cause damage to cellular DNA, thus causing lymphocyte dysfunction.

3.6.6. Equine Herpesvirus

Equine herpesviruses have been reported to infect monocytes and lymphocytes (Scott *et al.*, 1983). Both lymphocyte blastogenesis and ADCC are suppressed in these animals (Dutta *et al.*, 1980; Coignoul *et al.*, 1984). There did not seem to be a soluble factor responsible for the suppression of either response, and the role of infected macrophages in immune suppression was not identified.

3.6.7 Pseudorabies Virus

Pseudorabies virus (PRV) infects swine, causing a variety of disorders including Aujesky's disease, which involves the central nervous system. PRV infects pig lung macrophages *in vitro* (Smid *et al.*, 1981) and seems to infect macrophages *in vivo* with a resulting immune suppression (Babiuk *et al.*, 1987). PRV infection of swine renders them susceptible to secondary bacterial infection with *Pasturella multocida*. Macrophages infected *in vitro* cease function within 2 hr and lyse shortly thereafter but with only a few virus particles present, suggesting a toxic lysis rather than viral replication.

3.6.8. Marek's Disease Virus

The susceptibility of chicken macrophages to Marek's disease virus has been described (Jakowski *et al.*, 1970; Lee *et al.*, 1978). Lee and co-workers have suggested that splenic macrophages from infected birds are capable of suppressing mitogenic responses of lymphoid cells (1978). No mechanism of action was identified for these suppressor macrophages.

3.6.9. Infectious Laryngotracheitis Virus

Infectious laryngotracheitis virus (ILT) readily infects chicken macrophages *in vitro* (Calnek *et al.*, 1986), but infection of macrophages from infected animals has been difficult to assess and is not reported. Replication of ILT in bone marrow macrophages has been recorded, although replication is restricted. This appears to be a failure in assembly since large quantities of virus-directed macromolecules are detected but there is little or no mature virus. Curiously, attenuated vaccine strains replicate more efficiently than virulent virus strains in macrophages *in vitro*. This suggests that replication in macrophages is unrelated to virulence.

3.6.10. Guinea Pig Herpes-Like Virus

Guinea pig herpes–like virus, like most other herpesviruses, infects macrophages and lymphocytes during acute infection, often leading to latency (Gonzalez-Serva and Hsiung, 1978). There is a resulting lymphocytosis, mainly of T cells, but little else is reported regarding functional alterations.

3.7. PAPOVAVIRUSES

Macrophages appear to be resistant to papovaviruses. *In situ* studies of warts show an accumulation of macrophages at the base of the wart, but these cells do not infiltrate the wart per se (Oguchi *et al.*, 1981). These cells appear to be involved in wart repression, which suggests they may ingest and digest wart virus. Both monocytes and macrophages are also resistant to the BK virus, a polyomavirus. These cells have surface receptors for the virus and internalize the bound virus; however, there is no evidence of virus replication or persistence (Portolani *et al.*, 1985). Thus monocytes, regardless of their state of differentiation, are important for degradation of BK virus but not viral replication.

3.8. MYXOVIRUSES

The myxoviruses are composed of two major families, the Orthomyxoviruses and the Paramyxoviruses. Both families have representatives that replicate in leukocytes and are immunosuppressive.

3.8.1. Influenza Virus (Orthomyxovirus)

Macrophages of mice and humans have been reported to be infected by influenza virus both *in vivo* and *in vitro* (Kleinerman *et al.*, 1976; Warshauer *et al.*, 1977; Nugent and Pesanti, 1979). In certain instances, modulation of macrophage and lymphocyte functions is recorded but some activities are unaltered. Decreases in accessory cell function, chemotaxis, phagocytosis, and bactericidal activity are noted for macrophages (Kleinerman *et al.*, 1974, 1975, 1976; Warshauer *et al.*, 1977; Roberts and Steigbigel, 1978; Nugent and Pesanti, 1979; Roberts *et al.*, 1980; Gardner and Lawton, 1982). By contrast, mouse alveolar macrophages retained normal phagocytic function and human alveolar macrophages retained accessory cell functions (Nugent and Pesanti, 1979; Ettensohn and Roberts, 1984). Lymphocyte responses to mitogens were depressed in association with lost accessory cell function, but the response to influenza antigen was enhanced (Roberts, unpublished observations, 1982). These results suggest that the host may focus on selective antigens which challenge the integrity of the host at the expense of other immune functions. Roberts *et al.* (1979) refer to this as "immunofocusing."

Macrophages are not readily permissive for influenza, inactivating the virus *in vivo,* but they are infected *in vitro.* The immunosuppressive effects of the virus are not attributable to the virion, since inactivated virus does not induce suppression (Roberts and Steigbigel, 1978; Roberts *et al.,* 1980). Roberts and co-workers have eliminated prostaglandins (Ettensohn and Roberts, 1984) and IFN-α (Roberts and Steigbigel, 1978; Roberts *et al.,* 1979) as probable mediators of immune suppression. These investigators have detected both IL-1 and an IL-1 inhibitor in human macrophages infected with influenza virus *in vitro* (Roberts *et al.,* 1986). The role of this inhibitor as an immune suppressive factor in influenza and other virus infections remains to be determined.

3.8.2. Paramyxoviruses

The paramyxoviruses reported to infect macrophages include measles, mumps, respiratory synctial virus, and Sendai virus. Replication of measles virus can be demonstrated in human monocytes *in vitro,* but this is limited when compared to replication in lymphocytes (Joseph *et al.,* 1975; Sullivan *et al.,* 1975). Infected monocytes/macrophages can suppress mitogen-induced T-lymphocyte proliferation (Rinehart *et al.,* 1979). This suppression occurs most notably when there is direct lymphocyte–macrophage contact. Antibody production or B-cell mitogen stimulation is also depressed by measles-infected macrophages and this is related to cell contact and/or release of INF-α from macrophages (Kadish *et al.,* 1980; Harfast *et al.,* 1981).

Mumps virus has been demonstrated to infect monocytes/macrophages *in vitro;* however, the virus clearly infects lymphocytes more readily (Duc-Nguyen and Henle, 1966; Fleischer and Kreth, 1982). No relationship between infection of macrophages and alteration of immune function was suggested. Likewise, animal paramyxoviruses, canine distemper virus, and bovine rinderpest virus infect lymphocytes and macrophages (Appel, 1978; Rossiter and Wardley, 1985). These authors related ability to infect these cells with virulence, suggesting that restriction of replication inhibited disease.

Respiratory syncytial virus infection of macrophages is documented both *in vivo* and *in vitro* (Domurat *et al.,* 1985). Similar to influenza virus, RSV stimulates both IL-1 and IL-1 inhibitor (Roberts *et al.,* 1986). The levels of inhibitor are considerably greater with RSV than influenza infection; the significance of this finding is not known. Borysiewicz and co-workers (1985) describe an inhibition of IL-2 release by cytotoxic T cells during RSV infection; the relationship to the presence of IL-1 inhibitor is not reported.

Infection of mouse alveolar macrophages by Sendai virus resulted in decreased phagocytic function, even in the presence of immune serum (Jakab and Warr, 1981). In addition, Silverberg and co-workers (1979) have demonstrated an inhibition of phagosome–lysosome fusion in Sendai virus–infected macrophages. Sendai virus infection of macrophages does stimulate tumor necrosis factor production, signifying that these cells are activated (Aderka *et al.*, 1986). This type of activation, however, may enhance certain responses while limiting others. In this manner, Newcastle disease virus (NDV) infection results in a loss of cytotoxic T-cell activity associated with macrophage activation (Brenan and Zinkernagel, 1983). NDV is a potent IFN stimulant, and this and/or other soluble factors from macrophages could be responsible for the diminished T-cell activity.

3.9. PICORNAVIRUSES

The picornaviruses are comprised of the rhinoviruses, enteroviruses (polioviruses, coxsackieviruses, echoviruses, and hepatitis A virus), and cardioviruses. All of these have been demonstrated to modulate immune responsiveness to some extent (Garzelli *et al.*, 1987). Poliovirus has been demonstrated to infect human monocytes and mouse macrophages. The infected macrophages have been demonstrated to be responsible for inhibition of PHA-induced lymphocyte blastogenesis (Gresser and Chany, 1964; Kantoch and Dobrowolska, 1969; Soontiens and Van Der Veen, 1973; Van Loon *et al.*, 1979).

Murine peritoneal macrophages are resistant to encephalomyocarditis virus infection; however, they become susceptible *in vitro* when mice are treated with anti-IFN-α/β (Belardelli *et al.*, 1984). In addition, hepatitis A virus has been noted to primarily replicate in Kupffer cells (Mathiesen *et al.*, 1977) while feline picornavirus first infects kitten alveolar macrophages (Kahn and Gillespie, 1971). While these infections play an important part in the pathogenesis of the virus infections, their effect on the immune system is undetermined. Hepatitis A virus has been noted to depress immunity; thus the infection of Kupffer cells and possibly other macrophages may be important in this regard (Mella and Lang, 1967; Willems *et al.*, 1969; Aiuti *et al.*, 1973).

The most intensively studied picornaviruses relative to immunosuppression are the Coxsackie B viruses, especially B3 (CVB3) (Toniolo *et al.*, 1986). CVB3 infection of mouse spleen cells results in a depression of the antigen-presenting activity of macrophages (Matteucci *et al.*, 1985). It is not clear whether splenic or peritoneal macrophages can be infected by CVB3, and most evidence supports engulfment but not necessarily infection (Rager-Zisman and Allison, 1973; Gauntt *et al.*, 1979). Although the participation of macrophages in generalized immunosuppression by CVB3 is uncertain, it has been suggested that this is primarily due to the failure of antigen-presenting cells (APC). Splenic macrophages from CVB3-infected mice fail to restore APC responses to macrophage-depleted cultures, but normal macrophages can restore activity to infected cultures (Garzelli *et al.*, 1987).

3.10. POXVIRUSES

Smallpox (variola) and vaccinia have both been demonstrated to replicate within macrophages (Avila *et al.*, 1972; Buchmeier *et al.*, 1979; Moss, 1985). This infection *in vivo* is associated with the dissemination of infection and the generation of IFN by macrophages. Replication of vaccinia in rabbit alveolar macrophages is abortive in previously infected rabbits and fails to penetrate macrophages from immune rabbits. Evidence of macromolecular synthesis and a few viral particles in these infected cells suggest that the blockage occurs late in replication. McLaren *et al.* (1976) reported that when lymphoproliferative responses developed during acute vaccinia infection, depressed responses to Con A were seen. As with other viruses noted previously in this chapter, there seems to be a selective immunity against the virus and a sacrifice of generalized responsiveness. The ability to develop antivaccinia immunity was related to macrophage resistance. Resistance to ectromelia (mouse pox) was also related to macrophages and in this study linked to production of IFN (Tsuru *et al.*, 1983; Cohen *et al.*, 1984). One might speculate from this evidence that IFN might be important in vaccinia infection and the depression of Con A-induced lymphocyte blastogenesis.

Malignant fibroma virus (MV) of rabbits has been reported to generate suppressor factor(s) which, at least in part, are related to prostaglandins (PG) (Strayer, 1987). Since macrophages are an important source of PG and macrophages are noted to ingest virus-infected lymphocytes, these cells might contribute to the immune depression seen in MV. The evidence presented by Strayer (1987) supports this concept by indicating that most, but not all, suppressive activity is associated with nonadherent cells.

3.11. REOVIRUSES

Reovirus types 1 and 3 have been demonstrated to infect a mouse macrophage cell line, P388D1, in culture. Little else has been reported regarding the interaction of these viruses with macrophages, except that uptake and replication of reovirus by P388D1 cells was facilitated by nonneutralizing antibodies to the virus (Burstein *et al.*, 1983).

3.12. RHABDOVIRUSES

Vesicular stomatitis virus (VSV) infection in mice is the model usually employed to study rhabdoviruses. VSV has been demonstrated to replicate in mouse peritoneal macrophages, but only a small percentage of the cells become infected (Belardelli *et al.*, 1984). Injection of mice with anti-IFN prior to harvesting peritoneal macrophages allows the virus to replicate to higher titer. Thus IFN is seen to have a critical role in protection of the host from VSV via activation of macrophages. Macrophages or T lymphocytes infected with VSV are capable of completely eliminating proliferative responses of antigen-primed T cells, suggesting a defect in the macrophages as APC (Sy and Finberg, 1987). This defect in macrophage function is somewhat questionable since the macrophages may have merely carried virus to T cells, which were subsequently infected and rendered incompetent.

3.13. TOGAVIRUSES

Macrophages have been demonstrated to be infected permissively by a variety of togaviruses, including dengue virus, lactate dehydrogenase virus, rubella, West Nile virus, and yellow fever, as well as other togaviruses that result in persistent infections (Chaturvedi, 1987). Macrophages are the principal site of replication for dengue virus (Halstead *et al.*, 1977). *In vivo* studies reveal a decrease in the number of macrophages in mouse spleen and peritoneal cavity, and degenerative changes were noted in many of the macrophages present (Chaturvedi *et al.*, 1983a,b; Nath *et al.*, 1983). Phagocytic activity, attachment to a glass surface, and random migration were all inhibited *in vitro* (Gulati *et al.*, 1982; Chaturvedi *et al.*, 1983b). In contrast to these results, Japanese encephalitis virus infection of macrophages did not inhibit phagocytosis (Chaturvedi *et al.*, 1979).

The infection of mice by dengue virus results in the production of cytotoxic factors (CF) that appear to have a central role in immune suppression *in vivo* or cultures *in vitro* (Gulati *et al.*, 1983, 1986). A cytotoxic factor is generated by T cells infected *in vitro* by dengue virus; this CF subsequently destroys two-thirds of all macrophages. Surviving macrophages produce a second cytotoxic factor (CF2), which in turn is capable of killing lymphoid cells and rendering cultures or mice imcompetent immunologically. In addition, macrophage-produced CF2 mimics dengue virus by suppressing phagocytosis and E-rosette formation by human T cells (Chaturvedi *et al.*, 1982; Gulati *et al.*, 1984). There is also a pathway of formation of soluble suppressor factors which inhibits immune responses. Suppressor T cells produce a factor, SF, that is transmitted by macrophages to recruit a second T-cell subpopulation to produce a second suppressor factor, SF2. This signal transmission requires direct cell–cell contact (Chaturvedi and Shukla, 1981; Chaturvedi *et al.*, 1981; Shukla and Chaturvedi, 1981, 1983; Shukla *et al.*, 1982). This suppressor system is further characterized in the review by Chaturvedi (1987).

Lactate dehydrogenase virus replicates exclusively in macrophages (Herman *et al.*, 1966; Porter *et al.*, 1968). Only 5–20% of mouse macrophages are susceptible to LDV (Stueckemann *et al.*, 1982). The Ia antigen has been identified as the receptor for LDV, which may explain why only a limited number of macrophages (those expressing Ia) are susceptible (Inada and Mims, 1984, 1985). LDV is reported to severely diminish antigen-presenting activity of macrophages, thus resulting in decreased immune functions (Isakov *et al.*, 1982a,b). These same cells have normal phagocytic function. LDV has represented a particularly difficult problem in laboratories studying immune responsiveness. This virus often infects mice without causing any overt symptoms. Therefore, when animals are studied for immune responsiveness after tumors are induced or following infection by other viruses, LDV can complicate the interpretation of induced immune suppression (Riley, 1974). To avoid this complication, it is advised that mice should be periodically checked for LDV infection by testing the levels of lactate dehydrogenase in peripheral blood serum.

Rubella infection of human monocytes/macrophages has been readily demonstrated, as well as infection of lymphocytes (Chantler and Tingle, 1980; Van Der Logt *et al.*, 1980). The functional alterations due to macrophage infection by rubella are not delineated. Thus, it is possible that these infected cells may contribute to the suppression of migration inhibitory factor production, lymphocyte blastogenesis, and skin test reactions associated with rubella infection or vaccination (Midulla *et al.*, 1972; Ganguly *et al.*, 1976; Buimovici-Klein *et al.*, 1979).

3.14. NONONCOGENIC RETROVIRUSES

Infection of macrophages has been demonstrated for caprine/ovine lentiviruses and lentivirus infection of humans by human immunodeficiency virus (HIV), the etiologic agent of the acquired immune deficiency syndrome (AIDS). Caprine arthritis encephalitis virus (CAEV) replication does not occur within ovine monocytes; however, when they are activated to macrophages they are readily infected (Anderson *et al.*, 1983; Narayan *et al.*, 1983). Although macrophages readily replicate CAEV, the virus does not inhibit macrophage functions such as phagocytosis of latex particles, Fc, and complement receptors or production of hydrolytic enzymes (Narayan *et al.*, 1983). A second ovine lentivirus, visna-maedi, is also known to infect peripheral blood monocytes (Narayan *et al.*, 1982, 1985). The virus is reported to inhibit humoral and cell-mediated immune function of mouse lymphocytes (Svennerholm *et al.*, 1978). Narayan and co-workers (1985) have demonstrated a unique interferon produced in cultures of sheep lymphocytes and macrophages; however, its relationship to altered immune responsiveness is undetermined. Although there is no direct evidence that macrophages mediate immune suppression associated with ovine lentivirus infection, the fact that they are the primary (or only) infected cell suggests so.

HIV is known to be a lymphotropic virus; however, recent reports indicate that macrophages are also productively infected (Gartner *et al.*, 1986; Koenig *et al.*, 1986). Macrophage function in AIDS is normal until late in infection, when there is a degeneration of numerous host functions. More interesting is the role of macrophages as a disseminator of HIV. Recent evidence indicates that macrophages transmit HIV to the brain, with resulting neurologic infection (Koenig *et al.*, 1986). Thus, the failure of macrophages to control infection has a devastating influence on the degeneration of the host.

4. CONCLUSIONS

The importance of monocytes/macrophages in viral-induced immunosuppression is evident, as delineated in the preceding sections. The contribution to immune depression made by these phagocytic cells may be direct, e.g., loss of function, or it may be indirect, whereby macrophages render lymphocytes incompetent through cell–cell interactions or soluble factors. Tables 3 and 4 list the macrophage functions that can be altered by virus infection and some mechanisms by which this is known to occur.

TABLE 4. MECHANISMS OF ALTERATION OF MACROPHAGE FUNCTION BY VIRUSES

Infection of macrophage
Cell death due to productive infection
Persistence of virus—loss of function
Production of immunomodulatory substances by macrophages
IL-1
IL-1 Inhibitor
Interferon-α
Prostaglandins, especially PGE_2
Production of mediators by other cells that regulate macrophage function
Failure to produce lymphokines that regulate macrophage function

Studies on the contribution of macrophages to virus-induced immune suppression suggest that the host alterations and mechanisms by which they occur are almost as diverse as the number of viruses. However, much of the diversity may be attributed to the particular parameters of macrophage and/or lymphocyte functions tested. It is unfortunate that comprehensive studies have not been performed for many of these viruses regarding their effect on immune cell functions. Nevertheless, within the limitations of the studies performed, some generalizations can be deduced. For some virus infections, macrophage functions appear to be unaltered, but lymphoid cells are inhibited, clearly indicating that macrophages are only part of the suppressive mechanisms important in virus infections.

Accessory cell function, as either antigen-processing or -presenting activity, appears to be one of the most sensitive assays for loss of macrophage function. Several studies could demonstrate loss of antigen-presenting activity when phagocytosis was normal (Isakov *et al.*, 1982a,b; Krzystyniak and Dupuy, 1983; Sy and Finberg, 1987). Thus, it is important to test the accessory cell function of macrophages in any study that is valid, to ascertain that infection has "no effect" on macrophage function.

Another important conclusion from these diverse studies is the realization that several functions must be examined to determine whether viral infection is truly altering host function with regard to progression of the infection. Roberts *et al.* (1979) have clearly demonstrated that immunity, as measured by mitogen responses and antigenic stimulation unrelated to influenza, is depressed nonspecifically in influenza infection, while antiinfluenza responses are potent. The authors refer to this response as "immunofocusing." Certainly, the idea that the host concentrates its immune resources on the most threatening intruder for a short period of time until the "alien" is repelled is practical. This same phenomenon is seen for other virus infections, as indicated in Sections 3.4, 3.8.2, and 3.10. Also, the fact these depressed immune responses are transient and the host returns to normal supports this concept.

The immune system's ability to successfully repel virus infections is a good indication that to manipulate the immune system in any virus infection that is not life threatening might be unwise. Alteration of the delicate balance of host immunity and viral infection might favor the virus as opposed to the host. Since many viruses multiply in cells of the immune system, stimulating that system may provide a better milieu for virus replication. However, in situations such as AIDS, where HIV infection leads to immune cell destruction and ultimately death, immune therapy, perhaps in conjunction with antiviral drugs, seems warranted.

REFERENCES

Aderka, D., Holtmann, H., Toker, L., Hahn, T., and Wallace, D., 1986, Tumor necrosis factor induction by Sendai virus, *J. Immunol.* **136:**2938.

Aiuti, F., Ciarla, M. V., D'Asero, C., D'Amelio, R., and Garofalo, J. A., 1973, Surface markers on lymphocytes of patients with infectious diseases, *Infect. Immun.* **8:**110.

Allison, A., 1967, Lysosomes in virus-infected cells, in: *Perspectives in Virology,* Volume V (V. M. Pollard, ed.), pp. 29–62, Academic Press, New York.

Anderson, L. W., Klevjer-Anderson, P., and Liggitt, H. D., 1983, Susceptibility of blood derived monocytes and macrophages to caprine arthritis encephalitis virus, *Infect. Immun.* **41:**837.

Andiman, W. A., and Miller, G., 1982, Persistent infection with adenovirus types 5 and 6 in lymphoid cells from humans and woolly monkeys, *J. Infect. Dis.* **145:**83.

Appel, M. J. G., 1978, Reversion to virulence of attenuated canine distemper virus *in vivo* and *in vitro, J. Gen. Virol.* **41:**385.

Arneborn, P., and Biberfeld, G., 1983, T lymphocyte subpopulation in relation to immunosuppression in measles and varicella, *Infect. Immun.* **39:**29.

Avila, F. R., Schultz, R. M., and Tompkins, W. A. F., 1972, Specific macrophage immunity to vaccinia virus: Macrophage virus interaction, *Infect. Immun.* **6:**9.

Babiuk, L. A., Lawman, M. J. P., and Griebel, P. 1987, Immunosuppression by bovine herpesviruses and other selected herpesviruses, in: *Viral Induced Immunosuppression* (S. Specter, M. Bendinelli, and H. Friedman, eds.), Plenum Press, New York (in press).

Bang, F. B., and Warwick, A., 1960, Mouse macrophages as host cells for the mouse hepatitis virus and the genetic basis of their susceptibility, *Proc. Natl. Acad. Sci. USA* **46:**1065.

Belardelli, F., Vigmaux, F., Proietti, E., and Gresser, I., 1984, Injection of mice with antibody to interferon renders peritoneal macrophages permissive for vesicular stomatitis virus and encephalomyocarditis virus, *Proc. Natl. Acad. Sci USA* **81:**602.

Bendixen, P. H., Shewen, P. E., and Wilkie, B. N., 1981, The influence of bovine alveolar macrophages on the blastogenic response of peripheral blood mononuclear cells, in: *The Ruminant Immune System* (J. E. Butler, ed.), pp. 814–818, Plenum Press, New York.

Berensci, K., Bakay, M., and Kovacs, P., 1982, Effect of human adenovirus type 6 on the primary immune response in mice, *Acta Virol.* **26:**340.

Berensci, K., Bakay, M., and Beladi, I., 1985, The role of macrophages in adenovirus-induced immunosuppression in mice, *Acta Virol.* **29:**61.

Borysiewicz, L. K., Casali, P., Rogers, B., Morris, S., and Sissons, J. G. P., 1985, The immunosuppressive effects of measles virus on T cell function-failure to affect IL-2 release of cytotoxic T cell activity *in vitro, Clin. Exp. Immunol.* **59:**26.

Brenan, M., and Zinkernagel, R. M., 1983, Influence of one virus infection on a second concurrent primary *in vivo* antiviral cytotoxic T-cell response, *Infect. Immun.* **41:**470.

Bro-Jorgensen, K., 1978, The interplay between lymphocytic choriomeningitis virus, immune function and hemopoiesis in mice, *Adv. Virol. Res.* **22:**327.

Buchmeier, N. A., Gee, S. R., Murphy, F. A., and Rawls, W. E., 1979, Abortive replication of vaccinia virus in activated rabbit macrophages, *Infect. Immun.* **26:**328.

Buchmeier, M. J., Welsh, R. M., Dutko, F. J., and Oldstone, M. B. A., 1980, The virology and immunobiology of lymphocytic choriomeningitis virus infection, *Adv. Immunol.* **30:**275.

Buimovici-Klein, E., Lang, P. B., Ziring, P. R., and Cooper, L. Z., 1979, Impaired cell-mediated immune response in patients with congenital rubella: Correlation with gestational age at time of infection, *Pediatrics* **64:**620.

Burstein, S. J., Brandriss, M. W., and Schlesinger, J. J., 1983, Infection of a macrophage cell line, P388D1 with reovirus. Effects of immune ascitic fluids and monoclonal antibodies on the neutralization and on enhancement of viral growth, *J. Immunol.* **130:**2915.

Calnek, B. W., Fahey, K. J., and Bagust, T. J., 1986, *In vitro* infection studies with infectious laryngotracheitis virus, *Avian Dis.* **30:**327.

Carney, W. P., and Hirsch, M. S., 1981, Mechanisms of immunosuppression in cytomegalovirus mononucleosis. II. Virus-monocyte interactions, *J. Infect. Dis.* **144:**47.

Chantler, J. K., and Tingle, A. J., 1980, Replication and expression of rubella virus in human lymphocyte populations, *J. Gen. Virol.* **50:**317.

Chaturvedi, U. C., 1987, Togavirus induced immunosuppression, in: *Viral Induced Immunosuppression* (S. Specter, M. Bendinelli, and H. Friedman, eds.), Plenum Press, New York (in press).

Chaturvedi, U. C., and Shukla, M. I., 1981, Characterization of the suppressor factor produced in the spleen of dengue virus-induced mice, *Ann. Immunol. (Pasteur Inst.)* **132C:**245.

Chaturvedi, U. C., Mathur, A., Tandon, P., Natu, S. M., Rajvanshi, S., and Tandon, H. O., 1979, Variable effect on peripheral blood leukocytes during J E virus infection of man, *Clin. Exp. Immunol.* **38:**492.

Chaturvedi, U. C., Shukla, M. I., and Mathur, A., 1981, Thymus dependent lymphocytes of the dengue virus-infected mice spleen mediate suppression through prostaglandins, *Immunology* **42:**1.

Chaturvedi, U. C., Gulati, L., and Mathur, A., 1982, Inhibition of E-rosette formation and phagocytosis by human blood leukocytes after treatment with the dengue virus-induced cytotoxic factor, *Immunology* **45:**679.

Chaturvedi, U. C., Nagar, R., Gulati, L., and Mathur, A., 1983a, Macrophage functions in dengue virus infection, in: *Proc. Int. Conf. on Dengue/Dengue Haemorrhagic Fever, Kuala Lampur* (T. Pang and R. Pathmanathan, eds.), pp. 343–354, University of Malaya, Kuala Lampur.

Chaturvedi, U. C., Nagar, R., and Mathur, A., 1983b, Effect of dengue virus infection on Fc-receptor functions of mouse macrophages, *J. Gen. Virol.* **64:**2399.

Cohen, D. A., Morris, R. E., and Bubel, H. C., 1984, Abortive ectromelia virus infection in peritoneal macrophages activated by *Corynebacterium parvum*, *J. Leukocyte Biol.* **35:**179.

Cohen, P. L., Cross, S. S., and Mosier, D. E., 1975, Immunologic effects of neonatal infection with mouse thymic virus, *J. Immunol.* **115:**706.

Coignoul, F. L., Bertram, T. A., and Cheville, N. F., 1984, Functional and ultrastructural changes in neutrophils from mares and foals experimentally inoculated with a respiratory tract strain of equine herpesvirus-1, *Am. J. Vet. Res.* **45:**1977.

Cook, S. L., and Lewis, A. M., 1984, Differential NK and macrophage killing of hamster cells infected with non-oncogenic or oncogenic adenovirus, *Science* **224:**612.

Davis, G. L., Jicha, J. L., and Hoofnagle, J. H., 1984, Alpha and gamma interferon in patients with chronic type B, non-A non-B and delta hepatitis: Serum levels and *in vitro* production by lymphocytes, *Gastroenterology* **86:**1315 (Abstract).

Dinarello, C., 1984, Interleukin-1, *Rev. Infect. Dis.* **6:**52.

Domurat, F., Roberts, N. J., Jr., Walsh, E. E., and Dagan, R., 1985, Respiratory syncytial virus infection of human mononuclear leukocytes *in vitro* and *in vivo*, *J. Infect. Dis.* **152:**895.

Doyle, M. V., and Oldstone, M. B. A., 1978, Interactions between viruses and lymphocytes. I. *In vivo* replication of lymphocytic choriomeningitis virus in mononuclear cells during both acute and chronic viral infections, *J. Immunol.* **121:**1262.

Duc-Ngyuen, H., and Henle, W., 1966, Replication of measles virus in human leukocyte cultures, *J. Bacteriol.* **92:**258.

Dutta, S. K., Myrup, A., and Bumgardner, M. K., 1980, Lymphocyte responses to virus and mitogen in ponies during experimental infection with equine herpesvirus-1, *Am J. Vet. Res.* **41:**2066.

Einhorn, L., and Ost, A., 1984, Cytomegalovirus infection of human blood cells, *J. Infect. Dis.* **149:**207.

Ettensohn, D. B., and Roberts, N. J., Jr., 1984, Influenza virus infection of human alveolar and blood derived macrophages: Differences inaccessory cell function and interferon production, *J. Infect. Dis.* **149:**942.

Fleischer, B., and Kreth, H. W., 1982, Mumps virus replication in human lymphoid cell lines and in peripheral blood lymphocytes: Preference for T cells, *Infect. Immun.* **35:**25.

Forman, A. J., and Babiuk, L. A., 1982, Effect of infectious bovine rhinotracheitis virus infection on bovine alveolar macrophage function, *Infect. Immun.* **35:**1041.

Friedlander, A. M., Jahrling. P. B., Merrill, P., and Tobery, S., 1984, Inhibition of mouse peritoneal macrophages. DNA synthesis by infection with the arenavirus Pichinde, *Infect. Immun.* **43:**283.

Friedman, H., Szentivanyi, A., Specter, S., and Bendinelli, M., 1986, Virus interactions with the immune defense system, in: *Viruses, Immunity and Immunodeficiency* (A. Szentivanyi and H. Friedman, eds.), pp. 25–39, Plenum Press, New York.

Ganguly, R., Cusumano, C. L., and Waldman, R. H., 1976, Suppression of cell-mediated immunity after infection with attenuated rubella virus, *Infect. Immun.* **13:**464.

Gardner, I. D., and Lawton, J. W. M., 1982, Depressed human monocyte function after influenza infection *in vitro, J. Reticuloendothel. Soc.* **32:**443.

Gartner, S., Markovitz, P., Markovitz, D. M., Kaplan, M. H., Gallo, R. C., and Popovic, M., 1986, The role of mononuclear phagocytes in HTLV-III/LAV infection, *Science* **223:**215.

Garzelli, C., Basolo, F., Matteucci, D., Prabhakar, B. S., and Toniolo, A., 1987, Picornavirus induced immunosuppression, in: *Viral Induced Immunosuppression* (S. Specter, M. Bendinelli, and H. Friedman, eds.), Plenum Press, New York (in press).

Gauntt, C. J., Trousdale, M. D., LaBodie, D. R. L., Paque, R. E., and Nealon T., 1979, Properties of coxsackievirus B3 variants which are amyocarditic or myocarditic for mice, *J. Med. Virol.* **3:**207.

Gledhill, A. W., Bilbey, D. L. J., and Niven, J. S. F., 1965, Effect of certain murine pathogens on phagocytic activity, *Br. J. Pathol.* **46:**433.

Gonzalez, P. H., Lampuri, J. S., Coto, C. E., and Laguens, R. P., 1982, *In vitro* infection of murine macrophages with Junin virus, *Infect. Immun.* **35:**356.

Gonzalez-Serva, A., and Hsiung, G. D., 1978, Expression of herpesvirus in adherent bone marrow of latently infected cells from guinea pigs, *Am. J. Pathol.* **9:**483.

Gresser, I., and Chany, C., 1964, Multiplication of poliovirus type I in preparations of human leukocytes and its inhibition by interferon, *J. Immunol.* **92:**889.

Gulati, L., Chaturvedi, U. C., and Mathur, A., 1982, Depressed macrophage functions in dengue virus infected mice: Role of the cytotoxic factor, *Br. J. Exp. Pathol.* **63:**194.

Gulati, L., Chaturvedi, U. C., and Mathur, A., 1983, Dengue virus-induced cytotoxic factor induces macrophages to produce a cytotoxin, *Immunology* **49:**121.

Gulati, L., Chaturvedi, U. C., and Mathur, A., 1984, Effect of dengue virus-induced macrophage cytotoxin on functions of human blood leukocytes, *Indian J. Med. Res.* **79:**709.

Gulati, L., Chaturvedi, U. C., and Mathur, A., 1986, Production of dengue virus-induced macrophage cytotoxin, *Br. J. Exp. Pathol.* **67:**269.

Halstead, S. B., O'Rourke, E. J., and Allison, A. C., 1977, Dengue viruses and mononuclear phagocytes. II. Identity of blood and tissue leukocytes supporting *in vitro* infection, *J. Exp. Med.* **146:**218.

Harfast, B., Huddlestone, J. R., Casali, P., Merigan, T., and Oldstone, M. B. A., 1981, Interferon acts directly on human B lymphocytes to modulate immunoglobulin synthesis, *J. Immunol.* **127:**2146.

Herman, G., DuBuy, H. G., and Johnson, M. L., 1966, Studies on the *in vivo* and *in vitro* multiplication of the LDH virus of mice, *J. Exp. Med.* **123:**985.

Inada, T., and Mims, C. A., 1984, Mouse Ia antigens are receptors for lactate dehydrogenase virus, *Nature (London)* **308:**59.

Inada, T., and Mims, C. A., 1985, Ia antigens and Fc receptors of mouse peritoneal macrophages as determinants of susceptibility to lactic dehydrogenase virus, *J. Gen. Virol.* **66:**1469.

Isakov, N., Feldman, M., and Segal, S., 1982a, Acute infection of mice with lactic dehydrogenase virus (LDV) impairs the antigen presenting capacity of their macrophages, *Cell Immunol.* **66:** 317.

Isakov, N., Feldman, M., and Segal, S., 1982b, Lactic dehydrogenase virus (LDV) impairs the antigen presenting capacity of macrophages yet fails to affect their phagocytic activity, *Immunobiology* **162:**15.

Jakowski, R. M., Fredrickson, T. N., Chomiak, T. W., and Luginbuhl, R. E., 1970, Hematopoietic destruction in Marek's disease, *Avian Dis.* **14:**374.

Jacobs, R. P., and Cole, G. A., 1976, Lymphocytic choriomeningitis virus-induced immunosuppression: A virus-induced macrophage defect, *J. Immunol.* **117:**1004.

Jakab, G. J., and Warr, G. A., 1981, Immune enhanced phagocytic dysfunction in pulmonary macrophages infected with parainfluenza 1 (Sendai) virus 1–3, *Am. Rev. Respir. Dis.* **124:**575.

Joseph, B. S., Lampert, P. W., and Oldstone, M. B. A., 1975, Replication and persistence of measles virus in defined subpopulations of human leukocytes, *J. Virol.* **16:**1638.

Junker, A., Ochs, H. D., Clark, E. A., Puterman, M. L., and Wedgewood, R. J., 1986, Transient immune deficiency in patients with acute Epstein–Barr virus infection, *Clin. Immunol. Immunopathol.* **40:**436.

Kadish, A. S., Tansey, F. A., Yu, G. S. M., Doyle, A. T., and Bloom, B. R., 1980, Interferon as a mediator of human lymphocyte suppression, *J. Exp. Med.* **151:**637.

Kahn, D. E., and Gillespie, J. H., 1971, Feline viruses: Pathogenesis of picornavirus infection in the cat, *Am. J. Vet. Res.* **32:**51.

Kantoch, M., and Dobrowolska, H., 1969, Studies on the inheritance of the susceptibility to poliovirus of phytohemagglutinin transformed macrophages, *Acta Virol.* **13:**153.

Kleinerman, E. S., Snyderman, R., and Daniels, C. A., 1974, Depression of human monocyte chemotaxis by herpes simplex and influenza viruses, *J. Immunol.* **113:**1562.

Kleinerman, E. S., Sydnerman, R., and Daniels, C. A., 1975, Depressed monocyte chemotaxis during acute influenza infection, *Lancet* **2:**1063.

Kleinerman, E. S., Daniels, C. A., Polisson, R. P., and Synderman, R., 1976, Effect of virus infection on the inflammatory response: Depression of macrophage accumulation in influenza-infected mice, *Am. J. Pathol.* **85:**373.

Koenig, S., Gendelman, H. E., Orenstein, J. M., DalCanto, M. C., Pezeshkpour, G. H., Yungbluth, M., Janotta, F., Aksamit, A., Martin, M. A., and Fauci, A. S., 1986, Detection of AIDS virus in macrophages in brain tissue from AIDS patients with encephalopathy, *Science* **233:** 1089.

Kohl, S., Starr, S. E., Oleske, J., Shore, S. L., Ashman, R. B., and Nahmias, A. J., 1977, Human monocyte-macrophage-mediated antibody-dependent cytotoxicity to herpes simplex virus-infected cells, *J. Immunol.* **118:**729.

Korba, B. E., Wells, F., Tennat, B. C., Yoakum, G. H., Purcell, R. H., and Gerin, J. L., 1986, Hepadnavirus infection of peripheral blood lymphocytes *in vivo:* Woodchuck and chimpanzee models of viral hepatitis, *J. Virol.* **58:**1.

Kramp, W. J., Medveczky, P., Mulder, C., Hinze, H. C., and Sullivan, J. L., 1985, Herpes sylvilagus infects both B and T lymphocytes *in vivo*, *J. Virol.* **56:**60.

Krzystyniak, K., and Dupuy, J. M., 1983, Immunodepression of lymphocyte response in mouse hepatitis virus 3 infection, *Biomed. Pharmacol.* **37:**68.

Laguens, R. M., Avila, M. M., Samoilovich, S. R., Weissenbacher, M. C., and Laguens, R. P., 1983a, Pathogenicity of an attenuated strain ($XJCl_3$) of Junin virus. Morphological and virological studies in experimentally infected guinea pigs, *Intervirology* **20:**195.

Laguens, M., Chambo, J. G., and Laguens, R. P., 1983b, *In vivo* replication of pathogenic and attenuated strains of Junin virus in different cell populations of lymphatic tissue, *Infect. Immun.* **41:**1279.

Lee, L. F., Sharma, J. M., Nazerian, K., and Witter, R. L., 1978, Suppression and enhancement of mitogen induced proliferation of normal spleen cells by macrophages from chickens inoculated with Marek's disease virus, *J. Immunol.* **120:**1554.

Leray, D., Dupuy, C., and Dupuy, J. M., 1982, Immunopathology of mouse hepatitis virus type 3 infection. IV. MHV-3 induced immunosuppression, *Clin. Immunol. Immunopathol.* **23:**1457.

Levy-LeBlond, E., Oth, D., and Dupuy, J. M., 1979, Genetic study of mouse sensitivity to MHV-3 infection: Influence of the H-2 complex, *J. Immunol.* **112:**1359.

Lopez, C., 1975, Genetics of natural resistance to herpesvirus infections in mice, *Nature* **258:**152.

Lopez, C., and Dudas, G., 1979, Replication of herpes simplex virus type 1 in macrophages from resistant and susceptible mice, *Infect. Immun.* **23:**432.

Mathiesen, L. R., Feinstone, S. M., Purcell, R. H., and Wagner, J. O., 1977, Detection of hepatitis A antigen by immunofluorescence, *Infect. Immun.* **18:**524.

Matteucci, D., Toniolo, A., Conaldi, P. G., Basolo, F., Gori, Z., and Bendinelli, M., 1985, Systemic

lymphoid atrophy in coxsackievirus B3-infected mice: Effects of virus and immunopotentiating agents, *J. Infect. Dis.* **51**:1100.
McLaren, C., Cheng, H., Spicer, D. L., and Tompkins, W. A. F., 1976, Lymphocyte and macrophage responses after vaccinia virus infections, *Infect. Immun.* **14**:1014.
Mella, B., and Lang, D. J., 1967, Leukocyte mitosis: Suppression *in vitro* associated with acute infectious hepatitis, *Science* **155**:80.
Midulla, M., Businco, L., and Moschini, L., 1972, Some effects of rubella vaccination on immunologic responsiveness, *Acta Paediatr. Scand.* **61**:609.
Mims, C. A., 1964, Aspects of pathogenesis of virus diseases, *Bacteriol. Rev.* **28**:30.
Mims, C. A., and Wainwright, S., 1968, The immunodepressive action of lymphocytic choriomeningitis virus in mice, *J. Immunol.* **101**:717.
Mogensen, S. C., 1986, Macrophages and genetically determined natural resistance to virus infections, in: *Viruses, Immunity and Immunodeficiency* (A. Szentivanyi and H. Friedman, eds.), pp. 13–24, Plenum Press, New York.
Morahan, P. S., and Morse, S. S., 1979, Macrophage-virus interactions, in: *Virus–Lymphocyte Interactions: Implications for Disease* (M. R. Proffitt, ed.), pp. 17–35, Elsevier/North Holland Publishing Co., New York.
Morahan, P. S., Conner, J. R., and Leary, K. R., 1985, Viruses and the versatile macrophage, *Br. Med. Bull.* **41**:15.
Moss, B., 1985, Replication of poxviruses, in: Fundamental *Virology* (B. N. Fields, D. M. Knipe, R. M. Chanock, eds.), pp. 637–655, Raven Press, New York.
Murphy, F. A., Winn, W., Walker, D. H., Flemister, M. R., and Whitfield, S. G., 1976, Early lymphoreticular viral tropism and antigen persistence. Tamiami virus infection in the cotton rat, *Lab. Invest.* **34**:125.
Murphy, F. A., Buchmeier, M. J., and Rawls, W. E., 1977, The reticuloendothelium as the target in a virus infection: Pichinde virus pathogenesis in two strains of hamsters, *Lab. Invest.* **37**:502.
Narayan, O., Wolinsky, J. S., Clements, J. E., Shandberg, J. D., Griffin, D. E., and Cork, L. C., 1982, Slow virus replication: The role of macrophages in the persistence and expression of visna viruses of sheep and goats, *J. Gen. Virol.* **59**:345.
Narayan, O., Kennedy-Stoskopf, S., Sheffer, D., Griffin, D. E., and Clements, J. E., 1983, Activation of caprine arthritis-encephalitis virus expression during maturation from monocytes to macrophages, *Infect. Immun.* **41**:67.
Narayan, O., Sheffer, D., Clements, J. E., and Tennekoon, G., 1985, Restricted replication of lentiviruses: Visna viruses induce a unique interferon during interaction between lymphocytes and macrophages, *J. Exp. Med.* **162**:1954.
Nash, A. A., 1985, Tolerance and suppression in virus diseases, *Br. Med. Bull.* **41**:41.
Nath, P., Tandon, P., Gulati, L., and Chaturvedi, U. C., 1983, Histological and ultrastructural study of spleen during dengue virus infection of mice, *Indian J. Med. Res.* **77**:83.
Nick, S., Kampe, P., Knoblich, A., Metzger, B., and Falke, D., 1986, Suppression and enhancement of humoral antibody formation by herpes simplex virus types 1 and 2, *J. Gen. Virol.* **67**:1015.
Nugent, K. M., and Pesanti, E. L., 1979, Effect of influenza infection on the phagocytic and bactericidal activities of pulmonary macrophages, *Infect. Immun.* **26**:651.
Oguchi, M., Komura, J., Tagami, H., and Ofuji, S., 1981, Ultrastructural studies of spontaneously progressing plane warts. Macrophages attack verruca-epidermal cells, *Arch. Dermatol. Res.* **270**:403.
Oldstone, M. B. A., Tishon, A., Chiller, J. M., Weigle, W. O., and Dixon, F. J., 1973, Effect of chronic viral infection on the immune system. I. Comparison of the immune responsiveness of mice chronically infected with LCM virus with that of non-infected mice, *J. Immunol.* **110**:1268.
Porter, D. D., Porter, H. G., and Deerhale, B. B., 1968, Immunofluorescence assay for antigen and antibody in lactic dehydrogenase virus infection in mice, *J. Immunol.* **102**:431.
Portolani, M., Piani, M. Gazyenelli, G., Borgatti, A., Grossi, M. P., Corallini, A., and Barbanti-Brodano, G., 1985, Restricted replication of BK virus in human lymphocytes, *Microbiologica* **8**:59.
Rager-Zisman, B. R., and Allison, A. C., 1973, The role of anti body and host cells in the resistance of mice against infection by coxsackie B-3 virus, *J. Gen. Virol.* **19**:329.

Rhodes, J., 1985, Modulation of macrophage Fc receptor expression *in vitro* by insulin and cyclic nucleotides, *Nature* **257**:597.

Rice, G. P. A., Schrier, R. D., and Oldstone, M. B. A., 1984, Cytomegalovirus infects human lymphocytes and monocytes: Virus expression is restricted to immediate-early gene products, *Proc. Natl. Acad. Sci. USA* **81**:6134.

Riley, V., 1974, Erroneous interpretation of valid experimental observations through interference by the LDH virus, *J. Natl. Cancer Inst.* **52**:1673.

Rinaldo, C. R., Jr., Black, P. H., and Hirsch, M. S., 1977, Interactions of cytomegalovirus with leukocytes from patients with mononucleosis due to cytomegalovirus, *J. Infect. Dis.* **136**:667.

Rinaldo, C. R., Jr., Carney, W. P., Richter, B. S., and Hirsch, M. S., 1980, Mechanisms of immunosuppression in cytomegaloviral mononucleosis, *J. Infect. Dis.* **141**:488.

Rinehart, J. J., Orser, M., and Kaplan, M. E., 1979, Human monocytes and macrophage modulation of lymphocyte proliferation, *Cell Immunol.* **44**:131.

Roberts, N. J., Jr., 1982, Direct effect of influenza virus respiratory syncytial virus and Sendai virus on human lymphocytes and macrophages, *Infect. Immun.* **35**:142.

Roberts, N. J., Jr., and Steigbigel, R. T., 1978, Effect of *in vitro* virus infection on response of human monocytes and lymphocytes to mitogen, *J. Immunol.* **121**:1052.

Roberts, N. J., Jr., Douglas, R. G., Jr., Simons, R. L., and Diamond, M. E., 1979, Virus-induced interferon production by human macrophages, *J. Immunol.* **123**:365.

Roberts, N. J., Jr., Diamond, M. E., Douglas, R. G., Jr., Simons, R. L., and Stiegbigel, R. T., 1980, Mitogen responses and interferon production after exposure of human macrophages to infectious and inactivated influenza viruses, *J. Med. Virol.* **5**:17.

Roberts, N. J., Jr., Prill, A. H., and Mann, T. N., 1986, Interleukin 1 and interleukin 1 inhibitor production by human macrophages exposed to influenza virus or respiratory syncytial virus: Respiratory syncytial virus is a potent inducer of inhibitory activity, *J. Exp. Med.* **163**:511.

Rodgers, B. C., Scott, D. M., Mundin, J., and Sissons, J. G. P., 1985, Monocyte derived inhibitor of interleukin 1 induced by human cytomegalovirus, *J. Virol.* **55**:527.

Rossiter, P. B., and Wardley, R. C., 1985, The differential growth of rinderpest virus in bovine lymphocytes and macrophages, *J. Gen. Virol.* **66**:969.

Rothwell, D. J., 1975, Bone marrow granulomas and infectious mononucleosis, *Arch. Pathol.* **99**: 508.

Rouse, B. T., and Horohov, D. W., 1986, Immunosuppression in viral infections, *Rev. Infect. Dis.* **8**:850.

Santivatr, D., Maheswaran, S. K., Newman, J. A., and Pomeroy, B. S., 1981, Effect of infectious bursal disease virus infection on the phagocytosis of *Staphylococcus aureus* by mononuclear phagocytic cells of susceptible and resistant strains of chickens, *Avian Dis.* **25**:303.

Schwartz, R., Lohler, J., and Lehmann-Grube, F., 1978, Infection of cultivated mouse peritoneal macrophages with lymphocytic chroiomeningitis virus, *J. Gen. Virol.* **39**:565.

Scott, J. C., Dutta, S. K., and Myrup, A. C., 1983, *In vivo* harboring of equine herpesvirus-1 in leukocytes, populations and subpopulations and their quantitation from experimentally infected ponies, *Am. J. Vet. Res.* **44**:1344.

Sharma, J. M., and Lee, L. F., 1983, Effect of infectious bursal disease on natural killer cell activity and mitogenic response of chicken lymphoid cells: Role of adherent cells in cellular immune suppression, *Infect. Immun.* **42**:747.

Shukla, M. I., and Chaturvedi, U. C., 1981, Cycloheximide and mitomycin C treatment inhibits production of dengue virus-induced suppressor factor, *Indian J. Exp. Biol.* **19**:826.

Shukla, M. I., and Chaturvedi, U. C., 1983, Transmission of dengue virus-induced suppressor signal from macrophage to lymphocyte occurs by cell contact, *Br. J. Exp. Pathol.* **64**:87.

Shukla, M. I., Dalakoti, H., and Chaturvedi, U. C., 1982, Ly phenotype of T lymphocytes producing dengue virus-induced immunosuppressive factors, *Indian J. Exp. Biol.* **20**:525.

Silverberg, B. A., Jakab, G. J., Thompson, R. G., Warr, G. A., and Boo, K. S., 1979, Ultrastructural alterations in phagocytic functions of alveolar macrophages after parainfluenza virus infection, *J. Reticuloendothel. Soc.* **25**:405.

Sivanandan, V., and Maheswaran, S. K., 1980, Immune profile of infectious bursal disease. I. Effect of infectious bursal disease virus on peripheral blood T and B lymphocytes of chickens, *Avian Dis.* **24:**715.

Smid, B., Valick, L., and Sabol, A., 1981, Morphogenesis of Aujesky's disease virus in pig lung macrophages, *Acta Vet. Vysoka Skola Veterinarni Brno,* **50:**79.

Soontiens, F. J. C. J., and Van Der Veen, J., 1973, Evidence for a macrophage-mediated effect of poliovirus on the lymphocyte response to phytohemagglutinin, *J. Immunol.* **111:**1411.

Specter, S., and Friedman, H., 1978, Viruses and the immune response, *Pharmac. Ther. A.* **2:**595.

Stiehm, R. E., 1980, The human neonate as an immunocompromised host, in: *Pathogenesis, Prevention and Therapy* (J. Verhoef, P. K. Patterson, and P. G. Quie, eds.), pp. 77–91, Elsevier/North Holland Publishing Co., New York.

Strayer, D. S., 1987, Poxviruses, in: *Viral Induced Immunosuppression* (S. Specter, M. Bendinelli, and H. Friedman, eds.), *Plenum Press, New York (in press).*

Stueckemann, J. A., Ritzi, D. M., Holth, M., Smith, M. S., Swart, W. J., Cafruny, W. A., and Plagerman, P. G. W., 1982, Replication of lactate dehydrogenase virus in macrophages. I. Evidence for cytocidal replication, J. Gen. Virol. **59:**245.

Sullivan, J. L., Barry, D. W., Lucas, S. J., and Albrecht, P., 1975, Measles infection of human mononuclear cells. I. Acute infecfion of peripheral blood lymphocytes and monocytes, *J. Exp. Med.* **142:**773.

Svennerholm, B., Strannegard, O., and Lycke, E., 1978, Immune reactivity of visna virus–inoculated mice, *Infect. Immun.* **20:**412.

Sy, M. S., and Finberg, R., 1987, Rhabdoviruses: Effect of vesicular stomatitis virus (VSV) infection on the development and regulation of cell mediated and humoral immune responses, in: *Viral Induced Immunosuppression* (S. Specter, M. Bendinelli, and H. Friedman, eds.), Plenum Press, New York (in press).

Taguchi, F., Hirano, N., Kiuchi, Y., and Fujiwara, K., 1976, Difference in response to mouse hepatitis virus among susceptible mouse strains, *Jpn. J. Microbiol.* **20:**293.

Taguchi, F., Yamada, A., and Fujiwara, K., 1980, Resistance to highly virulent mouse hepatitis virus acquired by mice after low virulence infection: Enhanced antiviral activity of macrophages, *Infect. Immun.* **29:**42.

Taylor, C. F., Weiser, W. Y., and Bang, F. B., 1981, *In vitro* macrophage manifestation of cortisone-induced decrease in resistance to mouse hepatitis virus, *J. Exp. Med.* **153:**732.

Toniolo, A., Matteucci, D., Basolo, F., and Bendinelli, M., 1986, The immune system in experimental coxsackie virus B3 infection, in: *Viruses, Immunity and Immunodeficiency* (A. Szentivanyi and H. Friedman, eds.), pp. 101–108, Plenum Press, New York.

Tsuru, S., Kitani, H., Seno, M., Abe, M., Zinnaka, Y., and Nomoto, K., 1983, Mechanism of protection during the early phase of a generalized viral infection. I. Contribution of phagocytes to protection against ectromelia virus, *J. Gen. Virol.* **64:**2021.

Twomey, J. J., Gyorkey, F., and Norris, S. M., 1974, The monocyte disorder with herpes zoster, *J. Exp. Clin. Med.* **83:**768.

Van Der Logt, J. T. M., Van Loon, A. M., and Van Der Veen, J., 1980, Replication of rubella virus in human mononuclear blood cells, *Infect. Immun.* **27:**309.

Van Loon, A. M., Van Der Logt, J. T. M., and Van Der Veen, J., 1979, Poliovirus-induced suppression of lymphocyte-stimulation: A macrophage mediated effect, *Immunology* **37:**135.

Varesio, L., 1983, Suppressor cells and cancer: Inhibition of immune functions by macrophages, in: *The Reticuloendothelial System: A Comprehensive Treatise,* Volume 5, *Cancer* (R. B. Herberman and H. Friedman, eds.), pp. 217–252, Plenum Press, New York.

Virelizier, J. L., Dayan, A. D., and Allison, A. C., 1975, Neuropathologic effects of persistent infection of mice by mouse hepatitis virus, *Infect. Immun.* **12:**1127.

Warshauer, D., Goldstein, E., Akers, T., Lippert, W., and Kim, M., 1977, Effect of influenza viral infection on the ingestion and killing of bacteria by alveolar macrophages, *Am. Rev. Respir. Dis.* **115:**269.

Weiser, W., and Bang, F. B., 1976, Macrophages genetically resistant to mouse hepatitis virus concerted *in vivro* to susceptible macrophages, *J. Exp. Med.* **143:**732.

Weiser, W. Y., and Bang, F. B., 1977, Blocking of *in vitro* and *in vitro* susceptibility to mouse hepatitis virus, *J. Exp. Med.* **146:**1467.

Willlems, F. Th. C., Melnick, J. L., and Rawls, W. E., 1969, Viral inhibition of the phytohemagglutinin response of human lymphocytes and application to viral hepatitis, *Proc. Soc. Exp. Biol. Med.* **130:**652.

Woodruff, J. F., and Woodruff, J. J., 1975, The effect of viral infections on the function of the immune system, in: *Viral Immunology and Immunopathology* (A. L. Notkins, ed.), pp. 393–418, Academic Press, New York.

Yarborough, D. J., Meyer, O. T., Dannenberg, A. M., Jr., and Pearson, B., 1967, Histochemistry of macrophage hydrolases. III. Studies on β-glucuronidase and aminopeptides with indoyl and naphthyl substrates, *J. Reticuloendothel. Soc.* **4:**390.

11

The Role of the Reticuloendothelial System in Viral Hepatitis

FIORENZO PARONETTO

1. INTRODUCTION

In the last decade it became apparent that the liver can be attacked by a number of viruses that have affinity for it, that the host responds in a complex manner to the viral insult, and that the tissue damage or recovery depends to a great extent on the host's immune response. Thus, the reticuloendothelial system (RES) has a cardinal importance in the outcome of this disease. In this chapter the term *reticuloendothelial system* in not restricted to the mononuclear phagocyte system, but it is used in the more comprehensive sense to include a complex of macrophages, lymphocytes, and plasma cells, as well as cells involved not only in the clearance of infectious agents but also in the immune response (Friedman *et al.*, 1980).

Four major groups of viruses may affect the liver directly: hepatitis B virus (HBV), hepatitis A virus (HAV), hepatitis D virus (HDV or delta agent), and non-A, non-B (NANB) hepatitis virus. Papers on various aspects of the immunologic reaction in acute and chronic hepatitis have been written by Bianchi and Gudat (1979), Meyer zum Büschenfelde *et al.* (1979), Dienstag (1984), Levy and Chisari (1981), and Popper and Paronetto (1984). The purpose of this chapter is to update the present, still fragmentary, knowledge of the immune response to these viruses and to discuss the numerous problems still unresolved.

FIORENZO PARONETTO • Immunopathology Laboratory, Veterans Administration Medical Center, Bronx, New York 10468; and Department of Pathology, The Mount Sinai School of Medicine of the City University of New York, New York, New York 10029.

2. THE CELLULAR INFILTRATE IN THE VIRAL HEPATITIDES

2.1. MORPHOLOGIC CONSIDERATIONS

Although morphologic differentiation of the various forms of viral hepatitis is often difficult in liver biopsies, some distinguishing features may be helpful. The morphologic data are also corroborated by findings in experimentally transmitted hepatitis in nonhuman primates (Popper *et al.*, 1980).

HAV, a picornavirus recently designated enterovirus serotype 72, is probably responsible for 20–25% of clinical hepatitis. The virus elicits mainly an acute response that, as a rule, does not progress to chronic hepatitis or cirrhosis. The morphologic picture in humans is similar to that observed in chimpanzees with eosinophilic necrosis of hepatocytes and/or infiltrate mainly in the periportal area (Popper *et al.*, 1980). The cells in the infiltrates consist mainly of lymphocytes and histiocytes. By electron microscopy a large number of activated endothelial cells lining the sinusoids can be detected (Bardadin and Scheuer, 1984), and these cells have been considered to play a role in protecting hepatocytes from injury. These ultrastructural findings can be seen in acute viral hepatitis (AVH) induced by HBV and NANB viruses.

In AVH induced by HBV (AVH-B) the eosinophilic necrosis and cellular infiltrate are similar to those seen in AVH induced by HAV (AVH-A). The inflammatory reaction is particularly prominent around the central vein, and the extent of the inflammation in the portal tract varies from one case to another (Bianchi, 1983).

In chronic HBV hepatitis the cellular infiltrate is located in the portal tracts in the more mild form (portal hepatitis) and is characterized mainly by mononuclear cells. In the more severe form, chronic active hepatitis induced by HBV (CAH-B), the inflammatory cells spill over the limiting plate and surround hepatocytes in the periportal area (piecemeal necrosis) (Fig. 1). This form of CAH-B may progress to cirrhosis, especially when piecemeal necrosis is associated with focal parenchymal areas of necrosis and cellular infiltrates (Popper 1983).

HDV, an RNA-defective virus that requires HBV as a helper for replication and that has interfering properties on the expression of HBV antigen, was discovered by Rizzetto *et al.* (1977). It is observed in the Mediterranean areas and among drug addicts in North America (Ponzetto *et al.*, 1984b). It may superinfect or coinfect patients with AVH-B, CAH-B, or chronic carriers of HBV. The delta agent can be demonstrated immunologically in morphologically normal-appearing hepatocytes. The disease induces necrosis of hepatocytes without excessive lymphocyte accumulation but with marked activation of macrophages and sinusoidal cells (Rizzetto *et al.*, 1983). In patients with CAH-B delta agent superinfection induces more extensive inflammation in the periportal and portal areas and hepatocellular necrosis (Kanel *et al.*, 1984). Outbreaks of HDV infections were described among children in Venezuelan Indians in Brazil (Hadler *et al.*, 1984; Popper *et al.*, 1983; Popper, 1984). In these patients the inflammatory reaction in the liver was minimal and the changes were characterized by small-

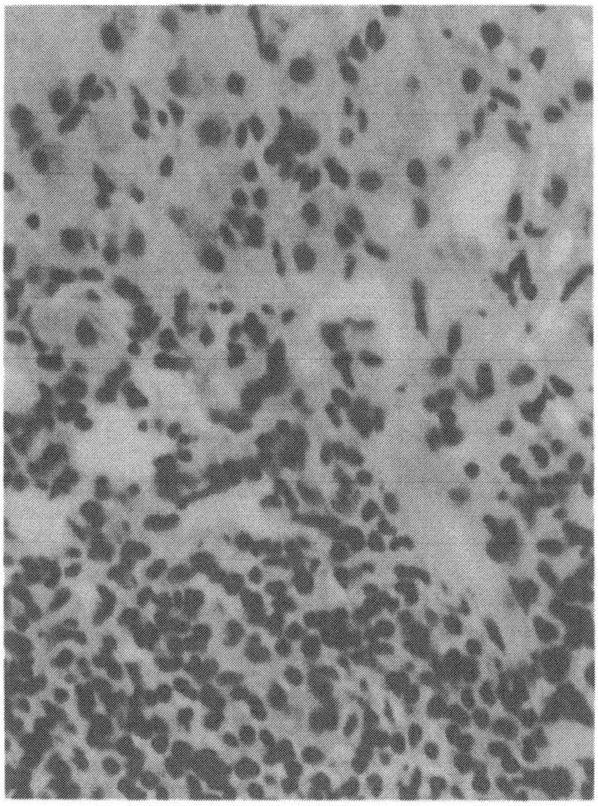

FIGURE 1. Figures 1–5 show cryostat sections from a patient with HBsAG-positive CAH. This is a cryostat section of liver showing marked inflammation in the portal tract and in the periportal area (piecemeal necrosis) with mononuclear cells surrounding isolated hepatocytes. H&E. ×250.

droplet steatosis of the hepatocytes, suggesting a cytotoxic rather than immunologic reaction.

The cellular reaction of NANB hepatitis varies and not all studies are comparable because the diagnosis of NANB hepatitis is not uniform owing to the lack of distinguishing markers. There is also evidence that more than one virus is involved, and each virus may elicit a different reaction (Rugge *et al.*, 1983). At present, there is no agreement as to the characteristic lesions of NANB hepatitis, but the consensus is that there is a marked activation of the sinusoidal cells and macrophages (Dienes *et al.*, 1982; Rugge *et al.*, 1983). In this respect the lesion is somewhat similar to that induced by HDV. Lymphocytes have been described as occasionally abundant but usually not in contact with hepatocytes, which have distinct cell borders, eosinophilic granular cytoplasm, and granular eosinophilic bodies without surrounding inflammatory cells (Dienes *et al.*, 1982).

2.2. THE PHENOTYPE OF THE CELLULAR INFILTRATE

In the last few years, monoclonal antibodies have allowed a better characterization of the mononuclear cell infiltrating the liver (Table 1). Most investiga-

TABLE 1. T-CELL SUBSETS IN LIVER OF PATIENTS WITH VIRAL HEPATITIS[a]

Disease	Prevalent cells in		Comment	Reference
	Portal tracts	Lobules		
AVH-A	CD8	CD8		Shibata *et al.*, 1984
AVH-B	0	CD8		Eggink *et al.*, 1984
	CD8	CD8		Govindarajan *et al.*, 1983
AVH-non-A, non-B	CD8	CD8		Govindarajan *et al.*, 1983
CPH-B	CD4	CD8		Eggink *et al.*, 1984
	CD8		CD4 cells were not studied	Pape *et al.*, 1983
CAH-B	CD4 and CD8	CD8		Colucci *et al.*, 1983
	CD4 and CD8	CD8	Many CD8 in PMN	Eggink *et al.*, 1984
	CD8	CD8	Lower CD4/CD8 ratio in HBeAg-positive patients	Montaño *et al.*, 1983
	CD8		Many CD8 in PMN	Pape *et al.*, 1983
	CD4 and CD8	CD4 and CD8	CD4/CD8 ratio lower in lobules than in portal tracts	Si *et al.*, 1983
CAH-non-A, non-B	CD8	CD4 and CD8		Si *et al.*, 1983

[a]PMN, areas of piecemeal necrosis; CPH-B, HBsAg-positive chronic persistent hepatitis; AVH, acute viral hepatitis; CAH, chronic active hepatitis.

tors have focused attention on CAH-B, and related knowledge in other conditions is fragmentary.

In AVH-A the lymphocytes in the portal tract are mostly T cells and carry the CD8 phenotype. By immunoelectron microscopy a close association between CD8 lymphocytes and hepatocytes can be observed, suggesting they participate in the hepatocellular injury (Shibata *et al.*, 1984).

In AVH-B the inflammatory cells are exclusively CD8 cells and are localized mainly in the lobular parencyhyma (Govindarajan *et al.*, 1983). In CAH-B both CD4 and CD8 cells are present in the portal and periportal lesions (Figs. 2 and 3) (Colucci *et al.*, 1983; Eggink *et al.*, 1984; Si *et al.*, 1983; Pape *et al.*, 1983), but the latter cells are predominant in the lobular parenchyma in areas of piecemeal necrosis (Fig. 4), especially in HBeAg-positive patients (Montaño *et al.*, 1983) (Fig. 5). The infiltrating lymphocytes express also DR antigens, suggesting that they are in the activated state. Monocytes and macrophages also carrying the DR antigens are present in the portal and periportal area, but cells reacting with monoclonal antibodies that identify NK/K cells are rare or absent (Colucci *et al.*, 1983). A recent report (Bernuau *et al.*, 1984) indicates that mononuclear phagocytes, characterized ultrastructurally by the presence of peroxidase-containing granules, are increased in the liver parenchyma of patients with highly active disease and are usually in contact with liver cells. This suggests that mononuclear phagocytes are involved in the lytic attack (Bernuau *et al.*, 1984).

Class I HLA (A, B, and C antigens) have been detected normally in sinusoidal cells and the bile duct epithelium (Barbatis *et al.*, 1981; Mantaño *et al.*, 1982) but not in the hepatocytes. In AVH-NANB the hepatocytes show a gran-

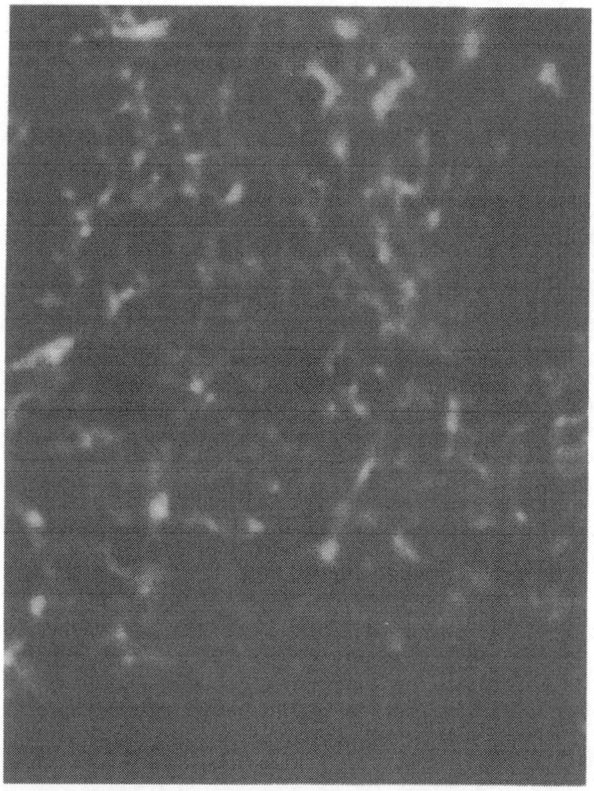

FIGURE 2. Figures 2–5 show cryostat sections treated with monoclonal antibodies to CD4 or CD8 followed by biotinylated antimouse Ig antiserum and fluoresceinated avidin. Here CD4 lymphocytes are scattered in the portal tract. ×400.

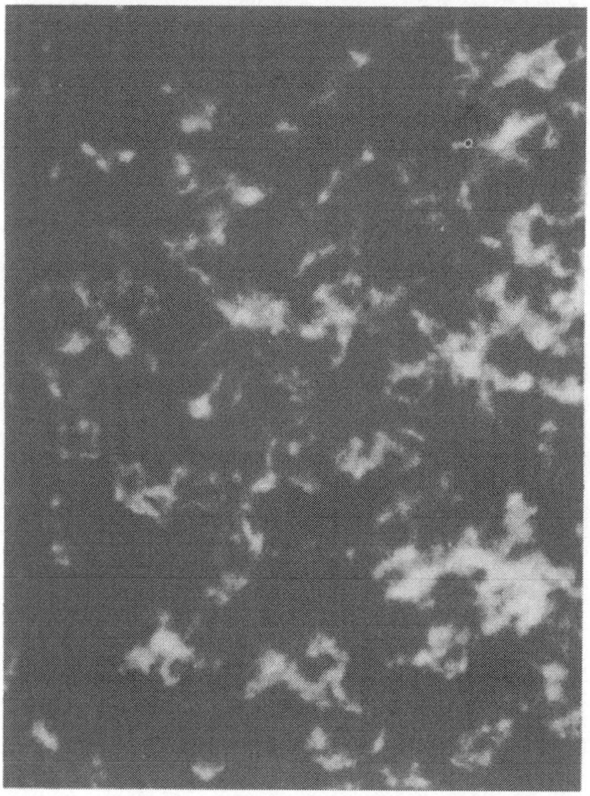

FIGURE 3. Numerous CD8 cells are seen in the portal tract. ×400.

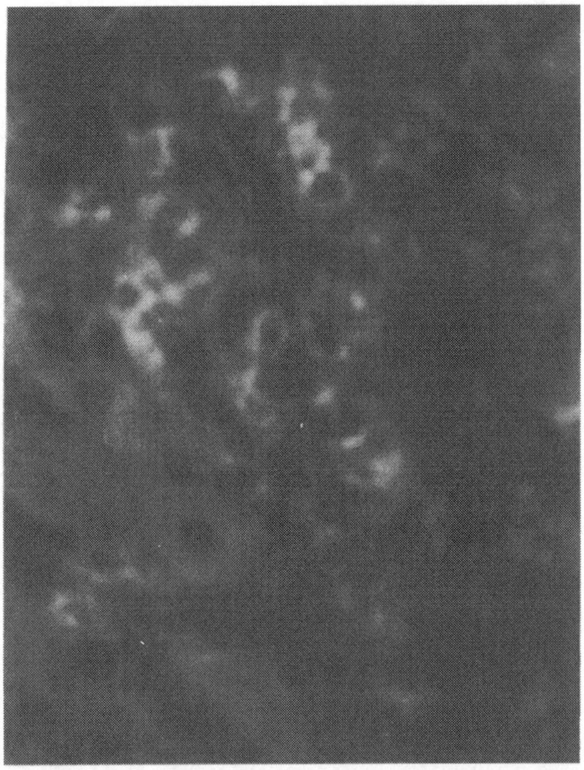

FIGURE 4. CD8 cells are observed in the periportal area adjacent to hepatocytes. ×400.

FIGURE 5. CD8 cells are the predominant cells in the sinusoid and lobular parenchyma. ×400.

ular cytoplasmic localization of class I HLA antigens, whereas in CAH-B the straining is more focal in the cell membrane (Barbatis *et al.*, 1981) and is more prominent in patients with antibodies to HBeAg, who have low levels of viral replication (Montaño *et al.*, 1982).

2.3. B CELLS AND ANTIBODIES

B cells are usually not abundant in AVH and CAH, but a variety of serum antibodies have been detected. The antibodies directed against viral antigens are, in general, considered to be protective or neutralizing. HAV antibodies of the IgM type usually indicate recent infection, while those of the IgG type suggest a previous exposure to the virus. In AVH-B, antibodies against HBsAg are considered protective and are usually not found in patients with CAH-B. Indeed, deficiency in the production of antibodies to HBsAg in the latter patients has been demonstrated *in vitro* (Dusheiko *et al.*, 1983). In contrast, antibodies to HBcAg are indicative of previous exposure to the virus and they are seen in both AVH-B and CAH-B. Antibodies against Dane particles are usually seen in the early stages of AVH-B and have been considered to play a role in the clearance of the virus (Alberti *et al.*, 1978). Antibodies to HBeAg are detected in the convalescent phase of AVH and in patients with other chronic liver disease. Antibodies to HBeAg seem to indicate a lack of viral replication in some areas—for example, in Europe and North America—while elsewhere, as in Greece and Africa, they may still be associated with an active HBV replication, as indicated by HBV-DNA in the blood (Thomas *et al.*, 1984; Hadziyannis *et al.*, 1983). While the production of antibodies to various antigenic components is important diagnostically and in monitoring the progression of the disease, the site of production is not certain.

Other antibodies of importance are those directed against liver-specific antigen (Meyer zum Büschenfelde *et al.*, 1979; Jensen *et al.*, 1978). These antibodies are primarily found in AVH and CAH, particularly in the late phase of the disease, suggesting to some investigators the presence of autoimmune reactions complicating chronic viral infection. These antibodies have been considered to mediate antibody-dependent cellular cytotoxicity, a phenomenon that has been demonstrated *in vitro* with peripheral blood mononuclear cells, but not yet convincingly at the site of liver lesions. A granular deposition of immunoglobulins has been observed on the surface of isolated hepatocytes of patients with CAH-B (Hopf *et al.*, 1976). This finding correlated with the localization of HBcAg in the nuclei of hepatocytes (Trevisan *et al.*, 1979). It has subsequently been demonstrated that immunoglobulins eluted from the liver react against core antigens (Trevisan *et al.*, 1982b), suggesting that the granular deposition is composed, at least in part, of antibodies against HBcAg. Antibodies to HBcAg have also been demonstrated in the nuclei of hepatocytes in patients with CAH-B (Arnold and Meyer zum Büschenfelde, 1977; Gerber *et al.*, 1976; Ray *et al.*, 1979; Rizzetto *et al.*, 1976) as well as in nonhuman primates with HDV superinfection (Rizzetto *et al.*, 1981). Although the role of these antibodies has not been elucidated so far, they do not seem to correlate with the progression or severity of the disease.

Antibodies to various viral and nonviral antigens may combine with the corresponding antigens and produce circulating immune complexes, which may deposit in a variety of organs, including the kidneys, joints, skin, and vascular system, producing different immune complex–mediated diseases (Nowoslawski, 1979). In some organs, but especially in the kidneys, antibodies to HBsAg and HBeAg have been demonstrated by immunofluorescent techniques (Hirose *et al.*, 1984; Ito *et al.*, 1981). Circulating immune complexes containing viral- and liver-specific antigens have been demonstrated in AVH-B, CAH-B, and NANB hepatitis. Prolonged high levels of these immune complexes seem to correlate with the severity of the liver disease (Sagnelli *et al.*, 1983).

2.4. FUNCTION OF CELLULAR INFILTRATE IN THE LIVER

The function of the cellular infiltrate has not been investigated mainly because of the technical difficulties in isolating cells from liver biopsies. Techniques that allow isolation of a few cells and their clonal expansion for functional studies have only recently been developed and may be useful.

2.5. FUNCTION OF PERIPHERAL BLOOD MONONUCLEAR CELLS

Although a great deal of emphasis has been placed on studies dealing with the function of peripheral blood mononuclear cells, results have often been controversial and different from one laboratory to another. However, the function of the mononuclear cells in the peripheral blood does not necessarily reflect the function of those in the liver. Indeed, the number and the function of the former may show a pattern complementary to that in the liver or may reflect a reaction unrelated to the hepatic damage.

It has been demonstrated in many laboratories over the last 10 years that lymphocytes of patients with AVH and CVH are cytotoxic against heterologous as well as autologous hepatocytes. Cytotoxicity against autologous hepatocytes has been particularly important because in this system target and effector cells are histocompatible (Paronetto and Vernace, 1975; Vergani-Mieli *et al.*, 1982; Stefanini *et al.*, 1983; Wands and Isselbacher, 1975). Two main types of cytotoxicity have been demonstrated: one mediated by T cells and the other by K cells. A variety of experiments point to HBcAg as a target of sensitized T cells (Mondelli *et al.*, 1982; Naumov *et al.*, 1984), based on the observation that HBcAg blocks T-cell cytotoxicity and is located not only in the nucleus but also in the cytoplasm of infected hepatocytes (Mondelli *et al.*. 1983). Also, the number of HBcAg-positive hepatocytes decreases when cells are incubated with sensitized lymphocytes. This cytotoxicity seems to be modulated by antibodies to HBcAg that may mask the relevant antigen and thereby render the target antigen unavailable to sensitized lymphocytes. Cytotoxic activity seems to be directed against liver-specific antigen and is observed mainly with a population of non-T lymphocytes, presumably K cells. This cytotoxicity occurs in some patients with CVH and may indicate the concomitant presence of autoimmune reactions.

The advent of monoclonal antibodies has focused interest on the number and phenotypes of lymphocytes and other mononuclear cells. In AVH, a decrease of the CD4/CD8 cell ratio with increase of CD8 cells has been described (Table 2). In chronic hepatitis the results are somewhat different, most likely related to the type of patients selected for investigation and their therapy. In patients with HBeAg chronic hepatitis there is a decreased ratio of helper to cytotoxic/suppressor cells. These studies are sharply limited because, at present, we do not know exactly the function of the cells identified by their phenotypes; a better delineation of helper and inducer or cytotoxic CD4 and suppressor or cytotoxic CD8 lymphocytes is not yet available.

But even when the function of the cells was investigated, there was no consensus. A number of techniques have been employed to evaluate spontaneous as well as mitogen- and prostaglandin-induced suppression. It is generallly agreed that in AVH suppressor cell function is defective (Chisari *et al.*, 1981; Kakumu *et al.*, 1980; Lemm *et al.*, 1983). Patients with CVH-B are heterogeneous, so that such defectiveness cannot be consistently demonstrated (reviewed by Dienstag, 1984).

Peripheral B lymphocytes from patients who are chronic carriers of HBV and have CAH-B, chronic persistent hepatitis B, or cirrhosis can produce anti-HBc but not anti-HBs antibodies. This alteration resides in B lymphocytes and, in some patients, in regulatory T lymphocytes (Dusheiko *et al.*, 1983).

In CAH-B patients the production of immunoglobulin-secreting cells or of immunoglobulins after mitogen stimulation is normal (Nardiello *et al.*, 1982; Triolo *et al.*, 1983). In the same patients, however, NK- and K-cell activities have generally been found to be decreased (Serdengecti *et al.*, 1981), but they were enhanced after stimulation with interleukin-2 and α-interferon (Nonomura *et al.*,

TABLE 2. T-CELL SUBSETS (EXPRESSED AS CD4/CD8 CELL RATIO) IN PERIPHERAL BLOOD OF PATIENTS WITH VIRAL HEPATITIS[a]

Disease	CD4/CD8	Reference
AVH-A	↓	Gocke *et al.*, 1981
AVH-B	↓	Gocke *et al.*, 1981; Thomas *et al.*, 1982; Klingenstein *et al.*, 1981
AVH non-A, non-B	↓	Gocke *et al.*, 1981
CPH	↓	Barnaba *et al.*, 1983
CAH-B	N	Klingenstein *et al.*, 1981; Regenstein *et al.*, 1983
	↓	Barnaba *et al.*, 1983; Carella *et al.*, 1982
	↑ N ↓	Gocke *et al.*, 1981; Lucas *et al.*, 1984
HBeAg+	↓	Barnaba *et al.*, 1983; Thomas *et al.*, 1982; Alexander *et al.*, 1983
HBeAg−	N	Thomas *et al.*, 1982; Alexander *et al.*, 1983
	↓	Barnaba *et al.*, 1983
HBsAg healthy carrier	↓	Gocke *et al.*, 1981; Regenstein *et al.*, 1983; Klingenstein *et al.*, 1981; Lucas *et al.*, 1984
	N	Thomas *et al.*, 1982

[a] ↓, decrease; ↑, increase; N, normal.

1984). By contrast, other investigators have reported an increased activity of NK cells in chronic HBV infection (Dienstag *et al.*, 1982).

3. NATURE OF THE ANTIGEN IN VIRAL HEPATITIS

A better understanding of the nature and function of the cellular infiltrate in the liver may be derived from a better knowledge of the location and nature of the viral antigens.

HAV has been detected in the cytoplasm of hepatocytes and Kupffer's cells of infected animals by immunofluorescence as a finely granular or diffuse pattern (Dienstag, 1979) for as long as 8 weeks after infection.

HBV antigens (HBsAg, HBcAg, or HBeAg) are usually undetectable in AVH-B. The assumption is that the cells that harbor these antigens are destroyed by the immune response (elimination phase, Gudat *et al.*, 1975). In CAH-B, HBsAg is seen in the cytoplasm of scattered hepatocytes. It is still debatable whether or not a membrane distribution of HBsAg is associated with a more severe form of HBV infection (Bianchi and Gudat, 1979; Ray *et al.*, 1976; Alberti *et al.*, 1976). The inconsistent results obtained from different laboratories may be due to technical variations or to different tissue-fixation methods. HBcAg is also seen in a few hepatocytes and is usually increased in patients undergoing immunosuppressive treatment (Gudat *et al.*, 1975). In the last few years circumstantial evidence has indicated that HBcAg is the target of an immunologic reaction mediated by lymphocytes (Mondelli *et al.*, 1982). Recent investigations indicated that in addition to the easily demonstrable nuclear localization, HBcAg can occasionally be seen in the cytoplasm (Mondelli *et al.*, 1983) and also near the surface of hepatocytes, a finding confirmed by electron microscopy (Gudat and Bianchi, 1977).

Like HBcAg, HBeAg has been demonstrated not only in the nuclei, but also in the cytoplasm of hepatocytes (Mondelli *et al.*, 1983), and the number of HBeAg-positive cells exceeds that of HBcAg-positive cells.

The recently developed technique of *in situ* hybridization has allowed the study of localization of HBV-DNA in patients with CAH-B (Burrell *et al.*, 1984; *Blum et al.*, 1983, 1984). Accordingly, HBV-DNA has been seen in the cytoplasm of hepatocytes with cytoplasmic damage (Burrell *et al.*, 1984).

HDV can be demonstrated in hepatocyte nuclei, although usually the cells are morphologically normal (Rizzetto *et al.*, 1983). The latter have not been found adjacent to inflammatory cells, suggesting that this agent does not elicit an intense immunologic reaction.

NANB hepatitis viral antigen has not been convincingly demonstrated so far. Viral particles frequently observed by electron microscopy in infected chimpanzees (reviewed by Gerety *et al.*, 1984), were not consistently detected in humans, and at present little is known about the target of the immune response.

Hepatocyte surface antigens (i.e., liver-specific and membrane) have been considered to be the site of immunologic attack (Meyer zum Büschenfelde *et al.*, 1979). While antibodies to liver membrane antigen are characteristic of autoimmune CAH, antibodies to liver-specific antigen are frequent in CAH-B (Jensen *et al.*, 1978; Behrens *et al.*, 1979). In addition, cytotoxicity against autologous or heterologous hepatocytes can be blocked by liver-specific antigen. Until now, however, no direct immunologic attack by sensitized mononuclear cells against these antigens has been demonstrated.

Other antigens that may be of importance in the immunologic attack are HLA class I and class II antigens. As indicated in Section 2.2, the former are increased in the liver of patients with AVH-B and CAH-B, especially in patients that have seroconverted to an anti-HBeAg status (Montaño *et al.*, 1982).

4. SERUM FACTORS AND THE IMMUNE RESPONSE

The activity and function of peripheral blood mononuclear cells can be influenced by a variety of serum factors. The previously mentioned immune complexes in patients with CVH may modify their number and function. In addition, the rosette inhibitory factor (RIF) (Chisari *et al.*, 1977), has been demonstrated in patients with AVH-A, AVH-B, and NANB hepatitis and it persists in patients who progress to CAH. Another factor, the serum inhibitory factor, was demonstrated in a variety of liver diseases including AVH. This factor has been considered to correlate with liver damage better than the RIF (Grauer *et al.*, 1984).

Because interferon has an inhibitory activity on viral replication and an enhancing activity on K cells, it has been considered important in CVH. Interferon-α, produced by lymphocytes and K cells, was depressed in CVH, but normal in those with AVH (A, B, and NANB) Kato *et al.*, 1982; Levin and Hahn, 1982), while interferon-γ, a product of T cells, was undetectable in asymptomatic HBV carriers, but normal in patients with CAH (Yousefi *et al.*, 1983).

5. THE RECEPTOR FOR HBV IN HEPATOCYTES

The hepatotropism of HAV, HBV, HDV, and NANB hepatitis virus has suggested the possibility that specific receptors are involved in the attachment of these viruses to the liver. Most of the work has been restricted to HBV. Polymerized human serum albumin (pHSA) has been postulated in the binding of HBV, because this protein binds to HBV and, according to some authors, also to human liver cells, bridging HBV and a putative receptor on the cell membrane. A number of investigations conclude that pHSA binds to HBV (Hansson and Purcell, 1979; Milich *et al.*, 1981; Pontisso *et al.*, 1983), but the presence of receptors on normal hepatocytes (Lenkei *et al.*, 1977b; Trevisan *et al.*, 1982a) is more controversial. A receptor in normal uninfected human hepatocytes has not been demonstrated by monoclonal antiidiotypes, which bear the internal image of pHSA (Colucci and Waksal, 1984). In addition, binding of fluoresceinated pHSA was found only in infected, but not in uninfected liver tissue (Thung and Gerber, 1981a). In addition, it has recently been demonstrated that cells other than hepatocytes (e.g., bone marrow, mononuclear, bile duct, and endothelial) may carry HBV-DNA or HBV antigens (Romet-Lemonne *et al.*, 1983; Shimoda *et al.*, 1981; Lie-Injo *et al.*, 1983; Blum *et al.*, 1983; Pontisso *et al.*, 1984), suggesting that hepatotropism is not absolute. Antibodies to pHSA have been demonstrated by a number of investigators (Lenkei *et al.*, 1977a; Mihaescu *et al.*, 1981; Thung and Gerber, 1981b), and it has been suggested that they may interfere with viral infectivity.

6. GENETIC FACTORS

The observation that only a minority of patients with AVH progress to chronic liver disease or become HBsAg carriers permanently has stimulated a number of investigations into the genetic factors for HBV infection. Sex-related factors have been considered since men more frequently than women become chronic carriers of HBV. Investigations of HLA A, B, and C have been inconclusive. Increased frequencies of HLA Bw 35 (Penner *et al.*, 1977) and of antigens of the 4C cross-reacting group (B5, B18, Bw35) (Paronetto *et al.*, 1982) seen in some populations have not been universally confirmed. Extensive investigations of the HLA DR antigens have not appeared as yet. A suggestion that DR4 is decreased in patients with CAH (Paronetto *et al.*, 1982) has recently been confirmed (Forzani *et al.*, 1984), suggesting that this antigen may be in some way related to the protection against HBV. It has also been reported that patients with chronic HBsAg-positive CAH have an increase of DR3. In contrast, patients with HDV infection have been reported to have an increase of DR2 antigen (Forzani *et al.*, 1984). These findings so far suggest that HBV-related chronic hepatitis may constitute a heterogeneous group of diseases with different pathogenetic mechanisms.

7. ANIMAL MODELS FOR HUMAN HEPATITIDES

Spontaneous viral hepatitis has been described in some animals, especially in mice (Sebesin and Koff, 1974a,b). Nude mice are particularly susceptible to mouse hepatitis virus, and acute necrotic lesions may progress to CVH (Tamura *et al.*, 1977). Genetic factors and the RES are of major importance in the susceptibility to mouse hepatitis (Weiser and Bang, 1978; Dupuy *et al.*, 1975; LePrevost *et al.*, 1975), but the significance of this disease as an animal model for human viral hepatitis is debatable.

New interest has been generated by the experimental transfer of HAV in marmosets (Holmes *et al.*, 1969) and HBV, HDV, and NANB hepatitis in chimpanzees (Alter *et al.*, 1977; Barker *et al.*, 1973; Maynard *et al.*, 1972; Purcell and Gerin, 1983; Rizzetto *et al.*, 1980; Tabor *et al.*, 1980; Wyke *et al.*, 1979). These experimental models offer the possibility of a more comprehensive observation of the role of the RES in viral hepatitides. A better insight on the interaction of viral hepatitis and RES should also come from the studies of other hepadna (hepatotropic DNA) viruses that have a remarkable morphologic and antigenic similarity with HBV (Summers, 1981). Woodchuck hepatitis virus induces an acute and chronic hepatitis in woodchucks (*Marmota monax*). Animals in the wild, as well as in captivity, injected with the virus develop hepatitis resembling AVH (Frommel *et al.*, 1984). In woodchucks, however, chronic hepatitis does not progress to cirrhosis but is frequently associated with hepatocellular carcinoma (Summers *et al.*, 1978; Popper *et al.*, 1981). Recently, woodchucks have been successfully infected with HDV (Ponzetto *et al.*, 1984a). Beechey ground squirrels may be infected by their respective hepadna virus (GSHBV) (Marion *et al.*, 1980), while in Pekin ducks another hepadna virus (DHBV) induces acute and chronic hepatitis which can be transferred to young animals by inoculation of eggs with the virus (Marion *et al.*, 1984; O'Connell *et al.*, 1983; Omata *et al.*, 1983).

The role of the RES in these experimental animal models has not been explored as yet, but fertile investigations can be foreseen.

8. SUMMARY

Most of our knowledge, still fragmentary, is derived from evaluation of the better-recognized and characterized HBV. There is a general consensus that in AVH-B CD8 (cytotoxic) lymphocytes are responsible for the focal hepatocellular damage. This hypothesis is based on the increase of these cells not only in the peripheral blood, but also at the site of lesions in close approximation with altered hepatocytes. Furthermore, in AVH, HLA class I antigens are increased in hepatocytes, supporting the finding in other systems that CD8 cells exert their cytotoxic activity in the context of class I HLA antigens. More difficult to explain is the chronicity of liver lesions and the lack of clearance of HBV. Most likely, various immunologic and nonimmunologic mechanisms may contribute. In the more active phase, in which there is evidence of replication of HBV, characterized by the presence in the serum of HBV-DNA, HBeAg, and DNA polymerase and by the presence of all viral antigens as well as HBV-DNA in the liver, CD8 cells have been considered important, but their efficiency in eliminating infected cells seems to be decreased. A number of factors may play a role: (1) the paucity of cells carrying HLA class I antigens, which may hinder a cytotoxic activity; (2) the increased number of suppressor cells that interfere with the activity of cytotoxic cells; (3) the presence or lack of certain modulating serum factors which may interfere with the activity of CD8 cells; (4) the cytopathic effect of the virus itself, which is supported by the finding that cells carrying HBV-DNA exhibit cytoplasmic swelling and vacuolization; and (5) the inability of CVH-B patients to produce protective antibodies against HBV.

In patients with little or no evidence of HBV replication, characterized by the lack of HBV-DNA and the presence of anti-HBeAg in the serum, different pathogenic mechanisms may be considered. Since HBV-DNA is integrated in the host DNA, these patients may still express viral markers in the hepatocytes and in the serum. The increased amount of HLA class I antigens and the presence of HBV markers on the hepatocyte surface may render the CD8 cells, still present in this phase of the disease, more efficient in eliminating infected cells. At this stage autoimmune processes directed against liver antigen may take place. CD4 lymphocytes present in the portal tract may be part of a delayed-type hypersensitivity reaction directed against autoantigens. Indeed, CD4 cells may also exert cytotoxicity expressed in the context of HLA class II antigens (Meuer *et al.*, 1982) that have been demonstrated in areas of piecemeal necrosis. An antibody-dependent cellular cytotoxicity mechanism has also been considered to play a role in this autoimmune attack. This hypothesis is based on the presence in the peripheral blood of cytotoxic non-T cells, identified as K cells, which have activity against autologous and heterologous hepatocytes. This cytotoxic activity can be blocked by liver-specific antigens. The paucity of K cells at the site of tissue damage, as demonstrated by specific antisera, has recently caused the role of this pathogenetic mechanism to be questioned. More recent ultrastructure investigations have suggested that mononuclear phagocytes, the most typical cells of the RES, may be important in the low-grade hepatocytic necrosis observed in certain forms of CVH. These mononuclear phagocytes do not seem to be involved in eliminating the virus, since they are not detected in AVH. Their activation, however, has been shown to produce cellular injury by releasing a variety of substances that may be responsible for cell injury and death.

Obviously, more investigations aimed at elucidating pertinent pathogenetic mechanisms in the various stages of HBV infection are needed, as well as a better understanding of the host–virus interactions in other forms of viral hepatitis, especially those associated with NANB hepatitis virus and superinfection or coinfection with HDV.

REFERENCES

Alberti, A., Realdi, G., Tremolada, F., and Spina, G. P., 1976, Liver cell surface localization of hepatitis B antigen and of immunoglobulins in acute and chronic hepatitis and in liver cirrhosis, *Clin. Exp. Immunol.* **25**:296–402.

Alberti, A., Diana, S., Scullard, G. H., Eddleston, A. L. W. F., and Williams, R., 1978, Detection of a new antibody system reacting with Dane particles in hepatitis B virus infection, *Br. Med. J.* **2**:1056–1058.

Alexander, G. J. M., Nouri-Aria, K. T., Eddleston, A. L. W. F., and Williams, R., 1983, Contrasting relations between suppressor-cell function and suppressor-cell number in chronic liver diseases, *Lancet* **1**:1291–1293.

Alter, H. J., Purcell, R. H., Gerin, J. L., London, W. T., Kaplan, P. M., McAuliffe, V. J., Wagner, J., and Holland, P. V., 1977, Transmission of hepatitis B to chimpanzees by hepatitis B surface antigen-positive saliva and semen. *Infect. Immun.* **16**:928–933.

Arnold, W., and Meyer zum Büschenfelde, K. H., 1977, Immune complexes of HBcAg/anti-HBc in nuclei of HBsAg positive chronic liver disease, *N. Engl. J. Med.* **296**:818.

Barbatis, C., Woods, J., Morton, J. A., Fleming, K. A., McMichael, A., and McGee, J. O. D., 1981, Immunohistochemical analysis of HLA (A,B,C) antigens in liver disease using a monoclonal antibody, *Gut* **22**:985–991.

Bardadin, K. A., and Scheuer, P. J., 1984, Endothelial cell changes in acute hepatitis. A light and electron microscopic study, *Pathology* **144**:213–220.

Barker, L. F., Chisari, F. V., McGrath, P. P., Dalgard, D. W., Kirschstein, R. L., Almeida, J. D.,

Edgington, T. S., Sharp, D. G., and Peterson, M. R., 1973, Transmission of type B viral hepatitis to chimpanzees, *J. Infect. Dis.* **127**:648–662.

Barnaba, V., Musca, A., Cordova, C., Levrero, M., Ruocco, G., Albertini-Petroni V., and Balsano, F., 1983, Relationship between T cell subsets and suppressor cell activity in chronic hepatitis B virus (HBV) infection, *Clin. Exp. Immunol.* **53**:281–288.

Behrens, U. J., Vernace, S., and Paronetto, F., 1979, Studies on "liver specific" antigens. II. Detection of serum antibodies to liver and kidney cell membrane antigens in patients with chronic liver disease, *Gastroenterology* **77**:1053.

Bernuau, D., Rogier, E., and Feldmann, G., 1984, *In situ* ultrastructural detection and quantitation of liver mononuclear phagocytes in contact with hepatocytes in chronic type B hepatitis, *Lab. Invest.* **51**:667–674.

Bianchi, L., 1983, Liver biopsy interpretation in hepatitis. Part II: Histopathology and classification of acute and chronic viral hepatitis/differential diagnosis, *Path. Res. Pract.* **178**:180–213.

Bianchi, L., and Gudat, F., 1979, Immunopathology of hepatitis B., in: *Progress in Liver Diseases*, Volume VI (H. Popper and F. Schaffner, eds.), pp. 371–424, Grune & Stratton, New York.

Blum, H. E., Stowring, L., Figus, A., Montgomery, C. K., Haase, A. T., and Vyas, G. V., 1983, Detection of hepatitis B virus DNA in hepatocytes, bile duct epithelium, and vascular elements by in situ hybridization, *Proc. Natl. Acad. Sci. USA* **80**:6685–6688.

Blum, H. E., Haase, A. T., and Vyas, G. N., 1984, Molecular pathogenesis of hepatitis B virus infection: Simultaneous detection of viral DNA and antigens in paraffin-embedded liver sections, *Lancet* **2**:771–775.

Burrell, C. J., Gowans, E. J., Rowland, R., Hall, P., Jilbert, A. R., and Marmion, B. P., 1984, Correlation between liver histology and markers of hepatitis B virus replication in infected patients: A study by *in situ* hybridization, *Hepatology* **4**:20–24.

Carella, G., Chatenoud, L., Degos, F., and Bach, M. A., 1982, Regulatory T cell–subset imbalance in chronic active hepatitis, *J. Clin. Immunol.* **2**:93–100.

Chisari, F. V., Routenberg, J. A., and Edgington, T. S., 1977, Mechanisms responsible for defective human T-lymphocyte sheep erythrocyte rosette function associated with hepatitis B virus infection, *J. Clin. Invest.* **59**:124–142.

Chisari, F. V., Castle, K. L., Xavier, C., and Anderson, D. S., 1981, Functional properties of lymphocyte subpopulations in hepatitis B virus infection. I. Suppressor cell control of T lymphocyte responsiveness, *J. Immunol.* **126**:38–44.

Colucci, G., and Waksal, S. D., 1984, Monoclonal anti-idiotypes recognize binding sites for human polymerized albumin on hepatitis B surface antigen (HBsAg), *Hepatology* **4**:1055.

Colucci, G., Colombo, M., DelNinno, E., and Paronetto, F., 1983, *In situ* characterization by monoclonal antibodies of the mononuclear cell infiltrate in chronic active hepatitis, *Gastroenterology* **85**:1138–1145.

Dienes, H. P., Popper, H., Arnold, W., and Lobeck, H., 1982, Histologic observations in human hepatitis non-A, non-B, *Hepatology* **2**:562–571.

Dienstag, J. L., 1979, The pathobiology of hepatitis A virus, *Int. Rev. Exp. Pathol.* **20**:1–42.

Dienstag, J. L., 1984, Immunologic mechanisms in chronic viral hepatitis, in: *Viral Hepatitis and Viral Disease* (G. N. Vyas, J. L. Dienstag, and J. H. Hoofnagle, eds.), pp. 135–166, Grune & Stratton, New York.

Dienstag, J. L., Savarese, A. M., and Bhan, A. K., 1982, Increased natural killer cell activity in chronic hepatitis B virus infection, *Hepatology* **2**:107S–115S.

Dupuy, J. M., Leblond-Levey, E., and LePrevost, C., 1975, Immunopathology of mouse hepatitis virus type 3 infection, *J. Immunol.* **114**:226–230.

Dusheiko, G. M., Hoffnagle, J. H., and Cooksley, W. G., 1983, Synthesis of antibodies to hepatitis B virus by cultured lymphocytes from chronic hepatitis B surface antigen carriers, *J. Clin. Invest.* **71**:1104–1113.

Eggink, H. F., Houthoff, H. J., Huitema, S., Wolters, G., Poppema, S., and Gips, C. H., 1984, Cellular and humoral immune reactions in chronic active liver disease. II. Lymphocyte subsets and viral antigens in liver biopsies of patients with acute and chronic hepatitis B, *Clin. Exp. Immunol.* **56**:121–128.

Forzani, B., Actis, G. C., Verme, G., Amoroso, A., Borelli, I., Curtoni, E. S., Rumi, M. G., Picciotto,

A., Marinucci, G., Freni, M. A., and Rizzetto, M., 1984, HLA-DR antigens in HBsAg-positive chronic active liver disease with and without associated delta infection, *Hepatology* **4**:1107–1110.

Friedman, H., Escobar, M. R., and Reichard, S. M., 1980, Preface, in: *The Reticuloendothelial System. A Comprehensive Treatise.* Volume I. *Morphology* (I. Carr, and W. T. Daems, eds.), pp. xi–xii, Plenum Press, New York.

Frommel, D., Crevat, D., Vitvitsky, L., Pichoud, C., Hantz, O., Chevalier, M., Grimaud, J. A., Lindberg, J., and Trépo, C. G., 1984, Immunopathologic aspects of woodchuck hepatitis, *Am. J. Pathol.* **115**:125–134.

Gerber, M. A., Sarno, E., and Vernace, V. J., 1976, Immune complexes in hepatocytic nuclei of HBAg-positive chronic hepatitis, *N. Engl. J. Med.* **294**:922–925.

Gerety, R. J., Tabor, E., Schaff, A., Seto, B., and Coleman, W. G. Jr., 1984, Non-A, non-B hepatitis agents, in: *Viral Hepatitis and Liver Disease* (G. N. Vyas, J. L. Dienstag, and J. H. Hoofnagle, eds.), pp. 23–47, Grune & Stratton, New York.

Gocke, D. J., Ackley, A. M., Kim, H. C., and Benson, G. D., 1981, Alterations in helper and suppressor T lymphocytes in hepatitis, *Hepatology* **1**:511.

Govindarajan, S., Uchida, T., and Peters, R. L., 1983, Identification of T lymphocytes and subsets in liver biopsy cores of acute viral hepatitis, *Liver* **3**:13–18.

Grauer, W., Brattig, N. W., Schomerus, H., Frosner, G., and Berg, P. A., 1984, Immunosuppressive serum factors in viral hepatitis. III. Prognostic relevance of rosette inhibitory factor and serum inhibition factor in acute and chronic hepatitis, *Hepatology* **4**:15–19.

Gudat, F., and Bianchi, L., 1977, Evidence for phasic sequences in nuclear HBcAg formation and cell membrane-directed flow of core particles in chronic hepatitis B, *Gastroenterology* **73**:1194–1197.

Gudat, F., Bianchi, L., Sonnabend, W., Thiel, G., Aenishaenslin, W., and Stalder G. A., 1975, Pattern of core and surface expression in liver tissue reflects state of specific immune response in hepatitis B, *Lab. Invest.* **32**:1–9.

Hadler, S. C., de Monzon, M., Ponzetto, A., Anzola, E., Rivero, D., Mondolfi, A., Bracho, A., Francis, D. P., Gerber, M. A., Thung, S., Gerin, J., Maynard, J. E., Popper, H., and Purcell, R. H., 1984, Delta virus infection and severe hepatitis: an epidemic of Yucpa Indians of Venezuela, *Ann. Intern. Med.* **100**:339–344.

Hadziyannis, S. J., Lieberman, H. M., Karvountzis, G. G., and Shafritz, D. A., 1983, Analysis of liver disease, nuclear HBcAg, viral replication, and hepatitis B virus DNA in liver and serum of HBeAg vs. anti-HBe-positive carriers of hepatitis B virus, *Hepatology* **3**:656–662.

Hansson, B. G., and Purcell, R. H., 1979, Sites that bind polymerized albumin on hepatitis B surface antigen particles: Detection by radioimmunoassay, *Infect. Immun.* **26**:125–130.

Hirose, H., Udo, K., Kojima, M., Takahashi, Y., Miyakawa, Y., Miyamoto, K., Yoshizawa, H., and Mayumi, M., 1984, Deposition of hepatitis B e antigen in membranous glomerulonephritis: Identification by F $(ab')_2$ fragments of monoclonal antibody, *Kidney Int.* **26**:338–341.

Holmes, A. W., Wolfe, L., Rosenblate H., and Deinhardt, F., 1969, Hepatitis in marmosets: Induction of disease with coded specimens from a human volunteer study, *Science* **165**:816–817.

Hopf, H., Meyer zum Büschenfelde, K. H., and Arnold, W., 1976, Detection of a liver membrane autoantibody in HBsAg-negative chronic active hepatitis, *N. Engl. J. Med.* **294**:578–582.

Ito, H., Hattori, S., Masuda, I., Amamiya, S., Hajikano, H., Yoshizawa, H., Miyakawa, Y., and Maxumi, M., 1981, Hepatitis B e antigen-mediated membranous glomerulonephritis: Correlation of ultrastructural changes with HBeAg in the serum and glomeruli, *Lab. Invest.* **44**:214–220.

Jensen, D., McFarlane, I. G., Portmann, B. S., Eddleston, L. W. F., and Williams, R., 1978, Detection of antibodies directed against a liver-specific membrane lipoprotein in patients with acute and chronic active hepatitis, *N. Engl. Med.* **299**:1–7.

Kakumu, S., Yata, K., and Kashio, T., 1980, Immunoregulatory T-cell function in acute and chronic liver disease, *Gastroenterology* **79**:613–619.

Kanel, G. C., Govindarajan, S., and Peters, R. L., 1984, Chronic delta infection and liver biopsy changes in chronic active hepatitis B, *Ann. Intern. Med.* **101**:51–54.

Kato, Y., Nakagawa, H., Kobayashi, K., Hattori, N., and Hatano, K., 1982, Interferon production by peripheral lymphocytes in HBsAg-positive liver disease, *Hepatology* **2**:789–790.

Klingenstein, R. J., Savarese, A. M., Dienstag, J. L., Rubin, R. H., and Bhan, A. K., 1981, Immunoregulatory T cell subsets in acute and chronic hepatitis, *Hepatology* **1**:523.

Lemm, G., Salzer, K., and Warnatz. H., 1983, Studies on immunoregulatory mechanisms in acute and chronic hepatitis B, *Clin. Exp. Immunol.* **52**:250–258.

Lenkei, R., Babes, V. T., Mustea, A., and Dobre, I., 1977a, Correlations between anti-albumin antibodies and HBsAg in hepatic patients, *J. Med. Virol.* **1**:29–34.

Lenkei, R., Onica, D., and Ghetie, V., 1977b, Receptors for polymerized albumin on liver cells, *Experientia* **33**:1046–1047.

LePrevost, C., Leblond-Levy, E., Virelizier, J. L., and Dupuy, J. M., 1975, Immunopathology of mouse hepatitis virus type 3 infection: Role of humoral and cell-mediated immunity in resistance mechanisms, *J. Immunol.* **114**:221–225.

Levin, S., and Hahn, T., 1982, Interferon system in acute viral hepatitis, *Lancet* **1**:592–594.

Levy, G. A., and Chisari, F. V., 1981, The immunopathogenesis of chronic HBV induced liver disease, *Springer Semin. Immunopathol.* **3**:439–459.

Lie-Injo, L. E., Balasegaram, M., Lopez, C. G., and Herrera, A. R., 1983, Hepatitis B virus DNA in liver and white blood cells of patients with hepatoma, *DNA* **2**:301–308.

Lucas, M., de Moura, M. C., and Victorino, R., 1984, Decreased helper/suppressor T-lymphocyte ratios in chronic hepatitis B virus (HBV) infection: Relation of liver histology and the hepatitis B e antigen (HBeAg) status, *Hepatology* **4**:1048.

Marion, P., Oshiro, L. S., Regnery, D. C., Scullard, G. H., and Robinson, W. S., 1980, A virus in Beechey ground squirrels that is related to hepatitis B virus of humans, *Proc. Natl. Acad. Sci. USA* **77**:2941–2945.

Marion, P. L., Knight, S. S., Ho, B. K., Robinson, W. S., and Popper, H., 1984, Liver disease associated with duck hepatitis B virus infection of domestic ducks, *Proc. Natl. Acad. Sci. USA* **81**:898–902.

Maynard, J., Berquist, K., Krushak, D. H., and Purcell, R. H., 1972, Experimental infection of chimpanzees with the virus of hepatitis B, *Nature* **237**:514–515.

Meuer, S. C., Schlossman, S. F., and Reinherz, E. L., 1982, Clonal analysis of human cytotoxic T lymphocytes: T4+ and T8+ effector T-cells recognize products of different major histocompatibility complex regions, *Proc. Natl. Acad. Sci. USA* **79**:4395–4399.

Meyer zum Büschenfelde, K. H., Hütteroth, T. H., Arnold, W., and Hopf, U., 1979, Immunologic liver injury: The role of hepatitis B virus antigens and liver membrane antigens as targets, in: *Progress in Liver Diseases*, Volume VI (H. Popper and F. Schaffner, eds.), pp. 407–424, Grune & Stratton, New York.

Mihaescu, S., Lenkei, R., and Ghetie, V., 1981, Radioimmunoassay of antialbumin autoantibodies in human sera, *J. Immunol. Methods* **42**:187–193.

Milich, D. R., Gottfried, T. D., and Vyas, G. N., 1981, Characterization of the interaction between polymerized human albumin and hepatitis B surface antigen, *Gastroenterology* **81**:218–225.

Mondelli, M., Vergani, G. M., Alberti, A., Vergani, D., Portmann, B., Eddleston, A. L. W. F., and Williams, R., 1982, Specificity of T lymphocyte cytotoxicity to autologous hepatocytes in chronic hepatitis B virus infection: Evidence that T-cells are directed against HBV core antigen expressed on hepatocytes, *J. Immunol.* **120**:2773–2778.

Mondelli, M., Alberti, A., Pontisso, P., Tedder, R. S., Cadrobbi, P., Busachi, C. A., Eddleston, A. L. W. F., Williams, R., and Realdi, G., 1983, Hepatitis B core (HBcAg) and e (HBeAg) antigens in the liver detected by immunofluorescence with monoclonal antibodies, *Gut* **24**:A995.

Montaño, L., Miescher, G. C., Goodall, A. H., Wiedmann, K. H., Janossy, G., and Thomas, H. C., 1982, Hepatitis B virus and HLA antigen display in the liver during chronic hepatitis B virus infection, *Hepatology* **2**:557–561.

Montaño, L., Aranguibel, F., Boffill, M., Goodall, A. H., Janossy, G., and Thomas, H. C., 1983, An analysis of the composition of the inflammatory infiltrate in autoimmune and hepatitis B virus–induced chronic liver disease, *Hepatology* **3**:292–296.

Nardiello, S., Schaffner, F., Vernace, S. J., and Paronetto, F., 1982, Pokeweed mitogen-induced immunoglobulin-secreting cells in hepatitis B surface antigen-positive-negative chronic active hepatitis: Evaluation by a protein A plaque assay, *Clin. Immunol. Immunopathol.* **22**:168–179.

Naumov, N. V., Mondelli, M., Alexander, G. J. M., Tedder, R. S., Eddleston, A. L. W. F., and Williams, R., 1984, Relationship between expression of hepatitis B virus antigens in isolated hepatocytes and autologous lymphocyte cytotoxicity in patients with chronic hepatitis B virus infection, *Hepatology* **4**:63–68.

Nonomura, A., Colucci, G., Vernace, S., Waksal, S. O., and Paronetto, F., 1984, Natural killer (NK) and antibody-dependent cellular cytotoxicity (ADCC) in chronic active hepatitis (CAH) and primary biliary cirrhosis (PBC): Effect of interleukin-2 (IL-2) and α interferon (α INF), *Hepatology* **2**:1087.

Nowoslawski, A., 1979, Hepatitis B virus-induced immune complex disease, in: *Progress in Liver Diseases*, Volume VI (H. Popper and F. Schaffner, eds.), pp. 393–406, Grune & Stratton, New York.

O'Connell, A. P., Urban, M. K., and London, W. T., 1983, Naturally occurring infection of Peking duck embryos by duck hepatitis B virus, *Proc. Natl. Acad. Sci. USA* **80**:1703–1706.

Omata, M., Uchiumi, K., Ito, Y., Yokosuka, 0., Mori, J., Terao, K., Wei-Fa, Y., O'Connell, A. P., London, W. T., and Okuda, K., 1983, Duck hepatitis B virus and liver diseases, *Gastroenterology* **85**:260–267.

Pape, G. R., Rieber, E. P., Eisenburg, J., Hoffmann, R., Balch, C. M., Paumgartner, G., and Riethmuller, G., 1983, Involvement of the cytotoxic/suppressor T-cell subset in liver tissue injury of patients with acute and chronic liver diseases, *Gastroenterology* **85**:657–662.

Paronetto, F., and Vernace, S. J., 1975, Immunological studies in patients with chronic active hepatitis: Cytotoxic activity of lymphocytes to autochtonous liver cells grown in tissue culture, *Clin. Exp. Immunol.* **19**:99–104.

Paronetto, F., Piccinino, F., Vernace, S., Sagnelli, E., Aloisio, V., Giusti, G., and Rubinstein, P., 1982, HLA antigens in HBsAg-positive chronic active hepatitis (CAH): Population and family study, in: *Viral Hepatitis. 1981 International Symposium* (W. Szmuness, H. J. Alter, and J. E. Maynard, eds.), pp. 728–729, The Franklin Institute Press. Philadelphia.

Penner, E., Grabner, G., Dittrich, H., and Mayr, W. R., 1977, HLA antigens in HBs antigen positive chronic active hepatitis, *Tissue Antigens* **10**:63–64.

Pontisso, P., Alberti A., Bortolotti, F., and Realdi, G., 1983, Virus-associated receptors for polymerized human serum albumin in acute and in chronic hepatitis B virus infection, *Gastroenterology* **84**:220–226.

Pontisso, P., Poon, M. C., Tiollais, P., and Brechot, C., 1984, Detection of hepatitis B virus DNA in mononuclear blood cells, *Br. Med. J.* **288**:1563–1566.

Ponzetto, A., Cote, P. J., Popper, H., Hoyer, B. H., London, W. T., Ford, E. C., Bonino, F., Purcell, R. H., and Gerin, J. L., 1984a, Transmission of the hepatitis B virus-associated delta agent to the eastern woodchuck, *Proc. Natl. Acad. Sci. USA* **81**:2208–2212.

Ponzetto, A., Seeff, L. B., Bales-Buskell, Z., Ishak, K. G., Hoofnagle, J. H., Zimmerman, H. J., Purcell, R. H., and Gerin, J. L., The Veterans Administration Hepatitis Cooperative Study Group, 1984b, Hepatitis B markers in United States drug addicts with special emphasis on the delta hepatitis virus, *Hepatology* **4**:1111–1115.

Popper, H., 1983, Changing concepts of the evolution of chronic hepatitis and the role of piecemeal necrosis, *Hepatology* **3**:758–762.

Popper, H., 1984, Pathologic observations on delta agent infection, in: *Viral Hepatitis and Liver Diseases* (G. N. Vyas, J. L. Dienstag, and J. H. Hoofnagle, eds.), pp. 381–384, Grune & Stratton, New York.

Popper, H., and Paronetto, F., 1984, Critical review of hepatic immunopathology, *Clin. Immunol. Newsletter* **5**:1–7.

Popper, H., Dienstag, J. L., Feinstone, S. M., Alter, H. J., and Purcell, R. H., 1980, The pathology of viral hepatitis in chimpanzees, Virchows *Arch. A Path. Anat. Histol.:* **387**:91–106.

Popper, H., Shih, J. W. K., Gerin, J. L., Wong, D. C., Hoyer, B. H., London, W. T., Sly, D. L., and Purcell, R. H., 1981, Woodchuck hepatitis and hepatocellular carcinoma: Correlation of histologic with virologic observations, *Hepatology* **1**:91–98.

Popper, H., Thung, S. N., Gerber, M. A., Hadler, S. C., de Monzon, M., Ponzetto, A., Anzola, E., Rivera, D., Mondolfi, A., Bracho, A., Francis, D. P., Gerin, J. L., Maynard, J. E., and Purcell, R. H., 1983, Histologic studies of severe delta agent infection in Venezuelan Indians, *Hepatology* **3**:906–912.

Purcell, R. H., and Gerin, J. L., 1983, Experimental transmission of the delta agent to chimpanzees, in: *Viral Hepatitis and Delta Infection* (G. Verme, F. Bonino and M. Rizzetto, eds.), pp. 79–89, Alan R. Liss, New York.

Ray. M. B., Desmet, V. J., Bradburne, A. F., Desmyter, J., Fevery, J., and de Groote, J., 1976, Differential distribution of hepatitis B surface antigen and hepatitis B core antigen in the liver of hepatitis B patients, *Gastroenterology* **71**:462–467.

Ray, M. B., Desmet, V. J., Bradburne, A. F., Desmyter, J., Fevery, J., and de Groote, J., 1979, Hepatitis B core antigen immune complexes in the liver of hepatitis B patients, *Clin. Exp. Immunol.* **37**:15–24.

Regenstein, F. G., Roodman, S. T., and Perrillo, R. P., 1983, Immunoregulatory T cell subsets in chronic hepatitis B virus infection: The influence of homosexuality, *Hepatology* **3**:951–954.

Rizzetto, M., Bonino, F., Crivelli, O., Canese, G., and Verme, G., 1976, Complement fixing hepatitis B core antigen immune complexes in the liver of patients with HBs antigen positive chronic disease, *Gut* **17**:837–843.

Rizzetto, M., Canese, M. G., Aricò, S., Crivelli, O., Trépo, C., Bonino, F., and Verme, G., 1977, Immunofluorescence detection of a new antigen-antibody system (delta/anti-delta) associated to hepatitis B virus in liver and serum of HBsAg carriers, *Gut* **18**:997–1003.

Rizzetto, M., Canese, M. G., Gerin, J. L., London, W. T., Sly, L. D.. and Purcell, R. H., 1980, Transmission of hepatitis B virus-associated delta antigen to chimpanzees, *J. Infect. Dis.* **141**:590–602.

Rizzetto, M., Canese, M. G., Purcell, R. H., London, W. T., Sly, L. D., and Gerin, J. L., 1981, Experimental HBV and infections of chimpanzees: Occurrence and significance of intrahepatic immune complexes of HBcAg and antigen, *Hepatology* **1**:567–574.

Rizzetto, M., Verme, G., Recchia, S., Bonino, F., Farci, P., Arico, S., Calzia, R., Picciotto, A., Colombo, M., and Popper, H., 1983, Chronic hepatitis in carriers of hepatitis B surface antigen, with intrahepatic expression of the delta antigen, *Ann. Intern. Med.* **98**:437–441.

Romet-Lemonne, J. L., McLane, M. F., Elfassi, E., Haseltine, W. A., Azocar, J., and Essex, M., 1983, Hepatitis B virus infection in cultured human lymphoblastoid cells, *Science* **221**:667–669.

Rugge, M., Vanstapel, M. J., Ninfo, V., Realdi, G., Tremolada, F., Montanari, P. G., van Damme, B., Fevery, J., de Groote, J., and Desmet, V., 1983, Comparative histology of acute hepatitis B and non-A, non-B in Leuven and Padova, *Virchows Arch. Path. Anat.* **401**:275–288.

Sagnelli, E., Felaco, F. M., Triolo, G., Vernace, S. J., Filippini, F. P., Piccinino, F., Behrens, U., and Paronetto, F., 1983, Circulating complement fixing immune complexes in chronic hepatitis. Use of anti-C_3 enzyme immunoassay to define antibody class and nature of antigen, *J. Clin. Lab. Immunol.* **12**:11–15.

Sebesin, S. M., and Koff, R. S., 1974a, Pathogenesis of experimental viral hepatitis (First of two parts), *N. Engl. J. Med.* **290**:944–950.

Sebesin, S. M., and Koff, R. S., 1974b, Pathogenesis of experimental viral hepatitis (Second of two parts), *N. Engl. J. Med.* **290**:996–1002.

Serdengecti, S., Jones, D. B., Holdstock, G., and Wright, R., 1981, Natural killer activity in patients with biopsy proven liver disease, *Clin. Exp. Immunol.* **45**:361–364.

Shibata, M., Kojima, T., Aoyama, K., Konda, T., Inoue, K., Sasaki, H., and Imoto, S., 1984, Morphological studies on the mechanism of hepatocellular injury in type A acute viral hepatitis. Ultrastructural investigation and identification of T-cell subsets using peroxidase antibody technique, *Acta Hepatol. Jpn.* **25**:1–11.

Shimoda, T., Shikata, T., Karasawa, T., Tsukagoshi, S., Yoshimura, M., and Sakurai, I., 1981, Light microscopic localization of hepatitis B virus antigens in the human pancreas, *Gastroenterology* **81**:998–1005.

Si, L., Whiteside, T. L., Schade, R. R., and Van Thiel, D. H., 1983, Studies of lymphocyte subpopulations in the liver tissue and blood of patients with chronic active hepatitis (CAH), *J. Clin. Immunol.* **3**:408–419.

Stefanini, G. F., Meliconi, R., Miglio, F., Mazzetti, M., Baraldini, M., Facchini, A., and Gasbarrini, G., 1983, Lymphocytotoxicity against autologous hepatocytes and membrane-bound IgG in viral and autoimmune chronic active hepatitis, *Liver* **3**:36–45.

Summers, J., 1981, Three recently described animal virus models for human hepatitis B virus, *Hepatology* **1**:179–193.

Summers, J., Smolec, J. M., and Snyder R., 1978, A virus similar to human hepatitis B virus associated with hepatitis and hepatoma in woodchucks, *Proc. Natl. Acad. Sci. USA* **75**:4533–4537.

Tabor, E., Peterson, D. A., April, M., Seeff, L. B., and Gerety, R. J., 1980, Transmission of human non-A, non-B hepatitis to chimpanzees following failure to transmit GB hepatitis, *J. Med. Virol.* **5**:103–108.

Tamura, T., Taguchi, F., Ueda, K., and Fujiwara, K., 1977, Persistent infection with mouse hepatitis virus of low virulence in nude mice, *Microbiol. Immunol.* **21**:683–691.

Thomas, H. C., Brown, D., Routhier, G., Janossy, G., Kung, P. C., Goldstein, F., and Sherlock, S., 1982, Inducer and suppressor T-cells in hepatitis B virus-induced liver disease, *Hepatology* **2**:202–204.

Thomas, H. C., Pignetelli, M., Goodall, A., Waters, J., Karayannis, P., and Brown, D., 1984, Immunologic mechanisms of cell lysis in hepatitis B virus infection, in: *Viral Hepatitis and Viral Disease* (G. N. Vyas, J. L. Dienstag, and J. H. Hoofnagle, eds.), pp. 167–178, Grune & Stratton, New York.

Thung, S. N., and Gerber, M. A., 1981a, Presence of receptors for polymerized albumin in HBsAg containing hepatocytes and hepatoma cell line, *Hepatology* **1**:132–136.

Thung, S. N., and Gerber, M. A., 1981b, HBsAg associated albumin receptors and antialbumin antibodies in sera of patients with liver disease, *Gastroenterology* **80**:260–264.

Trevisan, A., Realdi, G., Alberti, A., and Noventa, F., 1979, Relationship between membrane-bound immunoglobulin and viral antigens in liver cells from patients with hepatitis B virus infection, *Gastroenterology* **77**:209–214.

Trevisan, A., Gudat, F., Guggenheim, R., Krey, G., Durmuller, U., Luond, G., Duggelin, M., Landmann, J., Tondelli, P., and Bianchi, L., 1982a. Demonstration of albumin receptors on isolated human hepatocytes by light and scanning electron microscopy, *Hepatology* **2**:832–835.

Trevisan, A., Realdi, G., Alberti, A., Ongard, G., Pornaro, E., and Meliconi, R., 1982b, Core antigen-specific immunoglobulin G bound to the liver cell membrane in chronic hepatitis B, *Gastroenterology* **82**:218–222.

Triolo, G., Nardiello, S., Sagnelli, E., Schaffner, F., Vernace, S., and Paronetto, F., 1983, *In vitro* synthesis of IgG and IgM in patients with HBsAg-positive and HBsAg-negative chronic active hepatitis, *Liver* **3**:207–212.

Vergani-Mieli, G., Vergani, D., Portmann, B., White, Y., Lyon-Murray, I., Marigold, J. H., Woolf, I., Eddleston, A. L. F. W., and Williams, R., 1982, Lymphocyte cytotoxicity to autologous hepatocytes in HBsAg-positive chronic liver disease, *Gut* **23**:1029–1036.

Wands, J. R., and Isselbacher, K. J., 1975, Lymphocyte cytotoxicity to autologous liver cells in chronic active hepatitis, *Proc. Natl. Acad. Sci. USA* **72**:1301–1303.

Weiser, W., and Bang, F. B., 1978, Macrophages genetically resistant to mouse hepatitis virus. Converted in vitro to susceptible macrophages, *J. Exp. Med.* **143**:690–695.

Wyke, R. J., Tsiquaye, K. N., Thornton, A., White, Y., Portmann, B., Das, P. K., Zuckerman, A. J., and Williams, R., 1979, Transmission of non-A, non-B hepatitis to chimpanzees by factor IX concentrates: Fatal complications in patients with chronic liver disease, *Lancet* **1**:520–524.

Yousefi, S., Escobar, M. R., and Carithers, R. L., 1983, Immune interferon induction: A measure of cellular immune deficiency in hepatitis B carriers, *Hepatology* **3**:826.

12

Pathogenetic Mechanisms of the Acquired Immune Deficiency Syndrome

MARIO R. ESCOBAR

1. INTRODUCTION

1.1. STATEMENT OF THE PROBLEM

A new syndrome characterized by a rare malignancy, Kaposi's sarcoma, *Pneumocystis carinii* pneumonia, and other severe opportunistic infections was reported by the Centers for Disease Control (CDC), Atlanta, in 1981 (CDC, 1981a,b,c). The profound impairment of cellular immune function underlying these disorders led to their designation as the acquired immunodeficiency syndrome (AIDS). This sexually transmitted disease with a mortality close to 100% over time has been considered "the worst pandemic of the 20th century." Except for syphilis, no other infectious disease has depended so greatly on human behavior for its spread. In fact, in an attempt to communicate the threat of AIDS, researchers have sought to move away from placing those most at danger of infection in "risk groups" and have started to emphasize instead risk behavior. Since travel around the world has become faster, easier, and more frequent and because AIDS is becoming more of a heterosexually transmitted disease, it would be realistic to consider anyone as being at some degree of risk.

Although it has been assumed that AIDS originated in Africa, this may never be proven with certainty, as is still the case with syphilis nearly 5 centuries later. Nonetheless, both diseases have had a monumental psychosocial, political, and economic impact on the peoples of every country around the globe.

Every effort has been made to include the most recent information on AIDS to the time of submission of this chapter, since it is increasing logarithmically and quickly becoming outdated. This chapter focuses on the human immu-

MARIO ESCOBAR • Department of Pathology, Clinical Immunopathology and Virology Section, Medical College of Virginia, Virginia Commonwealth University, Richmond, Virginia 23298-0106.

nodeficiency virus (HIV), the etiologic agent of AIDS, the host responses to HIV infection, and the immunopathogenetic mechanisms involved. The clinical, epidemiologic, virologic, and immunodiagnostic aspects of AIDS are also covered, but more briefly.

1.2. CLINICAL PRESENTATION

1.2.1. Case Definitions

For the purposes of national reporting, a case definition of AIDS was first published by the CDC in 1982. A revision of this case definition was necessary following the identification of HIV antibodies and the culturing of infectious virions in 1985 (CDC, 1985). These two systems classify the manifestations of HIV infection into four mutually exclusive groups, designated by Roman numerals I through IV. Classification in a particular group is not explicitly intended to have prognostic significance or to imply severity of illness. However, classification into these four groups is hierarchic in that individuals classified in a particular group should not be reclassified in a preceding group if clinical findings resolve, because clinical improvement may not accurately reflect changes in the severity of the underlying disease. The definition of AIDS specifically lists the infections and malignancies that meet the criteria of the CDC case definition and the required diagnostic procedures that must be performed and the results that must be obtained in order for a patient to be diagnosed as having AIDS (Selik *et al.*, 1984). In general, the infections and malignancies include those that are usually seen in patients with primary or secondary immunodeficiency disorders or unusually severe or protracted disease with organisms that cause mild or subclinical infections in normal young adults. AIDS appears to include more than the CDC-defined opportunistic infections and malignancies. Thus, a persistent generalized lymphadenopathy and an AIDS prodrome referred to as AIDS-related complex (ARC) are syndromes thought to be associated with AIDS.

Another revision of the surveillance case definition was developed recently by CDC in collaboration with public health and clinical specialists and officially recommended for adoption by the Council of State and Territorial Epidemiologists (CDC, 1987a). Instead of the four sections included in the previous case definitions, the most recently revised version consists of three sections that depend on the status of laboratory evidence of HIV infection (e.g., HIV antibody). The major proposed changes apply to patients with laboratory evidence of HIV infection: (1) inclusion of HIV encephalopathy, HIV wasting syndrome, and a broader range of specific AIDS-indicative diseases (Section II.A); (2) inclusion of AIDS patients whose indicator diseases are diagnosed presumptively (Section II.B); and (3) elimination of exclusions due to other causes of immunodeficiency (Section I.A). Further details are provided in the CDC publication (1987a).

Application of the 1987 CDC case definition for children differs from that for adults in two ways. First, multiple or recurrent serious bacterial infections and lymphoid interstitial pneumonia/pulmonary lymphoid hyperplasia are accepted as indicative of AIDS among children but not among adults. Second, for children <15 months of age whose mothers are thought to have had HIV infection during the child's perinatal period, the laboratory criteria for HIV infection are more stringent, since the presence of HIV antibody in the child is, by itself, insufficient evidence for HIV infection because of the persistence of passively acquired maternal antibodies <15 months after birth.

Since the clinical spectrum of AIDS in Africa may differ substantially from that described in Europe and North America, WHO developed a separate clinical case definition for use in Africa. It has been tested on 174 inpatients at Mama Yemo Hospital in Kinshasa, Zaire. In this hospital population with a 34% infection rate of HIV, the clinical case definition had a specificity of 90%, a sensitivity of 59%, and a predictive value of 74% for HIV seropositivity. These results support the use of the WHO clinical definition for AIDS in Africa. However, since HIV prevalence and disease expression vary, similar evaluations should be carried out in different regions (Colebunders *et al.*, 1987).

1.2.2. Staging Classification

An alternative approach to the surveillance case definition as a method for the clinical staging of HIV infection was proposed by Redfield *et al.* (1986), who designated it as "the Walter Reed staging classification for HIV infection." This is an accurate method of clinically evaluating HIV infection, since the classification of symptoms in lymphadenopathy-associated syndrome/AIDS-related complex according to the CDC would not allow the clinical evolution to be detected in some of the patients (Terragna *et al.*, 1986). Table 1 provides in brief outline form a comparison between the latest CDC surveillance case definition and the Walter Reed staging classification.

TABLE 1. CLASSIFICATION SYSTEMS FOR HIV INFECTIONS

Centers for Disease Control	Walter Reed
Group I. Acute infection	1. (Seropositivity)
Group II. Asymptomatic seropositive	2. Persistent generalized lymphadenopathy
Group III. Persistent lymphadenopathy	3. $CD4^+$ T lymphocytes < 400
Group IV. Other diseases	4. $CD4^+$ T lymphocytes < 400 plus decreased delayed hypersensitivity
Subgroup: A. Constitutional disease	
B. Neurological disease	
C. Secondary infectious disease	5. $CD4^+$ T lymphocytes < 400 plus anergy and thrush
1. Listed in CDC definition of AIDS	
2. Other infections	6. $CD4^+$ T lymphocytes < 400 plus decreased delayed hypersensitivity plus opportunistic infections
D. Secondary cancer	
E. Other conditions	

2. EPIDEMIOLOGIC CONSIDERATIONS

2.1. WORLDWIDE DEMOGRAPHICS

It was estimated by the World Health Organization that in the middle of 1987 there were more than 51,000 cases of AIDS in 112 countries, a figure that may underestimate the true numbers by 50% (Fox, 1987). The number of individuals already infected by HIV worldwide is somewhere between 5 and 10 million, with about half that number in the African continent alone (Dickson, 1987). The incidence in the United States alone at that same time was over 36,000 cases of AIDS reported to the CDC, with nearly 21,000 deaths. It is expected that by 1991 in this country there will be 270,000 cumulative cases of AIDS and 197,000 deaths. Of the cases reported in the United States, about 36% are from New York City. However, AIDS cases have also been recorded in the majority of the states. The cities with the highest incidence of AIDS in the United States include Los Angeles, San Francisco, and Miami, such that 75% of the cases were reported to be residents of New York, New Jersey, California, Florida, and Texas (Selik *et al.*, 1984). Reports of AIDS cases outside the United States have been received from Europe (Gerstoft *et al.*, 1982; Brunet *et al.*, 1983; CDC, 1983b, 1984a,b; Clumeck *et al.*, 1984) and African countries (Anderson *et al.*, 1983; Piot *et al.*, 1984; Van De Perre *et al.*, 1984), as well as Canada (CDC, 1983c; Laveridiere *et al.*, 1983) and Australia (Cooper *et al.*, 1984).

The mortality rate for AIDS seems to be extremely high, but the exact numbers depend on clinical manifestations of the disease and the length of the follow-up (CDC, 1981c, 1982, 1983a; Jaffe *et al.*, 1983b; Rivin *et al.*, 1984). AIDS patients who develop opportunistic infections have the highest death rates. Two-to-three-year follow-up studies of patients diagnosed as having an opportunistic infection suggest that the mortality rate for AIDS with this clinical presentation is 80% or greater, eventually reaching 100%. Although many patients respond to specific therapy for microbial infections, all appear to eventually develop infections that are fatal (Welch *et al.*, 1984). Patients with Kaposi's sarcoma and opportunistic infections show slightly higher mortality than patients with the latter alone. On the other hand, patients with Kaposi's sarcoma alone demonstrate about half the mortality of patients with opportunistic infections, are less immunosuppressed, and have a better prognosis (Rivin *et al.*, 1984). The epidemiologic indices provided here refer almost exclusively to HIV-1. However, it was established in early 1987 that there are at least two distinct types of AIDS virus, designated HIV-1 and HIV-2 (Kanki *et al.*, 1987). The geographic distribution of HIV-2 infections is until now quite different than that of HIV-1. For instance, most cases of HIV-2 have been reported from countries in West Africa, including Senegal, Ivory Coast, Guinea-Bissau, and the Gambia. In contrast, HIV-1 is spread more widely in West, Central, and South Africa, although individuals with AIDS as well as carriers (e.g., seropositive) of HIV-1 have been found with the highest (and still increasing) incidence in discrete regions of Central Africa where infections with HIV-2 are virtually unknown. HIV-2 is primarily confined to West Africa, but the virus is gradually spreading eastward to other parts of Africa as well as into Europe (e.g., France,

West Germany, the United Kingdom, and Sweden) and South America (e.g., Brazil). The prevalence of HIV-2 is between 1 and 10% within the normal adult population, whereas it is between 15 and 65% in prostitutes from West Africa. It is remarkable that the development of AIDS among HIV-2-seropositive individuals is significantly lower than in those infected with HIV-1 (Marlink *et al.*, 1987). Double infections with both types of HIV are relatively infrequent. There are reports of cases that suggest that HIV-2 infections may have occurred in Africa even before the discovery of HIV-1. Ancelle *et al.* (1987) described the onset of AIDS in a heterosexual Portuguese couple after a long incubation period (16–19 years) for HIV-2. It is probable that the husband was infected between 1966 and 1969 through sexual contact in Guinea-Bissau. Saimot *et al.* (1987) reported another case of HIV-2 in a Portuguese man who died with AIDS in March 1980. He had served in Angola from 1968 to 1974. His serum collected in 1979 and stored frozen contained antibodies to the envelope glycoproteins of HIV-2 but not to those of HIV-1. This case strongly suggests that HIV-2 was present in Africa at least in the early 1970s. Of nine other AIDS cases in patients with antibodies to HIV-2 (of both sexes), seven were connected with Cape Verde, Guinea-Bissau, Mozambique, or Portugal. The apparent Africa/Portugal connection warrants further epidemiologic study. Possibly, this background information may also help to explain the spread of HIV-2 to Brazil, while sparing the United States. However, a number of putative HIV-2 isolates from Brazil were later confirmed as HIV-1.

2.2. AIDS RISK GROUPS

Of over 22,000 AIDS patients reported to the CDC by the middle of 1986, approximately 1.5% involved children. About 12,000 of these had died, including 71% of those diagnosed prior to July 1984. The number of cases continues to rise, although the doubling time has decreased from a mean of 6 months through 1983 to about 11 months in late 1985 (Goedert *et al.*, 1986). Homosexual and bisexual men continue to represent the dominant risk category, with 73% of the cases (Jaffe *et al.*, 1983a; CDC, 1984b; Selik *et al.*, 1984), but the susceptible population has expanded to include heterosexual intravenous-drug users (17%), recipients of blood or blood components within 5 years of diagnosis (1.5%), hemophiliacs treated with factor concentrates (0.7%), and the heterosexual partners of AIDS patients or of individuals at increased risk for AIDS (1.0%). The remaining 6.8% represent people born outside the United States, in countries such as Haiti, where many cases have not been linked to specific risk factors (Laurence, 1986).

2.3. MEANS OF TRANSMISSION

AIDS is most often transmitted by sexual contact. Since the first description of AIDS as a new disease entity in previously healthy homosexual men, promiscuous homosexual activity has been recognized as a major risk factor for acquisition of the disease. The risk increases in proportion to the number of sexual encounters with different partners. Particular sexual practices appear to be higher risk factors than others; there is a higher prevalence of AIDS among homosexual men who practice passive or receptive anal intercourse than among those who practice insertive anal intercourse. HIV has been isolated from blood, semen, vaginal secretions, saliva, tears, breast milk, cerebrospinal fluid, amniotic fluid, and urine, and it is likely to be isolated from other body fluids, secretions, and excretions. However, epidemiologic evidence has implicated only blood, semen, vaginal secretions, and possibly breast milk in transmission (CDC, 1987b).

Women may acquire the disease from bisexual men, and children of virus-infected mothers are increasingly being diagnosed. Pediatric AIDS cases, acquired from mothers in high-risk groups, usually present with clinical symptoms by 2 years of age.

Transfusion of infectious blood or blood products is an effective route for virus transmission. Thus, hemophiliacs who received contaminated clotting factor concentrates have been placed at high risk. Over 90% of recipients of factor VIII concentrates in the United States were reported to have developed antibodies to HIV by 1985. Intravenous-drug users are frequently infected through the use of contaminated needles (Jawetz *et al.*, 1987).

3. ETIOLOGIC AGENT

3.1. DESCRIPTION

Lentiviruses are a subgroup of retroviruses responsible for a wide range of diseases in animals (Bolton *et al.*, 1987). HIV, the etiologic agent of AIDS, was recently recognized as a novel member of this lentivirus subgroup (Sonigo *et al.*, 1985). This virus was first isolated in 1983 and called lymphadenopathy-associated virus (LAV) (Barre-Sinoussi *et al.*, 1983). Two other separate isolates of this virus were designated human T-lymphotropic virus type III (HTLV-III) (Popovic *et al.*, 1984) and AIDS-associated retrovirus (ARV) (Levy *et al.*, 1984). Numerous studies of the biologic and molecular characteristics of these three isolates have confirmed that they represent the same infectious agent for which the name HIV was later proposed by an international committee (Coffin *et al.*, 1986). Most recently, another pathogenic human retrovirus, termed HIV type 2 (formerly LAV-2), was recovered from West African patients with AIDS (Clavel *et al.*, 1986a,b). HIV-2 infection can be associated with an immunodeficiency syndrome clinically indistinguishable from that caused by the prototype AIDS virus, HIV-1 (Brun-Vezinet *et al.*, 1987; Clavel *et al.*, 1987). HIV-2 is related to, but distinct from, HIV-1 and from simian immunodeficiency virus (SIV or STLV-III), the etiologic agent of simian AIDS (Clavel *et al.*, 1986b). Extensive serologic cross-reactivity exists between HIV-2 and SIV, though cross-reactivity between HIV-2 and HIV-1 is limited to core antigens (Clavel *et al.*, 1986b; Brun-Vezinet *et al.*, 1987). An independent retroviral isolate called HTLV-IV (Barin *et al.*, 1985;

Kanki *et al.*, 1986, 1987) was found to be antigenically and genetically indistinguishable from HIV-2 (Barin *et al.*, 1985; Clavel *et al.*, 1986b; Kanki *et al.*, 1986).

3.2. MOLECULAR PROFILE

The HIV particle consists of a sphere that measures 100 nm across and is surrounded by a membrane comprised of two layers of lipid material. Studding the membrane are glycoproteins. Each glycoprotein has two components: a gp41, which spans the membrane (Veronese *et al.*, 1985), and a gp120 (Allan *et al.*, 1985), which extends beyond it. Enclosed by the membrane and protein envelope is a core that contains two major proteins, designated p24 and p17. There are slight ambiguities in the designation of the molecular weights of the HIV proteins, depending on the laboratory involved. The molecular weights provided here were based on a consensus from the recent literature. The 70S viral RNA is also present within the core of the enzyme reverse transcriptase, which catalyzes the assembly of the viral DNA (Steimer *et al.*, 1986; Veronese *et al.*, 1986).

The genome of HIV-1 contains at least eight genes (Wong-Staal *et al.*, 1987). In addition to the long terminal repeats and the *gag, pol,* and *env* genes, HIV has several genes that contribute to a system of genetic regulation far more complex than in most retroviruses (Knight *et al.*, 1987). Two of these are the *tat* and *art/trs* genes, both of which are required for viral replication *in vitro* (Dayton *et al.*, 1986; Fisher *et al.*, 1986; Sodroski *et al.*, 1986) and which code for "transactivators" that stimulate expression of other HIV genes (Knight *et al.*, 1987). The other three genes are *sor, R,* and *3'orf,* which have no known function. However, recent studies have suggested that although the *sor* gene is not absolutely required in HIV virion formation, it influences virus transmission *in vitro* and is crucial in the efficient generation of the infectious virus. It also appears that this gene influences virus replication at a novel, posttranslational stage and its action is independent of the regulatory genes *tat* and *trs* (Fisher *et al.*, 1987). A schematic representation of the HIV genome with its pertinent translational products and their location within the virion is shown in Fig. 1.

3.3. RELATIONSHIP TO OTHER RETROVIRUSES

Simian and human retroviruses are believed to have been derived from a common ancestor. Studies of the phylogenetic relationships among the known retroviruses indicate that they followed two evolutionary pathways: one including the transforming retroviruses and the other the cytopathic retroviruses. The former includes HTLV-I and HTLV-II as well as the mammalian and avian type C and bovine leukemia viruses. The latter pathway of evolution had as offshoots the equine infectious anemia, the caprine arthritis encephalitis, and the visna viruses on the one hand and the HIV, STLV-III, and HIV-2 on the other (Ho *et al.*, 1987). As shown in Fig. 2, despite the similarity of genomic organization,

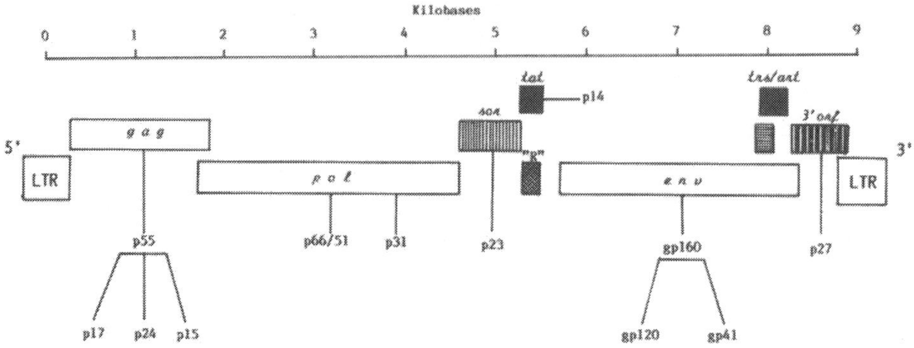

FIGURE 1. The HIV genome. LTR, long terminal repeat; *gag*, group antigen (core proteins); *pol*, polymerase (reverse transcriptase and endonuclease); *sor*, short open reading (functional); "R", region that codes for product of unknown function recognized by the immune system of HIV-infected persons; *tat*, transactivating; *env*, envelope; *trs/art*, posttranscriptional or transcriptional regulators; *3'orf*, open reading frame (functional, downregulates virus replication).

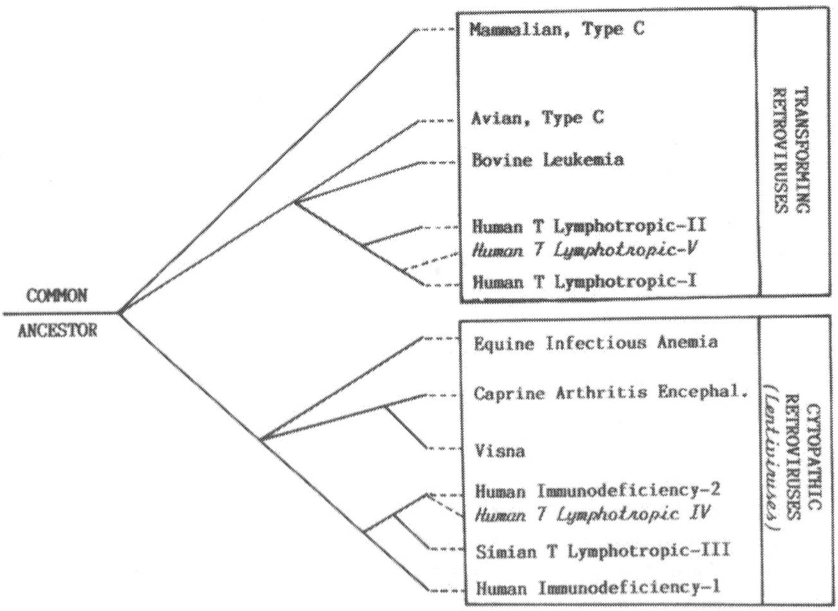

FIGURE 2. The genealogy of HIV-1, HIV-2, and HTLV-IV, outlining their evolutionary relationships with other transforming and cytopathic (e.g., lentiviruses) retroviruses (based on morphology, nucleotide sequence, and protein crossreactivity).

HTLV-I and HIV are only distantly related as determined by comparisons of nucleotide sequence and protein cross-reactivity. In contrast, HTLV-I and HTLV-II are very closely related and thus exhibit extensive serologic cross-reactivity and, unlike HIV, are not cytopathic but transforming viruses.

4. IMMUNOPATHOGENESIS

4.1. HOST–VIRUS INTERACTIONS

The role of HIV-1 as the etiologic agent of AIDS has been determined on the basis of seroepidemiologic studies, the cytopathogenicity of HIV for the $CD4^+$ subset of lymphocytes, the repetitive isolation of HIV from AIDS patients, and the transmission of HIV through infected blood and blood derivatives. In addition, voluminous amounts of data obtained from investigations in the fields of molecular and cell biology, as well as basic and clinical immunology, support such etiologic association. Despite the relatively precipitous advent of this wealth of information about a virus that was unknown only a few years ago, our understanding of the immunopathogenetic mechanisms responsible for the progression from asymptomatic infection to AIDS lags far behind (Klatzman and Gluckman, 1986; Ho *et al.*, 1987). Thus, a number of fundamental questions still remain unanswered concerning the mechanisms of pathogenesis in AIDS. Although HIV produces clinically apparent infection in a much greater number of infected individuals than is the case with other viral agents, for example, poliovirus, Epstein–Barr virus (EBV), herpes simplex, and cytomegalovirus (CMV), still there are a significant number of seropositive healthy carriers of HIV who may be infectious. Epidemiologic studies and observations derived from animal models have revealed that a number of factors (e.g., genetic constitution, congenital or acquired immunologic defects, nutritional imbalance, environmental agents, life-style), relative to the set of circumstances of exposure to the infecting agent, may play an important role in the outcome of the natural infection (Bendinelli and Ceccherini-Nelli, 1988). With particular regard to AIDS, these interactions may apply in view of the frequency of concurrent opportunistic infections as well as the presence of other potential sources of repeated antigenic stimulation or immunomodulation (e.g., sperm, drugs of abuse). There is sufficient evidence from *in vitro* studies that immunologic activation results in viral expression, cytolysis, and the release of large quantities of virus by lymphocytes and monocytes that would otherwise remain latently infected. Such interpretation suggests that opportunistic infections and neoplasia, which signal the onset of full-blown AIDS, might be the sequelae of the immunodeficiency state and/or the precipitating events that lead to the general impairment of the immune system by HIV. The role of a number of viruses as possible cofactors in the development of AIDS has been seriously considered.

Evidence in favor of CMV as a cofactor in immune suppression, in the pathogenesis of *Pneumocystis carinii* pneumonia, and in the development of Kaposi's sarcoma has been reviewed recently by Drew (1987). It has been suggested that CMV infection may be a necessary cofactor for the full expression of infection by HIV. Regarding the cofactor role of CMV in *P. carinii* pneumonia, CMV infection of alveolar macrophages interferes with their ability to interact with *P. carinii* by impairing cell-mediated immune responses. With reference to Kaposi's sarcoma, whether or not CMV is a cofactor in the pathogenesis of AIDS, the evidence is compelling that CMV does contribute to the development of Kaposi's sarcoma. However, definitive proof of an etiologic role for CMV in the genesis of this malignancy requires further investigation and eventually the demonstration of protection by a vaccine.

It has been proposed that EBV may be associated with AIDS, neoplasia, and oral hair leukoplakia (Escobar, 1987). There is little evidence to date that EBV is involved in the marked lymphadenopathy characteristic of ARC patients. Although the role of EBV in the AIDS-associated B-cell lymphoma is still under investigation, the number of reports of EBV-associated neoplasms is increasing steadily. There are indications, both *in vitro* and *in vivo,* that EBV may interact with HIV at least at the level of the B lymphocytes. This interaction(s) may not be necessarily direct, since EBV affects the humoral and HIV the cellular limb of the immune system. In EBV infections, the T suppressor cells are required to maintain the equilibrium between latency and activation. Therefore, perturbations of T suppressor cells resulting from HIV infection could lead to enhanced EBV replication. As the T-cell population diminishes, the effects of EBV-induced B-cell activity become more pronounced. In the final stage of T-cell depression, B-cell dysfunction may lead to lymphoproliferative growth and, most prominently, to EBV-associated neoplasia. The discovery of another B-lymphotropic virus, which is also a member of the herpesvirus group, was recently reported (Salahuddin *et al.*, 1986; Josephs *et al.*, 1986). This virus was called human B-lymphotropic virus (HBLV). Although this virus may also infect HIV-seropositive individuals and AIDS patients, it is not exclusively an AIDS-associated agent (Salahuddin *et al.*, 1986).

The putative etiologic role of hepatitis B virus (HBV) in AIDS was suggested initially, but was dismissed following the discovery of HIV. However, ongoing studies were designed to determine what role, if any, HBV plays as a cofactor in the etiology of AIDS. If it could be found that HBV does indeed play a role, the elimination of blood containing HBV and HBV-like agents should lead to incidental reduction of transfusion-associated AIDS (Escobar, 1987). Prospective studies of chronic and asymptomatic carriers of HBV as well as of immune and healthy individuals, who later become infected with HIV, would be valuable in order to determine if there are any differences among the four groups in terms of their progression from asymptomatic infection to full-blown AIDS. Conversely, one could also ask if HIV infection has any influence on the natural history of HBV infection. Krogsgaard *et al.* (1987) studied 33 male homosexual HBsAg carriers in order to answer this question. The observed HBV DNA clearance rates in the HIV-infected versus those without HIV infection were significantly different. Disease activity, as determined by transaminase levels, was significantly lower in HIV-infected individuals. Infection with HIV appeared to extend the period of active viral replication or even trigger HBV reactivation. It also seemed to diminish inflammatory disease activity in chronic HBsAg carriers.

4.2. TARGETS OF HIV ATTACK

4.2.1. T Helper Lymphocytes

HIV is characteristically a lymphotropic virus. This fact led to its initial identification in lymphocyte cultures and also to its original name. In contrast to HTLV-I and HTLV-II, HIV does not induce malignant transformation of infected cells but produces cytopathic effects in cell culture manifested by either cell lysis or the formation of multinucleated giant cells and syncytia. Consequently, the immediate effect of HIV in humans may consist of the impairment of the immune system by destruction of certain lymphocyte subpopulations. The latter have a central role in immunoregulation so that their dysfunction would of necessity have a universal impact on the immune system. A mechanism to explain the lack of antigen-induced T helper cell responses in AIDS patients, not mutually exclusive with depletion of antigen-reactive cells, is that the latter are viable but functionally impaired (Fauci, 1987). For example, one or more of the following functional defects have been demonstrated in HIV-infected individuals: decreased or absent delayed hypersensitivity responses (Gottlieb *et al.*, 1981; Masur *et al.*, 1981; Mildvan *et al.*, 1982; Schroff *et al.*, 1983), decreased mitogen- and antigen-induced blastogenesis (Fauci *et al.*, 1984; Ciobanu *et al.*, 1983; Gupta and Safai, 1983; Lane *et al.*, 1985), decreased lymphokine production (Ciobanu *et al.*, 1983; Murray *et al.*, 1984), decreased cytotoxicity (Rook *et al.*, 1983), decreased allogenic (Masur *et al.*, 1981; Ammann *et al.*, 1983; Cunningham-Rundles *et al.*, 1983) and autologous (Gupta and Safai, 1983) mixed lymphocyte reactions, decreased T-cell help for antibody production (Benveniste *et al.*, 1983; Lane *et al.*, 1983), and decreased interleukin-2 receptor expression (Munn *et al.*, 1984) and generation (Ciobanu *et al.*, 1983; Kirkpatrick *et al.*, 1985; Murray *et al.*, 1985).

A recent prospective study of patients with ARC found that progression to AIDS was more highly correlated with lack of antigen-induced lymphokine release than with numbers of T helper cells (Murray *et al.*, 1985b). The most likely explanation for this selective loss of antigen specific T-cell function is the relative or absolute depletion of the antigen-reactive T helper–inducer cell population. The latter has been reported to bear the 4B4 antigen (Morimoto *et al.*, 1985a), whereas the complementary suppressor–inducer subset of T helper cells expresses the 2H4 antigen (Morimoto *et al.*, 1985b). Information on whether there is a depletion of $T4^{+}/4B4^{+}$ cells in individuals infected with HIV should be available soon.

Lymphatic tissues may reveal follicular disruption, a decrease of T helper cells in the paracortical area (T zone), and a reactive polyclonal B-cell hyperplasia (Krueger *et al.*, 1985; Racz *et al.*, 1986). Respective lymphocyte changes in the peripheral circulation are similar. Such lesions correlate with deranged antigen processing, insufficient production of specific antibodies and cytotoxic T lymphocytes, as well as with progressive hypergammaglobulinemia (Seyda and Krueger, 1987). This concept received further acceptance when the CD4 antigen, which is present on the surface of T helper cells and monocytes, was shown to function as the receptor for HIV attachment. Notwithstanding, a rather surprising finding was that only very few T lymphocytes from infected individuals (e.g., $1/10^4$–$1/10^5$ T cells) expressed HIV antigen or contained HIV DNA by *in situ* hybridization. The small number of such cells did not coincide with the advanced degree of immunodeficiency and T helper cell loss. That discrepancy prompted, therefore, the search for additional mechanisms responsible for HIV immunopathogenesis.

Among the mechanisms that have been proposed to explain the destruction of T lymphocytes by HIV (Margolick *et al.*, 1987) are (1) direct cytopathic effect, as discussed earlier; (2) selective depletion of a $CD4^+$ T-cell subset or a $CD4^+$ nonlymphoid cell target, which is critical to the propagation of the entire T-cell pool and indirectly leads to the attrition; there appears to exist a good precedent for that because T helper cells and their factors are responsible for many myelopoietic functions and can maintain the integrity of the myeloid series; (3) induction by HIV of soluble substances with toxic effects on T helper cells, a mechanism that is presently under investigation; (4) autoimmune phenomena, although the available data have not demonstrated that this is a major factor; (5) induction of syncytia and cell death in other neighboring T helper cells as a result of a high level of HIV envelope gene expression in infected cells; that is, the infected cell that expresses the envelope antigen binds to the surrounding noninfected cells that possess the CD4 receptor on their surface, hence the formation of the so-called syncytium; or (6) some combination of the mechanisms described.

4.2.2. T Suppressor/Cytotoxic Lymphocytes

There is no evidence that HIV infects T suppressor lymphocytes (Zagury *et al.*, 1986). However, these cells are stimulated by the infectious process. Therefore, one of the earliest manifestations of HIV infection appears to be an increase in the number and percentage of circulating T suppressor cells. This has been found to occur in the mononucleosislike syndrome that characterized the acute phase (i.e., during the period lasting 6–8 weeks after exposure) of HIV infection (Cooper *et al.*, 1985). An increase in the level of circulating T suppressor cells was observed in apparently healthy homosexual men early in the epidemic (Stahl *et al.*, 1983; Kornfeld *et al.*, 1982; Hersh *et al.*, 1984; Fahey *et al.*, 1983) and in asymptomatic HIV-seropositive individuals in other groups at high risk for AIDS (Luben *et al.*, 1983). An increase in the number of cells with the T suppressor phenotype early in infection is a normal response of the immune system. Thus, stimulated by HIV-infected cells, T helper cells induce the differentiation of T cytotoxic cells whose function is to kill virus-infected cells. These T cytotoxic lymphocytes display the same surface markers as T suppressor lymphocytes, so that an increase in cells expressing the surface markers of T suppressor/cytotoxic lymphocytes is recorded (Cooper *et al.*, 1985). Although this type of immune response is also observed after infection with a number of other viruses, such as EBV (DeWaele *et al.*, 1981) and CMV (Carney *et al.*, 1981), in the latter cases the T suppressor lymphocyte count is restabilized within several months after the initial infection. In many HIV-seropositive individuals, both T helper and T suppressor counts remain normal for prolonged periods; however, in those individuals in whom clinical manifestations develop, the number of T suppressor cells, like the number of T helper cells and the total number of lymphocytes, eventually falls dramatically (Zolla-Pazner, 1986).

Functional and phenotypic abnormalities of suppressor–cytotoxic ($T8^+$ or $CD8^+$) cell function in AIDS patients have been reported. The proportion of such cells expressing the HLA-DR and T10 markers has been reported to be increased (Gottlieb *et al.*, 1981; Schroff *et al.*, 1983). This was originally interpreted as suggesting an increased number of activated cells, although more recent evidence has shown that the cells with these markers are immature rather than activated cells. In fact, their presence in the circulation correlates with the degree of depletion of $CD4^+$ T cells (Salazar-Gonzalez *et al.*, 1985). In consonance with this idea, $T8^+$ cell function is decreased in patients with AIDS. For instance, specific *in vitro* killing of CMV-infected target cells by $T8^+$ cells from AIDS patients was markedly decreased compared to controls (Rook *et al.*, 1983). It is remarkable that this functional defect was partially reversed by the addition of IL-2 to the cultures, demonstrating that cytotoxic functional potential was present in the cells from AIDS patients but required the presence of an inductive signal normally provided by the $CD4^+$ T cells. It is also of interest that similar results were obtained with natural killer cells (Rook *et al.*, 1983). This finding is described in further detail in Section 4.2.4.

4.2.3. B Lymphocytes

Hypergammaglobulinemia was one of the first effects of HIV on the immune system of AIDS patients that was described (Friedman-Kien *et al.*, 1982; Masur *et al.*, 1981; Siegel *et al.*, 1981; Gottlieb *et al.*, 1981). Specifically, IgG, IgA, and IgM were later reported to be two to three times the normal levels (El-Sadr *et al.*, 1984) and IgD was found to be increased up to 10 times (Chess *et al.*, 1984; Zolla-Pazner *et al.*, 1986). These findings appeared to be inconsistent with the destruction of the T helper cell subset by HIV, which is normally required for the production of antibodies to T-cell-dependent antigens. On the other hand, this hypergammaglobulinemia could have been explained by the presence of immune complexes and autoantibodies in those AIDS patients (McDougal *et al.*, 1985; Williams *et al.*, 1984; Tomar *et al.*, 1985; Walsh *et al.*, 1984) reflecting an intense polyclonal activation of their B cells (Lane *et al.*, 1983; Pahwa *et al.*, 1984). Furthermore, it was also determined subsequently that B-cell function was quite abnormal (Stahl *et al.*, 1982) and that this abnormality appeared to antedate the impairment of the T-cell proliferative response. Later studies (Lane *et al.*, 1985) revealed that B-cell responses to soluble antigens were more profoundly suppressed than was the response to B-cell mitogens (e.g., pokeweed) and that the proliferative response of the AIDS patients' B cells to pokeweed mitogen was more severely impaired than was the response of T cells to PHA (Stahl *et al.*, 1982).

The mechanisms responsible for this intense state of B-cell activation and the scenario of events observed in AIDS patients are still unclear. Nonetheless, many possible mechanisms have been considered, including (1) induction or transformation of B cells by B lymphotropic viruses, such as EBV or HBLV, both of which can be detected frequently in AIDS patients and which induce hypersecretion of immunoglobulins by B cells *in vitro;* (2) a direct effect of HIV on B cells (Montagnier *et al.*, 1984); HIV can infect B cells under certain conditions (e.g., following transformation by EBV); (3) absence of normal regulation of B cells by functional T cells, with special reference to newly encountered antigens (memory B cells are capable of mounting a secondary response to newly encountered antigens without the assistance of T helper cells); (4) excessive secretion of B-cell–stimulating factors by virally transformed B cells (Salahuddin *et al.*, 1984), such as with HTLV-I; or (5) some combination of these mechanisms. That at least some of these mechanisms may operate in AIDS has been suggested by the results of a number of studies (Lane *et al.*, 1985; Pahwa *et al.*, 1985; Schnittman *et al.*, 1986; Ammann *et al.*, 1984). In any case, one of the practical implications of these studies regarding the interpretation of serologic test results is that the use of the conventional serologic criteria for the diagnosis of infection may be unreliable when dealing with AIDS patients.

Whatever the mechanisms, it has been established that B-cell activation plays an important role in many of the clinical manifestations of AIDS. The B-cell lymphomas, seen at increased frequency in AIDS patients (Birx *et al.*, 1986; Ziegler *et al.*, 1984), appear in part to result from the increased EBV activity and from the uncontrolled activity of the B-cell population.

4.2.4. Antigen-Presenting Cells

The first step in the induction of an adaptive, cell-mediated immune response is antigen processing and presentation by dendritic cells and macrophages to T cells (Mackaness, 1970; Steinman and Nussenweig, 1980). Langerhans cells (LC) seem to be closely related to dendritic cells and function as antigen-presenting cells in the skin (Silberberg-Sinakin *et al.*, 1980). Since a deficiency of antigen-presenting cells might also be central to an acquired cell-mediated immune deficiency, this issue was also addressed by a number of investigators (Belisto *et al.*, 1984). Belisto *et al.* (1984) demonstrated that AIDS patients with opportunistic infections had a severe deficiency of Ia^+ LC in skin biopsy specimens. AIDS patients with Kaposi's sarcoma had less severely diminished numbers. The marked deficiency detected in AIDS patients with opportunistic infections was also seen in individuals with ARC and, therefore, at high risk of developing an opportunistic infection. Tschachler *et al.* (1987) demonstrated by electron microscopy in one of several biopsies of HIV-infected patients the presence of LC-associated viral particles characteristic of HIV. They also detected cytopathic changes in approximately one-third of the LC population. These results strongly suggest the LC may harbor HIV. The infection of these cells with HIV may have deleterious consequences for the immunologic functions of this antigen-presenting cell system and thus may contribute to both the acquisition of immunodeficiency and the infectious and neoplastic complications of AIDS.

Other studies have suggested a functional deficiency of antigen-presenting cells in AIDS patients. Belisto *et al.* (1984) found a functional deficiency as well as subnormal numbers of antigen-presenting cells in AIDS patients. They performed experiments with cells from HLA-identical, MLC-nonreactive siblings, one with and the other without AIDS. The mononuclear cells of the AIDS patient were unable to present antigen to his own or his sibling's T cells. However, the sibling's mononuclear cells functioned normally with the T cells from either subject.

4.2.5. Mononuclear Phagocytes

It has been shown that monocytes and macrophages (Armstrong *et al.*, 1985) as well as macrophage cell lines (Levy *et al.*, 1985) can be infected by HIV. In various functional assays, these cells from AIDS patients display a diversity of activities that range from normal to abnormal (Zolla-Pazner, 1986). For example, Murray and co-workers (1984) found several physiologic functions of monocytes from AIDS patients to be within normal limits. These included oxidative burst, antimicrobial activity, and phagocytosis. In addition, stimulation of these cells with a mixture of lymphokines, or with either partially purified or recombinant interferon-γ, resulted in normal expression of the activated state as measured by augmented hydrogen peroxide release and enhanced oxygen-dependent and oxygen-independent antimicrobial activity.

On the other hand, other investigators have reported that although AIDS patients generally have normal numbers of circulating monocytes, these cells display a number of *in vitro* functional abnormalities. Prince *et al.* (1985) found a defect in monocytes of lymphadenopathy syndrome (LAS) patients leading to a poor response in a monocyte-dependent T-cell proliferation assay. As suggested earlier, many of these functional defects appear to be due to a lack of interferon-γ. Thus they can be reversed by the addition of this lymphokine to the cultures. For instance, defective extracellular killing of *Toxoplasma gondii in vitro* by monocytes from AIDS patients was enhanced by the addition of interferon-γ to the cultures (Murray *et al.*, 1984). AIDS monocytes express lower levels of class II HLA antigens than normal monocytes (Belisto *et al.*, 1984; Heagy *et al.*, 1984; Murray *et al.*, 1985a) and this can also be reversed by exogenous interferon-γ (Heagy *et al.*, 1984; Murray *et al.*, 1985a). Moreover, monocytes from AIDS patients increase hydrogen peroxide release and cytotoxicity normally in response to interferon-γ (Murray *et al.*, 1984). These data also suggest that functional abnormalities of mononuclear phagocytes from AIDS patients are a result, at least in part, of a lack of inductive signals from helper/inducer T lymphocytes. Whether this defect accounts for other *in vitro* and *in vivo* functional abnormalities of these cells, such as spontaneous secretion of interleukin-1 and prostaglandin E2 (Smith *et al.*, 1984), diminished response to inducers of interleukin-1 release, diminished chemotactic responses (Smith *et al.*, 1984), and impaired Fc receptor–mediated clearance (Bender *et al.*, 1985), remains to be elucidated. Other possibilities are a lack of regulatory signals other than interferon-γ from T cells, direct HIV infection of mononuclear phagocytes, as well as secondary effects from opportunistic infections in AIDS patients.

One of the difficulties in the interpretation of the data described here is the uncertainty of whether the different assays used in the various studies are actually measuring different subsets of mononuclear phagocytes at different stages of activation or whether, alternatively, various functions of these cells are affected differently by the infectious process (Zolla-Pazner, 1986).

The immunopathogenetic mechanisms involving monocytes/macrophages in AIDS patients have eluded us for some time, but are becoming gradually more clear. As stated earlier, HIV can infect monocytes without killing them, perhaps very early during infection (Levy *et al.*, 1985; Armstrong *et al.*, 1985). Like T helper cells, monocytes/macrophages, including dendritic reticular cells of the lymphoid follicle and, apparently, certain thymic stromal cells (Seyda and Krueger, 1987), as well as glial cells in the brain (Koenig *et al.*, 1986), can also express the CD4 receptor and CD4 messenger RNA. An important question is why is HIV cytocidal for T helper cells, but not for mononuclear and antigen-presenting cells? It now appears that at least monocytes play an important role as the reservoir of infectious virus (Rosenberg and Fauci, 1988). The difficulty in obtaining pure populations of monocytes, without contaminating T helper cells, was overcome by Fauci's group at the National Institutes of Health (Folks *et al.*, 1987). They infected a promonocyte cell line, strain U937, which was subsequently cloned and subcloned repeatedly to produce a chronically HIV-infected cell line designated U1. This U1 cell line could not produce spontaneously any virus in detectable amounts. However, exposure of these cells to a PHA supernatant (medium alone) brought about a remarkable up-regulation of virus expression. Antibodies to interleukin-1β completely obviated the induction capability of the supernatant, whereas antibodies to interleukin-1α, tumor necrosis factor-α, lymphotoxin, and normal rabbit serum had no effect. Although interleukin-1β, by itself, did not induce up-regulation of virus expression, it was shown that there were other factors that induced a membrane-bound interleukin-1β that was then responsible for this up-regulation of virus expression. The biologic relevance of this finding is that one does not necessarily have to have infection of the cell. Thus, an activation signal—whatever that may be (e.g., another microbe or a cytokine, such as interleukin-1)—can induce up-regulation of virus expression of infected monocytes which are intimately interacting with T helper cells, since these are monocyte-dependent responses. Consequently, the T helper cell becomes secondarily infected and, unlike the monocyte, develops the cytopathic effect.

4.2.6. Natural Killer Cells

Generally, it has been determined that natural killer cell activity is significantly impaired in AIDS patients (Rook *et al.*, 1983; Gold *et al.*, 1985), although it has also been found to be normal in a small group of AIDS patients (Schroff *et al.*, 1983). *In vitro*, natural killer cell activity could be completely restored by a 72-hr pretreatment with interleukin-2. This seems to suggest that the well-documented defect in interleukin-2 production by T cells from infected patients (Hersh *et al.*, 1984; Lane *et al.*, 1985) may play a particularly crucial role in the natural killer cell defect observed in AIDS (Rook *et al.*, 1983; Zolla-Pazner, 1986).

5. IMMUNOLOGIC MARKERS OF HIV INFECTION AND THEIR EVALUATION

5.1. ANTIGEN DETECTION

HIV has been isolated from cells of a large number of AIDS and ARC patients at all stages of the disease and from some clinically normal individuals (Schupbach *et al.*, 1986). The virus was isolated from several tissue sources including peripheral blood and bone marrow mononuclear cells, lymph nodes, saliva, urine, semen, cerebrospinal fluid, neural tissue, and plasma. Isolation studies initiated in 1982 and procedures established for the isolation of HTLV-I and HTLV-II were modified for the isolation of HIV (Popovic *et al.*, 1984). Virus was isolated directly from peripheral blood and bone marrow mononuclear cells and from lymph node and brain cells established in culture. However, its isolation from saliva and plasma was achieved only by transmission or cocultivation of lymphocytes with mitogen-stimulated normal human lymphocytes (pretreated with polybrene or DEAE dextran) or permissive T-cell lines, like subclones of the human T-cell line HT, which is resistant to the cytopathic effect of HIV and is a good virus producer after infection. Various criteria have been utilized to identify HIV isolates, such as release of particulate reverse transcriptase activity, visualization of viral particles by electron microscopy, indirect immunofluorescence for HIV proteins using monoclonal antibodies, or transmission of virus to cell cultures resulting in both cytopathic effect and virus release. Unfortunately, these methods are labor- and instrument-intensive and demand considerable expertise.

Recent technical innovation has provided simpler procedures than those described earlier. They are based on immunoassay methodology (using either enzyme or radioisotope labels) not only for detection of HIV antigen, but also for detection of the HIV genome itself (i.e., using molecular probes). Several assay kits are currently available commercially for research purposes that detect the presence of HIV protein antigen (p24). At least two of these assays are designed to detect HIV antigen in human serum or plasma, as well as cell cultures. However, these immunoassays detect solubilized viral proteins and not intact, viable virus. Increases, therefore, in the quantity of antigen detected may indicate, but not prove, an increase in viral replication. Positive HIV culture is still the only procedure that can demonstrate definitively the presence of intact virus in the patient specimen capable of replication (Lombardo, 1987). The prevalence of HIV antigen is high in culture-positive samples. Very few HIV antigen–positive sera are culture negative. The presence of HIV antigen, as detected by immunoassay methods, most likely indicates active viral replication. On the other hand, lymphocyte culture may reflect both replicating HIV and its proviral form (Dittel *et al.*, 1987). The clinical situations where HIV antigen detection is useful are (1) in early diagnosis of HIV infection during the first 2 weeks to 3 months after exposure (Allain *et al.*, 1986; Kessler *et al.*, 1987); (2) in the diagnosis of HIV infection in individuals with common variable immunodeficiency; (3) in young infants whose mothers are infected with HIV, since maternal antibody usually confuses the interpretation of serologic test results; (4) in the prognosis of HIV-infected individuals, since the disappearance of the antibody to the p24 core antigen within 18 months prior to the reappearance of HIV antigen indicates progression to AIDS (Allain *et al.*, 1987); and (5) in monitoring therapy with certain antiviral drugs (Chaisson *et al.*, 1987).

5.2. ANTIBODY DETECTION

Serologic tests for the detection of antibodies to the core and envelope proteins of HIV are the mainstay of the immunodiagnosis of HIV infection, even though antigen detection is critical in the clinical situations indicated in Section 5.1. It should be stressed, however, that these tests were, and still are, only licensed to screen blood products for the presence of HIV antibodies. They are not meant to screen patients or populations for AIDS. AIDS is a diagnosis made only when specific clinical criteria have been fulfilled (see Section 1.2). The presence of HIV antibodies alone is insufficient to diagnose AIDS.

5.2.1. Conventional Tests

At this time there are seven commercial sources of kits for the detection of antibodies to HIV (Abbott, DuPont, Electro-Nucleonics, Genetics Systems, Organon Teknika, Ortho Diagnostics, and Wellcome). Five of these made their appearance for the first time in the spring of 1985, and only one (Wellcome) is based on a competitive binding assay. All derive their source of antigen from the American strain of HIV-1 (HTLV-III) propagated in the T-lymphocyte cell line H9, with two exceptions: (1) Genetics System and Wellcome use the CEM cell line to grow the virus, and (2) Genetics System utilizes the French strain of HIV-1 (LAV) as the antigen source. These tests vary to different degrees from one another in terms of sensitivity and specificity.

5.2.2. Second-Generation Tests

A major problem with the conventional assays was that, because the antigen was derived from HIV grown in human lymphocyte cultures, they produced a significant number of false positive results. This was often the case in multiparous women who developed antibodies to the HLA antigens of their husbands to whom they were immunized through pregnancy. The second-generation tests were thus made available in 1986 especially to resolve this problem. These new methods are competitive immunoassays using recombinant DNA antigens (CIA-RA) where patient antibodies directed to the specific antigen (either p24 core or gp41 envelope) compete with labeled polyclonal antibodies.

5.2.3. Tests under Evaluation

Other immunoassays are being developed, but none are yet approved for the detection of HIV antibodies in donor plasma or serum. To date, there are the following commercial sources: (1) Chiron strip enzyme-linked immunosorbent assay (ELISA), which uses the ARV strain of HIV-1; (2) Cambridge Diagnostics latex agglutination test sensitized with recombinant DNA antigen; and (3) Fujirebion Serodia-HIV, which utilizes a gelatin particle carrier sensitized with inactivated HIV-1 antigen.

Regardless of their Food and Drug Administration status, none of the conventional, second-generation, or newer tests listed here are adequate to screen for antibodies directed against HIV-2. It is expected that tests for HIV-2 antibodies will also become available soon in anticipation of the penetration of HIV-2 into the United States.

5.2.4. Supplementary Tests

Grave concerns have been expressed over the erroneous labeling of a false positive specimen as a true positive specimen. This problem led, in part, to the need for further confirmation of a positive ELISA result by (1) repeating the ELISA test in duplicate and (2) testing those specimens that repeat positive by methods that are more qualitative in detecting the presence of antibodies but still use HIV-1 antigens derived from lymphocyte culture, e.g., Western blot or immunofluorescence (CDC, 1986). A new *in vitro* competitive-binding ELISA is already commercially available (Abbott) which uses recombinant DNA antigens on two separate beads, one coated with gp41 and gp120 and the other with p24 and p55. The test is performed in duplicate, and either the envelope or core antigen-coated bead, or both, must be reactive in duplicate before the test result is considered conclusively reactive. This test may perhaps be proven to be adequate as a confirmatory test and replace the more cumbersome and expensive Western blot. A more detailed and recent discussion on the interpretation of HIV test results is available (Hausler, 1987).

5.3. OTHER IMMUNOLOGIC PARAMETERS

Serologic assays for the detection of antibodies to HIV-1 can generally determine only exposure to HIV, but do not provide information about the development of AIDS or the extent of the disease. A number of bioassays have been proposed as useful markers for disease progression. These, however, are expensive tests that are not readily available outside a research setting. Nonetheless, in addition to the *in vitro* tests described earlier, certain assays are used by the majority of those involved in the management of AIDS, including a complete blood cell count with a differential white-blood-cell-count; enumeration of total lymphocytes, of circulating T helper (CD4) and T suppressor (CD8) cells and their ratio; and quantitation of serum immunoglobulins and β-$_2$ microglobulin.

6. CONCLUDING REMARKS

Physicians and scientists engaged in the diagnosis and management of patients with LAS/ARC/AIDS and Kaposi's sarcoma have become all too familiar with the relentless progression of these disorders. Epidemiologists around the world are spending more and more of their valuable time to contain the spread of HIV infection. Researchers of many countries are eagerly drawn into ways to gain a better understanding of the mechanisms of this disease. Although major advances have been made in terms of prevention and treatment, their effectiveness has yet to catch up with their success in slowing down the rate of HIV infection or improving survival. The therapeutic shortcomings of contemporary medicine, however, should not be viewed with frustration or pessimism. In only 6 years, medicine and biotechnology working side by side have permitted the identification and classification of AIDS as an entirely new disorder, elucidation of its pathophysiology, and the discovery of HIV as its etiological agent. Progress should continue uninterrupted, like our victory over smallpox, to conquer one of the largest threats to humankind.

REFERENCES

Allain, J-P., Laurian, Y., Paul, D. A., and Senn, D., 1986, Serological markers in early stages of HIV infection in hemophilia, *Lancet* **2:**1233–1236.

Allain, J-P., Laurian, Y., Paul, D. P., Verroust, F., Leather, M., Gazengel, C., Senn, D., Larrieu, M-J., and Bosser, C., 1987, Long-term evaluation of HIV antigen and antibodies to p24 and gp41 in patients with hemophilia: Potential clinical importance, *N. Engl. J. Med.* **317:**1114–1121.

Allan, J. S., Coligan, J. E., Barin, F., McLane, M. F., Sodroski, J. G., Rosen, C., Haseltine, W. A., Lee, T. H., and Essex, M., 1985, Major glycoprotein antigens that induce antibodies in AIDS patients are encoded by HIV, *Science* **228:**1091–1094.

Ammann, A. J., Abrams, D., Conant, M., Dhudwin, D., Cowan, M., Volberding, P., Lewis, B., and Casavant, C., 1983. Acquired immune dysfunction in homosexual men: Immunologic profiles, *Clin. Immunol. Immunopathol.* **27:**315–325.

Ammann, A. J., Schiffman, G., Abrams, D., Voldberding, P., Ziegler, J., and Conant, M., 1984, B-cell immunodeficiency in acquired immune deficiency syndrome, *JAMA* **251:**1447–1449.

Ancelle, R., Bletry, O., Baglin, A. C., Brun-Vezinet, F., Rey, M. A., and Godeau, P., 1987, Long incubation period for HIV-2 infection, *Lancet* **1**:688–689.

Anderson, R., Prozesky, O. W., Eftychis, H. A., van der Merwe, M. F., Swanevelder, C., and Simson, I. W., 1983, Immunological abnormalities in South African homosexual men, S. A. *Mediese Tydskrif.* **64**:119.

Armstrong, J. A., Dawkins, R. L., and Horne, R., 1985, Retroviral infection of accessory cells and the immunological paradox in AIDS, *Immunol. Today* **6**:121–122.

Barin, F., Denis, F., Allan, J. S., M'Boup, S., Kanki. P., Lee, T. H., and Essex, M., 1985, Serological evidence for virus related to simian T-lymphotropic retrovirus III in residents of West Africa, *Lancet* **2**:1387–1389.

Barre-Sinoussi, F., Chermann, J. C., Rey, F., Nugeyre, M. T., Chamaret, S., Gruest, J., Dauguet, C., Axter-Blin, C., Brun-Vezinet, F., Rouzioux, C., Rozenbaum, W., and Montagnier, L., 1983, Isolation of a T lymphotropic retrovirus from a patient at risk for acquired immune deficiency syndrome (AIDS), *Science* **220**:868–871.

Belisto, C. V., Sanchez, M. R., Baer, R. L., Valentine, R., and Thorbecke, G. J., 1984, Reduced Langerhans' cell Ia antigen and ATPase activity in patients with the acquired immunodeficiency syndrome, *N. Engl. J. Med.* **310**:1279–1282.

Bender, B. S., Frank, M. M., Lawley, T. J., Smith, W. J., Brickman, C. M., and Quinn, T. C., 1985, Defective reticuloendothelial system Fc-receptor function in patients with acquired immunodeficiency syndrome, *J. Infect. Dis.* **152**:409–412.

Bendinelli, M., and Ceccherini-Nelli, L., 1988, Mechanisms of retrovirus pathogenicity, *Clin. Immunol. Newsletter*, **9**(4): 57–60.

Benveniste, E., Schroff, R., Stevens, R. H., and Gottlieb, M. S., 1983, Immunoregulatory T cells in men with a new acquired immunodeficiency syndrome, *J. Clin. Immunol.* **3**:359–367.

Birx, D. L., Redfield, R. R., and Tosato, G., 1986, Defective regulation of Epstein-Barr virus infection in patients with acquired immunodeficiency syndrome (AIDS) or AIDS-related disorders, *N. Engl. J. Med.* **314**:874–879.

Bolton, V., Pedersen, N. C., Higgins, J., Jennings, M., and Carlson, J., 1987, Unique p24 epitope marker to identify multiple human immunodeficiency virus variants in blood from the same individuals, *J. Clin. Microbiol.* **8**:1411–1415.

Brunet, J. B., Bouvet, E., Leibowitch, J., Chaperon, J., Mayaud, C., Gluckman, J. C., Picard, O., Kernbaum, S., Revuz, J., Klatzmann, D., Rozenbaum, W., Lachiver, D., Villalonga, J., and Wesselberg, C., 1983, Acquired immunodeficiency syndrome in France, *Lancet* **1**:700–701.

Brun-Vezinet, F., Kaltama, C., Roulot, D., Lenoble, L., Alizon, M., Madgar, J. J., Rey, M. A., Girard, P. M., Yeni, P., Clavel, F., Gadelle, S., and Harzic, M., 1987, Lymphadenopathy-associated virus type-2 in AIDS and AIDS-related complex: Clinical and virological features in four patients, *Lancet* **1**:128–132.

Carney, W. P., Rubin, R. H., Hoffman, R. A., Hansen, W. P., Healey, K., and Hirsh, M. S., 1981, Analysis of T lymphocyte subsets in cytomegalovirus mononucleosis, *J. Immunol.* **126**:2114–2116.

CDC, 1981a, *Pneumocystis* pneumonia—Los Angeles, *MMWR* **30**:250.

CDC, 1981b, Kaposi's sarcoma and *Pneumocystis* pneumonia among homosexual men—New York City and California, *MMWR* **30**:305.

CDC, 1981c, Follow-up on Kaposi's sarcoma and *Pneumocystis* pneumonia, *MMWR* **30**:409.

CDC, 1982, Update on acquired immune deficiency syndrome (AIDS)—United States, *MMWR* **31**:507–514.

CDC, 1983a, Update: Acquired immunodeficiency syndrome (AIDS)—United States, *MMWR* **32**:389.

CDC, 1983b, Acquired immunodeficiency syndrome (AIDS)—Europe, *MMWR* **32**:610.

CDC, 1983c, Acquired immunodeficiency syndrome (AIDS)—Canada, *MMWR* **32**:635.

CDC, 1984a, Update: Acquired immunodeficiency syndrome (AIDS)—Europe, *MMWR* **33**:607.

CDC, 1984b, Update: Acquired immunodeficiency syndrome (AIDS)—United States, *MMWR* **33**:661.

CDC, 1985, Revision of the case definition of acquired immunodeficiency syndrome for national reporting—United States, *MMWR* **34**:373–375.

CDC, 1986, Enzyme-Linked Immunoelectrotransfer Blot Technique (Western Blot) for Human T Lymphotropic Virus Type III/Lymphadenopathy-Associated Virus *HTLV-III/LAV Antibodies),* U.S. Dept. of Health and Human Serv., Washington, DC.

CDC, 1987a, Revision of the CDC surveillance case definition for acquired immunodeficiency syndrome, *MMWR* (Suppl. No. 1S):3S–15S.

CDC, 1987b, Recommendations for prevention of HIV transmission in health-care settings, *MMWR* (Suppl. No. 2S):35–185.

Chaisson, R. E., Allain, J. P., Leuther, M., Parks, W., Lehrman, S., and Volberding, P., 1987, Decline in serum HIV p24 antigen (Ag) in patients treated with AZT, III International Conference on AIDS, Washington, DC, Abstract #T.8.1, p. 58.

Chess, Q., Daniels, J., North, E., *et al.*, 1984, Serum immunoglobulin elevations in the acquired immunodeficiency syndrome (AIDS): IgG, IgA, IgM and IgD, *Diag. Immunol.* **2:**148–153.

Ciobanu, N., Welk, K., Kruger, G., Venuta, S., Gold, J., Feldman, S., Wang, C. Y., Koziner, B., Moore, M. A. S., Safai, B., and Mertelsmann, R., 1983, Defective T-cell responses to PHA and mitogenic monoclonal antibodies in male homosexuals with acquired immune deficiency syndrome and its *in vitro* correction by interleukin-2, *J. Clin. Immunol.* **3:**332–340.

Clavel, F., Brun-Vezinet, F., Guetard, D., Chamaret, S., Laurant, A., Rouzioux, C., Rey, M., Katlama, C., Rey, F., Champelinaud, J. L., and Montagnier, L., 1986a, LAV type II: A second retrovirus associated with AIDS in West Africa, *C.R. Acad. Sci.* (III) **302:**485–488.

Clavel, F., Guetard, D., Brun-Vezinet, F., Chamaret, S., Rey, M-A., Santos-Ferreira, M. O., Lauret, A. G., Dauguet, C., Katlama, C., Rouzioux, C., Klatzman, D., Champalimaud, J. L., and Montagnier, L., 1986b, Isolation of a new human retrovirus from West African patients with AIDS, *Science* **233:**343–346.

Clavel, F., Mansinho, K., Chamaret, S., Guetard, D., Favier, V., Nina, J., Santos-Ferreira, M-O., Champalimaud, J-L., and Montagnier, L., 1987, Human immunodeficiency virus type 2 infection associated with AIDS in West Africa, *N. Engl. J. Med.* **316:**1180–1185.

Clumeck, N., Sonnet, J., Taelman, H., Mascart-Lemone, F., De Bruyere, M., Vandeperre, P., Dasnoy, J., Marcelis, L., Lamy, M., Jonas, C., Eyckmann, L., Noel, H., Vanhaeverbeek, M., and Butzler, J. P., 1984, Acquired immunodeficiency syndrome in African patients. *N. Engl. J. Med.* **310:**492–497.

Coffin, J., Hause, A., Levy, J. A., Montagnier, L., Orozlan, S., Teich, N., Temin, H., Toyoshima, K., Varmus, H., Vogt, P., and Weiss, R., 1986, Human immunodeficiency viruses, *Science* **232:**697.

Colebunders, R. H., Francis, H., Izaley, L., Kabasele, K., Nzilambi, N., Van der Groen, G., Vercauteren, B., Mann, J. M., Bila, K., Kakonde, N., Ifoto, L., Quinn, T. C., Curran, J. W., and Piot, P., 1987, Evaluation of a clinical case definition of acquired immunodeficiency syndrome in Africa, *Lancet* **1:**492–494.

Cooper, D. A., Gold, J., May, W., Kaminsky, L. S., Penny, R., and Levy, J. A., 1984, Contact tracing in the acquired immune deficiency syndrome (AIDS). Evidence for transmission of virus and disease by an asymptomatic carrier, *Med. J. Aust.* **141:**579.

Cooper, D. A., Maclean, P., Finlayson, R., Michelmore, H. M., Gold, J., Donovan, B., Barnes, T. G., Brooke, P., and Penny, R., 1985, Acute AIDS retrovirus infection. Definition of a clinical illness associated with seroconversion, *Lancet* **1:**537–540.

Cunningham-Rundles, S., Michelis, M. A., and Masur, H., 1983, Serum suppression of lymphocyte activation *in vitro* in acquired immunodeficiency disease, *J. Clin. Immunol.* **3:**156–165.

Dayton, A. I., Sodroski, J. G., Rosen, C. A., Goh, W. C., and Haseltine, W. A., 1986, The *trans*-activator gene of the human T cell lymphotropic virus type III is required for replication, *Cell* **44:**941–947.

DeWaele, M., Theilemans, C., and Van Camp, B. K. G., 1981, Characterization of immunoregulatory T cells in Epstein-Barr virus-induced infectious mononucleosis by monoclonal antibodies, *N. Engl. J. Med.* **304:**460–462.

Dickson, D., 1987, Africa begins to face up to AIDS, *Science* **238:**605–607.

Dittel, B., Falk, L., Paul, D., Spear, J., Kessler, H., and Landay, A., 1987, Correlation of serum HIV antigen detection with isolation of HIV from patients with AIDS and patients at risk for AIDS, III International Conference on AIDS, Washington, DC, Abstract #WP.9, p. 111.

Drew, W. L., 1987, Is cytomegalovirus a cofactor in AIDS? *Clin. Immunol. Newsletter* **8**(8):117–120.

El-Sadr, W., Stahl, R., Sidhu, G., *et al.*, 1984, The acquired immunodeficiency syndrome: Laboratory findings, clinical features, and leading hypotheses, *Diag. Immunol.* **2**:73–85.
Escobar, M. R., 1987, AIDS: Current status and research perspectives, *Clin. Immunol. Newsletter* **8**(6):91–94.
Fahey, J. L., Detels, R., and Gottlieb, M., 1983, Immune augmentation (with altered T-subset ratio) is common in healthy homosexual men, *N. Engl. J. Med.* **308**:842–843.
Fauci, A. S., 1987, AIDS: Immunopathogenic mechanisms and research strategies, *Clin. Res.* **35**:503–510.
Fauci, A. S., Macher, A. M., Longo, D. I., Lane, H. C., Rook, A. H., Masur, H., and Gelmann, E. P., 1984, Acquired immunodeficiency syndrome: Epidemiologic, clinical, immunologic, and therapeutic considerations, *Ann. Intern. Med.* **100**:92–106.
Fisher, A. G., Feinberg, M. B., Joseph, S. F., Harper, M. E., Marselle, L. M., Reyes, G., Gonda, M. A., Adovini, A., Debouk, C., Gallo, R. C., and Wong-Staal, F., 1986, The *trans*-activator gene of HTLV-III is essential for virus replication, *Nature*·**320**:367–371.
Fisher, A. G., Ensoli, B., Ivanoff, L., Chamberlain, M., Petteway, S., Ratner, L., Gallo, R. C., and Wong-Staal, F., 1987, The *sor* gene of HIV-1 is required for efficient virus transmission *in vitro*, *Science* **237**:888–893.
Folks, T. M., Justement, J., Kinter, A., Dinarello, C. A., and Fauci, A. S., 1987, Cytokine-induced expression of HIV in a chronically infected promonocyte cell line, *Science* **238**:800–802.
Fox, J. L., 1987, AIDS: The public health–public policy crisis, *ASM News* **58**(8):426–430.
Friedman-Kien, A. E., Laubenstein, L. J., Rubenstein, P., Buimovici-Klein, E., Marmor, M., Stahl, R., Spigland, I., Kim, K. S., and Zolla-Pazner, S., 1982, Kaposi's sarcoma in homosexual men, *Ann. Intern. Med.* **96**:693–700.
Gerstoft, J., Malchow-Moller, A., Bygbjerg, I., Dickmeiss, E., Enk, C., Halberg, P., Haahr, S., Jacobsen, M., Jensen, J., Mejer, J., Neilsen, J. O., Thomsen, H. K., Sondergaard, J., and Lorenzen, I., 1982, Severe acquired immunodeficiency in European homosexual men. *Br. Med. J.* **285**:17–19.
Goedert, J. J., Biggar, R. J., Weiss, S. H., Eyster, M. E., Melbye, M., Wilson, S., Ginzburg, H. M., Grossman, R. J., DiGiola, R. A., Sanchez, W. C., Giron, J. A., Ebbesen, P., Gallo, R. C., and Blattner, W. A., 1986, Three-year incidence of AIDS in five cohorts of HTLV-III-infected risk group members, *Science* **231**:992–995.
Gold, J. W. M., Weike, C. S., Godbold, J., Garcia, C., Urmacher, C., Cunningham-Rundles, S., Koziner, B., Pollack, M., Gallo, R. C., Sarngadharan, M. G., and Armstrong, D., 1985, Unexplained persistent lymphadenopathy in homosexual men and the acquired immune deficiency syndrome, *Medicine* **64**:203–213.
Gottlieb, M. S., Schroff, R., Schander, H. M., Weisman, D. O., Fan, P. T., Wolf, R. A., and Saxon, A., 1981, *Pneumocystis carinii* pneumonia and mucosal candidiasis in previously healthy homosexual men: Evidence for a new acquired cellular immunodeficiency, *N. Engl. J. Med.* **305**:1425–1431.
Gupta, S., and Safai, B., 1983, Deficient autologous mixed lymphocyte reaction in Kaposi's sarcoma associated with deficiency of Leu-3 positive responder cells, *J. Clin. Invest.* **71**:296–300.
Hausler, W. J., Jr., 1987, Second state laboratories consensus conference on HIV testing, *Clin. Immunol. Newsletter* **8**(10):156–158.
Heagy, W., Kelley, V. E., Strom, T. B., Mayer, K., Shapiro, H. M., Mandel, R., and Finberg, R., 1984, Decreased expression of human class II antigens on monocytes from patients with acquired immune deficiency syndrome: Increased expression with interferon-gamma, *J. Clin. Invest.* **74**:2989–2096.
Hersh, E. M., Reuben, J. M., Mansell, P. W. A., Rios, A., Gutterman, J. U., Munn, G., Murray, J. L., Spector, S. P., Goldstein, A. L., and Newell, G. R., 1984, Immunologic studies of AIDS: Relationship of immunodeficiency to extent of disease, *Ann. NY Acad. Sci.* **437**:364–372.
Ho, D. D., Pomerantz, R. J., and Kaplan, J. C., 1987, Pathogenesis of infection with human immunodeficiency virus, *N. Engl. J. Med.* **317**:278–286.
Jaffe, H. W., Bregman, D. J., and Selik, R. M., 1983a, Acquired immune deficiency syndrome in the United States: The first 1,000 cases, *J. Infect. Dis.* **148**:339–345.
Jaffe, H. W., Choi, K., Thomas, P. A., Haverkos, H. W., Auerbach, D. M., Guinan, M. E., Rogers,

M. F., Spira, T. J., Darrow, W. W., Kramer, M. A., Friedman, S. M., Monroe, J. M., Friedman-Kien, A. E., Laubenstein, L. J., Marmor, M., Safai, B., Dritz, S. K., Crispi, S. J., Fannin, S. L., Orkwis, J. P., Kelter, A., Rushing, W. R., Thacker, S. B., and Curran, J. W., 1983b, National case-control study of Kaposi's sarcoma and *Pneumocystis carinii* pneumonia in homosexual men: Part 1, Epidemiologic results, *Ann. Intern. Med.* **99:**145–151.

Jawetz, E., Melnick, J. L., and Adelberg, E. A., 1987, Acquired immune deficiency syndrome (AIDS), in: *Review of Medical Microbiology*, Appleton & Lange, Norwalk, CT, pp. 533–537.

Josephs, S. F., Salahuddin, S. Z., Ablashi, D. V., Schachter, F., Wong-Staal, F., and Gallo, R. C., 1986, Genomic Analysis of the human B-lymphotropic virus (HBLV), *Science* **234:**601–603.

Kanki, P. J., Barin, F., M'Boup, S., Allan, J. S., Romet-Lemonne, J. L., Marlink, R., McLane, M. F., Lee, T-H., Arbeille, B., Denis, F., and Essex, M., 1986, New human T-lymphotropic retrovirus related to simian T-lymphotropic virus type III ($STLV\text{-}III_{AGM}$), *Science* **232:**238–243.

Kanki, P. J., M'Boup, S., Ricard, D., Barin, F., Denis, F., Boye, C., Sangare, L., Travers, K., Albaum, M., Marlink, R., Romet-Lemonne, J. L., and Essex, M., 1987, Human T-lymphotropic virus type 4 and the human immunodeficiency virus in West Africa, *Science* **236:**827–831.

Kessler, H. A., Blaauw, B., Spear, J., Paul, D. A., Falk, L. A., and Landay, A., 1987, Diagnosis of human immunodeficiency virus infection in seropositive homosexuals presenting with an acute viral syndrome, *JAMA* **9:**1196–1199.

Kirkpatrick, C. H., Davis, K. C., Horsburgh, C. R., Jr., Cohn, D. L., Penley, K., and Judson, F. N., 1985, Interleukin-2 production by persons with the generalized lymphadenopathy syndrome or the acquired immune deficiency syndrome, *J. Clin. Immunol.* **5:**31–37.

Klatzmann, D., and Gluckman, J. C., 1986, HIV infection: Facts and hypotheses, *Immunol. Today* **7**(10):291–296.

Knight, D. M., Flomerfelt, F. A., and Ghrayeb, J., 1987, Expression of the art/trs protein of HIV and study of its role in viral envelope synthesis, *Science* **236:**837–840.

Koenig, S., Gendelman, H. E., Orenstein, J. M., Dal-Canto, M. C., Pezeshkpour, G. H., Yungbluth, M., Janotta, F., Aksamit, A., Martin, M. A., and Fauci, A. S., 1986, Detection of AIDS virus in macrophages in brain tissue from AIDS patients with encephalopathy, *Science* **233:**1089–1093.

Kornfeld, H., Vande Stouwe, R. A., Lange, M., Reddy, M. M., and Grieco, M. H., 1982, T lymphocyte subpopulations in homosexual men, *N. Engl. J. Med.* **307:**729–731.

Krogsgaard, K., Lindhardt, B. O., Neilsen, J. O., Andersen, P., Kryger, P., Aldershvile, J., Gerstoft, J., and Pedersen, C., 1987, The influence of HTLV-III infection on the natural history of hepatitis B virus infection in male homosexual HBsAg carriers, *Hepatology* Vol. 7 (1):37–41.

Krueger, G. R. F., Otten, J. K., Ortmann, M., Mueller, W., Haussermann, L., Rasokat, H., Degenhart, S., Hehlmann, R., Wolf, H., and Auffermann, W., 1985, Immunopathology of AIDS (acquired immune deficiency syndrome). Observations in 75 patients with preAIDS and AIDS, *Path. Res. Pract.* **180:**463–480.

Lane, H. C., Masur, H., Edgar, L. C., Whalen, G., Rook, A. H., and Fauci, A. S., 1983, Abnormalities of B lymphocyte activation and immunoregulation in patients with the acquired immunodeficiency syndrome, *N. Engl. J. Med.* **309:**453–458.

Lane, H. C., Depper, J. M., Greene, W. S., Whalen, G., Waldmann, T. A., and Fauci, A. S., 1985, Qualitative analysis of immune function in patients with the acquired immunodeficiency syndrome: Evidence for a selective defect in soluble antigen recognition, *N. Engl. J. Med.* **313:**79–84.

Laurence, J., 1986, AIDS: Definition, epidemiology, and etiology, *Lab. Med.* **11:**659–663.

Laveridiere, M., Tremblay, J., Lavallee, R., Bonny, Y., Lacombe, M., Boileau, J., Lachapelle, J., and Lamoureux, C., 1983, AIDS in Haitian immigrants and in a caucasian woman closely associated with Haitians, *Can. Med. Assoc. J.* **129:**1209–1212.

Levy, J. A., Hoffman, A. D., Kramer, S. M., Landis, J. A., Shimabukuro, J., and Mro, L. S., 1984, Isolation of lymphocytopathic retroviruses from San Francisco patients with AIDS, *Science* **225:**840–842.

Levy, J. A., Shimabukurd, J., McHugh, T., Casavant, C., Stites, D., and Oshiro, L., 1985, AIDS-associated retrovirus (ARV) can productively infect other cells besides human T helper cells, *Virology* **147:**441–448.

Lombardo, J. M., 1987, HIV-I testing: An overview, *Am. Clin. Prod. Rev.* **11**:10–19.

Luben, N. L. C., Kelleher, J. F., Jr., and Reaman, G. H., 1983, Altered distribution of T lymphocyte subpopulations in children and adolescents with haemophilia, *Lancet* **1**:503–505.

Mackaness, G. B., 1970, The monocyte in cellular immunity, *Semin. Hematol.* **7**:172.

Margolick, J. B., Lane, H. C., and Fauci, A. S., 1987, Immunopathogenesis of HTLV-III/LAV infection, in: *Viruses and Human Cancer* (R. C. Gallo, W. Haseltine, G. Klein, and H. zur Hausen, eds.), Alan R. Liss, New York, pp. 59–79.

Marlink, R. G., Ricard, D., M'Boup, S., Kanki, P., Romet-Lemonne, J. L., and Essex, M., 1987, Clinical, hematological and immunological evaluation of individuals exposed to human T-lymphotropic virus type IV (HTLV-IV), III International Conference on AIDS, Washington, DC, Abstract #THP.7, p. 164.

Masur, H., Michelis, M. A., Greene, J. B., Onorato, I., Vande-Stouwe, R. A., Holzman, R. S., Wormser, G., Brettman, L., Lange, M., Murray, H. W., and Cunningham-Rundles, S., 1981, An outbreak of community-acquired *P. carinii* pneumonia: Initial manifestation of cellular immune dysfunction, *N. Engl. J. Med.* **305**:1431–1438.

McDougal, J. S., Hubbard, M., Nicholson, J. K. A., Jones, B. M., Holman, R. C., Roberts, J., Fishbein, C. B., Jaffe, H. W., Kaplan, J. E., Spira, T. J., and Evatt, B. L., 1985, Immune complexes in the acquired immunodeficiency syndrome (AIDS): Relationship to disease manifestation, risk group, and immunologic deficit, *J. Clin. Immunol.* **5**:130–138.

Mildvan, D., Mathur, U., Enlow, R. W., Roman, P. L., Winchester, R. J., Cop, C., Singman, H., Adelsberg, B. R., and Springland, I., 1982, Opportunistic infections and immunodeficiency in homosexual men, *Ann. Intern. Med.* **96**:700–704.

Montagnier, L., Gruest, J., Chamaret, S., Dauguet, C., Axler, C., Guetard, D., Nugeyre, M. T., Barre-Sinoussi, F., Chermann, J-C., Brunet, J. B., Klatzmann, D., and Gluckman, J. C., 1984, Adaptation of lymphadenopathy associated virus (LAV) to replication in EBV-transformed B lymphoblastoid cell lines, *Science* **225**:63–66.

Morimoto, C., Letvin, N. L., Boyd, A. W., Hagan, M., Brown, H. M., Kornacki, M. M., and Schlossman, S. F., 1985a, The isolation and characterization of the human helper inducer T cell subset, *J. Immunol.* **134**:3762–3769.

Morimoto, C., Letvin, N. L., Distaso, J. A., Aldrich, W. A., and Schlossman, S. F., 1985b, The isolation and characterization of the human suppressor inducer T cell subset, *J. Immunol.* **134**:1508–1515.

Munn, C. G., Reuben, J. M., and Hersh, E. M., 1984, Interleukin-2 receptor expression on lymphocytes of patients with AIDS during *in vitro* mitogen stimulation, *Lymphokine Res.* **3**:97–101.

Murray, H. W., Rubin, B. Y., Masur, H., and Roberts, R. B., 1984, Impaired production of lymphokines and immune (gamma) interferon in the acquired immunodeficiency syndrome, *N. Engl. J. Med.* **310**:883–889.

Murray, H. W., Gellene, R. A., Libby, D. M., Rothermel, C. D., and Rubin, B. Y., 1985a, Activation of tissue macrophages from AIDS patients: In vitro response of AIDS alveolar macrophages to lymphokines and interferon-gamma, *J. Immunol.* **135**:2374–2377.

Murray, J. L., Hillman, J. K., Rubin, B. Y., Kelly, C. D., Jacobs, J. L., Tyler, L. W., Donelly, D. M., Carriero, S. M., Godbold, J. H., and Roberts, R. B., 1985b, Patients at risk for AIDS-related opportunistic infections: Clinical manifestations and impaired gamma interferon production, *N. Engl. J. Med.* **313**:1504–1510.

Murray, J. L., Hersh, E. M., Reuben, J. M., Munn, C. G., and Mansell, P. W. M., 1985, Abnormal lymphocyte response to exogenous interleukin-2 in homosexuals with acquired immune deficiency syndrome (AIDS) and AIDS related complex (ARC), *Clin. Exp. Immunol.* **60**:25–30.

Pahwa, S., Pahwa, R., Saxinger, C., Gallo, R. C., and Good, R. A., 1985, Influence of the human T-lymphotropic virus/lymphadenopathy-associated virus on functions of human lymphocytes: Evidence for immunosuppressive effects and polyclonal B cell activation by banded viral preparations, *Proc. Natl. Acad. Sci. USA* **82**:8198–8202.

Pahwa, S. G., Quilop, M. T. J., Lange, M., Pahwa, R. N., and Grieco, M. H., 1984, Defective B-lymphocyte function in homosexual men in relation to the acquired immunodeficiency syndrome, *Ann. Intern. Med.* **101**:757–763.

Piot, P., Quinn, T. C., Taelman, H., Feinsod, F. M., Minlangu, K. B., Wobin, O., Mbendi, N., Mazebo, P., Ndangi, K., Stevens, W., Kalambayi, K., Mitchell, S., Bridts, C., and McCormick, J. B., 1984, Acquired immunodeficiency syndrome in a heterosexual population in Zaire, *Lancet* **2**:65–69.

Popovic, M., Sarngadharan, M. G., Read, E., and Gallo, R. C., 1984, Detection, isolation and continuous production of cytopathic retrovirus (HTLV-III) from patients with AIDS and pre-AIDS, *Science* **224**:497–500.

Prince, H. E., Moody, D., Shubin, B., and Fahey, J. L., 1985, Defective monocyte function in AIDS: Evidence from a monocyte-dependent T cell proliferative system, *J. Clin. Immunol.* **5**:21–25.

Racz, P., Tenner-Racz, K., Kahl, C., Feller, A. C., Kern, P., and Dietrich, M., 1986, Spectrum of morphologic changes of lymph nodes from patients with AIDS or AIDS-related complexes, *Prog. Allergy* **37**:81–181.

Redfield, R. R., Wright, D. C., and Tramont, E. C., 1986, The Walter Reed staging classification for HTLV-III/LAV infection, *N. Engl. J. Med.* **314**:131–132.

Rivin, B. E., Monroe, J. M., Hubschman, B. P., and Thomas, P. A., 1984, AIDS outcome: A first follow-up [Letter], *N. Engl. J. Med.* **311**:857.

Rook, A. H., Masur, H., Lane, H. C., Frederick, W., Kasahara, T., Macher, A. M., Djeu, J. Y., Manischewitz, J. F., Jackson, L., Fauci, A. S., and Quinnan, G. V., Jr., 1983, Interleukin-2 enhances the depressed natural killer and cytomegalovirus-specific cytotoxic activities of lymphocytes from patients with the acquired immune deficiency syndrome, *J. Clin. Invest.* **72**:398–403.

Rosenberg, Z. F., and Fauci, A. S., 1988, Immunopathogenesis of human immunodeficiency virus infection, *Clin. Immunol. Newsletter* **9**(1):1–3.

Saimot, A. G., Coulaud, J. P., Mechali, D., Materon, S., Dazza, M. C., Rey, M. A., Brun-Vezinet, F., and Leibowitch, J., 1987, HIV-2/LAV-2 in Portuguese man with AIDS (Paris, 1978) who had served in Angola in 1968–1974, *Lancet* **1**:688.

Salahuddin, S. Z., Markham, P. D., Lindner, S. G., Gootenberg, J., Popovic, M., Hemmi, H., Sarin, P. W., and Gallo, R. C., 1984, Lymphokine production by cultured human T cells transformed by human T cell leukemia lymphoma virus I, *Science* **223**:703–707.

Salahuddin, S. Z., Alblashi, D. V., Markham, P. D., Josephs, S. F., Sturzenegger, S., Kaplan, M., Halligan, G., Biberfield, P., Wong-Staal, F., Kramarsky, B., and Gallo, R. C., 1986, Isolation of a new virus, HBLV, in patients with lymphoproliferative disorders, *Science* **234**:596–601.

Salazar-Gonzalez, J. F., Moody, D. J., Giorgi, J. V., Martinez-Maza, O., Mitsuyasu, R. T., and Fahey, J. L., 1985, Reduced ecto-5'-nucleotidase activity and enhanced OKT10 and HLA-DR expression of CD8 (T suppressor/cytotoxic) lymphocytes in the acquired immune deficiency syndrome: Evidence of CD8 cell immaturity, *J. Immunol.* **135**:1778–1785.

Schnittman, S., Lane, H. C., Higgins, S. E., Folks, T., and Fauci, A. S., 1986, Direct polyclonal activation of human B lymphocytes by the acquired immunodeficiency syndrome virus, *Science* **233**:1084–1086.

Schroff, R. W., Gottlieg, M. S., Prince, H. E., Chai, L. L., and Fahey, J. L., 1983, Immunological studies of homosexual men with immunodeficiency and Kaposi's sarcoma, *Clin. Immunol. Immunopathol.* **27**:300–314.

Schupbach, J., Veronese, F., and Gallo, R. C., 1986, Human retroviruses, in: *Clinical Virology Manual* (S. Specter and G. J. Lancz, eds.), Elsevier, New York, pp. 451–490.

Selik, R. M., Haverkos, H. W., and Curran, J. W., 1984, Acquired immune deficiency syndrome (AIDS) trends in the United States, 1978–1982, *Am. J. Med.* **76**:493–500.

Seyda, M., and Krueger, G. R. F., 1987, Complex infectious copathogenesis of AIDS in HIV-positive individuals, *Clin. Immunol. Newsletter* **8**(6):81–86.

Siegel, F. P., Lopez, C., Hammar, G. S., Brown, A. E., Kornfeld, S. J., Gold, J., Hassett, J., Hirschman, S. Z., Cunningham-Rundles. C., Adelberg, B. R., Parham, D. M., Siegel, M., Cunningham-Rundles, S., and Armstrong, D., 1981, Severe acquired immunodeficiency in male homosexuals, manifested by chronic perianal ulcerative herpes simplex lesions, *N. Engl. J. Med.* **305**:1439–1444.

Silberberg-Sinakin, I., Gigli, I., Baer, R. L., and Thorbecke, G. J., 1980, Langerhans cells: Role in contract hypersensitivity and relationship to lymphoid dendritic cells and to macrophages, *Immunol. Rev.* **53**:203.

Smith, P. S., Ohura, K., Masur, H., Lane, H. C., Fauci, A. S., and Wahl, S. M., 1984, Monocyte function in the acquired immune deficiency syndrome: Defective chemotaxis, *J. Clin. Invest.* **74**:2121–2128.

Sodroski, J., Goh, W. C., Rosen, C., Dayton, A., Terwilliger, E., and Haseltine, W. A., 1986, A second post-transcriptional *trans*-activator gene required for HTLV-III replication, *Nature* **321**:412–416.

Sonigo, P., Alizon, M., Staskus, K., Klatzmann, D., Cole, S., Danos, O., Retzel, E., Tiollais, P., Haase, A., and Wain-Hobson, S., 1985, Nucleotide sequence of the visna lentivirus: Relationship to the AIDS virus. *Cell* **42**:369–382.

Stahl, R. E., Friedman-Kien, A. E., Dubin, R., Marmor, M., and Zolla-Pazner, S., 1982, Immunologic abnormalities in homosexual men: Relationship to Kaposi's sarcoma, *Am. J. Med.* **73**:171–178.

Stahl, R. E., Friedman-Kien, A., and Zolla-Pazner, S., 1983, AIDS: A medical conundrum, *J. Cut. Pathol.* **10**:550–558.

Steimer, K. S., Higgins, K. W., Powers, M. A., Stephans, J. C., Gyenes, A., George-Nascimento, C., Luciw, P. A., Barr, P., Hallewell, R. A., and Sanchez-Peseador, R., 1986, Recombinant polypeptide from the endonuclease region of the acquired immune deficiency syndrome retrovirus polymerase (pol) gene detects serum antibodies in most individuals, *J. Virol.* **58**:9–16.

Steinman, R. M., and Nussenweig, M. C., 1980, Dendritic cells: Features and functions, *Immunol. Rev.* **53**:127.

Terragna, A., Dodi, F., Anselmo, M., Del Bono, V., Damasio, E., and Vimercati, R., 1986, The Walter Reed staging classification in the follow-up of HIV infection, *N. Engl. J. Med.* **315**:1355–1356.

Tomar, R. H., John, P. A., Hennig, A. K., and Kloster, B., 1985, Cellular targets or antilymphocyte antibodies in AIDS and LAS, *Clin. Immunol. Immunopathol.* **37**:37–47.

Tschachler, E., Groh, V., Popovic, M., Mann, D. L., Konrad, K., Safai, B., Eron, L., Veronese, F. M., Wolff, K., and Stingl, G., 1987, Epidermal Langerhans cells: A target for HTLV-III/LAV infection, *J. Invest. Dermatol.* **88**(2):233–237.

Van De Perre, P., Rouvroy, D., Lepage, P., Bogaerts, J., Kestelyn, P., Kayihigi, J., Hekker, A. C., Butzler, J. P., and Clumeck, N., 1984, Acquired immunodeficiency syndrome in Rwanda, *Lancet* **2**:62–65.

Veronese, F., Copeland, T., Oroszlan, S., Gallo, R. C., and Sarngadharan, M. G., 1985, Characterization of gp41 as the transmembrane protein coded by the HIV/LAV envelope gene, *Science* **228**:291–294.

Veronese, F., DeVico, A., Copeland, T., DeVico, A., Rahman, R., Oroszlan, S., Gallo, R. C., and Sarngadharan, M. G., 1986, Characterization of highly immunogenic p66/p51 as the reverse transcriptase of HIV/LAV, *Science* **231**:1289–1291.

Walsh, C. M., Mardi, M. A., and Karpatkin, S., 1984, On the mechanism of thrombocytopenia in sexually active homosexual men, *N. Engl. J. Med.* **311**:635–639.

Welch, K., Finkbeiner, W., Alpers, C. E., Blumenfeld, W., Davis, R. L., Smuckler, E. A., and Beckstead, J. H., 1984, Autopsy findings in the acquired immune deficiency syndrome, *JAMA* **252**:1152–1159.

Williams, R. C., Jr., Masur, H., and Spira, T. J., 1984, Lymphocyte-reactive antibodies in acquired immune deficiency syndrome, *J. Clin. Immunol.* **4**:118–123.

Wong-Staal, F., Chanda, P., and Ghrayeb, J., 1987, Human immunodeficiency virus: The eighth gene, *AIDS Res. Hum. Retrovirol.* **3**:33–39.

Zagury, D., Bernard, J., Leonard, R., Cheynier, R., Feldman, M., Sarin, P. S., and Gallo, R. C., 1986, Long-term cultures of HTLV-III-infected T cells: A model of cytopathology of T-cell depletion in AIDS, *Science* **231**:850–853.

Ziegler, J. L., Beckstead, J. A., Volberding, P. A., Abrams, D. I., Levine, A. M., Lukes, R. J., Gill, P. S., Burkes, R. L., Meyer, P. R., Metroka, C. E., Mouradian, J., Moore, A., Riggs, S. A., Butler, J. J., Cabanillas, F. C., Hersh, E., Newell, G. R., Laubenstrin, L. J., Knowles, D., Odajnyk, C., Raphael, B., Koziner, B., Urmacher, C., and Clarkson, B. D., 1984, Non-Hodgkin's lymphoma in 90 homosexual men: Relation to generalized lymphadenopathy syndrome, *N. Engl. J. Med.* **311**:565–570.

Zolla-Pazner, S., 1986, Immunologic abnormalities in infections with the human immunodeficiency virus, *Lab. Med.* **17**(11):685–689.

Zolla-Pazner, S., Mizuma, H., El-Sadr, W., Sharpe, S., and Litwin, S., 1986, IgD secretion as a marker of B cell activation in AIDS, *J. Cell. Biochem.* **10A**(suppl):107.

13

The Reticuloendothelial System in Infection with RNA Tumor Viruses

MAURO BENDINELLI

1. INTRODUCTION

The interactions between the reticuloendothial system (RES) and RNA tumor viruses (or oncoviruses or oncogenic retroviruses, terms used interchangeably throughout the chapter) are of interest for a number of reasons. First, such viruses represent the largest group of agents associated with neoplasia. They affect such divergent species as birds, mice, cats, cattle, subhuman primates, and humans and are etiologically linked to a large variety of tumors both under laboratory conditions and in the natural environment. Second, few viruses proliferate and persist in their hosts to the extent of retroviruses. This is related to an unequaled genetic intimacy with the host and to the compatibility of their replication with host cell survival (Section 2) and also to the ability, well documented in several systems, to infect and impair the RES. RNA tumor viruses—the exogenous members at least—are primarily infectious agents, whose perpetuation depends on host invasion and consequently on a bidirectional interaction with the RES and its multiple functions (Section 3). Third, oncovirus-induced tumorigenesis often involves cellular compartments—lymphoid cells, for example—closely related with the RES both anatomically and functionally. Expansion of the initially transformed cells into a conspicuous mass may, thus, be under the influence of microenvironmental factors of RES origin, both cellular and humoral. Moreover, certain oncoviruses transform cells of the monocyte–macrophage lineage itself (Section 4). Finally, the RES is an important ancillary and regulatory component of the immune response system and may be implicated in the more or less pronounced immunodeficiency associated with many oncoviral infections (Section 5).

MAURO BENDINELLI • Institute of Epidemiology, Hygiene and Virology, University of Pisa, I-56100 Pisa, Italy.

2. GENERAL PROPERTIES OF RNA TUMOR VIRUSES

A comprehensive discussion of RNA tumor viruses is beyond the scope of this chapter. The broad overview provided focuses on the aspects that may help in understanding the information in subsequent sections. For a wider discussion or where specific references are not given, the reader is referred to extensive reviews and books that have recently appeared in all areas of oncovirology (Hanafusa, 1977; Klein, 1980, 1982; Stephenson, 1980; Weiss *et al.*, 1982; Vande Woude *et al.*, 1984).

All RNA tumor viruses are classified in the family Retroviridae, the only group of RNA viruses known to replicate by means of a DNA provirus integrated in the host cell genome. From the standpoint of structure, chemical composition, genetic organization, and strategy of replication, they are rather homogeneous, but individual members differ in biology and pathogenetic potential. Some including agents of chronic degenerative diseases such as visna-maedi of sheep and infectious anemia of horses, and apparently nonpathogenic agents such as the foamy viruses, are not noticeably tumorigenic and will not be considered here. Oncogenic members are mostly associated with leukemias, lymphomas, and fibrosarcomas, but may induce many other types of tumors.

Oncogenic retroviruses are grouped in the subfamily Oncoviridae and further classified according to morphology, maturation characteristics, and natural host species (Table 1). Additional criteria of subdivision are antigenic markers, cross-interference, experimental host range, and other biologic and pathogenic properties. For example, with reference to replication *in vitro,* retroviruses are distinguished as ecotropic (growth only in cells of the species of origin), xenotropic (efficient growth only in cells of species other than that of origin), and amphitropic (growth in both kinds of cells). A distinction based on modalities of

TABLE 1. THE SUBFAMILY ONCOVIRIDAE OF RETROVIRUSES

Genus	Subgenus	Major species
Type C oncovirus	Reptilian	Reptilian oncoviruses
	Avian	Avian sarcoma and leukemia viruses
		Avian reticuloendotheliosis viruses
		Pheasant oncoviruses
	Mammalian	Murine sarcoma and leukemia viruses
		Rat leukemia virus
		Guinea pig oncovirus
		Feline sarcoma and leukemia viruses
		Porcine oncovirus
		Bovine leukemia virus
		Woolly monkey sarcoma virus
		Baboon type C oncovirus
		Gibbon ape leukemia virus
		Human T lymphotropic virus
Type B oncovirus		Mouse mammary tumor virus
Type D oncovirus		Mason–Pfizer monkey virus

transmission is discussed in the following section. The key subdivision is, however, based on the length of latency for tumor induction. Differences between viruses which induce macroscopic neoplasms within a few days or weeks (acute oncoviruses) and those which cause neoplasia after a latency of several months or more (slow oncoviruses) have taken long to emerge (probably because viral preparations are frequently mixtures of different viruses), but are now known to involve important features: while slow oncoviruses are replication-competent and do not cause readily detectable transformation in culture, acute oncoviruses (there are about 30 characterized viruses of this type isolated from birds, rodents, cats, and primates) are generally replication-defective and tranform cells efficiently *in vitro*. Moreover, the two groups differ markedly in genomic organization and transformation mechanisms (see Section 2.2 and 2.4). Table 2 lists the major viruses mentioned in this chapter, giving some of their main features and the abbreviations used.

2.1. DIFFUSION AND MEANS OF TRANSMISSION

Retroviral genomes are ubiquitous as proviruses integrated in host's chromosomes in a stable form. Such endogenous retroviruses have been identified in divergent species and may be common to all animals. They are present in various numbers of copies in all the cells of the host, including the germ line, and are transmitted from one generation to another as mendelian traits. Many are xenotropic, and, their horizontal transmission seems to occur according to current evidence, rarely if ever. Thus, endogenous retroviruses may be considered basically as cellular genes with the unique ability to generate virions.

Most endogenous retroviruses do not appear to be associated with any form of disease. Their remarkable evolutionary conservation has led to speculation about possible beneficial effects for the host (Aaronson and Stephenson, 1976), though it also seems possible that their generally low level of expression and frequent defectiveness render selective pressure against them minimal. However, endogenous retroviruses share many important characteristics and may be progenitors of exogenous oncoviruses. They can be activated by various treatments (including exogenous infections) and, if passaged in suitable hosts, may exhibit some degree of oncogenicity, as exemplified by a number of leukemia viruses, isolated from x-irradiation-induced tumors. Moreover, some members appear to be directly related to cancer; classical examples are the endogenous ecotropic retroviruses of AKR and C58 mice that are associated with a frequent spontaneous T-cell lymphocytic lymphoma and leukemia. Indeed, the relationship of endogenous retroviruses to various naturally occurring neoplastic and nonneoplastic (e.g., autoimmune) affections remains a promising avenue of research.

Most oncogenic retroviruses are exogenous. Some are broadly distributed, infecting many individuals of their respective hosts' species, even though they generally cause tumors in a minority. Usually they are not transmitted genetically and circulate horizontally or vertically in a manner identical to other infec-

TABLE 2. ABBREVIATIONS AND MAIN BIOLOGIC FEATURES OF VIRUSES MENTIONED IN THE TEXT[a]

Abbreviation	Virus[b]	Replication	Tumorigenesis	Predominant tumors	Cells transformed *in vitro*	Other features
AEV	Avian erythroblastosis	d	a	Erythroid L, fibrosarcoma	Erythroid, fibroblasts	
ALV	Avian leukosis	c	s	B-cell Ly		Broad spectrum of tumors
AK-MuLV	AKR murine L	c	s	T-cell Ly		Endogenous
A-MuLV	Abelson murine L	d	a	B-cell Ly, plasmocytoma	B cells, fibroblasts	
AMbV	Avian myeloblastosis	d	a	Myeloblastic L	Myelomonocytic	
AMcV	Avian myelocytomatosis	d	a	Myelocytic L	Myelomonocytic, fibroblasts	
AOV	Avian osteopetrosis	c	s	B-cell Ly, osteoblastoma		Derived from ALV; osteopetrosis is nonmalignant
ASV	Avian sarcoma[c]	d[d]	a	Fibrosarcoma	Fibroblasts	
Cas-MuLV	Lake–Casitas murine L	c	s	Hemopoietic (various)		
BoLV	Bovine L	c	s	B-cell Ly		
FeLV	Feline L	c	s	T-cell Ly and L		Broad spectrum of tumors

FeSV	Feline sarcoma	d	a	Fibrosarcoma, melanoma	Fibroblasts	
FLV	Friend murine L	d	a	Erythroid L	Hemopoietic precursors	F-SFFV (spleen focus forming virus) + F-MuLV (helper)
G-MuLV	Gross murine L		Same as AK-MuLV, but exogenous			
Mo-MuLV	Moloney murine L	c	s	T-cell L		
MuSV	Murine sarcoma[c]	d	a	Fibrosarcoma	Fibroblasts	
MuMTV	Murine mammary tumor	c	s	Mammary adenocarcinoma		
Rad-MuLV	Radiation murine L	c	s	T-cell Ly		
REV	Avian reticuloendotheliosis	d	a	B-cell Ly		
Rg-FLV			Same as FLV, but the leukemia tends to regress spontaneously			
RLV	Rauscher murine L	d	a	Erythroid L	Hemopoietic precursors	Similar to FLV
SSV	Simian sarcoma (woolly monkey)	d	a	Fibrosarcoma, fibroma, glioma	Fibroblasts, other cells	Closely related to gibbon ape L virus

[a] a, acute; c, replication-competent; d, helper-dependent; L, leukemia; Ly, lymphoma; s, slow.
[b] A precise taxonomical nomenclature is not yet available; several viruses are named after the scientist who first isolated them.
[c] There are several isolates containing distinct oncogenes; unless specified, the results discussed in the text have been obtained with the Rous strain of ASV and the Moloney strain of MuSV.
[d] Some strains are competent.

tious agents. Feline leukemia virus (FeLV) is spread via saliva; avian leukosis (ALV), through the egg albumin or, less frequently, by contact. Murine mammary tumor virus (MuMTV) is spread via lactation and also by mating; bovine leukemia virus (BoLV) is shed with milk but more effectively transmitted horizontally; gibbon ape leukemia virus is excreted in saliva and urine. Transpecies transmission has also been documented (Todaro, 1980). Genetic and extragenetic ways of transmission, however, are not mutually exclusive. Strains of MuLV usually transmitted genetically may also be passed to the progeny in milk (Squartini *et al.*, 1973), and, conversely, distinct strains of MuMTV are transmitted genetically (Bentvelzen and Daams, 1969). Endogenization of exogenous viruses has also been documented (Jaenish, 1976). Modalities of transmission may influence considerably the consequences of infection in terms of host invasion, tumorigenesis, induction of immunity, and chances of diffusion (Herberman, 1982).

Several exogenous retroviruses are highly oncogenic under natural conditions and leading causes of death in animal species ranging from chickens to cats and cattle. Many viruses in experimental research, however, are laboratory-derived strains deliberately selected over numerous passages by artificial routes. Typical examples are the acute oncoviruses. Although this view has recently been challenged (Neil *et al.*, 1984), these viruses do not appear to circulate significantly in nature, and presumably have been modified considerably by the selective pressures imposed by investigators (several have been isolated from animals bearing tumors induced by slow viruses). Considering how they are generated (see Section 2.2), it seems likely that they exist ephemerally and perish together with the tumor-bearing host, unless isolated in time in the laboratory. Yet, their study is furnishing important clues to the mode of action of naturally occurring oncoviruses and to carcinogenesis in general.

2.2. STRUCTURE AND GENOMIC ORGANIZATION

Mature oncoviruses appear in this sections as spherical particles, 80–120 nm in diameter, composed of an outer envelope, an inner coat, and an electron-dense nucleoid about 70 nm in diameter (bull's-eye shape). In the majority the nucleoid is located in the center (type C particles), but in a few it is eccentric (B and D particles). More or less prominent, loosely attached spikes and knobs are regularly scattered over the outer envelope surface. The internal structure, still poorly resolved, is thought to consist of an icosahedral capsid containing a possibly helical ribonucleoprotein. Type A particles are found only intracellularly. Whereas the role of those located in the cisterna is unknown, intracytoplasmic forms probably represent immature virions.

The major internal structural proteins are a group of nonglycosylated proteins whose molecular weights range from 10 to 30 kilodaltons. They mainly contain group-specific and interspecies-specific antigens (that is, shared by retroviruses infecting the same or related species, respectively) and are designated *gag* proteins. The envelope is composed of lipids derived from the plasma membrane of the host cell and by two to three species of virus-coded proteins (*env* proteins), of which at least one is glycosylated (gp70 in many mammalian viruses, gp37 and gp85 in avian members). Envelope proteins contain type-specific antigens that differentiate closely related viruses and are important protective antigens. Another internal protein is reverse transcriptase (about 30 molecules per virion), the polymerase used to synthesize the DNA intermediate, complementary to virion RNA, needed for replication and which also exhibits exonuclease activity. This key enzyme, which in different viruses consists of one or two polypeptide chains, carries group-specific and type-specific antigenic determinants.

Virion RNA (60–70S) is noninfectious and composed of two usually identical, single-stranded subunits, which in nondefective replication-competent members are 9 kilobases long (sufficient to code for 300 kilodaltons of protein) and in defective members (see Section 2.2) may be considerably shorter. Each subunit carries a terminal repetition 20–50 nucleotides long and, similar to cellular mRNA, is polyadenylated at the 3′ end and capped at the 5′ end. Moreover, a tRNA of cellular origin that serves as primer for reverse transcriptase is noncovalently bound to the 5′ terminus, and the two subunits are themselves joined by base pairing of the 5′ ends.

Thus, retroviruses are the only known disploid viruses, but the implications of this property are not well understood, and it is not even known whether the two strands are both functional. Replication-competent members have at least three genes (defined as the region translated into a primary protein product) that furnish the information essential for the production of infectious progeny: *gag* codes for the core structural proteins, *pol* for reverse transcriptase, and *env* for the envelope proteins. A fourth region *c* (for common) with unknown functions is present in some viruses only. In all the slow retroviruses studied the map reads 5′-TR-gag-pol-env-(c)-TR-3′. Acute oncoviruses contain additional sequences that confer on them the ability to induce neoplasms with short latency (viral oncogenes, Section 2.4) and that appear to have been acquired by recombination with host cell DNA (possibly due to diploidy, retroviruses show a high frequency of recombination with other endogenous or exogenous retroviruses and with cellular genes; Coffin, 1979). Usually, such additional information replaces viral sequences (e.g., within the *env* region in certain strains of ASV, within the replicative genes in AMcV and mammalian sarcoma viruses), rendering the virus replication-defective.

2.3. REPLICATION

Retroviruses enter cells by interaction of their surface glyocoproteins with specific cell receptors and subsequent fusion of the virion envelope with the plasma membrane. In the cytoplasm, viral RNA is rapidly transcribed by the reserve transcriptase carried in the virion into double-stranded DNA. In the process long terminal repeats (LTR), composed of a few hundred nucleotides and derived from the fusion of sequences from the 3′ and 5′ termini, are generated at both ends of the transcript. The DNA intermediate is soon found in the nucleus, where it becomes covalently integrated into the cell chromosomal DNA. This step is considered essential for replication and thought to occur by illegitimate recombination, though molecular events and enzymatic requirements for integration are still poorly defined (Hughes, 1983). The viral DNA is now used by host RNA polymerase II as a template for repeated synthesis of viral RNA destined to serve partly as messenger and partly as genome in progeny virions. Viral RNA is then spliced and/or processed into forms resembling cellular mRNA and transported to the cytoplasm. Here, viral protein synthesis occurs (the translational products are mostly polyproteins that require cleavage), and nucleocapsids assemble in proximity of the plasma membrane. Release occurs by a budding process during which virions acquire the outer envelope.

Although there are exceptions (Ziegler *et al.*, 1981; Popovic *et al.*, 1984), retrovirus replication does not generally lead to cytopathology or death of the host cell. When certain conditions pertaining to the virus and to the cell occur (Sections 2.4 and 4.1), host cells become transformed, but otherwise there often are no obvious deleterious effects on cell physiology and behavior, despite large quantities of progeny virus continuously being produced. The proviral DNA is now a stable component of the cell chromosome, where it is replicated and inherited in concert with flanking cell DNA. Here, its transcription is partly self-governed (e.g., LTRs appear to function in enhancement, promotion, and initiation of transcription as well as in polyadenylation of transcripts) but is also regulated by cellular mechanisms (Hughes, 1983). Thus, the levels of expression of individual proviruses may exhibit considerable differences and be modulated by a number of genetic and epigenetic (hormonal, chemical, immunologic, etc.) factors (Weinberg and Steffen, 1981). As a rule, however, less than 1% of RNA and protein synthesis is of viral origin (Coffin, 1979).

2.4. MECHANISMS OF CELL TRANSFORMATION AND TUMOR INDUCTION

Like other forms of neoplastic transformation, oncovirus-induced transformation consists of a heritable acquisition of abnormal growth properties, which are usually accompanied by a variable spectrum of morphologic, physiologic, and biochemical changes collectively referred to as "transformed phenotype." Transformation of fibroblasts and other cells by oncoviruses has been actively studied in recent years in the effort to gain access to the genetic and biochemical events responsible for neoplasia. Indeed, recent progresses in this area may represent an important basis for the understanding of the molecular basis of carcinogenesis.

As currently understood, transformation by RNA tumor viruses is correlated with the existence in normal cells of genes, whose deregulation and/or alteration can transform cultured cells and elicit tumor formation in animals (Bishop, 1982; Enrietto and Wike, 1983). Known as protooncogenes or c-*onc*, such loci appear to be present in the genetic endowment of many, probably all, cells as sequences entirely unlinked to endogenous proviruses. Presently the subject of intensive investigation, cellular oncogenes were initially recognized by hybridization experiments due to their extensive omology with the transforming sequences of acute oncoviruses (v-*onc*, see below). They may also be demonstrated by the ability to induce morphologic transformation, when transfected into suitable cells (such as the murine permanent NIH 3T3 line), an assay that has so far allowed the detection of over 20 such genes in DNAs extracted from nonviral tumors, tumorigenic lines, and, under special conditions of test, normal cells. In normal uninfected cells oncogenes are generally expressed at low levels, a fact that, together with the high degree of phylogenetic conservation, has suggested that their physiologic role might be confined to critical stages of cell differentiation or division (Coll *et al.*, 1983; Goyette *et al.*, 1983). As indicated by accumulating data, in many oncovirus-transformed cells oncogenes are overexpressed (Hayward *et al.*, 1982), deregulated, or structurally altered in subtle critical ways by mutation or other DNA rearrangements, such as insertion, amplification, or translocation (Duesberg, 1983), a series of possibilities that are now being actively considered also for the genesis of nonviral tumors (Weinberg, 1982; Vande Woude *et al.*, 1984).

RNA tumor viruses appear to influence oncogene expression by at least two mechanisms. The first is exploited by acute oncoviruses, whose rapid tumorigenic action has been clearly associated with the activity of specific sequences present in the viral genome (mutations or deletions within such genes abrogate the rapid transforming activity) known as v-*onc* (Bishop, 1982). Twenty v-*onc* are presently recognized: some are found in multiple, independent, retroviral isolates and for each a similar, but not necessarily identical, counterpart, from which they presumably derive, has been detected in normal cells. Thus, infection by acute oncoviruses leads to the transduction into cells of one oncogene (only a few retroviruses contain two distinct v-*onc*), which is immediately and actively expressed under the control of the viral LTR and this, in turn, rapidly causes transformation, no matter where the provirus is integrated (Bishop, 1982). Consistent with this mode of action (1) most acute oncovirus-induced tumors are, at least initially, polyclonal; (2) oncogene-carrying viruses do not need to replicate in order to transform; cells infected with retroviruses, whose replication is restricted (because they are defective or for other reasons), give rise to transformed "nonproducer" cells, from which the transforming virus can be rescued by superinfecting with suitable helper viruses. However, there are also oncoviruses that appear not to contain a cell-derived oncogene but nevertheless induce rapid tumors. Notable examples are FLV and RLV, whose acute leukemogenic activity has been related to products of viral structural genes (Ruta *et al.*, 1983; Machida *et al.*, 1984).

The second mechanism whereby RNA tumor viruses alter oncogene expression in infected cells has been termed "promoter insertion" and appears to be used by slow oncoviruses. These viruses lack demonstrable v-*onc* and display properties, that suggest that transformation is associated with relatively rare events occurring after prolonged extensive growth. For some members of this group, such as ALV and MuMTV, there is evidence that in tumor cells the viral genome is integrated in the vicinity of a c-*onc* (Hayward *et al.*, 1982; Nusse and Varmus, 1982). As a result, transcription of the c-*onc* is believed to be no longer tuned by normal cellular mechanisms and to be activated by the provirus owing to its strong promoters and enhancers (insertion of only a viral LTR may suffice to enhance the expression of flanking genes; Blair and Vande Woude, 1982) or by less well-defined mechanisms (Nusse *et al.*, 1984).

Since possible sites of integration for slow viruses are numerous, perhaps random, insertion at a location that would activate a cellular oncogene is a rare event. This may explain, at least partly, the long latency for tumor appearance *in vivo* and the failure to transform cells in culture. Alternative or additional factors that have been invoked include the following: (1) Oncovirus integration and c-*onc* activation may represent only the initiating events of the chain of reactions ultimately leading to the neoplastic phenotype (Cullen *et al.*, 1984; Peters *et al.*, 1984). Indeed, there are several viruses that initiate a premalignant tissue-specific hyperplasia, which progresses to an increasingly malignant cancer. Carcinogenesis by MuMTV is a known example. (2) Recombinants must generate within the host before the oncogenic potential of the infecting virus is fully expressed. Dualtropic *env* gene recombinants with endogenous xenotropic viruses, capable of accelerating spontaneous lymphoma when injected into newborns, are regularly found in AKR mice during the late preleukemic period and similar recombinants have been implicated in tumorigenesis by other oncoviruses (Famulari, 1983; Herr and Gilbert, 1984). (3) Owing to their monoclonal nature, the tumors induced by slow viruses may be more susceptible to host defense mechanisms.

Considerable uncertainty also exists on how altered expression of oncogenes leads to the pleiomorphic cellular changes associated with transformation. The translational products of several oncogenes have been identified, but how they are linked to transformation is still essentially unknown. Some possess protein phosphokinase activity with an unusual specificity for tyrosine residues (the p60src phosphoprotein of ASV and the p120 protein of A-MuLV are well-known examples) an enzymatic activity which, in principle at least, is capable of affecting many cellular properties, but whose relation to transformation is still largely speculative (Cooper and Hunter, 1983). Others, including the p21 of Harvey and Kirstein MuSV and p110 of AMbV, have no demontrable tyrosine phosphokinase activity, even though some may be structurally related to the above products (the protein encoded by the AEV v-*onc*, for instance). Still others have been linked to normal cell growth factors or their receptors (Robbins *et al.*, 1983). These data, together with the different subcellular localization of different v-*onc* products, suggest that biochemical mechanisms leading to transformation might be multiple.

3. RES INVOLVEMENT IN HOST INVASION BY RNA TUMOR VIRUSES

As discussed earlier, exogenous RNA tumor viruses enter their hosts through routes similar to those of nononcogenic viruses. Though virtually nothing is known of these aspects, it seems reasonable to assume that the paths used by oncoviruses to reach susceptible cells within the host organism do not differ substantially from those used by other viruses. At least following exogenous infection, host invasion is an essential prerequisite for oncoviruses to induce tumors, and many such viruses must replicate extensively in their hosts before the critical event(s) for neoplastic transformation is generated. In this section existing evidence on the interplay of oncoviruses with the RES during the preneoplastic stages of infection is reviewed. Before doing so, it will be informative to recapitulate briefly the most salient conclusions drawn from studies on the interactions between the RES and viruses in general.

3.1. RES INVOLVEMENT DURING VIRAL INFECTIONS IN GENERAL

Extensive investigations have shown that animal viruses establish a complex series of interactions with the RES of their hosts (Table 3). Although RES involvement varies considerably as a function of a number of factors, some of which are nature and virulence of the virus, extension of infection, genetic constitution, and age of the host, the following general considerations are possible:

1. The RES is crucially implicated in resistance to viruses, possibly more than to other infectious agents. Thus, for example, by using substances selectively toxic for macrophages, it has been shown in several acute viral infections that (during the early stages when specific immune effectors have not yet formed or are still scarce) the RES probably represents the major barrier against the virus by affecting both its dissemination from the initial site of entry and its spread within target organs. The participation of macrophages to the afferent and efferent limbs of virus-specific immune responses is also fairly well documented.

2. If exposed to relatively high multiplicities *in vitro*, mononuclear phagocytes are efficiently entered by widely different viruses. This may occur via specific receptors and also through phagocytosis, a pathway that for certain viruses is facilitated by trace amounts of antiviral antibody. The events that ensue after viral entry into macrophages vary. The virus may be more or less rapidly inactivated, may undergo an aborted cycle of replication with synthesis of generally low amounts of viral macromolecules followed by scanty or no release of mature virions, or may replicate with the production of abundant progeny. As a result of viral invasion, macrophages may either continue an apparently normal life or show overt cytopathology and die, depending largely on the virus–host interaction. The outcome is also strongly influenced by factors such as differentiation or functional stage and anatomic origin of the cell, physiologic and pathologic state of the cell donor, and the presence of soluble lymphokines. Although the relevance of such *in vitro* findings to the *in vivo* context of infection is not entirely clear (for example, the number of viruses which have been shown to invade mononuclear phagocytes in the intact host is limited), in several instances a close correlation between virus ability to kill macrophages *in vitro* and severity of infection *in vivo* has been observed.

TABLE 3. FUNCTIONS ASCRIBED TO MACROPHAGES AND CHANGES OF MACROPHAGE BEHAVIOR OBSERVED IN THE COURSE OF VIRAL INFECTIONS[a]

A. Functions
1. Virus adsorption and internalization
2. Virus replication
3. Dissemination of replicated virus
4. Inactivation and degradation of phagocytized virus
5. Direct control of extracellular and cell-to-cell virus spread
6. Accumulation at sites of virus multiplication
7. Protection of adjacent cells from viral attack
8. Selective lysis/cytostasis of virus-infected cells
9. Interferon-α production
10. Potentiation of interferon-γ production by virus-sensitized lymphocytes
11. Participation to the antiviral effects of interferon, interferon inducers, and immunomodulators
12. Potentiation of viral antigen immunogenicity
13. Cooperation with specific effectors of antiviral immunity in
—delayed hypersensitivity
—lysis of infected cells
—control of cell-to-cell spread
—immune phagocytosis
—passive protection from infection

B. Behavior changes
1. Reduced circulation in peripheral blood
2. Decreased motility and/or chemotaxis
3. Decreased or enhanced attachment/spreading onto surfaces
4. Decreased or enhanced phagocytosis
5. Impaired phagosome–lysosome fusion
6. Decreased or enhanced digestion of ingested microorganisms
7. Enhanced lysis/cytostasis of suitable target cells
8. Decreased or enhanced response to interferon inducers
9. Enhanced retention of membrane-bound antigen
10. Reduced expression of Ia antigen
11. Impaired antigen presentation to lymphocytes
12. Reduced support to lymphocyte blastogenesis
13. Production of cytotoxins
14. Suppression of lymphocyte responsiveness

[a]Supporting evidence reviewed in Allison, 1974a, b; McFarland and Johnson, 1975; Denman *et al.*, 1976; Morahan and Morse, 1979; Mogensen, 1981; Bendinelli, 1984.

3. Viruses that persistently infect their hosts are often capable of noncytocidal replication in macrophages. Such correlation has been observed with diverse viruses such as arenaviruses, herpesviruses, parvoviruses, togaviruses, and nononcogenic retroviruses. Consequently, it has been repeatedly suggested, though never formally proved, that replication in macrophages is instrumental for viruses to persist.

4. In virus-infected individuals RES function may be perturbed. Certain alterations are observed also under conditions of *in vitro* infection of macrophages and are apparently due to direct contact with the virus or viral prod-

ucts, but others may be regular adaptive responses to infection (Bendinelli *et al.*, 1982).

5. Relatively little attention has been paid to the participation of macrophages in the generation of tissue lesions and other pathologic effects of viral infections. However, macrophages are often the predominant cell type recruited by the inflammatory response at sites of virus multiplication, and it is well known that such response may be either beneficial or harmful to the host depending on many variables (Escobar and Swenson, 1983). In addition, macrophages have been pinpointed as key cells in the immunosuppressed state and in the reduced resistance to superinfections induced by many viral infections.

3.2. REPLICATION OF RNA TUMOR VIRUSES IN MONONUCLEAR PHAGOCYTES

The range of cell types supporting oncovirus replication varies widely with the virus/host combination and may be entirely unrelated to the cell type(s) eventually transformed. Thus, Rad-MuLV, AK-MuLV, Mo-MuLV, and FeLV are all associated with the induction of T-cell lymphomas. However, whereas Rad-MuLV appears to replicate only in the T-cell lineage (Haran-Ghera, 1980), the other three viruses may grow in many cell types and organs throughout the majority of the preleukemic period (Gisselbrecht *et al.*, 1978; Horak *et al.*, 1981). Similarly, bursal cells are the principal transformation target for ALVs, but these viruses multiply extensively in many tissues before tumors develop (Beard, 1980). Even MuMTV, which exclusively transforms mammary cells in susceptible mice, has an extremely wide tissue distribution (Nandi and McGrath, 1973).

It has been confirmed in the monocyte–macrophage lineage that cell targets for oncovirus replication versus oncovirus-induced transformation may be distinct and occasionally divergent. Cells in this lineage may be productively infected by many oncoviruses and, nevertheless, may maintain a nontransformed phenotype (Table 4). Evidence dealing with RNA tumor virus replication in nontransformed macrophages is discussed here. Since, at least *in vitro*, transformed and nontransformed cells may show different permissiveness (Horak *et al.*, 1981), replication in transformed macrophages is considered separately (Section 4.2).

As early as the mid-twenties, ASV was reported to grow in buffy coat macrophage cultures of chickens (Carrel and Ebeling, 1924). Though questioned on the ground of possible fibroblastic contamination by the unfiltered embryonic extracts used for the cultures (Sanford *et al.*, 1952), these findings were essentially confirmed by later studies showing that ASV generated a low-grade, persistent infection in avian macrophage cultures (Rangan and Bang, 1967). More recently, embryonic and adult peripheral blood macrophages of chickens were challenged *in vitro* with various helper-dependent and -independent strains, pseudotypes and recombinants of ASV and ALV belonging to different subgroups of envelope specificity. Subgroups B and C replicated efficiently. In contrast, subgroups A and D–G grew little in spite of the fact that macrophages

TABLE 4. RNA TUMOR VIRUSES WHOSE ABILITY TO REPLICATE IN AND TO TRANSFORM CELLS OF THE MONOCYTE–MACROPHAGE LINEAGE HAS BEEN DOCUMENTED[a]

Host	Virus	Cell source[b]	Replication		Transformation	
			In vivo	*In vitro*	*In vivo*	*In vitro*
Chicken	AEV	BM, E		+		
	ALV	BM, E, PB	+	+		
	AMbV	BM, E, PB	+	+	+	+
	AMcV	BM, E, PB	+	+	+	+
	ASV	E, PB, T	+	+		
Quail	AEV	BM, PB	+	+	+	+
	AMcV	PB	+	+	+	+
	ASV	PC		+		+
Mouse	A-MuLV	BM, S[c]	+[c]	+	+[c]	+
	Balb/c:virus-1	BM		+		
	Cas-MuLV				+	
	F-MuLV	BM, PB, S	+	+		+
	F-SFFV	S	+			
	G-MuLV	BM	+	+		+
	Mo-MuLV	S	+			
	MuSV	BM		+		+
	RLV	BM		+		
Cat	FeLV	BM, LN, PB, PC, S	+	+		

[a]References discussed in Sections 3.2 and 4.1.
[b]BM, bone marrow; E, embryo; LN, lymph node; PB, peripheral blood; S, spleen; T, virus-induced tumor.
[c]Uncertain.

were from animals whose fibroblasts were fully permissive. This epigenetic restriction was not further investigated, but its dependence on viral envelope properties was considered an indication that virus penetration or uncoating was the discriminating step. Embryonic macrophages proved more susceptible than the adult counterpart, but produced significantly less virus than embryonic fibroblasts. Productively infected macrophages exhibited morphologic and metabolic changes (clumping, enhanced polykaryocytosis, increased medium acidification, sugar uptake, and protein synthesis), but an unaltered potential of growth (Gazzolo *et al.*, 1974, 1975). Later studies confirmed the failure of ASV to transform chicken macrophages and found no changes whatsoever on infected cells (Durban and Boettiger, 1981b). Producer lines of ASV-transformed Japanese quail macrophages have, however, been described (Yamanouchi *et al.*, 1979).

Invasion of mononuclear phagocytes by avian oncoviruses occurs also *in vivo* (Table 4). Peripheral blood monocytes of susceptible chickens, which had been infected at 1–2 months of age, harbored infectious ALV for up to 3 years, albeit neutralizing antibodies were present, and no virus was detectable in plasma. ALV-positive cells ranged between 0.001 and 0.7% of monocytes, depending on the infecting virus and the individual animal tested, a proportion significantly higher than that found in lymphocytes. Despite the use of sensitive coculture techniques, ASV was not detected in circulating monocytes of sarcoma-bearing chickens, thus indicating that macrophage permissiveness *in vitro* and in the host may differ considerably (Gazzolo *et al.*, 1979a). Intratumoral macrophagelike cells have, however, been found to replicate ASV efficiently (Wainberg *et al.*, 1983).

Oncovirus replication in the monocyte–macrophage lineage has been investigated also in the murine system. Early electron microscopic investigations evidenced mature C-type particles within Kupffer cells and other macrophages of mice infected with several MuLVs. The presence of budding particles suggested that such cells were actually replicating virus (Yumoto *et al.*, 1966; Feldman *et al.*, 1967).

More conclusive data were obtained using cell culture methods. In an early study, starch-elicited peritoneal exudate macrophages from FLV-infected mice released virus in culture for several weeks. Moreover, peritoneal macrophages from normal mice, when exposed to a large dose of FLV *in vitro*, remained virus-positive for several days, though a degree of inconsistency was encountered (Odaka and Köhler, 1965). Later studies with Rg-FLV (Section 4.2.2) confirmed these results partially: (1) resident peritoneal macrophages produced infectious F-MuLV and expressed the viral gp70 on their surface, only if harvested after at least 20–25 days of infection; (2) the proportion of virus-producing cells was low, ranging from 0.6% to less than 0.001%; (3) the peritoneal cells of some mice remained constantly virus-negative; (4) attempts to infect peritoneal macrophages *in vitro* were unsuccessful. These results led to speculation that FLV may infect precursors rather than mature macrophages and that the delayed appearance of infected cells in the peritoneal cavity may reflect the time required for differentiation and peripherization of infected precursors (Marcelletti and Furmanski, 1978).

Evidence supporting such a concept was obtained by the same workers. First, in accordance with the requirement for ongoing DNA synthesis in order that retroviruses replicate (Temin, 1967), a direct relationship was observed between productive Rg-FLV infection of peritoneal macrophages and DNA synthesis by these cells. Resident macrophages did not incorporate [^{3}H-]thymidine and were refractory to infection also following transfer into viremic mice. Starch- and thyioglycolate-elicited cells contained a few DNA-synthesizing cells and supported some viral replication. Treatment with colony-stimulating factor enhanced both DNA synthesis and virus production. Second, a high proportion of bone marrow macrophage precursors could be infected *in vitro*. Finally, in infected mice such precursors were virus-positive considerably earlier than peritoneal macrophages. Another interesting observation was that peritoneal macrophages could remain permanently virus-negative, even in animals whose bone marrow yielded infected myeloid precursors (Marcelletti and Furmanski, 1979). The bone marrow seems a favorable environment for initial replication of, and target cell infection by, a number of oncoviruses, including FeLV (Rojko *et al.*, 1979) and several MuLVs (Haran-Ghera, 1980).

Studies with conventional strains of FLV gave somewhat different results.

For example, resident peritoneal (Levy and Wheelock, 1975) and spleen (Toniolo *et al.*, 1980) macrophages were found to be productively infected as early as 5 and 3 days postinfection, respectively. Virus release by splenic macrophages was found to depend on cellular viability and protein synthesis and to encompass both the components of FLV. Though the spleen has been reported to contain numerous macrophage precursors (Yochuchi *et al.*, 1980), attempts to infect splenic macrophages *in vitro* gave inconsistent results (unpublished data). Whether the discrepancies encountered reflect differences in the FLV strains (Furmanski *et al.*, 1980) or in methodology is unclear.

Replication of FeLV in macrophages has also been studied. FeLV-specific antigen and/or infectivity was intermittently detectable in peripheral blood monocytes of cats during the first 2 weeks after oral–nasal infection, irrespective of whether the animals eventually progressed or regressed (Rojko *et al.*, 1978), and in lymph node, spleen, and peritoneal macrophages at later times (Rojko *et al.*, 1979, 1981; Hoover *et al.*, 1981). Moreover, *in vitro* thyioglycolate-elicited peritoneal macrophages infected *in vitro* proved permissive to FeLV, though the proportion of virus-positive cells remained low (less than 0.3% in kitten cells, which proved significantly more susceptible than adult cells). Peritoneal macrophages harvested from viremic animals also yielded low proportions of infected cells. The latter finding was unexpected, because FeLV replicates extensively in myeloid precursors of viremic cats (Rojko *et al.*, 1979), but was reminiscent of the already discussed results in mice infected with Rg-FLV and interpreted as indicating that the majority of mature peritoneal macrophages either derive from uninfected precursors or abort infection during the process of differentiation.

Although low as compared to that of fibroblasts, macrophage permissiveness to FeLV *in vitro* was still considerably greater than that of lymphocytes, which represent the ultimate transformation target for the virus (Rojko *et al.*, 1979). Also, in cocultures of infected or noninfected peritoneal macrophages with precultured peripheral blood lymphocytes (precultivation was a prerequisite for FeLV replication in lymphocytes), the bulk of virus was invariably produced by adherent cells, irrespective of the cell type initially exposed to the virus. In such cocultures lymphocytes transferred the infection efficiently to normal macrophages, whereas macrophages failed to transfer the virus to lymphocytes (Hoover *et al.*, 1981; Rojko *et al.*, 1981).

Cells in the monocyte–macrophage lineage may also produce endogenous retroviruses. For example: (1) macrophages of NZB mice were shown to release xenotropic and ecotropic viruses (Levy *et al.*, 1981); (2) macrophage precursors were found to represent at least one-third of the AK-MuLV-producing cells in the bone marrow of young AKR mice (Godard *et al.*, 1983); (3) the endogenous virus *ev*-3 was activated in chicken macrophages following exogenous infection with ASV, though less efficiently than in fibroblasts (Gazzolo *et al.*, 1975); (4) macrophages infiltrating spontaneous lymphoproliferative diseases of turkeys evidenced intracytoplasmic clusters of type A particles, which were attributed to activation of endogenous virus by the disease (Perk *et al.*, 1979); (5) mouse peritoneal macrophages were shown to release endogenous viruses following fusion with human cells (Huebner and Croce, 1976) or spontaneously (Sher *et al.*, 1983). In the latter report endogenous retrovirus expression by macrophages was suppressed coincident with the development of Rad-MuLV-induced leukemia.

In conclusion, though in most cases their permissiveness has ranked low as compared to other susceptible cell types, nontransformed cells of the monocyte–macrophage lineage can serve as sites of replication for several oncoviruses. Possible implications for host invasion and for persistence are discussed below.

By analogy with nononcogenic virus infections (Section 3.1), it is reasonable to infer that oncovirus replication in cells of the monocyte–macrophage lineage is a critical event both for the initial invasion of the host and for viral persistence (Gazzolo *et al.*, 1979a). Evidence that this assumption is correct stems from the already cited studies with Rg FLV and FeLV. In either system a close correlation was observed between productive infection of mature peritoneal macrophages and failure of leukemia to regress. Indeed, presence or absence of macrophage infection by Rg-FLV was an excellent indicator of subsequent progression or regression, respectively (Marcelletti and Furmanski, 1979; Johnson *et al.*, 1980). The outcome of FeLV may vary in individual cats; some animals become persistently viremic, excrete large amounts of FeLV for prolonged periods, and experience a high incidence of leukemia, while others eliminate the infection rapidly (Essex and Worley, 1981). Though no differences were observed in macrophage permissiveness between progressor and regressor adult cats, the mean permissiveness of kitten macrophages was five times that of adult macrophages, a difference that correlated with the greater susceptibility of weanling animals to infection and tumorigenesis.

The finding that the macrophage is one of the cell types where FeLV infection is amplified in progressing infections is also consistent with the observation that hydrocortisone, which greatly enhances cat susceptibility to FeLV, increases the permissiveness of macrophage cultures up to 1000-fold and renders them as permissive as embryonic cells. Since cellular DNA synthesis was only doubled by the treatment, such striking increase of macrophage permissiveness could not be explained by the higher numbers of cells entering mitosis (Hoover *et al.*, 1981). Corticosteroids also induced FeLV production by otherwise virus-negative macrophages and macrophage precursors of latently infected cats (Rojko *et al.*, 1982). Recent circumstantial evidence indicates that replication in macrophages contributes an important source of virus and facilitates tumor progression also in the ASV system (Wainberg *et al.*, 1983).

3.3. CHANGES OF RES FUNCTIONS DURING RNA TUMOR VIRUS INFECTIONS

This aspect was first approached by following the removal of foreign particulate material from the circulation of mice infected with acute leukemia viruses, such as FLV and RLV. Histologically, carbon-laden macrophages were progressively fewer in spleens and either decreased or increased in livers (Old *et al.*,

1961; Elliott and Schloss, 1971; Seidel and Nothdurft, 1976). Uptake of xenogeneic red cells by FLV-infected spleens was also reduced, especially when calculated on an organ weight basis (Wedderburn and Salaman, 1968). However, in the course of ALV infections—due to early neoplastic infiltration of spleen and liver—particle removal from the blood is an unsatisfactory parameter of general RES function. Moreover, most murine oncovirus preparations used in early studies were probably contaminated by LDH virus, which is known to affect RES function (Rowson and Mahy, 1975). Macrophages crammed with cell debris and conferring a "starry sky" appearance onto lymphoid tissue, originally considered characteristic of FLV- and RLV-induced leukemias, were subsequently related to LDH virus (Proffitt *et al.*, 1972). In spontaneous lymphoma-bearing mice carbon clearance was unaffected (Boggs and Schwartz, 1978).

Subsequent investigations focused on macrophage functions as studied *in vitro*. Interestingly, phagocytosis was effected normally or near normally by productively infected (Cox and Keast, 1973; Gazzolo *et al.*, 1974, 1975; Durban and Boettiger, 1981b; Bendinelli *et al.*, 1980a; Price and Smith, 1982) and also by transformed (Raschke *et al.*, 1978) macrophages even after years of propagation *in vitro* (Ralph and Nakoinz, 1981). Properties such as motility, ability to spread, and chemotactic responses were more promptly and consistently impaired, and this often occurred early after infection (Friedman and Ceglowski, 1971; Bendinelli, 1973a; Komitowski and Kurka, 1974; Fimmel *et al.*, 1975; Levy and Wheelock, 1975; Marcelletti and Furmanski, 1978; Normann *et al.*, 1981).

Other alterations of mononuclear phagocyte functions described in conjunction with oncovirus infections include impaired killing of ingested bacteria (in RLV-infected mice; Cox and Keast, 1973), enhanced production of granulocyte–macrophage colony-stimulating factor(s) (following infection with a variety of leukemogenic and sarcomagenic murine viruses; Greenberger *et al.*, 1980), decreased responsiveness to soluble mediators (in A-MuLV-infected mice; Broxmeyer *et al.*, 1980), increased natural cytotoxicity (in MuMTV-infected mice; Tagliabue *et al.*, 1980), and reduced antibody-dependent cytoxicity (in mice infected with FLV; Genovesi *et al.*, 1982). Quantitative and qualitative perturbations of bone marrow myelomonocytic precursors such as augmented responsiveness to granulocyte–macrophage colony-stimulating factor(s) were also described (Lieberman and Sachs, 1979; Klein *et al.*, 1982). The finding that macrophages of oncovirus-infected animals may suppress lymphocyte functions or fail to cooperate normally with lymphoid cells in the generation of immune responses is discussed in Section 5.

The mechanisms whereby macrophages are driven to change functional behavior during oncovirus infections are not known. At least some modifications may be due to activation by the concomitant antiviral immune response (Tagliabue *et al.*, 1980) or by the release of lymphokines (David and Remold, 1976). A more direct involvement of the infecting virus or its products is also possible. A correlation between alteration of macrophages and oncovirus replication in these cells has repeatedly been noted (Marcelletti and Furmanski, 1978; Greenberger *et al.*, 1980; Biasi *et al.*, 1983). Moreover, injecting mice with nanogram amounts of extracts prepared from FLV, RLV, and Mo-MuLV caused a generalized inhibition of macrophage accumulation at sites of inflammation, and the activity could be reproduced with the purified protein p15(E). Similar chemotaxis inhibitors were detected also in extracts of several primary and transplantable tumors but not in extracts of MuMTV, FeLV, and other oncoviruses (Cianciolo *et al.*, 1980a, b). Thus, macrophage alterations may also be secondary to tumor induction (Normann *et al.*, 1981) as well as to damages of other cell types (Pack and Chapman, 1980).

Collectively, these observations show that in RNA tumor virus–infected animals the monocyte–macrophage lineage may present considerable, though as yet ill defined, alterations. The impact of such changes on host invasion and pathogenesis is poorly understood. Rg-FLV infected mice, whose peripheral macrophages were normal-functioning at 1 month of infection, eventually regressed, whereas those with impaired macrophages invariably progressed. Inoculation of normal macrophages led to regression of otherwise progressing FLV-infected animals (Marcelletti and Furmanski, 1978). Intratumoral macrophages derived from regressing MuSV were more cytotoxic for various neoplastic cells *in vitro* as compared to macrophages from progressing tumors (Taniyama and Holden, 1979a). These and other results (Genovesi *et al.*, 1982) are suggestive of a causal correlation between altered macrophage function and progression of oncoviral infections and tumors. The role macrophages play in oncovirus-induced immusuppression (Section 5) may also be important.

3.4. ROLE OF RES IN HOST RESISTANCE TO INVASION BY RNA TUMOR VIRUSES

Studies in birds and mammals have shown that resistance to RNA tumor viruses is multifactorial in nature and strongly influenced by the genetic constitution of the host (Meruelo and Bach, 1983). A number of loci that govern resistance to different oncoviruses have been identified, both within and outside the major histocompatibility complex. A fairly well-defined example is the control of FLV in mice. One gene (*Fv*-1) regulates cell permissiveness to the helper F-MuLV (and to a number of other retroviruses), one (*Fv*-2) affects the defective component (Lilly and Pincus, 1973), one (*Fv*-3) governs host sensitivity to the immunodepressive effects of the virus (Kumar and Bennett, 1981), at least three genes (*Rfv*) control regression of advanced leukemia (Chesebro and Wehrly, 1979), and still others regulate additional aspects of infection (Shibuya and Mak, 1982; Silver and Fredrickson, 1983).

The mechanisms of genetically governed resistance are still poorly understood. One phenotypic trait of utmost importance is certainly the restriction of viral replication by host cells, which may occur either at the cell surface or at various intracellular steps. For example, chicken susceptibility to ALV is mainly determined by the availability of appropriate receptors for virus penetration (Beard, 1980), while *Fv*-1-controlled resistance of mice to strains of MuLV of different tropism is probably regulated at the stage of viral DNA integration (Chinski *et al.*, 1984), and ASV restriction in mammalian cells results from incomplete transcription of the integrated provirus (Varmus *et al.*, 1974). Inheritance and expression of endogenous proviral genes can also reduce susceptibility to selected exogenous retroviruses (Robinson *et al.*, 1981; Kozak *et al.*, 1984). However, resistance can also be expressed at several supracellular levels, including the control of viral dissemination and the ability to mount natural or postexposure immune responses strong enough to control efficiently infection or tumor progression (Nowinski, 1976; Lilly and Mayer, 1980). Epigenetic factors such as age, sex, and hormonal status are also important and may influence host susceptibility due to their influence on the immune system (Ruscetti *et al.*, 1982) or by more direct routes (as well exemplified by MuMTV oncogenesis; Ringold, 1983).

It is commonly accepted that mechanisms of resistance to neoplasia and to viruses are largely superimposable. In the case of nonviral tumors this is essentially an act of faith, but in the case of oncovirus-induced tumors there is solid evidence favoring such a view. It is well established, for example, that immunologic surveillance against such tumors is primarily directed against virus-specified antigens and may actually be regarded as a special case of antiviral immunity (Herberman, 1982). Furthermore, in tumor virus infections both antiviral and antitumoral defenses eventually represent a barrier against tumorigenesis. Artificial as it may be, keeping separate the two groups of mechanisms may, however, be conceptually and operationally very difficult. In this section discussion is focused on data implicating macrophages in the mechanisms that combat progression of oncovirus infections. The functions ascribed to RES in resistance to oncovirus-induced tumors are reviewed in Section 4.2. Unfortunately, in most experiments resistance has been evaluated solely by using tumor parameters (incidence, latency, and so forth), making it difficult to identify its site of action.

3.4.1. Innate Nonspecific Resistance

Oncovirus effects may manifest themselves long after infection. Since antiviral immune responses also may take weeks or months to develop, the intervention of constitutive broad-range defenses against pathogens must be postulated. The cellular compartment of such first-line defense is presently thought to be composed of mononuclear phagocytes, polymorphonuclear leukocytes, and various types of natural killer (NK) cells (Mizuno *et al.*, 1982). Preexisting humoral factors and antibody triggered by cross-reacting endogenous retroviruses may also be involved (Bendinelli *et al.*, 1974, 1980b; Nowinski, 1976).

Evidence for the participation of RES in defense against oncoviral infections has mainly been obtained by using macrophage-impairing agents. Antimacrophage serum, known to abate resistance to several nononcogenic viruses, gave conflicting results. For example, it enhanced leukemia induction by FLV in mice, but this effect was reproduced by control rabbit serum (Bendinelli, 1973b), possibly also due to altered RES function (Larson *et al.*, 1972). RES blockade with colloidal gold or carbon also gave conflicting information; for example, it inhibited FLV (Friedman and Ceglowski, 1973), but reportedly enhanced the related RLV (Buchman *et al.*, 1973; Knyszynski and Danon, 1977).

Convincing data were, instead, gathered by using more potent and specific macrophage poisons, such as microcrystalline silica or carrageenan. These substances enhanced tumorigenesis by acute and slow oncoviruses, as exemplified respectively by RLV or FLV in susceptible mouse strains (Larson *et al.*, 1972; Knyszynski and Danon, 1977) and FeLV in cats (Hoover *et al.*, 1981). The bone marrow-seeking isotope ^{89}Sr, which supposedly interferes with macrophage renewal, and carrageenan even abrogated the marked resistance to FLV exhibited by mice resistant at the *Fv-2* locus (Kumar and Bennett, 1981; Ostertag and Pragnell, 1981). The antimacrophage action of silica and carrageenan lasts for a few days only (Collins and Snyderman, 1980; Schwartz and Catanzaro, 1981). This, coupled with the observation that such substances, to be effective, had to be given just before or soon after virus inoculation, clearly indicated that RES function is essential for the early containment of oncoviruses.

Ninety-two percent of silica-treated FeLV-infected cats developed persistent viremia as compared to 9% of untreated controls, despite equivalent immune responses to the virus and to the cell-associated FOMCA antigen (Hoover *et al.*, 1981). Moreover, silica also potentiated a strain of F-MuLV that proved nonpathogenic for athymic mice, as shown by generation of dualtropic viruses and leukemogenesis (Ruscetti *et al.*, 1982). It seems, thus, possible that the RES functions required for early containment of RNA tumor viruses are independent from the generation of virus-specific immunity. However, even the selectivity of silica and carrageenan is not absolute. Both have been seen to affect lymphocyte and NK cell functions (Levy and Wheelock, 1975; Cudkowicz and Hochman, 1980) and also stimulate hematopoiesis, which might induce an accumulation of virus-susceptible cycling cells (Ruscetti *et al.*, 1982).

Less direct evidence for RES involvement in nonspecific resistance to RNA tumor viruses is that preinfection with LDH virus, an agent that selectively replicates in, and impairs, macrophages (Rowson and Mahy, 1975), increased mouse susceptibility to RLV (Hanna *et al.*, 1970). Relevant also is the observation that ^{59}Fe-labeled red cells from RLV-infected mice, once transfused into normal recipients, disappeared from the circulation and accumulated in the spleen and liver faster than normal red cells, suggesting that an unstimulated RES is capable of recognizing cell surface changes induced by RLV (Knyszynski *et al.*, 1978).

3.4.2. Nonspecifically Stimulated Resistance

Numerous immunomodulators and biologic response modifiers reduce the incidence and severity of neoplasia induced by oncoviruses and increase host survival (Table 5). In the absence of reported attempts to clarify whether such effects were due to reduced viral replication, inhibition of target cell transformation, or suppression of tumor growth, the fact that most treatments afforded resistance only when given prophylactically or initiated soon after infection suggests a primarily antiviral action.

Immunostimulants exert a myriad of effects (on immune responses, on the production of regulatory molecules, on cell cycling, and so forth), but a common denominator is that they stimulate the RES. Macrophage activation has often been invoked to explain the beneficial effects of various agents against oncoviral

TABLE 5. EFFECT OF SOME IMMUNOMODULATORS AND BIORESPONSE MODIFIERS ON RNA VIRUS-INDUCED TUMORIGENESIS[a]

Virus	Host	Inhibition	Enhancement
ASV	Chicken	Freund's complete adjuvant	Freund's complete adjuvant
		Shark liver extract	
AK-MuLV	Mouse	BCG	
		Besnoitia jellisoni (chronic infection)	
		Bordetella pertussis	
		Corynebacterium parvum[b]	
		Freund's complete adjuvant	
		Pyran copolymer	
		Toxoplasma gondii (chronic infection)	
FLV	Mouse	Ammonium 5-tungstate-2-antimoniate[c]	Bacterial lipopolysaccharide
		BCG	BCG
		Besnoitia jellisoni (chronic infection)	Leukogenol
		Bovine serum albumin[d]	Poly I:C
		Coenzyme Q10	Pyran copolymer (intravenous)
		Corynebacterium granulosum	
		Double-stranded RNA (viral and synthetic)	
		Freund's complete adjuvant	
		Keyhole limpet hemocyanin[d]	
		Poly A:U[d]	
		Poly I:C	
		Polymaleic anhydride[e]	
		Pyran copolymer (intraperitoneal)	
		Statolon	
		Streptococcus hemolyticus, killed[f]	
		Thyoglicollate	
RLV	Mouse	BCG	Freund's complete adjuvant
		Phytohemagglutinin	Freund's incomplete adjuvant
		Pyran copolymer	
		Tuftsin[g]	
MuSV	Mouse	*Corynebacterium parvum* and fractions	Poly I:C
Kirsten-MuLV	Rat	Poly I:C	

[a] Reviews Marx and Wheelock, 1976; Strauss *et al.*, 1976; Schuller and Morahan, 1979; Bendinelli *et al.*, 1980a.
[b] Chany *et al.*, 1983.
[c] Larnicol *et al.*, 1981.
[d] Nahon-Merlin *et al.*, 1979.
[e] Goodell *et al.*, 1978.
[f] Tsuruo *et al.*, 1980.
[g] Knyszyanski *et al.*, 1983.

infections. In most instances supporting evidence was, at best, correlative. For example, the increased resistance of mice bearing chronic protozoan infections to FLV and AK-MuLV was paralleled by a potentiation of the typically macrophage-dependent resistance to *Listeria monocytogenes* (Hibbs *et al.*, 1971), and the synthetic polyanionic product pyran copolymer was shown also to protect mice that had been immunologically depleted by thymectomy and antilymphocytic serum (Hirsch *et al.*, 1973). However, in a few cases, direct evidence for an involvement of RES also exists. Administration of silica reduced or abolished the protection against FLV conferred on mice by pyran and statolon (Wirth *et al.*, 1976; Schuller and Morahan, 1977). Peritoneal macrophages from donors treated with pyran, BCG, or killed Streptococcus protected untreated mice, inhibiting viremia as well as foci of spleen transformation and leukemia progression. In similar experiments, resident and thioglycolate-elicited macrophages were ineffective, suggesting the existence of a critical level of activation (Schuller and Morahan, 1977; Goodell *et al.*, 1978; Tsuruo *et al.*, 1980). It is obvious, however, that immunoenhancing agents may affect oncoviral infections by many diverse means, including modulation of defenses other than macrophages, reversal of virus-induced immunodeficiency (Section 5), and modification of endogenous retrovirus expression (Nahon-Merlin *et al.*, 1979).

As shown in Table 5, certain immunomodulators also increased the severity of oncoviral infections and, in some instances, the same substance ameliorated or aggravated the course of infection and disease depending on variables such as dose, route of inoculation, timing, and animal strain or age. The possible mechanisms for such paradoxic effects include influences on selected immune and RES compartments shifting the balance toward less effective resistance, or expansion of target cell populations for oncogenesis (Schuller and Morahan, 1979). Reportedly, the mere administration of an antigen such as heterologous red cells potentiates viral leukemogenesis (Kouttab and Jutila, 1972).

3.4.3. Immune Responses

As mentioned earlier, certain resistance genes do specify events that occur after a given oncoviral infection is fully established. They succeed in eliminating the infection or in reducing the viremia and prolonging tumor latency. Since the resistance they govern can often be modulated by manipulating the host's immune system, most such genes are believed to regulate immunologic responses to oncovirus-associated antigens. In addition to the *Rfv* loci already cited, examples in mice include loci that appear to control antibody responses to Mo-MuLV (Boyer *et al.*, 1980), the *Rgv*-1 locus that influences susceptibility to spontaneous leukemia and to Rad-MuLV, G-MuLV, and MuMTV (Lilly and Pincus, 1973), and the loci *Av*-1 and *Av*-2 that confer resistance against A-MuLV (Risser *et al.*, 1978).

A number of studies have investigated immune responses to oncoviruses and oncovirus-infected and transformed cells. The virion antigens involved have been briefly outlined (Section 2.2). Determinants expressed on the surface of oncovirus-infected cells include virion proteins and glycoproteins and their precursors. Animals immunized with oncoviruses or oncovirus-infected cells may develop antiviral antibodies (detectable in neutralization, radioimmunoassays, complement-dependent cytoxicity, and so forth) and sensitized T lymphocytes (demonstrable in cytotoxic, blastogenic, MIF, and other *in vitro* assays as well as in adoptive transfer). Under conditions of preimmunization these responses have proved capable of protecting animals against several oncoviruses and also against the graft of oncovirus-transformed cells, but their relative roles are still not clearly defined.

During the course of active infections in unimmunized susceptible hosts, immune effectors are elicited, but often remain feeble and fail to eliminate the infection or to cope with appearance and progressive growth of transformed cells. Nevertheless, such effectors contribute to overall host resistance, as shown by the increased frequency of progressing infections induced by immunosuppressive treatments, which do not ablate target cells. However, the action of immune factors is not one way. Immunologic potentiation of oncovirus-induced tumors has been described (Herberman, 1982), and suggestions have been put forward that chronic immune stimulation by oncoviral antigens might be instrumental in permitting the critical interaction for leukemogenesis between certain slow oncoviruses and target cells (Ihle and Lee, 1982). The role played by cells of the monocyte–macrophage series in this complex context is far from clear.

Macrophages exert critical functions in the early activation of lymphoid cells by stimulants in general (Unanue, 1981). The finding that macrophages are essential also to the inductive phases of immune responses against oncoviruses is supported by elegant experiments showing their requirement for the *in vitro* induction of T-lymphocyte responses against oncoviral antigens. Nonsensitized mouse spleen cells could be stimulated to generate a primary cytotoxic T-cell response against the FMR antigen (a marker shared by FLV, Mi-MuLV, and RLV) only when incubated with FMR-positive stimulator cells in the presence of spleen or peritoneal Ia-positive macrophages (Gomard *et al.*, 1981a). Metabolically active macrophages were also needed for triggering the secondary cytotoxic T-cell response against cell-free and cell-associated antigens of oncoviruses such as Rad-MuLV (Treves *et al.*, 1976) and MSV (Taniyama and Holden, 1979b; Biasi *et al.*, 1983). A double path to T-cell stimulation with different requirements for presenting macrophages has also been postulated (Wybier-Franqui *et al.*, 1982).

Macrophages were also needed for the production of lymphokines and for the inhibition of leukocyte adherence by immune rodent lymphoid cells stimulated by oncoviral antigens *in vitro* (Landolfo *et al.*, 1977; Sharma *et al.* 1979; Koppi and Halliday, 1982), although divergences have been noted between cell-associated and cell-free antigens (Landolfo *et al.*, 1978). In contrast to most antigen-induced blastogenic responses, the *in vitro* proliferation of T lymphocytes from Mo-MuLV- and MuSV-infected mice in response to the gp71 and p12 components of the infecting virus proved, instead, largely macrophage-independent (Enjuanes *et al.*, 1981).

Precisely what function(s) macrophages perform as accessory cells for the induction of immune reactions in general is still essentially unknown, and responses to oncovirus antigens are no exception. The usual possibilities (improvement of culture conditions, uptake of antigen/virus from stimulator cells or medium and its processing and presentation, and so forth) have been suggested. In one study, macrophages did not appear to interact with precursors of cytotoxic T cells directly but through intermediate helper T cells (Gomard *et al.*, 1981b). Even the need for a genetic compatibility between interacting cells and the role of soluble mediators in such responses is controversial (Ihle and Lee, 1982; Biasi *et al.*, 1983).

There are no data on the role of the RES in the efferent arm of immune mechanisms to oncoviruses. As reported, experiments in this area have focused on events occurring after tumor onset. However, the results contain elements that may implicate macrophages, not only in regression or amelioration of the induced tumors, but also in immunologically mediated containment of established oncoviral infections (see Section 4.2.2).

4. RES INVOLVEMENT IN ONCOGENESIS BY RNA TUMOR VIRUSES

As extensively reviewed in volume 5 of this series (Herberman and Friedman, 1983), RES involvement in naturally occurring and experimental neoplasia is manyfold. Briefly, the RES may represent the site of origin and secondary implantation of tumors, may be affected in its functioning by the tumorous process, and may constitute a barrier against tumor development and spread. It is possible that during growth of oncovirus-induced tumors, macrophages may experience a similarly wide range of involvement. However, only those aspects that appear unique to, or strongly influenced by, the oncoviral etiology of the tumor are discussed here.

4.1. THE MONOCYTE–MACROPHAGE LINEAGE AS TARGET FOR TRANSFORMATION

As discussed in Section 3.2, the range of cell types that are transformed by oncoviruses is usually narrower than the spectrum of cells supporting their replication. Moreover, the induced tumors generally present a remarkable specificity and uniformity: they are usually composed of cells belonging to a single lineage (predominantly hemopoietic or fibroblastic, though *in vivo* the precise origin may be difficult to assess) and displaying a defined phenotype of differentiation that may occasionally present minor differences in individual tumors (Boyer *et al.*, 1980). In fact, transformation by oncoviruses appears to depend on the expression of specific differentiated functions by the infected cell, though the mechanisms controlling such selectivity are essentially unknown. It has been suggested that slow oncoviruses integrate near a particular cellular oncogene only in cells where such a gene is actively expressed (Hayward *et al.*, 1982). However, this explanation may not hold true for acute oncoviruses, whose oncogenes are expressed also in infected nontransformed cells, thus confirming that transformation occurs only when viral products interact with specific phenotypic programs of the cell (Graf *et al.*, 1976). It is also still unclear whether cells can serve as transformation targets at various stages, or only at a particular point in differentiation (Whitlock *et al.*, 1983).

These and related general problems are well illustrated also by oncoviruses that transform cells in the monocyte–macrophage series (Moscovici and Gazzolo, 1982). As summarized in Table 4, there are several viruses that infect, but do not appear to transform, such cells. For example, productive AEV infection *in vitro* did not transform chicken macrophages, despite the fact that the v-*onc* product was expressed, thus clearly showing that transformation of this cell type also results from the interaction between viral and cell lineage–specific functions (Durban and Boettiger, 1981a). In keeping with this conclusion, selected AMcV mutants with deletions in the *myc* oncogene had a drastically reduced ability to transform macrophages while retaining the capacity to transform fibroblasts *in vitro* and *in vivo*. On passage through chick embryo cells, a revertant was obtained that had reacquired the missing sequence and transformed macrophages efficiently, even though in susceptible animals, it induced a spectrum of tumors different from that caused by the wild type (Bister *et al.*, 1982; Ramsay *et al.*, 1982; Enrietto *et al.*, 1983).

Neoplasms induced by oncoviruses can vary depending on the animal host species. For example, in chickens AEV inoculation induced a pure erythroleukemia, but in quails it caused a leukemia involving both erythroid and myeloid cells (Moscovici *et al.*, 1981). Strains of AMcV showed differences in the inability to transform chicken and quail macrophages (Linial, 1982). Microenvironmental factors are also important. Bone marrow cells infected *in vitro* with AMbV strain E26 showed an outgrowth of transformed erythroblasts or myeloblasts, depending on conditions of culture and on the presence of appropriate growth factors (Radke *et al.*, 1982; Moscovici *et al.*, 1984).

Usually, RNA virus-transformed cells present a mosaic of differentiation markers, which diverges from that of mature cells and, more or less faithfully, resembles that of normal differentiating cells. Thus, chicken mononuclear cells transformed *in vivo* or *in vitro* by AMcV exhibit properties of mature macrophages (morphology, surface antigens, Fc and C3 receptors, immune and nonimmune phagocytosis) but, like precursor cells, proliferate rapidly and adhere loosely to surfaces. Neoplastic cells that accumulate in the blood of AMbV-diseased chickens and cells transformed *in vitro* by this virus are more immature, since they engulf latex particles and express Fc receptors, but are nonadherent, lack C3 receptors and immune phagocytosis, and display enzymatic and antigenic markers typical of precursors (Gazzolo *et al.*, 1979a; Durban and Boettiger, 1981a). When treated with a tumor promoter, these cells differentiate further along the macrophage pathway (Pessano *et al.*, 1979), as did cells transformed by *ts* mutants of AMbV, once shifted to the nonpermissive temperature (Graf *et al.*, 1983; Moscovici and Moscovici, 1983).

It has been generally assumed that the immature phenotype of transformed cells is due to recruitment of target cells among precursors whose maturation is then blocked and/or deranged. Nonetheless, pure cultures of functionally mature chicken macrophages were efficiently transformed by AMbV and AMcV, thus implying that the immature phenotype may also result from dedifferentiation or deregulation of fully differentiated cells (Linial, 1982; Moscovici and Gazzolo, 1982). Though it was argued that in such experiments transformed cells originated from contaminating immature myeloid cells (Graf and Beug, 1978; Graf *et al.*, 1981), it seems likely that all stages of macrophage differentiation, from the committed progenitor to the mature macrophage, may serve as targets for infection by AMbV and give rise to transformed progeny cells (Boettiger and Durban, 1984). However, treatment with a tumor promoter causes chick embryo macrophages to become activated (tumor promoters may affect various properties of mononuclear phagocytes with disparate consequences; Keller *et al.*, 1982) and resistant to transformation by AMbV (Pessano *et al.*, 1979). Efficiency of transformation appears proportional to the DNA-synthesizing activity of cultures at the time of infection (Durban and Boettiger, 1981a).

The transforming potential of oncoviruses includes cells expressing markers of the monocyte–macrophage lineage more frequently than previously believed, on the basis of the histotype of tumors that develop *in vivo*. This is the unexpected conclusion of recent experiments using long-term murine bone marrow cultures (Dexter and Testa, 1976). Thus FLV infection, whose transformation targets *in vivo* are erythroid precursors (Tambourin *et al.*, 1979), led to the establishment of cell lines exhibiting morphologic, functional, and cytochemical characteristics typical of mononuclear macrophages (Greenberger *et al.*, 1979; Testa *et al.*, 1980). Similar findings were reported with A-MuLV (Greenberger *et al.*, 1979), MuSV (various strains and pseudotypes; Greenberger, 1979), and G-MuLV (Hays *et al.*, 1980). It is unclear, however, whether these results are due to the *in vitro* environment favoring transformation of particular cell types or to unmasking of events which *in vivo* remain obscured by overgrowth of other cells. It is also conceivable that the transformed cells belong to a given pathway of differentiation, but are led to express traits that are normally depressed. In keeping with this possibility, FLV-transformed mouse bone marrow cells concomitantly exhibited myeloid and erythroid markers, whose expression could be independently modulated (Revoltella *et al.*, 1979), and one A-MuLV-transformed B lymphocytic line could be induced to acquire macrophagelike properties (Boyd and Schrader, 1982).

These results have led to reconsideration of the significance of the myelomonocytic leukemias and cell lines obtained from mice infected with viruses that usually induce other types of leukemia, such as FLV or RLV, under conditions that retard or prevent leukemia development (Fredrickson *et al.*, 1972; Clarke *et al.*, 1976; Shibuya and Mak, 1982; Heard *et al.*, 1983; Nobis and Löhler, 1983), or following *in vitro* passage of A-MuLV (Raschke *et al.*, 1978; Hines and Gragowski, 1983) and Mo-MuLV (Devani *et al.*, 1983) lymphomas. However, several monocytic lines obtained from RNA virus-infected bone marrow cultures were not malignant, in that they never gave rise to *in vivo* transplantable tumors or to granulocyte–macrophage colony-stimulating factor-independent immortalized lines (Heard *et al.*, 1981). Infection by a number of oncoviruses can enhance the production of this growth factor by macrophages (Greenberger *et al.*, 1980) and fibroblasts (Koury and Pragnell, 1982), and this effect may play a part in the establishment of malignant and nonmalignant monocytic lines (Heard *et al.*, 1983).

Generally, the transformed mononuclear phagocytes released the transforming oncovirus permanently, but replication-defective viruses also generated nonproducer lines from which they could be rescued by superinfection with appropriate helper viruses (Raschke *et al.*, 1978; Graf *et al.*, 1979). Chicken macrophages transformed *in vitro* by AMcV strain E26 produced ratios of defective-to-helper virus higher than transformed fibroblasts, owing to enhanced production of the defective component (Hu *et al.*, 1978, 1979).

In conclusion, macrophages or their precursors are the primary transformation target for certain avian acute oncoviruses, but there is evidence suggesting that they may be transformed at relatively low frequency also by viruses that usually induce tumors of other histotypes. A murine oncovirus (Cas-MuLV) that appears to cause a wide spectrum of hematopoietic tumors, including monocytic leukemias and macrophage tumors, has recently been described (Fredrickson *et al.*, 1984). Macrophage-related cells, such as glial cells, can also be transformed by RNA tumor viruses (Bigner and Palgram, 1976). Macrophages may also be involved in events leading to transformation of other cell types (Boniver *et al.*, 1981; Johnson *et al.*, 1984).

4.2. ROLE OF RES IN RESISTANCE TO RNA VIRUS-INDUCED TUMORS

Studies on host resistance to neoplasia have led to the realization that development and spread of tumor cells are subject to the influence of various defense factors. These include RES-mediated mechanisms, which, although still under scrutiny, are believed capable of discriminating malignant from normal cells (Herberman and Friedman, 1983). Although in nonviral tumors there are difficulties in identifying markers, which may serve for recognition by such resistance mechanisms, virus-induced tumors have conspicuous antigenic and biochemical surface changes, which may represent elements useful for discrimination. Nonetheless, information concerning RES participation in resistance against oncovirus-induced neoplasms does not go very far beyond data derived from a number of virus–host systems showing that macrophages are, indeed, important effectors of resistance.

4.2.1. Nonspecific Resistance

In recent years, studies on resistance to tumors have switched emphasis from specific, adaptive immune mediators to nonspecific effectors and have implicated macrophages as important cellular elements, which might kill tumor cells or prevent their growth. *In vitro*, macrophages can be nonspecifically activated to destroy tumor cells in a way that spares normal cells, and although the relevance of this phenomenon to tumor control *in vivo* and the biochemical steps involved still need definition, there is circumstantial evidence for a major role in host defense against primary and metastatic tumors (Evans and Haskill, 1983; Stern, 1983; Zacharchurk *et al.*, 1983). In contrast to the exquisite specificity of sensitized cytolytic T cells, activated macrophages are cytotoxic to a wide range of tumor cells, like NK cells, even though their modes of action appear to differ (Wiltrout *et al.*, 1982).

There is little evidence implicating macrophages in nonspecific defenses against primary oncovirus-induced tumors. Whether this reflects macrophage defects caused by the infecting virus (Section 3.3), inappropriate investigation, or masking of the generally feeble natural defenses by immunologically specific mechanisms—much more vigorous in viral than in nonviral tumors—is not known. Large numbers of mononuclear macrophages have been shown to infiltrate oncoviral tumors from the early phases of growth (Metcalf *et al.*, 1967; Stanton *et al.*, 1968; Siegler, 1970; Puccetti and Holden, 1979). Macrophage accumulation occurs also in immunocompromised animals (Becker and Haskill, 1980), indicating that recruitment is not necessarily immunologically stimulated, whereas the cytotoxic capacity of macrophages accumulated by such hosts is still controversial (Taniyama and Holden, 1979a; Becker and Haskill, 1980). Functions and mechanisms of attraction of intratumoral macrophages in general are still a matter of debate (Evans and Haskill, 1983).

Oncoviral infection might affect RES-mediated nonspecific mechanisms in several ways, including influences on the susceptibility to lysis of target cells and modulation of macrophage antitumoral activity. Virus infection has been shown to increase the sensitivity of cultured cells to macrophage cytostasis (Goldman and Hogg, 1978), but there is no convincing evidence that oncoviral structures serve as recognition sites for macrophages and other effectors of natural antitumor activity, or that the oncoviral etiology correlates with tumor cell sensitivity to killing or cytostasis. However, one cannot be certain that a given cell does not express any endogenous virus marker (Croker *et al.*, 1979). As for the second possibility, during oncoviral infections macrophages may present traits typical of the activated state (Section 3.3), and their antitumoral activity might, therefore, be concomitantly potentiated. However, at least *in vitro*, the tumoricidal phenotype of activated macrophages persists for a few days only and then becomes refractory to restimulation (Poste and Kirsch, 1979). The antitumor activity of interferon, one of the factors that might contribute to macrophage activation, appears to be mediated by host cells, but is little affected by treating the host with silica (Belardelli *et al.*, 1982).

4.2.2. Immune Resistance

RNA virus-induced tumors are strongly immunogenic, expressing a complex mosaic of antigens including virus components, derepressed cellular antigens and other determinants (the putative tumor- and/or transplantation-specific antigens described in several systems, for example), whose origin and relation to the former molecules are ill understood (Chandra, 1983). These antigens may serve as targets for an array of specific effector mechanisms: T-dependent cytolytic, cytostatic, and graft responses, antibody and complement, antibody-dependent cell cytotoxicity, and so forth. As a result, immunologic mechanisms acting directly against transformed cells contribute significantly to overall resistance to oncovirus-induced tumorigenesis (Herberman, 1982). The importance of immunologic factors in this context is revealed, among other evidence, by the spontaneous regression of certain established oncoviral tumors and by instances of successful immunotherapy, two situations in which direct RES contribution is documented.

While most oncoviral tumors grow progressively until the host succumbs, others may regress with a frequency that varies depending on virus and host factors. One system where the effectors of regression have been studied is the erythroleukemia induced by Rg-FLV. This disease develops progressively for several weeks, but then in a proportion of animals may be followed by a gradual return to normality of spleen mass, histology, and other pathologic manifestations. Even though the virus persists and relapses may occur, regression is accompanied by a marked reduction of viral levels in tissues (Rich *et al.*, 1969; Dawson and Fieldsteel, 1971). A great deal of evidence indicates that immune responses are of primary importance in determining the process; for example, immunosuppressive treatments convert regressing into lethal forms (Furmanski *et al.*, 1979), and remission is accompanied by recovery of general immunologic competence (Cerny and Essex, 1979) as well as by the appearance of virus-specific antibodies and cytotoxic lymphocytes (Chesebro and Wehrly, 1976; Marcelletti and Furmanski, 1978; Johnson *et al.*, 1980).

Several pieces of evidence indicate that macrophages play a central role in regression of leukemia induced by Rg-FLV: (1) A single dose of silica at 4 weeks from infection delayed time and frequency of regression, and the effect was reproduced by a battery of other macrophage-toxic agents, including antimacrophage serum, carrageenan, trypan blue, and ^{89}Sr, and prevented by the macrophage-protecting substance poly-2-vinylpyridine-*N*-oxide. (2) Macrophage infection and impairment correlated with progression. Apart from the slow-acting ^{89}Sr, to affect regression, macrophage poisons had to be given late in the course of disease, thus suggesting that macrophage activity was critical at an advanced stage of disease and presumably worked in concert with immune mechanisms. (3) Some progressors developed vigorous antiviral immune responses, indicating that immunity is not in itself sufficient for regression to occur. (4) Inoculation of normal resident peritoneal macrophages achieved regression in a significant proportion of otherwise progressing mice infected with conventional FLV (Marcelletti and Furmanski, 1978; Furmanski *et al.*, 1979). Since relatively few macrophages were sufficient to exert this effect, an amplification mechanism such as reversal of virus-induced immunosuppression (Bendinelli *et al.*, 1975) was considered feasible. Macrophages were ineffective in a strain of mice inherently incapable of limiting Rg-FLV and possibly of mounting efficient anti-FLV immune responses (Chesebro and Wehrly, 1976).

Other RNA virus-induced tumors that can regress are the fibrosarcomas induced by MuSV. Regression is the usual outcome of these tumors, if infection is initiated in immunocompetent adult mice (Levy and Leclerc, 1977). In this system, T-cell responsiveness is essential (Collavo *et al.*, 1980), but macrophages also appear to be important. First, macrophages infiltrating such sarcomas are particularly numerous (Holden *et al.*, 1976). Second, macrophages were more numerous and cytotoxic in regressing than in progressing tumors (Russell *et al.*, 1976; Taniyama and Holden, 1979a). Further, there are indications that the failure of mice neonatally infected with Mo-MuLV to have a regression of MSV-induced tumors may be partly due to a macrophage deficit (Biasi *et al.*, 1983).

Evidence is scanty for a role of macrophages in successful immunotherapy of oncovirus-induced tumors. The antileukemic effect obtained by grafting syngeneic immune spleen cells to FLV-infected mice was markedly reduced by administering silica to the treated animals (Wirth *et al.*, 1976). An intact RES was also essential for therapy of oncovirus-induced tumors with a virus-specific homoantiserum (Gerlier *et al.*, 1979), though not for curing a FLV-induced leukemia with a xenoantiserum (Collins and Snyderman, 1980). In a chemoimmunotherapy protocol of Friend leukemias, both transplanted and virus-induced, macrophage-rich peritoneal cells from immunized mice were substantially more active than spleen and lymph node cells derived from the same donors (unpublished results).

5. RES INVOLVEMENT IN RNA TUMOR VIRUS-INDUCED PATHOLOGY OTHER THAN TUMORIGENESIS

RNA tumor viruses may cause a variety of nonneoplastic manifestations. In the avian system examples are thrombocytopenia, anemia, bone marrow and lymphoid tissue atrophy, stunting, alterations of metabolism, obesity, ataxia, and osteopetrosis caused by strains of ALV (Smith, 1982; Smith and Schmidt, 1982; Carter *et al.*, 1983). Severe erythroid aplasia and thymic atrophy occur frequently also in FeLV-infected cats, which may also exhibit abortions, runting, and glomerulonephritis (Essex and Worley, 1981). Nonneoplastic disorders associated with oncoviruses in mice include hyperplasia of lymphoid organs, induced for example by F-MuLV (Bendinelli and Nardini, 1973a) and A-MuLV (Siegler *et al.*, 1972), hematocrit changes caused by FLV and RLV (Ostertag and Pragnell, 1981), and neurogenic paralysis associated with retroviral infections of wild mice (Gardner, 1978; Fredrickson *et al.*, 1984). FLV also produced runting, autoimmune hemolytic anemia, and glomerulonephritis in rats (Kuzumaki *et al.*, 1974).

The mechanisms underlying oncovirus-induced nonmalignant affections have been little studied, but are probably multiple. Although the broad range of pathology occurring in infections (e.g., ALV and FeLV) may be due to coexistence of different etiologic agents, some cloned oncoviruses continued to cause both neoplastic and nonneoplastic diseases (Smith and Morgan, 1982). Some manifestations are secondary to tumor development. Examples are the abnormal gait and paraplegia due to malignant infiltration of paravertebral muscles seen in A-MuLV-infected mice (Rosenberg, 1982) and the end-stage terminal anemias found in leukemic cats. Other manifestations, including early anemias observed in FeLV-infected cats (Essex and Worley, 1981) and mammary hyperplastic nodules caused by MuMTV (Cardiff and Young, 1980), may instead be considered preneoplastic changes. Still others, such as the renal lesions in various virus/host combinations, are probably immunopathologic (Fleissner and Snyder, 1982; Sanfilippo *et al.*, 1982). However, the only nonneoplastic manifestation in which the possibility of a direct RES involvement has been investigated to some extent is immunodeficiency.

As first observed in mice inoculated with FLV (Old *et al.*, 1961), many RNA tumor viruses adversely affect the immunologic homeostasis of the host (Table 6). The effect, which has attracted much attention, is not an experimental artifact. Cats infected with FeLV under natural conditions exhibit a severe immunosuppressive syndrome, as shown by reduced responses to mitogens, antigens, and allografts and depressed levels of peripheral blood lymphocytes (Essex, 1977). Similarly, mice genetically infected with MuLV show immunodepressive changes, which become more evident in the preleukemic stages (Hatten and Dunton, 1978). Oncoviruses have also been implicated in the etiology of a severe acquired immunodeficiency in humans (Gallo *et al.*, 1983; Popovic *et al.*, 1984) and other primates (Stromberg *et al.*, 1984).

Retrovirus-infected animals are at risk of developing superinfections and of succumbing to microorganisms easily controlled by uninfected animals. Indeed, it was suggested that in free-living cats FeLV causes higher mortality by func-

TABLE 6. RNA TUMOR VIRUSES THAT HAVE BEEN ASSOCIATED WITH IMMUNODEFICIENCY

Host	Acute viruses	Slow viruses
Chicken	AEV	AEV-associated virus
	AMbV	AMbV-associated virus
	AMcV	AMcV-associated virus
	ASV	ASV-associated virus
	REV	REV-associated virus
Mouse	FLV	AK-MuLV
	MuSV	DMBA-MuLV
	RLV	F-MuLV
		Graffi-MuLV
		G-MuLV
		Mo-MuLV
		MuMTV
		Rad-MuLV
		Urethane-MuLV
		334C-MuLV
Rat		Mo-MuLV
Cat		FeLV
Baboon		Baboon endogenous virus
Macaque		Mason–Pfizer monkey virus
		Other type-D viruses
Humans		Human T lymphotropic virus III
		Human immunodeficiency virus

tioning as an immunodepressant than by acting as an oncogenic agent (Cerny and Essex, 1979), and for years in the laboratory the immunodeficiencies caused by oncoviruses have been regarded as a nuisance, because infected animals died of infections before they could develop tumors (Weiss, 1984). On the other hand, the fact that immunodepression often precedes the development of recognizible neoplasia has suggested that it may facilitate the full expression of the pathogenic potential of RNA tumor viruses and help explain the paradox of tumors that are highly immunogenic and, nevertheless, often grow to kill the host (Specter and Friedman, 1978; Cerny and Essex, 1979; Genovesi *et al.*, 1982; Rup *et al.*, 1982).

The wide range of immune functions that have been found suppressed in oncoviral-infected animals have been extensively reviewed (Dent, 1972; Specter and Friedman, 1978; Cerny and Essex, 1979). Briefly, the range of humoral and cell-mediated parameters affected, as well as the extent and rapidity of suppression, varied considerably with the virus type and variant, with the characteristics of the host (e.g., species, strain, age), and with modalities of infection (e.g., exogenous versus endogenous, timing, dosage). With few exceptions, such effects were paralleled by a number of alterations of immunocompetent cells as detected in a variety of *in vivo* and *in vitro* assays. For example, modifications of T and B lymphocytes encompassed such diverse aspects as availability of precursors (Friedman and Ceglowski, 1968), plasma membrane architecture (Farber *et al.*, 1975), surface antigens and functions (Bendinelli and Friedman, 1976), and circulatory ability (Bainbridge and Bendinelli, 1972).

Many of the aforementioned changes occurred independent of tumorigenic activity, thus confirming that oncoviruses have direct immunodepressive properties. No definite answers have, however, been forthcoming on the mechanisms whereby they are generated. For example, FLV-infected mice represent the model most extensively investigated (Bendinelli, 1981), but even in this system it has as yet been impossible to pinpoint the primary immunocompetent cell alteration(s) sustaining the immunodeficiency. The problem is compounded, because different oncoviruses appear to immunodepress by different routes, and the overall immune deficit caused by a given virus may result from the combined action of multiple independent mechanisms. Evidence implicating macrophages in the genesis of immunosuppression is discussed in Sections 5.1 and 5.2.

5.1. IMPAIRMENT OF MACROPHAGE ACCESSORY FUNCTIONS

At the time of early studies on oncovirus-induced immunodepression the accessory functions of macrophages in the inductive phases of immune responses were just beginning to be delineated, and their alteration was soon suggested as a likely mechanism (Ceglowski and Friedman, 1968; Dracott *et al.*, 1972). The first experimental evidence was, however, obtained when spleen cell cultures prepared from F-MuLV-infected mice reacquired near-normal antibody responses to sheep erythrocytes, once supplemented with peritoneal exudate cells from normal syngeneic donors. The restoring cells had properties (morphology, adherence, phagocytosis, presence of Fc receptors, absence of T-cell markers) typical of macrophages (Bendinelli *et al.*, 1975, 1976a). Since F-MuLV-infected spleen cell populations showed no numerical deficiencies of total adherent phagocytic cells, these results clearly indicated that macrophages of infected animals were qualitatively deficient.

The functions attributed to macrophages in antibody responses *in vitro* are several, including promotion of lymphocyte viability and proliferation, antigen processing or presentation, and mediation of T cell–B cell cooperation (Unanue, 1981). Restoration of F-MuLV-infected spleen cell cultures required low numbers of viable macrophages (0.5–1% responder cells) and within limits (when the added cells exceeded 10–20% little or no restoration was observed) was dose-dependent. The degree of macrophage activation was critical: spleen and resident peritoneal macrophages were consistently less active than adequately elicited peritoneal macrophages, and the nature of the eliciting stimulus was also important. Ultraviolet rays at doses known to block cell division did not reduce, and in certain instances enhanced, the beneficial activity of macrophages, whereas treatments known to interfere with protein synthesis were inhibitory. Also, it was crucial that macrophages were added during the early inductive phases of the response. Thus, collectively these results indicated that the macrophage function required for restoring F-MuLV-immunodepressed spleen cells (and presumably the macrophage deficit of the latter cells) was located in the events needed for an efficient antigen presentation (Bendinelli *et al.*, 1980a).

This interpretation was corroborated by corollary findings. Agents substituting for the lymphocyte viability/proliferation-promoting action of macrophages, such as 2-mercaptoethanol and peritoneal exudate cell-conditioned media, or agents substituting for T-cell help, such as concanavalin A–conditioned medium, failed to overcome the hyporesponsiveness of F-MuLV-infected spleen cultures. In contrast, factors known to curtail the antigen-presenting function of macrophages, such as increased antigen dosages and addition of *Escherichia coli* lipopolysaccharide or allogeneic-conditioned medium, partially circumvented the need for added normal macrophages (Bendinelli *et al.*, 1976a, 1980a). Moreover, *in vivo* antibody responses to relatively T-cell- and macrophage-independent antigens were little affected by F-MuLV (Bendinelli *et al.*, 1978), and the response to sheep red cells was markedly restored by injecting intravenously as few as 2×10^5 syngeneic peritoneal exudate cells at the time of immunization (Bendinelli *et al.*, 1976b). The fact that responses achieved by administering optimal numbers of exogenous macrophages averaged about 80% *in vitro* and 40% *in vivo* relative to uninfected counterparts was, however, consistent with the multifactorial nature of F-MuLV-induced immunodeficiency (Bendinelli and Friedman, 1976; Bendinelli *et al.*, 1979).

Further evidence implicating inefficient macrophage help was obtained in similar experiments on mice infected with the entire FLV complex. In this system, however, restoration by macrophages *in vitro* was generally of lower degree than in F-MuLV-infected cultures and *in vivo* failed completely (Ceglowski and

Friedman, 1975; Specter *et al.*, 1976, 1978; Bendinelli *et al.*, 1980a). Recent data also indicate that a deficit of soluble factors produced by macrophages and needed for optimal antibody responses may be present in FLV-infected mice (Butler *et al.*, 1983). Since helper viruses contribute significantly to the immunosuppression caused by acute oncoviruses (Bendinelli and Nardini, 1973b; Smith and van Eldik, 1978; Rup et al., 1982), the moderate immunologic restoration effected by macrophages in FLV infection may be due to reversal of the aliquot of immunodepression sustained by F-MuLV. Addition of normal macrophages also reversed the *in vitro* suppression of lymphocyte responses exerted by established lines of F-MuLV-producing tumor cells. Though it was considered unlikely that this immunodepression was virus-mediated (Ting and Rodrigues, 1979, 1982), recent studies have shown that the suppressive effects of transplantable tumors may be mediated by oncoviral products (Bluestone and Lopez, 1982).

That impaired macrophage accessory functions participate in RNA tumor virus-induced immusuppression was recently confirmed in two distinct systems. AOV inoculation of chick embryos led to a dramatic loss of lymphoid tissue mass and to generalized immunodeficiency (Smith and van Eldik, 1978). The mitogenic response of such infected birds to concanavalin A was restored by spleen or blood macrophages from normal chickens, indicating that their T lymphocytes were intrinsically normal-responsive, and that the deficit was located at the level of spleen macrophages, although these cells did not present gross alterations in Fc-receptor content, Fc-dependent phagocytosis, and staining for nonspecific esterase (Price and Smith, 1982). In the second system, Ia-positive accessory cells of neonatally Mo-MuLV-infected mice were unable to support the generation of a secondary cytotoxic T-lymphocyte response by Ia-depleted uninfected spleen cell cultures. In the latter study the macrophage deficit was selective for the antigens induced by the infecting virus (Biasi *et al.*, 1983). Possibly relevant also is the failure to cooperate with lymphocytes exhibited by adherent spleen cells of NZB mice (McCombs *et al.*, 1975), a strain that harbors a highly expressed endogenous retrovirus (Morse and Hartley, 1981).

5.2. SUPPRESSOR MACROPHAGES

Suppressor cells are considered important determinants of immunodeficiency (Oehler *et al.*, 1978; Broder *et al.*, 1978). Cells capable of specifically or nonspecifically inhibiting afferent or efferent arms of immune responses, belonging to various cell classes, and requiring or not requiring direct contact with responder cells have been described in diverse natural and experimental situations. Macrophages are key cells in this context, since they may acquire autonomous suppressive properties and also act as essential intermediaries in both regulation and expression of soluble factors and of other suppressor cells or networks (Aune and Pierce, 1981; Holt *et al.*, 1981).

Nonspecific suppressor cells have also been described in oncovirus-infected animals, including mice infected with MuSV, Mo-MuLV, FLV, F-MuLV, Rad-

MuLV, and MuMTV, cats infected with FeLV, and chickens infected with REV and other avian retroviruses (Cerny and Essex, 1979; Specter and Friedman, 1978; Rup *et al.*, 1982; Yefenof and Zilcha, 1982; Langweiler *et al.*, 1983), and were variously characterized as neoplastic cells, T and B lymphocytes, macrophages, and macrophagelike cells.

An early indication that in the course of oncovirus infections macrophages may exert suppression was obtained while investigating the antigen-independent differentiation into antierythrocyte antibody-forming cells that occurs when peritoneal lymphocytes of mice are released from restraints operating *in vivo*. The phenomenon, which has attracted attention because of its possible links with autoimmunity (Bussard and Pages, 1976), was inhibited by FLV infection of donor animals. However, when depleted of macrophages, lymphocytes of infected mice differentiated as efficiently as similarly treated normal lymphocytes, thus showing that in diseased animals macrophages were inhibiting lymphocyte activity (Bendinelli, 1968, 1970).

Subsequently, nonspecific suppressor macrophages were described in mice bearing MuSV-induced tumors. Addition of spleen cells from these animals to functionally competent lymphoid cultures inhibited mitogenic and lymphokine responses to antigens and lectins as well as to tumor-specific reactions, and responsible cells, initially identified as B cells (Gorczynski, 1974), were later definitely characterized as macrophages, which had infiltrated spleen and sarcomas (Varesio, 1983).

Other systems in which suppressor macrophages have been described include mice infected with DMBA-MuLV (a virus isolated from a thymic lymphoma induced by dimethylbenzanthracene; Roder *et al.*, 1978) and MuMTV (Tagliabue *et al.*, 1980) and murine lymphoid cultures exposed to various oncoviruses (Israel and Wainberg, 1981). More recently, macrophagelike cells of FLV-infected mice were also shown to suppress *in vitro* cytotoxic T-cell responses (Garaci *et al.*, 1981) and NK activity (Bendinelli *et al.*, 1981; Migliorati *et al.*, 1983; Moody *et al.*, 1984). Since previous studies in the same system had failed to detect macrophages capable of inhibiting elicited antibody responses (Kamo *et al.*, 1977; Bendinelli *et al.*, 1980a), cytotoxic cells may be especially prone to macrophage suppression (Brunda *et al.*, 1983).

Thus, during infection by RNA tumor viruses recruitment and activation of suppressor macrophages occur frequently. However, a number of questions remain unanswered. For example, it is not clear whether the phenomenon is caused by infection directly or by the virus-induced tumors. Suppressing macrophages and macrophagelike cells are regarded as one major component of immunosuppression in animals bearing transplantable tumors, whether originally oncovirus-induced or not (Bluestone and Lopez, 1979; Mortensen and Elson, 1980; Varesio, 1983). In DMBA-MuLV-infected mice suppressor macrophages appeared at the site of tumor growth earlier than in the spleen, raising the possibility that they were induced within the tumors and then migrated to the spleen (Roder *et al.*, 1978). In MuSV-infected mice suppressor macrophage activity and tumor size peaked together (Kirchner *et al.*, 1975). Also, the fact that no suppressor macrophages have hitherto been described in infections by slowly transforming oncoviruses unless the infected animals were overtly tumorous might not be simply coincidental.

On the other hand, it has been noted that inoculation of immunogens as diverse as parasites, bacteria, foreign or haptenated cellular antigens, and proteins also leads to activation of suppressor cells, either T lymphocytes or macrophages, and many stimuli that activate macrophages also induce suppressor activity by these cells (Oelher *et al.*, 1978). Hence, suppressing macrophages in oncovirus-infected hosts might also result from activation of feedback mechanisms known to regulate immune and other adaptive responses to infection. A further possibility is that viral invasion of immunocompetent cells deranges immunoregulatory circuit. Many RNA tumor viruses replicate in immunocytes, which may well be incompatible with the normal function of the immune system; for example, incorporation of viral proteins onto the plasma membrane, an essential step in oncovirus replication, might not only derange infected cells but also disturb interacting cells (Bendinelli, 1984).

Also unknown are the means whereby oncovirus-induced suppressor macrophages inhibit lymphocyte responses. Macrophages produce a variety of bioactive materials that inhibit immune responses (arginase, complement components, polyamine oxidase, prostaglandins, and so forth; Allison, 1978; Hume and Gordon, 1982) and may activate inhibitory substances produced by other cells (Aune and Pierce, 1981). Suppression of NK activity by macrophagelike cells of FLV-infected mice was reversed by indomethacin, indicating that the effect was mediated by endogenous production of prostaglandins (Friedman *et al.*, unpublished observations). Disorders of prostaglandin synthesis have been associated with virus-induced leukemogenesis (Karmali *et al.*, 1984). However, such substances did not appear to be responsible for the activity of MuSV-induced suppressor macrophages (Varesio and Holden, 1980). It is also possible that productively infected macrophages shed immunosuppressive products of viral origin (Israel and Wainberg, 1981).

A final point is the role of suppressor macrophages in infection and tumorigenesis by oncoviruses. A general underlying assumption is that they participate in the genesis of the overall immunodeficiency, thus contributing to prevention of the development of immune responses that might effectively control virus and tumor proliferation. However, findings that suppressor cells may function *in vitro*, but not display appreciable activity *in vivo*, warn against overestimating their significance (Hanna and Kripke, 1979; Webb *et al.*, 1980; Bendinelli *et al.*, 1982; Forni *et al.*, 1982).

6. CONCLUSIONS

This chapter has set into a coherent frame existing evidence on the interactions of the host's RES with infecting RNA tumor viruses and with the cells invaded and transformed by such viruses. Undoubtedly, the resulting picture is complex and unilluminated areas still represent a large part of the scene. This reflects (1) the broad biologic and pathogenetic pleiomorphism exhibited by RNA tumor viruses; (2) the wide number of functions currently ascribed to macrophages in viral infections as well as in tumor development and growth; and, possibly, in greater measure, (3) the many uncertainties existing about the real significance of many such functions under *in vivo* conditions. Nonetheless, the facts that have been reasonably documented are largely sufficient to establish the centrality of the RES and its functions in oncovirus infections and oncovirus-induced pathology, encouraging further research in the field. It seems feasible that many aspects will soon be elucidated when progress in technology can allow the separate analysis of cells belonging to different sublines and stages of differentiation within the monocyte–macrophage lineage.

ACKNOWLEDGMENTS. I gratefully acknowledge the colleagues who have collaborated in the experiments described in this chapter, including Drs. Herman Friedman, Donatella Matteucci, Steven Specter, and Antonio Toniolo. A special acknowledgment is due to my wife for her careful secretarial help. This work was partially supported by CNR, Special Projects "Control of Neoplastic Growth" and "Oncology," and by the Italian Ministry of Education.

REFERENCES

Aaronson, S. A., and Stephenson, J. R., 1976, Endogenous type-C RNA viruses of mammalian cells, *Biochim. Biophys. Acta,* **458:**323.

Allison, A. C., 1974a, On the role of mononuclear phagocytes in immunity against viruses, *Prog. Med. Virol.* **18:**15.

Allison, A. C., 1974b, Interaction of antibodies, complement components and various cell types in immunity against viruses and pyogenic bacteria, *Transplant. Rev.* **19:**3.

Allison, A. C., 1978, Mechanisms by which activated macrophages inhibit lymphocyte responses, *Immunol. Rev.* **40:**3.

Aune, T. M., and Pierce, C. W., 1981, Mechanism of action of macrophage-derived suppressor factor produced by soluble immune response suppressor-treated macrophages, *J. Immunol.* **127:**368.

Bainbridge, D. R., and Bendinelli, M., 1972, Circulation of lymphoid cells in mice infected with Friend leukemia virus, *J. Natl. Cancer Inst.* **49:**773.

Beard, J. W., 1980, Biology of avian oncornaviruses, in: *Viral Oncology* (G. Klein, ed.), p. 55, Raven Press, New York.

Becker, S., and Haskill, S., 1980, Non-T-cell-mediated cytotoxicity in MSV tumor bearing mice. III. Macrophage-mediated cytotoxicity against autocthonous MSV tumor-isolated target cells. *Int. J. Cancer.* **25:**535.

Belardelli, F., Gresser, I., Maury, C., and Maunoury, M-T., 1982, Antitumor effects of interferon in mice injected with interferon-sensitive and interferon-resistant Friend leukemia cells. II. Role of host mechanisms, *Int. J. Cancer* **30:**821.

Bendinelli, M., 1968, Haemolytic plaque formation by mouse peritoneal cells, and the effect on it of Friend virus infection, *Immunology* **14:**837.

Bendinelli, M., 1970, *In vitro* studies on the immunodepressive effect of Friend virus, in: *Immunity and Tolerance in Oncogenesis* (L. Severi, ed.), p. 623, Division of Cancer Research, Perugia.

Bendinelli, M., 1973a, Immunodepression by Rowson-Parr virus, in: *Virus Tumorigenesis and Immunogenesis* (W. S. Ceglowski and H. Friedman, eds.), p. 181, Academic Press, New York.

Bendinelli, M., 1973b, Effetto del siero antimacrofagi sull'infezione da virus del complesso della leucemia di Friend, *G. Batteriol. Virol. Immunol.,* **64:**1.

Bendinelli, M., 1981, Rapporti tra sistema immunitario e virus di Friend, in: *Virus Oncogeni ad RNA* (C. De Giuli Morghen, ed.), p. 318, Piccin, Padova.

Bendinelli, M., 1984, Immunomodulation in viral infections: Virus or infection-induced? in: *Immunomodulation: New Frontiers and Advances* (H. H. Fudenberg, H. D. Whitten, and F. Ambrogi, eds.), p. 161, Plenum Press, New York.

Bendinelli, M., and Friedman, H., 1976, Immunodepression by Rowson-Parr virus in mice: Lymphocyte markers and capping response of spleen and lymph node cells after infection, *Infect. Immun.* **14**:613.

Bendinelli, M., and Nardini, L., 1973a, Immunodepression by Rowson-Parr virus in mice. I. Growth curves of Rowson-Parr virus and immunological relationships with Friend virus, *Infect. Immun.* **7**:152.

Bendinelli, M., and Nardini, L., 1973b, Immunodepression by Rowson–Parr virus in mice. I. Effect of Rowson–Parr virus infection on the antibody response to sheep red cells *in vivo* and *in vitro, Infect. Immun.* **7**:160.

Bendinelli, M., Nardini, L., and Campa, M., 1974, Neutralization of Friend leukaemia virus by sera of unimmunized animals, *J. Gen. Virol.* **22**:207.

Bendinelli, M., Kaplan, G. S., and Friedman, H., 1975, Reversal of leukemia virus-induced immunosuppression *in vitro* by peritoneal macrophages, *J. Natl. Cancer Inst.* **55**:1425.

Bendinelli, M., Toniolo, A., and Friedman, H., 1976a, Reversal of immunodepression induced by murine leukemia virus, *Ann. NY Acad. Sci.* **276**:431.

Bendinelli, M., Toniolo, A., and Friedman, H., 1976b, Reversal of a leukemia virus-induced immunodepression by macrophages *in vitro* and *in vivo, Adv. Exp. Biol. Med.* **66**:507.

Bendinelli, M., Campa, M., Toniolo, A., and Garzelli, C., 1978, An analysis of the role of lymphatic leukaemia virus in the immunodepression exerted by Friend complex in leukaemia-resistant C57BL/6 mice, *J. Gen. Virol.* **39**:243.

Bendinelli, M., Matteucci, D., Toniolo, A., and Friedman, H., 1979, Suppression of *in vitro* antibody response by spleen cells of mice infected with Friend-associated lymphatic leukemia virus, *Infect. Immun.* **24**:1.

Bendinelli, M., Matteucci, D., Toniolo, A., and Friedman, H., 1980a, Macrophage involvement in leukemia virus-induced tumorigenesis, *Adv. Exp. Biol. Med.* **121B**:493.

Bendinelli, M., Toniolo, A., Matteucci, D., Celeste, F., and Revoltella, R., 1980b, Immunoglobulins in human serum reactive with Friend leukaemia virus, *J. Gen. Virol.* **46**:311.

Bendinelli, M., Cox, W. I., Specter, S., and Friedman, H., 1981, Interferon-induced augmentation of natural killer activity by splenocytes from leukemia virus-immunosuppressed mice and dimethyl sulfoxide-induced enhancement of Friend leukemia cell susceptibility to natural cytolysis, *Curr. Chemother. Immunother,* in: *Proceedings of the Twelfth International Congress of Chemotherapy,* pp. 1121, Florence.

Bendinelli, M., Matteucci, D., Toniolo, A., Patanè. A. M., and Pistillo, M. P., 1982, Impairment of immunocompetent mouse spleen cell functions by infection with cosxackievirus B3, *J. Infect. Dis.* **146**:797.

Bentvelzen, P., and Daams, J. H., 1969, Hereditary infections with mammary tumor viruses in mice, *J. Natl. Cancer Inst.* **43**:1025.

Biasi, G., Collavo, D., and Chieco-Bianchi, L., 1983, Selective failure of accessory cell function in Moloney murine leukemia virus tolerant mice, *J. Immunol.* **131**:16.

Bigner, D. D., and Pelgram, C. N., 1976, A review of virus-induced experimental brain tumors and of putative associations of viruses with brain human tumors, *Adv. Neurol.* **15**:57.

Bishop, J. M., 1982, Retroviruses and cancer genes, *Adv. Cancer Res.* **37**:1.

Bister, K., Ramsay, G. M., and Hayman, M. J., 1982, Deletions within the transforming specific RNA sequences of acute leukemia virus MC29 give rise to partially transformation-defective mutants, *J. Virol.* **41**:754.

Blair, D. G., and Vande Woude, G. F., 1982, Moloney sarcoma virus: A model for transformation by endogenous cellular genes, *Adv. Viral Oncol.* **1**:189.

Bluestone, J. A., and Lopez, C., 1979, Suppression of the immune response in tumor-bearing mice. I. Response to virus-producing tumor cells and non-virus-producing tumor cells, *J. Natl. Cancer Inst.* **63**:1215.

Bluestone, J. A., and Lopez, C., 1982, Induction of gp70-specific suppressor T-cells in mice inoculated with virus-producing tumor cells, *J. Natl. Cancer Inst.* **69**:953.

Boettiger, D., and Durban, E., 1984, Target cells for avian myeloblastosis virus in embryonic yolk sac and relationship of cell differentiation to cell transformation, *J. Virol.* **49:**841.

Boggs, S. S., and Schwartz, G. N., 1978, Increased lethality after endotoxin in old or leukemic AKR mice. *Proc. Soc. Exp. Biol. Med.* **157:**424.

Boniver, J., Declève, A., Dayley, M. O., Honsik, C., Lieberman, M., and Kaplan, H. S., 1981, Macrophage and lymphocyte-depleted thymus reticuloepithelial cell cultures: Establishment and functional influence on T-lymphocyte maturation, C-type virus expression and lymphomatous transformation *in vitro, Thymus* **2:**193.

Boyd, A. W., and Schhrader, J. W., 1982, Derivation of macrophage-like lines from the pre-B lymphoma ABLS 8.1 using 5-azacytidine, *Nature* **297:**691.

Boyer, B., Gisselbrecht, S., Debre, P., McEnzie, I., and Levy, J. P., 1980, Genetic control of sensitivty to Moloney leukemia virus in mice. IV. Phenotypic heterogeneity of the leukemic mice, *J. Immunol.* **125:**1415.

Broder, S., Muul, L., and Waldmann, T. A., 1978, Suppressor cells in neoplastic disease, *J. Natl. Cancer Inst.* **61:**5.

Broxmeyer, H. E., Ralph, P., Gilbertson, S., and Margolis, V. B., 1980, Induction of leukemia-associated inhibitory activity and bone marrow granulocyte-macrophage progenitor cell alterations during infection with Abelson virus, *Cancer Res.* **40:**3928.

Brunda, M. J., Taramelli, D., Holden, H. T., and Varesio, L., 1983, Suppression of *in vitro* maintenance and interferon-mediated augmentation of natural killer cell activity by adherent peritoneal cells from normal mice, *J. Immunol.* **130:**1974.

Buchman, V. M., Radzichovskaya, R. M., Svet-Moldavsky, G. J., Kozlova, M. D., Mikaelian, S. E., and Levin, V. J., 1973, Loading of the reticuloendothelial system: Effect on the induction, development and chemotherapy of Rauscher virus leukemia, *Biomedicine* **19:**335.

Bussard, A. E., and Pages, J., 1976, The peritoneal cell phenomenon: Its general significance in cellular immunology, *Ann. Immunol.* **127C:**583.

Butler, R. C., Frier, J. M., Chapekar, M. S., Graham, M. O., and Friedman, H., 1983, Role of antibody response helper factors in immunosuppressive effects of Friend leukemia virus, *Infect. Immun.* **39:**1260.

Cardiff, R. D., and Young, L. J. T., 1980, Mouse mammary tumor biology: A new synthesis, in: *Viruses in Naturally Occurring Cancers* (M. Essex, G. Todaro, and H. zur Hausen, eds.), p. 1105, Cold Spring Harbor Laboratory, Cold Spring Harbor, New York.

Carrel, A., and Ebeling, A. H., 1924, The transformation of monocytes into fibroblasts through the action of Rous virus, *J. Exp. Med.* **43:**461.

Carter, J. K., Ow, C. L., and Smith, R. E., 1983, Rous-associated virus type 7 induces a syndrome in chickens characterized by stunting and obesity, *Infect. Immun.* **39:**410.

Ceglowski, W. S., and Friedman, H., 1968, Immunosuppression by leukemia virus. I. Effect of Friend disease virus on cellular and humoral hemolysin responses of mice to a primary immunization with sheep erythrocytes, *J. Immunol.* **101:**594.

Ceglowski, W. S., and Friedman, H., 1975, Failure of peritoneal exudate macrophages to reverse immunologic impairment by Friend leukemia virus, *Proc Soc. Exp. Biol. Med.* **148:**808.

Cerny, J., and Essex, M., 1979, Mechanisms of immunosuppression by oncogenic RNA viruses, in: *Naturally Occurring Biological Immunosuppressive Factors and their Relationship to Disease* (R. Neubeneur, ed.), p. 233, CRC Press, Cleveland.

Chandra, P., (ed.), 1983, *Biochemical and Biological Markers of Neoplastic Transformation,* pp. 654, Plenum Press, New York.

Chany, C., Cerutti, I., and Mace, B., 1983, Effect of coördinated therapeutic assay using *C. parvum,* interferon and arginine butyrate on spontaneous disease and survival of AKR mice. *Int. J. Cancer* **32:**379.

Chesebro, B., and Wehrly, K., 1976, Studies on the role of the host immune response in recovery from Friend virus leukemia. II. Cell-mediated immunity, *J. Exp. Med.* **143:**85.

Chesebro, B., and Wehrly, K., 1979, Identification of a non-H-2 gene (*Rfv*-3) influencing recovery from viremia and leukemia induced by Friend virus complex, *Proc. Natl. Acad. Sci. USA* **76:**425.

Chinski, J., Soeiro, R., and Kopchick, J., 1984, *Fv*-1 host cell restriction of Friend leukemia virus: Microinjection of unintegrated viral DNA. *J. Virol.* **50:**271.

Cianciolo, G. J., Herberman, R. B., and Snyderman, R., 1980a, Depression of murine macrophage accumulation by low-molecular-weight factors derived from spontaneous mammary carcinomas, *J. Natl. Cancer Inst.* **65**:829.

Cianciolo, G. J., Matthews, T. J., Bolognesi, D. P., and Snyderman, R., 1980b, Macrophage accumulation in mice is inhibited by low molecular weight products from murine leukemia viruses, *J. Immunol.* **124**:2900.

Clarke, B. J., Axelrad, A. A., and Housman, D., 1976, Friend spleen focus-forming virus production *in vitro* by a nonerythroid cell line, *J. Natl. Cancer Inst.* **57**:853.

Coffin, J. M., 1979, Structure, replication and recombination of retrovirus genomes: some unifying hypothesis, *J. Gen. Virol.* **42**:1.

Coll, J., Saule, S., Martin, P., Raes, M. B., Lagrou, C., Graf, T., Beug, H., Simon, I. E., and Stehelin, D., 1983, The cellular oncogenes c-*myc*, c-*myb* and c-*erb* are transcribed in defined types of avian hematopoietic cells, *Exp. Cell Res.* **149**:151.

Collavo, D., Zanovello, P., Leuchars, E., Davies, A. J. S., Chieco-Bianchi, L., and Biasi, G., 1980, Moloney murine sarcoma virus oncogenesis in T lymphocyte-deprived mice. Biological and immunological studies, *J. Natl. Cancer Inst.* **64**:97.

Collins, J. J., and Snyderman, R., 1980, Absence of macrophage involvement in the passive serum therapy of Friend leukemia virus-induced disease, *Cancer Res.* **40**:557.

Cooper, J. A., and Hunter, T., 1983, Regulation of cell growth and transformation by tyrosine-specific protein kinases: The search for important cellular substrate proteins, *Curr. Top. Microbiol. Immunol.* **107**:125.

Cox, K. O., and Keast, D., 1973, Rauscher virus infection, erythrocyte clearance studies, and autoimmune phenomena, *J. Natl. Cancer Inst.* **50**:941.

Croker, B. P., Zinkernagel, R. M., and Dixon, F. J., 1979, Natural cytotoxic cells in NZB mice: Spontaneous *in vivo* and *in vitro* primed activity and structural characteristics, *Clin. Immunol. Immunopathol.* **12**:410.

Cudkowicz, G., and Hochman, P. S., 1980, Carrageenan-induced decline of natural killer activity. I. *In vitro* activation of adherent non-T-suppressor cells, *Cell. Immunol.* **53**:395.

Cullen, B. R., Lomedico, P. T., and Ju, G., 1984, Transcriptional interference in avian retroviruses—Implications for the promoter insertion model of leukaemogenesis, *Nature* **307**:241.

David, J. R., and Remold, H. G., 1976, Macrophage activation by lymphocyte mediators and studies on the interaction of macrophage inhibitory factor with its target cell, in: *Immunobiology of the Macrophage* (D. S. Nelson, ed.), p. 401, Academic Press, New York.

Dawson, P. J., and Fieldsteel, A. H., 1971, The influence of age on chronic remittent Friend disease, *Cancer Res.* **31**:974.

Denman, A. M., Pelton, B. K., Appleford, D., and Kinsley, M., 1976, Virus infections of lymphoreticular cells and auto-immune diseases, *Transplant. Rev.* **31**:79.

Dent, P. B., 1972, Immunodepression by oncogenic viruses, *Prog. Med. Virol.* **14**:1.

Devani, G. L., Fioretti, A., Isetta, A. M., and Trizio, D., 1983, Macrophage characteristics of a cell line isolated *in vitro* from a Moloney lymphoma, *Tumori* **69**:177.

Dexter, T. M., and Testa, N. G., 1976, Differentiation and proliferation of hemopoietic cells in culture, *Meth. Cell. Biol.* **14**:387.

Dracott, B. N., Wedderburn, N., and Salaman, M. H., 1972, *In vitro* immune response of spleen cells from Friend virus infected mice, *J. Gen. Virol.* **14**:77.

Duesberg, P. H., 1983, Retroviral transforming genes in normal cells? *Nature* **304**:219.

Durban, E. M., and Boettiger, D., 1981a, Differential effects of transforming avian RNA tumor viruses on avian macrophages, *Proc. Natl. Acad. Sci. USA* **78**:3600.

Durban, E. M., and Boettiger, D., 1981b, Replicating, differentiated macrophages can serve as *in vitro* targets for transformation by avian myeloblastosis virus, *Virology* **37**:488.

Elliott, S. C., and Schloss, G. T., 1971, Effect of 7, 12-dimethylbenz(a)anthracene on phagocytosis and antibody formation in Friend virus leukemia, *Infect. Immun.* **3**:217.

Enjuanes, L., Lee, J. C., and Ihle, J. N., 1981, T cell recognition of Moloney sarcoma virus proteins during tumor regression. I. Lack of a requirement for macrophages and the role of blastogenic factors in T cell proliferation, *J. Immunol.* **126**:1478.

Enrietto, P. J., and Wike, J. A., 1983, The pathogenesis of oncogenic avian retroviruses, *Adv. Cancer Res.* **39**:269.

Enrietto, P. J., Hayman, M. J., Ramsay, G. M., Wyke, J. A., and Payne, L. N., 1983, Altered pathogenicity of avian myelocytomatosis (MC29) viruses with mutations in the v-*myc* gene, *Virology* **124**:164.

Escobar, M. R., and Swenson, P. D., 1983, Mechanisms of viral immunopathology, in: *The Reticuloendothelial System. A Comprehensive Treatise.* Vol. 4. *Immunopathology* (Rose, N. R. and Siegel, B. V., eds.), p. 201, Plenum Press, New York.

Essex, M., 1977, Immunity to leukemia, lymphoma and fibrosarcoma in cats: A case for immunosurveillance, *Contemp. Top. Immunobiol.* **6**:71.

Essex, M., and Worley, M., 1981, Naturally occurring retroviruses of animals and birds, in: *Comparative Diagnosis of Viral Diseases.* Part IV. *Vertebrate Animal and Related Viruses* (E. Kurstak and C. Kurstak, eds.), p. 553, Academic Press, New York.

Evans, R., and Haskill, S., 1983, Activities of macrophages within and peripheral to the tumor mass. in: *The Reticuloendothelial System. A Comprehensive Treatise.* Vol. 5. *Cancer* (R. B. Herberman and H. Friedman, eds.), p. 155, Plenum Press, New York.

Famulari, N. G., 1983, Murine leukemia viruses with recombinant *env* genes: A discussion of their role in leukemogenesis, *Curr. Top. Microbiol. Immunol.* **103**:76.

Farber, P., Specter, S., and Friedman, H., 1975, Leukemia virus-induced immunosuppression: Scanning electron microscopy of infected spleen cells, *Science* **190**:469.

Feldman, D. G., Dreyfuss, Y., and Gross, L., 1967, Electron microscopic study of mouse leukemia virus (Gross) in organs of mouse embryos from virus-injected and normal C3Hf parents. *Cancer Res.* **27**:1792.

Fimmel, P. J., Cox, K. O., and Keast, D., 1975, Inhibition of migration of murine spleen cells by Rauscher leukemia virus-infected syngeneic cells, *J. Natl. Cancer Inst.* **54**:263.

Fleissner, E., and Snyder, H. W., 1982, Oncoviral proteins as cellular antigens, *Curr. Top. Microbiol. Immunol.* **98**:37.

Forni, G., Giovarelli, M., Lanfrancone, L., and Varesio, L., 1982, Suppressor macrophages in tumor-bearing mice. Inconsistency between *in vivo* and *in vitro* findings? *Int. J. Cancer* **29**:695.

Fredrickson, T. N., Lobue, J., Alexander, P., Schultz, E. F., and Gordon, A. S., 1972, A transplantable leukemia from mice inoculated with Rauscher leukemia virus, *J. Natl. Cancer Inst.* **48**:1597.

Fredrickson, T. N., Langdon, W. Y., Hoffman, P. M., Hartley, J. W., and Morse, H. C., 1984, Histologic and cell surface antigen studies of hematopoietic tumors induced by Cas-Br-M murine leukemia virus, *J. Natl. Cancer Inst.* **72**:447.

Friedman, H., and Ceglowski, W. S., 1968, Cellular basis for the immunosuppressive properties of a leukaemogenic virus, *Nature* **218**:1232.

Friedman, H., and Ceglowski, W. S., 1971, Leukemia virus-induced immunosuppression. VIII. Rapid depression of *in vitro* leukocyte migration after infection of mice with Friend leukemia virus, *J. Immunol.* **107**:1637.

Friedman, H., and Ceglowski, W. S., 1973, Virus tumorigenesis and immunity: Influence of immunostimulation and immunodepression, in: *The Role of Immunological Factors in Viral and Oncogenic Processes* (R. F. Beers, R. C. Telghman, and E. G. Bassett, eds.), p. 187, Johns Hopkins University Press, Baltimore.

Furmanski, P., Dietz, M., Fouchey, S., Hall, L., Clymer, R., and Rich, M. A., 1979, Spontaneous regression of Friend murine leukemia virus-induced erythroleukemia. IV. Effects of radiation and athymia on leukemia regression in mice, *J. Natl. Cancer Inst.* **63**:449.

Furmanski, P., Longley, C., Bolles, C. S., Hines, D. L., and Dietz, M., 1980, Spontaneous regression of Friend virus-induced erythroleukemia. VI. Structural and antigenic differences between the regressing and the conventional strains of virus, *J. Virol.* **33**:1083.

Gallo, R., Sarin, P. S., Gelmann, E. P., Robert-Guroff, M., Richardson, E., Kayanaraman, V. S., Mann, D., Sidhu, G. D., Stahl, R. E., Zolla-Pazner, S., Leibowitch, J., and Popovic, M., 1983, Isolation of human T-cell leukemia virus in acquired immune deficiency syndrome, *Science* **220**:865.

Garaci, E., Migliorati, G., Jezzi, T., Bartocci, A., Gioia, L., Rinaldi C., and Bonmassar, E. 1981, Impairment of *in vitro* generation of cytotoxic or T suppressor lymphocytes by Friend leukemia virus infection in mice, *Int. J. Cancer* **28**:367.

Gardner, M. B., 1978, Type C viruses of wild mice: Characterization and natural history of amphotropic, ecotropic and xenotropic MuLV, *Curr. Topics Microbiol. Immunol.* **79**:215.

Gazzolo, L., Moscovici, M. G., and Moscovici, C., 1974, Replication of avian sarcoma viruses in chicken macrophages, *Virology* **58**:514.

Gazzolo, L., Moscovici, M. G., Moscovici, C., and Vogt, P. K., 1975, Susceptility and resistance of chicken macrophages to avian RNA tumor viruses, *Virology* **67**:553.

Gazzolo, L., Moscovici, C., and Moscovici, M. G., 1979a, Persistence of avian oncoviruses in chicken macrophages, *Infect. Immun.* **23**:294.

Gazzolo, L., Moscovici, C., and Moscovici, M. G., and Samarut, J., 1979b, Response of hemopoietic cells to avian acute leukemia viruses: Effects on the differentiation of the target cells, *Cell* **16**:627.

Genovesi, E. V., Livnat, D., and Collins, J. J., 1982, Immunotherapy of murine leukemia. VII. Prevention of Friend leukemia virus-induced immunosuppression by passive serum therapy, *Int. J. Cancer* **30**:609.

Gerlier, D., Avice, T., and Doré, J. F., 1979, Specific passive immunotherapy of experimental lymphoma. Effects of macrophage toxic drugs, *Biomedicine* **31**:106.

Gisselbrecht, C., Blaineau, D., Hurot, M., Pozo, F., and Levy, J. P., 1978, Prevalence of non-T-cells in the replication of the *N*-tropic type C virus of young AKR mice, *Cancer. Res.* **38**:939.

Godard, C. M., Augery, Y., and Jasmin, C., 1983, Expression of ecotropic murine leukaemia virus in haemopoitic cells of AKR mice during the embryonic and neonatal period. *J. Gen. Virol.* **64**: 2051.

Goldman, R., and Hogg, N., 1978, Enhanced susceptibility of virus-infected fibroblasts to cytostasis mediated by peritoneal exudate cells, *J. Immunol.* **14**:1657.

Gomard, E., Wybier-Franqui, J., and Levy, J. P., 1981a, Accessory cells in the *in vitro* generation of type C viruses-specific T killer lymphocytes. I. Role of macrophages in primary anti-FMR reaction, *J. Immunol.* **126**:891.

Gomard, E., Wybier-Franqui, J., Simmler, M. C., McKenzie, I. F. C., and Levy, J. P., 1981b, Accessory cells in the *in vitro* generation of type C virus-specific T killer lymphocytes. II. Demonstration of a T helper cell in the spleen of *in vivo* primed mice, *J. Immunol.* **127**:2291.

Goodell, E. M., Ottenbrite, R. M., and Munson, A. E.. 1978, Polymaleic anhydride: Effects on the immune system and Friend leukemia disease of mice, *J. Reticuloendothel. Soc.* **23**:183.

Gorczynski, R. M., 1974, Immunity to murine sarcoma virus-induced tumors. II. Suppression of T cell-associated immunity by cells from progressor animals, *J. Immunol.* **112**:1826.

Goyette, M., Petropoulos, C. J., Shank, P. R., and Fausto, N., 1983, Expression of a cellular oncogene during liver regeneration, *Science* **219**:510.

Graf, T., and Beug, H., 1978, Avian leukemia viruses: interaction with their target cells *in vivo* and *in vitro*, *Biochim. Biophys. Acta* **516**:269.

Graf, T., Royer-Pokora. B., Schubert, G. E., and Beug, H., 1976, Evidence for the multiple oncogenic potential of cloned leukemia virus: *In vitro* and *in vivo* studies with avian erythroblastosis virus, *Virology* **71**:423.

Graf, T., Oker-Bloom, N., Tudorov, T. G., and Beug, H., 1979, Transforming capacities and defectiveness of avian leukemia viruses OK10 and E26, *Virology* **99**:431.

Graf, T., von Kirchbach, A., and Beug, H., 1981, Characterization of the hematopoietic target cells of AEV, MC29 and AMV avian leukemia viruses, *Exp. Cell Res.* **131**:331.

Greenberger, J. S., 1979, Phenotypically distinct target cells for murine sarcoma virus and murine leukemia virus marrow transformation *in vitro*, *J. Natl. Cancer Inst.* **62**:337.

Greenberger, J. S., Davisson, P. B., Gans, P. J., and Moloney, W. C., 1979, *In vitro* induction of continuous acute promyelocytic leukemic cell lines by Friend or Abelson murine leukemia virus, *Blood* **53**:987.

Greenberger, J. S., Wroble, L. M., and Sakakeeny, M. A., 1980, Murine leukemia viruses: Induction of macrophage production of granulocyte-macrophage colony-stimulating factor *in vitro*, *J. Natl. Cancer Inst.* **65**:841.

Hanafusa, H., 1977, Cell transformation by RNA tumor viruses, *Comprehen. Virol.* **10**:401.

Hanna, M. G., Walburg, H. E., Tyndall, R. L., and Snodgrass, M. J., 1970, Histoproliferative effect of Rauscher leukemia virus on lymphatic tissue. II. Antigen-stimulated germfree and conventional BALB/c mice, *Proc. Soc. Exp. Biol. Med.* **134**:1132.

Hanna, N., and Kripke, L. M., 1979, Immunological significance of nonspecific suppressor cells in spleen of tumor-bearing mice, *Cell. Immunol.* **43**:293.

Haran-Ghera, N., 1980, Pathogenesis of murine leukemia, in: *Viral Oncology* (G. Klein, ed.), p. 161, Raven Press, New York.

Hatten, B. A., and Dunton, H., 1978, Age-related humoral antibody responses of AKR/J mice to T-cell dependent and independent antigens, *Immunology* **35**:707.

Hays, E. F., Warren, J., Hale, L., and Rodensky, D., 1980, AKR murine leukemia viruses and long-term bone marrow cultures from AKR and SJL mice, *Exp. Hematol.* **8**:134.

Hayward, W. S., Neel, B. G., and Astrin, S. M., 1982, Avian leukosis viruses: Activation of cellular "oncogenes," *Adv. Viral. Oncol.* **1**:207.

Heard, J. M., Fichelson, S., Gisselbrecht, S., and Varet, B., 1981, Comparative effects of polycythemia-inducing Friend leukemia virus and Moloney leukemia virus on hematopoietic progenitors in murine long-term bone marrow cultures, *Cancer Res.* **41**:3260.

Heard, J. M., Fichelson, S., Choppin, J., and Varet, B., 1983, Autocrine function of murine F-MuLV induced myeloblastic cell lines, *Int. J. Cancer* **31**:337.

Herberman, R. B., 1982, Immunology of oncornaviruses, in: *Immunology of Human Infection. Part II. Viruses and Parasites; Immunodiagnosis and Prevention of Infectious Diseases,* (A. J. Nahmias and F. J. O'Reilly, eds.), p. 273, Plenum Press, New York.

Herberman, R. B., and Friedman, H., 1983. *The Reticuloendothelial System. A Comprehensive Treatise. 5. Cancer*, pp. 356, Plenum Press, New York.

Herr, W., and Gilbert, W., 1984, Free and integrated recombinant murine leukemia virus DNAs appear in preleukemic thymuses of AKR/J mice, *J. Virol.* **50**:155.

Hibbs, J. B., Lambert, L. H., and Remington, J. S., 1971, Resistance to murine tumors conferred by chronic infection with intacellular protozoa, *Toxoplasma gondi* and *Besnoitia jellisoni, J. Infect. Dis.* **124**:587.

Hines, D. L., and Gragowski, L., 1983, Abelson murine leukemia virus: Effect of helper virus from regressing Friend virus on leukemia development, *Leukemia Res.* **7**:251.

Hirsch, M. S., Black, P. H., Wood, M. L., and Monaco, A. P., 1973, Effects of pyran copolymer on leukemogenesis in immunosuppressed AKR mice, *J. Immunol.* **111**:91.

Holden, H. T., Haskill, J. S., Kirchner, H., nd Herberman, R. B., 1976, Two functionally distinct antitumor effector cells isolated from primary murine sarcoma virus-induced tumor, *J. Immunol.* **117**:440.

Holt, P. G., Warner, L. A., and Mayrhofer, G., 1981, Macrophages as effectors of T suppression: T-lymphocyte-dependent macrophage-mediated suppression of mitogen-induced blastogenesis in the rat, *Cell. Immunol.* **63**:57.

Hoover, E. A., Rojko, J. L., Wilson, P. L., and Olsen, R. G., 1981, Determinants of susceptibility and resistance to feline leukemia virus infection. I. Role of macrophages, *J. Natl. Cancer Inst.* **67**:889.

Horak, I., Enjuanes, L., Lee, J. C., and Ihle, J. N., 1981, Resistance of cultures of normal T cells to infection with murine type C viruses, *J. Virol.* **37**:483.

Hu, S. S. F., Duesberg, P. H., Lai, M. M. C., and Vogt, P. K., 1979, Avian oncovirus MH2: Preferential growth in macrophages and exact size of the genome, *Virology* **96**:302.

Hu, S. S. F., Moscovici, C., and Vogt, P. K., 1978, The defectiveness of Mill Hill 2, a carcinoma-inducing avian oncovirus, *Virology* **89**:162.

Huebner, K., and Croce, C. M., 1976, Production of B-tropic murine leukemia virus by somatic cell hybrids between mouse peritoneal macrophages and simian virus 40-transformed human cells, *J. Virol.* **18**:1143.

Hughes, S. H., 1983, Synthesis, integration, and transcription of the retroviral provirus, *Curr. Top. Microbiol. Immunol.* **103**:23.

Hume, D. A., and Gordon, S., 1982, Macrophage biochemistry, *Life Chem. Rep.* **1**:1.

Ihle, J. N., and Lee, J. C., 1982, Possible immunological mechanisms in C-type viral leukemogenesis in mice, *Curr. Top. Microbiol. Immunol.* **98**:85.

Israel, E., and Wainberg, M. A., 1981, Viral inhibition of lymphocyte mitogenesis: The role of macrophages as primary targets of virus-cell interactions, *J. Reticuloendothel. Soc.* **29**:105.

Jaenish, R., 1976, Germ line integration and Mendelian transmission of the endogenous Moloney leukemia virus, *Proc. Natl. Acad. Sci. USA* **73**:1260.

Johnson, C. S., Fouchey, S. P., and Furmanski, P., 1980, Spontaneous regression of Friend murine leukemia virus-induced erythroleukemia. V. Cell-mediated antivirus immunity in mice undergoing erythroleukemia regression and in leukemic mice, *J. Natl. Cancer Inst.* **64**:645.

Johnson, C. S., Marcelletti, J., and Furmanski, P., 1984, Stimulation of normal erythropoiesis *in vitro* by macrophages from erythroleukemic mice, *J. Natl. Cancer Inst.* **72**:491.

Kamo, I., and Friedman, H., 1977, Immunosuppression and the role of suppressive factors in cancer, *Adv. Cancer Res.* **25**:271.

Karmali, R. A., Wustrow, T., Thaler, H. T., and Good, R. A., 1984, Prostaglandins in cells of the lymphoid system in AKR leukemia, *Cancer Res.* **44**:467.

Keller, R., Keist, R., and Hecker, E., 1982, Tumor-promoting diterpene esters prevent macrophage activation, *Br. J. Cancer* **45**:144.

Kirchner, H., Muchmore, A. V., Chused, T. M., Holden, H. T., and Herberman R. B., 1975, Inhibition of proliferation of lymphoma cells and T lymphocytes by suppressor cells from spleens of tumor-bearing mice, *J. Immunol.* **114**:206.

Klein, B., Le Bousse-Kerdiles, C., Smadja-Joffe, F., Pragnell, I., Ostertag, W., and Jasmin, C., 1982, A study of added GM-CSF independent granulocyte and macrophage precursors in mouse spleen infected with myeloproliferative sarcoma virus (MPSV), *Exp. Hematol.* **10**:373.

Klein, G. (ed.), 1980, *Viral Oncology*, Raven Press, New York.

Klein, G. (ed.), 1982, *Advances in Viral Oncology*. Vol. 1. *Oncogene Studies, pp. 262, Raven Press, New York.*

Knyszynski, A., and Danon, D., 1977, The role of macrophages in defense against the development of Rauscher virus leukemia. *J. Reticuloendothel. Soc.* **22**:341

Knyszynski, A., Leibovich, S. J., and Danon, D., 1978, Phagocytosis of red blood cells from Rauscher leukemia virus-infected mice by macrophages from normal syngeneic mice, *J. Reticuloendothel. Soc.* **23**:243.

Knyszynski, A., Gottlieb, P., and Fridkin, M., 1983, Inhibition by tuftsin of Rauscher virus leukemia development in mice, *J. Natl. Cancer Inst.* **71**:87.

Komitowski, D., and Kurka, J., 1974, Ergebnisse des makrophagen-migration-testes wahrend der fruhpase der Rauscher-leukamie, *Experientia* **30**:648.

Koppi, T. A., and Halliday, W. J., 1982, Cell-mediated immunity to viral proteins in mice bearing Moloney sarcomas, *Immunol. Lett.* **4**:135.

Koury, M. J., and Pragnell, I. B., 1982, Retroviruses induce granulocyte–macrophage colony stimulating activity in fibroblasts, *Nature* **299**:638.

Kouttab, N. M., and Jutila, J. W., 1972, Friend leukemia virus infection in germfree mice following antigen stimulation, *J. Immunol.* **108**:591.

Kozak, C. A., Gromet, N. J., Ikeda, H., and Buckler, C. E., 1984, A unique sequence related to the ecotropic murine leukemia virus is associated with the *Fv-4* resistance gene. *Proc. Natl. Acad. Sci USA* **81**:834.

Kumar, V., and Bennett, M., 1981, Genetic resistance to Friend virus-induced erythroleukemia and immunosuppression, *Curr. Top. Microbiol. Immunol.* **92**:65.

Kuzumaki, N., Kodama, T., Takeichi, N., and Kobayashi, H., 1974, Friend lymphatic leukemia virus-induced autoimmune hemolytic anemia with glomerulonephritis in the rat, *Int. J. Cancer* **14**:483.

Landolfo, S., Herberman, R. B., and Holden, H. T., 1977, Stimulation of mouse migration inhibitory factor (MIF) production from MSV-immune lymphocytes by soluble tumor associated antigen: Requirement for histocompatible macrophages, *J. Immunol.* **118**:1244.

Landolfo, S., Herberman, R. B., and Holden, H. T., 1978, Macrophage–lymphocyte interaction in migration inhibition factor production against soluble or cellular tumor-associated antigens. I. Characteristics and genetic control of two different mechanisms of stimulating MIF production, *J. Immunol.* **121**:695.

Langweiler, M., Cockerell, G. L., and deNoronha F., 1983, Role of suppressor cells in feline leukemia virus-associated immunosuppression, *Cancer. Res.* **43**:1957.

Larnicol, N., Augery, Y., Le Bousse-Kerdiles, C., Degiorgis, V., Chermann, J. C., Teze, A., and Jasmin, C., 1981, *In vivo* effect of a new mineral condensed ion (HPA 39) on murine Friend leukaemia, *J. Gen. Virol.* **55**:17.

Larson, C. L., Ushijima, R. N., Baker, R. E., Baker, M. B., and Gillespie, C. A., 1972, Effect of normal serum and antithymocyte serum on Friend disease of mice, *J. Natl. Cancer Inst.* **48**:1403.

Levy, J. A., Barrett, S. G., Leong, J. C., and Dirksen, E. R., 1981, Transformation of macrophages from NZB hybrid mice by simian virus 40, *J. Reticuloendothel. Soc.* **29**:35.

Levy, M. H., and Wheelock, E. F., 1975, Impaired macrophage function in Friend virus leukemia: Restoration by statolon, *J. Immunol.* **114**:962.

Levy, P., and Leclerc, J. C., 1977, The murine sarcoma virus-induced tumor: Exception or general model in tumor immunology? *Adv. Cancer Res.* **24**:1.

Lieberman, D., and Sachs, L., 1979, Increase of normal myeloblast viability and multiplication without blocking differentiation by type C RNA virus from myeloid leukemic cells, *Proc. Natl. Acad. Sci. USA* **76**:3353.

Lilly, F., and Mayer, A., 1980, Genetic aspects of murine type-C viruses and their hosts in oncogenesis, in: *Viral Oncology* (G. Klein, ed.), p. 89, Raven Press, New York.

Lilly, F., and Pincus, T., 1973, Genetic control of murine viral leukemogenesis, *Adv. Cancer Res.* **17**:231.

Linial, M., 1982, Two retroviruses with similar transforming genes exhibit differences in transforming potential, *Virology* **119**:382.

Machida, C. A., Bestwick, R. K., and Kabat, D., 1984, Reduced leukemogenicity caused by mutations in the membrane glycoprotein gene of Rauscher spleen focus-forming virus, *J. Virol.* **49**:394.

Marcelletti, J., and Furmanski, P., 1978, Spontaneous regression of Friend virus-induced erythroleukemia. III. The role of macrophages in regression, *J. Immunol.* **120**:1.

Marcelletti, J., and Furmanski, P., 1979, Infection of macrophages with Friend virus: Relationship to the spontaneous regression of viral erythroleukemia, *Cell* **16**:649.

Marx, P. A., and Wheelock, E. F., 1976, Influence of immune stimulators on viral leukemogenesis, *Ann. NY Acad. Sci.* **276**:502.

McCombs, C., Hom., J., Talal, N., and Mishell, R. I., 1975, Functional deficiency of splenic adherent cells in New Zealand Black mice, *J. Immunol.* **115**:1695.

McFarland, H. F., and Johnson, R. T., 1975, The role of the inflammatory response in viral infections, in: *Viral Immunology and Immunopathology* (A. L. Notkins, ed.), p. 137, Academic Press, New York.

Meruelo, D., and Bach, R., 1983, Genetics of resistance to virus-induced leukemias, *Adv. Cancer. Res.* **40**:107.

Metcalf, D., Ishidata, M., and Brumby, M., 1967, Proliferation and behavior of phagocytic cells in mouse lymphoma tissue, *J. Natl. Cancer Inst.* **38**:527.

Migliorati, G., Jezzi, T., Frati, L., Bonmassar, E., Rossi, G. B., Garaci, E., and Riccardi, C., 1983, Modulation of natural killer cell activity during FLV-P virus infection of mice, *Int. J. Cancer* **31**:81.

Mizuno, D., Cohn, Z. A., Takeya, K., and Ishida, N. (eds.), 1982, *Self-defense mechanisms. Role of macrophages.*, pp. 340, Elsevier Biomedical, Amsterdam.

Mogensen, S. C., 1981, Viral interactions with phagocyte functions, in: *Microbial Perturbation of Host Defences* (F. O'Grady and H. Smith, eds.), p. 165, Academic Press, London.

Moody, D. J., Specter, S., Bendinelli, M., and Friedman, H., 1984, Suppression of natural killer cell activity by Friend murine leukemia virus, *J. Natl. Cancer. Inst.* **72**:1349.

Morahan, P. S., and Morse, S. S., 1979, Macrophage-virus interactions, in: *Virus-Lymphocyte Interactions: Implications for Disease* (M. R. Proffitt, ed.), p. 17, Elsevier/North Holland, New York.

Morse, H. C., and Hartley, J. W., 1981, Expression of xenotropic murine leukemia viruses, *Curr. Top. Microbiol. Immunol.* **98**:17.

Mortensen, R. F., and Elson, L. M., 1980, Suppression of the leukocyte adherence inhibition response to murine-sarcoma-virus-induced tumors by tumor-associated suppressor cells, *Oncology* **37**:406.

Moscovici C., and Gazzolo, L., 1982, Transformation of hemopoietic cells with avian leukemia viruses, *Adv. Viral. Oncol.* **1**:83.

Moscovici, M. G., and Moscovici, C., 1983, Isolation and characterization of a temperature-sensitive mutant of avian myeloblastosis virus, *Proc. Natl. Acad. Sci. USA* **80**:1421.

Moscovici, C., Samarut, J., Gazzolo, L., and Moscovici, M. G., 1981, Myeloid and erythroid neoplastic responses to avian defective leukemia viruses in chickens and in quail, *Virology* **113**:765.

Moscovici, M. G., Jordic, P., Samarut, J., Gazzolo, L., Mura, C. V., and Moscovici C., 1984, Characterization of the hemopoietic target cells for the avian leukemia virus E26, *Virology* **129**:65.

Nahon-Merlin, E., Lacour, F., Friend, C., and Revoltella, R., 1979, Protective effect of immunization with nonviral antigens against Friend leukemia virus in mice, *Proc. Natl. Acad. Sci. USA* **76**:2018.

Nandi, S., and McGrath, C. M., 1973, Mammary neoplasia in mice, *Adv. Cancer Res.* **17**:353.

Neil, J. C., Hughes, D., McFarlane, R., Wilkie, N. M., Onions, D. E., Lees, G., and Jarrett, O., 1984, Transduction and rearrangement of the *myc* gene by feline leukaemia virus in naturally occurring T-cell leukaemias, *Nature* **308**:814.

Nobis, P., and Löhler, J., 1983, Induction of histiocytomas by pristane treatment of mice chronically infected with Moloney murine leukemia virus, *Haematol. Blood Transf.* **28**:280.

Normann, S. J., Schardt, M., and Sorkin, E., 1981, Alteration of macrophage function in AKR leukemia, *J. Natl. Cancer Inst.* **66**:157.

Nowinski, R. C., 1976, Genetic control of natural immunity to ecotropic mouse leukemia viruses. II. Immune response genes, *Infect. Immun.* **13**:1098.

Nusse, R., and Varmus, H. E., 1982, Many tumors induced by the mouse mammary tumor virus contain a provirus integrated in the same region of the host genome, *Cell* **31**:99.

Nusse, R., van Ooyen, A., Cox, D., Fung, Y. K. T., and Varmus, H., 1984, Mode of proviral activation of a putative mammary oncogene (*int*-1) on mouse chromosome 15, *Nature* **307**:131.

Odaka, T., and Köhler, K., 1965, Uber aufnahme und vermehrung des Friend-leukamie-virus in macrophagen, *Z. Naturforsch.* **20b**:473.

Oehler, J. R., Herberman, R. B., and Holden, H. T., 1978, Modulation of immunity by macrophages, *Pharmacol. Ther. A* **2**:551.

Old, L. J., Benaceraff, B., Clarke, D. A., Carswell, E. A., and Stockert, E., 1961, The role of the reticuloendothelial system in the host reaction to neoplasia, *Cancer Res.* **21**:1281.

Ostertag, W., and Pragnell, I. B., 1981, Differentiation and viral involvement in differentiation of transformed mouse and rat embryo cells, *Curr. Top. Microbiol. Immunol.* **94**:143.

Pack, F. D., and Chapman, W. L., 1980, Light and electron microscopic evaluation of thymuses from feline leukemia virus-infected kittens, *Exp. Pathol.* **18**:96.

Perk, K., Ianconescu, M., Yanlv, A., and Zimber, A., 1979, Morphologic characterization of proliferative cells and virus particles in turkeys with lymphoproliferative disease, *J. Natl. Cancer Inst.* **62**:1483.

Pessano, S., Gazzolo, L., and Moscovici, C., 1979, The effect of a tumor promoter on avian leukemic cells, *Microbiologica* **2**:379.

Peters, G., Lee, A. E., and Dickson, C., 1984, Activation of cellular gene by mouse mammary tumor virus may occur early in mammary tumor development, *Nature* **309**:273.

Popovic, M., Sargadharan, M. G., Read, E., and Gallo, R. C., 1984, Detection, isolation, and continuous production of cythopatic retroviruses (HTLV-III) from patients with AIDS and pre-AIDS, *Science* **224**:497.

Poste, G., and Kirsch, R., 1979, Rapid decay of tumoricidal activity and loss of responsiveness to lymphokines in inflammatory macrophages, *Cancer Res.* **39**:2582.

Price, J. A., and Smith, R. E., 1982, Ihibition of concanavalin A response during osteopetrosis virus infection, *Cancer Res.* **42**:3617.

Proffitt, M. R., Congdon, C. C., and Tyndall, R. L., 1972, The combined action of Rauscher leukemia virus and lactic dehydrogenase virus on mouse lymphatic tissue, *Int. J. Cancer* **9**:193.

Puccetti, P., and Holden, H. T., 1979, Cytolytic and cytostatic anti-tumor activities of macrophages from mice injected with murine sarcoma virus, *Int. J. Cancer* **23**:123.

Radke, K., Beug, H., Kornfeld, S., and Graf, T., 1982, Transformation of both erythroid and myeloid cells by E26, an avian leukemia virus that contains the *myb* gene, *Cell* **31**:643.

Ralph, P., and Nakoinz, I., 1981, Differences in antibody-dependent cellular cytotoxicity and activated killing of tumor cells by macrophage cell lines, *Cancer Res.* **41**:3546.

Ramsay, G. M., Enrietto, P. J., Graf, T., and Hayman, M. J., 1982, Recovery of *myc*-specific sequences by a partially transformation-defective mutant of avian myelocytomatosis virus, MC29, correlates with the restoration of transforming activity, *Proc. Natl. Acad. Sci. USA* **79**:6885.

Rangan, S. R. S., and Bang, F. B., 1967, Rous sarcoma virus infection of chicken buffy coat macrophages, *Proc. Soc. Exp. Biol. Med.* **125**:593.

Raschke, W. C., Baird, S., Ralph, P., and Nakoinz, I., 1978, Functional macrophage cell lines transformed by Abelson leukemia virus, *Cell* **15**:261.

Revoltella, R. P., Bertolini, L., Fioritoni, G., and Tortolano, G., 1979, Specific esterase activity induced in mouse Friend erythroleukemia cells by dimethylsulfoxide, *J. Cell. Physiol.* **99**:461.

Rich, M. A., Siegler, R., Karl, S., and Clymer, R., 1969, Spontaneous regression of virus induced murine leukemia. I. Host virus system, *J. Natl. Cancer Inst.* **43:**559.

Ringold, G. M., 1983, Regulation of mouse mammary tumor virus gene expression by glucocorticoid hormones, *Curr. Top. Microbiol. Immunol.* **406:**79.

Risser, R., Potter, M., and Rowe, W. P., 1978, Abelson virus-induced lymphomagenesis in mice, *J. Exp. Med.* **148:**714.

Robbins, K. C., Antoniades, H. N., Devare, S. G., Hunkapiller, M. W., and Aaranson, S. A., 1983, Structural and immunological similarities between simian sarcoma virus gene product(s) and human platelet-derived growth factor, *Nature* **305:**605.

Robinson, H. L., Astrin, S. M., Senior, A. M., and Salazar, F. H., 1981, Host susceptibility to endogenous viruses: Defective, glycoprotein-expressing proviruses interfere with infections, *J. Virol.* **40:**745.

Roder, J. C., Tyler, L., Ball, J. K., and Singhal, S. K., 1978, The immunocompetence of tumor T cells and their role in generalized immunosuppression and immunostimulation following inoculation of dimethylbenzanthracene-induced leukemia virus in mice, *Cell. Immunol.* **36:**128.

Rojko, J. L., Hoover, E. A., Mathes, L. E., Hause, W. R., Schaller, J. P., and Olsen, R. G., 1978, Detection of feline leukemia virus in tissues of cats by a paraffin embedding immunofluorescence procedure, *J. Natl. Cancer Inst.* **61:**1315.

Rojko, J. L., Hoover, E. A., Mathes, L. E., Olsen, R. G., and Schaller, J. P., 1979, Pathogenesis of experimental feline leukemia virus infection, *J. Natl. Cancer Inst.* **63:**759.

Rojko, J. L., Hoover, E. A., Finn, B. L., and Olsen, R. G., 1981, Determinants of susceptibility and resistance to feline leukemia virus infection. II. Susceptibility of feline lymphocytes to productive feline leukemia virus infection. *J. Natl. Cancer Inst.* **67:**899.

Rojko, J. L., Hoover, E. A., Quackenbush, S. L., and Olsen, R. G., 1982, Reactivation of latent feline leukaemia virus infection, *Nature* **298:**385.

Rosenberg, N., 1982, Abelson leukemia virus, *Curr. Top. Microbiol. Immunol.* **101:**95.

Rowson, K. E. K., and Mahy, B. W. J., 1975, Lactic dehydrogenase virus, in *Virol. Monogr. 13,* Springer-Verlag, Wien.

Rup, B. J., Hoelzer, J. D., and Bose, H. R., 1982, Helper viruses associated with avian acute leukemic viruses inhibit the cellular immune response, *Virology* **116:**61.

Ruscetti, S., Feild, J., Davis, L., and Oliff, A., 1982, Factors determining the susceptibility of NIH Swiss mice to erythroleukemia induced by Friend murine leukemia virus, *Virology* **117:**357.

Russell, S. W., Doe, W. F., and Cochrane, C. G., 1976, Number of macrophages and distribution of mitotic activity in regressing and progressing Moloney sarcomas, *J. Immunol.* **116:**164.

Ruta, M., Bestwick, R., Machida, C., and Kabat, D., 1983, Loss of leukemogenicity caused by mutations in the membrane glycoprotein structural gene of Friend spleen focus-forming virus, *Proc. Natl. Acad. Sci. USA* **80:**4704.

Sanfilippo, F., Genovesi, E. V., and Collins, J. J., 1982, Immunotherapy of murine leukemia. V. Protection against Friend leukemia virus-induced immune complex glomerulonephritis by passive serum therapy, *J. Natl. Cancer Inst.* **67:**703.

Sanford, K. K., Likely, G. D., Bryan, W. R., and Earle, W. R., 1952, The infection of cells in tissue culture with Rous sarcoma virus, *J. Natl. Cancer Inst.* **12:**1317.

Schuller, G. B., and Morahan, P. S., 1977, Cellular and serum involvement in protection against Friend leukemia virus, *Cancer Res.* **37:**4064.

Schuller, G. B., and Morahan, P. S., 1979, Enhancement of erythroid target cells for Friend murine leukemia virus by intravenous pyran treatment, *J. Natl. Cancer Inst.* **62:**1257.

Schwartz, H. J., and Catanzaro, P. J., 1981, The use of carrageenan in determining immunological functions of macrophages, in: *Manual of Macrophage Methodology* (H. B., Herscowitz, H. T. Holden, J. A. Bellanti, and A. Ghaffar, eds.), p. 145, Marcel Dekker, New York.

Seidel, H. J., and Nothdurft, W., 1976, The phagocytic activity of the reticuloendothelial system of mice infected with Rauscher leukemia virus, *J. Reticuloendothel. Soc.* **19:**173.

Sharma, J. M., Herberman, R. B., Djeu, J. Y., and Nunn, M. E., 1979, Production of migration inhibition factor by spleen cells of normal rats upon culture *in vitro* with tumor cells and cells expressing endogenous virus, *J. Immunol.* **123:**222.

Sher, T., Haas, M., and Feldman, M. 1983, Elimination of endogenous xenotropic retroviruses from peritoneal macrophages in virus-induced T cell lymphoma. *J. Gen. Virol.* **64**:421.

Shibuya, T., and Mak, T. W., 1982, Host control of susceptibility to erythroleukemia and to the types of leukemia induced by Friend murine leukemia virus: Initial and late stages, *Cell* **31**:483.

Siegler, R., 1970, Pathogenesis of virus-induced murine sarcoma. I. Light microscopy, *J. Natl. Cancer Inst.* **45**:135.

Siegler, R., Zajdel, S., and Lane, I., 1972, Pathogenesis of Abelson-virus-induced murine leukemia, *J. Natl. Cancer Inst.* **48**:189.

Silver, J. E., and Fredrickson, T. N., 1983, A new gene that controls the type of leukemia induced by Friend murine leukemia virus. *J. Exp. Med.* **158**:493.

Smith, R. E., 1982, Avian osteopetrosis, *Curr. Top. Microbiol. Immunol.* **101**:75.

Smith, R. E., and Morgan, J. H., 1982, Identification of plaque isolates of an avian retrovirus causing rapid and slow onset osteopetrosis, *Virology* **119**:488.

Smith, R. E., and Schmidt, E. V., 1982, Induction of anemia by avian leukosis viruses of five subgroups, *Virology* **119**:516.

Smith, R. E., and van Eldik, L. J., 1978, Characterization of the immunosupsion accompanying virus-induced avian osteopetrosis, *Infect. Immun.* **22**:452.

Specter, S., and Friedman, H., 1978, Viruses and the immune response, *Pharmacol. Ther. A* **2**:595.

Specter, S., Patel, N., and Friedman, H., 1976, Restoration of leukemia virus-suppressed immunocytes *in vitro* by peritoneal exudate cells, *Proc. Soc. exp. Biol. Med.* **151**:163.

Specter, S., Bendinelli, M., Ceglowski, W. S., and Friedman, H., 1978, Macrophage-induced reversal of immunosuppression by leukemia viruses, *Fed. Proc.* **37**:97.

Squartini, F., Bolis, G. B., and Rossi, G., 1973, BALB/cfRIII: A new high mammary tumor and high leukemia-mouse strain, *J. Natl. Cancer Inst.* **52**:1581.

Stanton, L. F., Law, L. W., and Ting, R. C., 1968, Some biologic, immunologic, and morphological effects on mice after infection with a murine sarcoma virus. I. Morphologic studies, *J. Natl. Cancer Inst.* **40**:1968.

Stephenson, J. R. (ed.), 1980, *Molecular Biology of RNA Tumor Viruses*, Academic Press, New York.

Stern, K., 1983, Control of tumors by the RES, in: *The Reticuloendothelial System. A Comprehensive Treatise*. Vol. 5. *Cancer* (R. B. Herberman, and H. Friedman, eds.), p. 56, Plenum Press, New York.

Strauss, R. R., Mills, L., and Patel, N., 1976, Nonspecific activation of macrophages for antitumor immunity, *Ann. NY Acad. Sci.* **276**:520.

Stromberg, K., Benveniste, R. E., Arthur, L. O., Rabin, H., Giddens, W. E., Ochs, H. D., Morton, W. R., and Tsa, C-C., 1984, Characterization of exogenous type D retrovirus from a fibroma of a macaque with simian AIDS and fibromatosis, *Science* **224**:289.

Tagliabue, A., Boraschi, D., and McCoy, J. L., 1980, Development of cell-mediated antiviral immunity and macrophage activation in C3H/HeN mice infected with mouse mammary tumor virus, *J. Immunol.* **124**:2203.

Tambourin, P. E., Wendling, F., Jasmin, C., and Smadia-Joffe, F., 1979, The physiopathology of Friend leukemia, *Leukemia Res.* **3**:117.

Taniyama, T., and Holden, H. T., 1979a, Cytolytic activity of macrophages isolated from primary murine sarcoma virus-induced tumors, *Int. J. Cancer* **24**:151.

Taniyama, T., and Holden, H. T., 1979b, Requirement of histocompatible macrophages for the induction of a secondary cytotoxic response to syngeneic tumor cells *in vitro, J. Immunol.* **123**:43.

Temin, H. M., 1967, Studies on carcinogenesis by avian sarcoma virus. V. Requirement for new DNA synthesis and for cell division, *J. Cell Physiol.* **69**:53.

Testa, N. G., Dexter, T. M., Scott, D., and Teich N. M., 1980, Malignant myelomonocytic cells after *in vitro* infection of marrow cells with Friend leukemia virus, *Br. J. Cancer* **41**:38.

Ting, C-C., and Rodrigues, D., 1979, Reversal by peritoneal adherent cells of tumor cell suppression of T cell-mediated immunity, *J. Immunol.* **123**:801.

Ting, C-C., and Rodrigues, D., 1982, Tumor cell-triggered macrophage-mediated suppression of the T-cell cytotoxic response to tumor-associated antigens. I. Characterization of the cell components for induction of suppression, *J. Natl. Cancer Inst.* **69**:867.

Todaro, G. J., 1980, Interspecies transmission of mammalian retroviruses, in: *Viral Oncology* (G. Klein, ed.), p. 291, Raven Press, New York.

Toniolo, A., Matteucci, D., Pistillo, M. P., Gori, Z., and Bendinelli, M., 1980, Early replication of Friend leukaemia viruses in spleen macrophages, *J. Gen. Virol.* **49**:203.

Treves, A. J., Schechter, B., Cohen, I. R., and Feldman, M., 1976, Sensitization of T lymphocytes *in vitro* by syngeneic macrophages fed with tumor antigens, *J. Immunol.* **116**:1059.

Tsuruo, T., Iida, H., Tsukagoshi, S., and Sakurai, Y., 1980, Suppression of Friend leukemia virus by bacillus Calmette-Guérin and a streptococcal preparation, OK-432, *Int. J. Cancer* **25**:131.

Unanue, E. R., 1981, The regulatory role of macrophages in antigenic stimulation. Part two: Symbiotic relationship between lymphocytes and macrophages, *Adv. Immunol.* **31**:1.

Vande Woude, G. F., Levine, A. J., Topp, W. C., and Watson, J. D. (eds.), 1984, *Cancer Cells.* Vol 2. *Oncogenes and Viral Genes,* pp. 657, Cold Spring Harbor Laboratory, Cold Spring Harbor, New York.

Varesio, L., and Holden, H. T., 1980, Suppression of lymphokine production. I Macrophage–mediated inhibition of MIF production, *Cell. Immunol.* **56**:16.

Varesio, L., 1983, Suppressor cells and cancer: Inhibition of immune functions by macrophages, in: *The Reticuloendothelial System. A Comprehensive Treatise.* Vol. 5. *Cancer* (R. B. Herberman and H. Friedman, eds.), p. 217, Plenum Press, New York.

Varmus, H. E., Guntala, R. V., Deng, C. T., and Bishop, J. M., 1974, Synthesis, structure and function of avian sarcoma virus-specific DNA in permissive and nonpermissive cells, *Cold Spring Harbor Symp. Quant. Biol.* **39**:987.

Wainberg, M. A., Beiss, B., Fong, H., Beaupré, S., and Menezes, J. 1983, Involvement of macrophage-like cells in growth of tumors induced by avian sarcoma virus, *Cancer Res.* **43**:1550.

Webb, P. J., Brooks, L. G., and Baldwin, R. W., 1980, Macrophage-like suppressor cells in rats. II. Evidence for a quantitative rather than qualitative change in tumor-bearer animals, *Cell. Immunol.* **52**:381.

Wedderburn, N., and Salaman, M. H., 1968, The immunodepressive effect of Friend virus. II. Reduction of splenic haemolysin-producing cells in primary and secondary responses, *Immunology* **15**:439.

Weinberg, R. A., 1982, Oncogenes of spontaneous and chemically induced tumors, *Adv. Cancer Res.* **36**:149.

Weinberg, R. A., and Steffen, D. L., 1981, Regulation of expression of the integrated retrovirus genome, *J. Gen. Virol.* **54**:1.

Weiss, R. 1984, Retroviruses linked with AIDS, *Nature* **309**:12.

Weiss, R., Teich, N., Varmus, H., and Coffin, J. (eds.), 1982, *Molecular Biology of Tumor Viruses: RNA Tumor Viruses,* 2nd ed., pp. 1396, Cold Spring Harbor Laboratory, Cold Spring Harbor, New York.

Whitlock C. A., Ziegler, S. F., Treiman, L. J., Stafford, J. I., and Witte, O. N., 1983, Differentiation of cloned populations of immature B cells after transformation with Abelson murine leukemia virus, *Cell* **32**:903.

Wiltrout, R. H., Brunda, M. J., and Holden, H. T., 1982, Variation in selectivity of tumor cell cytolysis by murine macrophages, macrophage-like cell lines and NK cells, *Int. J. Cancer* **30**:335.

Wirth, J. J., Levy, M. H., and Wheelock, E. F., 1976, Use of silica to identify host mechanisms involved in suppression of established Friend virus leukemia, *J. Immunol.* **117**:2124.

Wybier-Franqui, J., Gomard, E., and Levy, J. P., 1982, Accessory cells in the *in vitro* generation of type C virus-specific T-killer lymphocytes. III. Two methods of antigen presentation to cytolytic T-lymphocyte precursors, *Cell. Immunol.* **68**:287.

Yamanouchi, K., Yoshikawa, Y., Nayami, H., Nishiyama, H., and Boschek, C. B., 1979, Continuous lines of RSV-transformed embryo cells and peritoneal macrophages of Japanese quails, *Jpn. J. Med. Sci. Biol.* **32**:19.

Yefenof, E., and Zilcha, D., 1982, High- and low-leukemogenic variants of the radiation leukemia virus differ in ability to induce tumor-specific immune suppression, *Int. J. Cancer* **30**:355.

Yochuchi, T., Nakashima, I., and Kato, N., 1980, Further studies on generation of macrophages in *in vitro* cultures of mouse spleen cells and its inhibition by the capsular polysaccharide of *Klebsiella pneumoniae, J. Gen. Virol.* **49**:203.

Yumoto, T., Recher, L., Sykes, J. A., and Dmochowski, L., 1966, Morphology and development of some murine leukemia virus, *Natl. Cancer Inst. Mon.* **22**:107.
Zacharchunk, C. M., Drysdale, B.-E., Mayer, M. M., and Shin, H. S., 1983, Macrophage-mediated cytotoxicity: Role of a soluble macrophage cytotoxic factor similar to lymphotoxin and tumor necrosis factor, *Proc. Natl. Acad. Sci. USA* **80**:6341.
Ziegler, S. F., Whitlock, C. A., Goff, S. P., Gifford, A., and Witte, O. N., 1981, Lethal effects of Abelson murine virus transforming gene product, *Cell* **27**:477.

Index

GPSR Compliance

The European Union's (EU) General Product Safety Regulation (GPSR) is a set of rules that requires consumer products to be safe and our obligations to ensure this.

If you have any concerns about our products, you can contact us on ProductSafety@springernature.com

In case Publisher is established outside the EU, the EU authorized representative is:

Springer Nature Customer Service Center GmbH
Europaplatz 3
69115 Heidelberg, Germany

MIX
Papier | Fördert gute Waldnutzung
FSC® C083411

Kolibri 360 GmbH
Ferdinand-Jühlke-Straße 7
99095 Erfurt, Deutschland
produktsicherheit@kolibri360.de